CAMBRIDGE MONOGRAPHS ON PHYSICS

Introduction to the theory of solid surfaces

Introduction to the theory of solid surfaces

FEDERICO GARCÍA-MOLINER AND
FERNANDO FLORES

Institute of Solid State Physics
Spanish Research Council and Autonomous
University of Madrid

CAMBRIDGE UNIVERSITY PRESS
CAMBRIDGE
LONDON · NEW YORK · MELBOURNE

CAMBRIDGE UNIVERSITY PRESS
Cambridge, New York, Melbourne, Madrid, Cape Town, Singapore, São Paulo, Delhi

Cambridge University Press
The Edinburgh Building, Cambridge CB2 8RU, UK

Published in the United States of America by Cambridge University Press, New York

www.cambridge.org
Information on this title: www.cambridge.org/9780521114356

First published 1979
This digitally printed version 2009

A catalogue record for this publication is available from the British Library

Library of Congress Cataloguing in Publication data

García-Moliner, Federico.
Introduction to the theory of solid surfaces.
(Cambridge monographs on physics)
Bibliography: p.
Includes index.
1. Surface (Physics) 2. Metallic surfaces.
3. Semiconductors. I. Flores, Fernando, joint
author. II. Title.
QC173.4.S94G37 530.4′1 78-17617

ISBN 978-0-521-22294-5 hardback
ISBN 978-0-521-11435-6 paperback

To Carmen and Araceli

Contents

Preface

Surface physics has advanced very rapidly in the last few years, largely based on an impressive range of new and powerful experimental techniques. Empirical progress seems to have stimulated a great deal of theoretical work in this field, or it might be that perhaps some people felt that the more traditional (bulk) solid state theory was about to die on its feet. Or perhaps both. Certainly there are attractive new scientific challenges to be met in trying to understand bulk condensed matter other than the traditional archetypes of crystalline bulk solids. But, let us be frank, none of these fields has – for the time being – anything like the immense potential for technological applications that surface physics has (to the extent and degree of immediacy that even theoreticians and administrators can see it).

As people who have been concerned with teaching, we would like to say that surface physics, besides presenting interesting challenges to the professional researcher, also has a high intrinsic value to basic knowledge. It is the physical materialisation of the mathematical principle which says that a new boundary condition changes the solutions of a given differential equation. Take any simple classical problem, introduce a surface and consider it again. The surprises are numerous and very instructive; a real, deep and thorough understanding of the new situation requires a serious effort of critical rethinking which always proves highly rewarding. It is this intrinsic educational value that we had in mind when we embarked upon the work presented here, and we believe that it is in itself a good justification for a book of this sort.

The above remarks should not be taken to imply that this is an undergraduate text. What we have tried to write is a selfcontained book which can be read from cover to cover without urgent need for interruption to consult the references: a book which by starting in each case from first principles, may lead the reader from an elementary analysis for beginners in this field to the level of current research papers, to which of course he or she should go for further details, factual information, and everything else of interest to a professional or would-be professional in the areas touched on here. It follows that this is most definitely not a treatise on the subject, in which one would find emphasis on its empirical basis and on descriptive material related to the particular properties of given surfaces or interfaces. This is simply a book of theory.

The choice of material is avowedly subjective. Surface theory is currently in the making as a result of the work of many people and it is impossible to do justice to everything and everybody in a reasonably sized volume. In some particular areas it might even be premature to try and construct at this stage a logically coherent account of the theoretical basis behind seemingly unconnected research papers. At any rate we did not feel competent to attempt it, and this is ultimately the best and only genuine justification for everything humans do not do; life is short and we all have our limitations. But we do feel that there is some coherence in the theoretical discourse embodied in this volume. We try to emphasise basic principles, and to demonstrate their application for simple tractable models, and we hope that the recurrence of general common concepts – surface modes, matching, Green functions, phase shifts, etc. – provides the connection between different parts of the book, thereby enabling the reader to perceive some logical continuity. In order to avoid breaking this up we have not inserted the references in the text. We give them instead (by chapter) at the end of the book, indicating the sections to which they apply.

As regards references we should warn the reader that there are very few. There is no pretence of 'covering' the field here, or synthetising everything that is known about a given problem. We refer only to the literature that we have actually used while working on this book, but we try to give proper credit whenever this is due. If a given section is based on some particular papers, this is pointed out in the references for that chapter. Although it has been our policy not to discuss detailed calculations for particular models and their comparison with experimental data, sometimes the treatment would be too dry if it were simply presented in the form of equations, while the illustrative power of a few appropriate numerical results is much too great to be neglected. In these cases we include the numerical results in the form of tables or figures, and the original references are given in the legend.

Concerning notation we hope no reader is misled by the use of the same symbol to denote different things. This is always done either in different chapters or in different contexts and should cause no confusion, while avoiding the use of too many uncommon symbols.

Finally, we wish to express our gratitude both to Mrs Vergés, whose technical skill has been a great asset in the preparation of this book, and to our many friends and colleagues at the Solid State Theory Section of the

Institute of Solid State Physics of the Spanish Research Council and the Solid State Department of the Autonomous University of Madrid. This has been our home for several years, during which we have enjoyed their company and collaboration. We are glad, in particular, to express our thanks to Enrique Louis, Carlos Tejedor, Luis Garrido, Víctor R. Velasco, Rosa Monreal and Fernando Franco for their help in preparing some of the illustrative material included in this book.

Some colleagues have been very kind in maintaining a correspondence on their work which has proved very useful to us. We gratefully acknowledge our gratitude to Drs R. Monnier, J. P. Perdew, J. R. Smith and H. B. Shore. We are also grateful to all the authors who gave us their kind permission to adapt some of their figures or numerical results for use in this book.

Madrid
February 1978

F. GARCÍA-MOLINER
AND F. FLORES

1

Classical electrodynamics of surface systems

1.1. Maxwell's equation in a polarisable medium

In the electromagnetic (e.m.) theory of surfaces we recognise the two typical elements of every surface problem. First, we study any field obeying certain differential equations: in this case Maxwell's equations. Secondly, out of all possible solutions we must seek those satisfying some specific *boundary conditions* due to the surface.

Let us consider the first question. We regard matter as a polarisable system, i.e. as composed of elementary charges e_i which, under the forces of the e.m. field, can undergo displacements $\boldsymbol{r}_i$, thus producing a *total* polarisation field,

$$\boldsymbol{P} = \sum_i e_i \boldsymbol{r}_i. \tag{1.1}$$

We know the e.m. field satisfies Maxwell's equations but, how do we write these down in order to describe field and matter in interaction? Since $\boldsymbol{P}$ must be related to the forces causing the motion of the elementary charges we must introduce some appropriate coefficients to describe the response of the medium to external stimuli. In a linear theory these coefficients will be related to the linear response functions obtained from perturbation theory. At the same time they must have a definite phenomenological meaning. Like all matters of definitions and conventions this can be done in different ways. We shall discuss two alternative schemes which we shall call the (L, T) and the (ε, μ) schemes. The use of either one is purely optional and depends on practical expediency, but it is convenient to have a clear understanding of the connection between them.

Let us go back to (1.1). In solid matter we adopt the convention of separating the elementary bound charges, which contribute to the polarisation in the usual dielectric sense, from the free, i.e. charge carriers. Thus we rewrite (1.1) as

$$\boldsymbol{P} = \boldsymbol{P}_b + \sum_i^f e_i \boldsymbol{r}_i. \tag{1.2}$$

Then in the time derivative ($\dot{\boldsymbol{r}}_i = \boldsymbol{v}_i$)

$$\dot{\boldsymbol{P}} = \dot{\boldsymbol{P}}_\text{b} + \sum_i^\text{f} e_i \boldsymbol{v}_i = \dot{\boldsymbol{P}}_\text{b} + \boldsymbol{J}_\text{f}, \tag{1.3}$$

we separate the *displacement current* $\dot{\boldsymbol{P}}_\text{b}$ from the (free) *induced current* associated with the motion of the charge carriers. Now *define the displacement field* $\boldsymbol{D}$ by

$$\boldsymbol{D} = \boldsymbol{E} + 4\pi \boldsymbol{P}, \tag{1.4}$$

where $\boldsymbol{E}$ is the *electric field.* Thus the dielectric polarisation, by definition, gives rise to

$$\boldsymbol{D}_\text{b} = \boldsymbol{E} + 4\pi \boldsymbol{P}_\text{b}. \tag{1.5}$$

The fields defined in this scheme satisfy the equations

$$\left.\begin{aligned} \nabla \cdot \boldsymbol{H} &= 0, & \nabla \wedge \boldsymbol{E} &= -\frac{1}{c}\frac{\partial \boldsymbol{H}}{\partial t}, \\ \nabla \cdot \boldsymbol{D} &= 4\pi\sigma_\text{ext}, & \nabla \wedge \boldsymbol{H} &= \frac{1}{c}\frac{\partial \boldsymbol{D}}{\partial t} + \frac{4\pi}{c}\boldsymbol{J}_\text{ext}, \end{aligned}\right\} \tag{1.6}$$

where we regard the external charge σ_ext and current $\boldsymbol{J}_\text{ext}$ as the sources of the e.m. field. We stress that our definition of $\boldsymbol{D}$ (1.2)–(1.4) includes the effects associated with *both* free and bound charges.

Now we introduce linear response theory. Even for an isotropic medium the vectors $\boldsymbol{E}$, $\boldsymbol{P}$, $\boldsymbol{D}$ will be related through tensors. Consider, for example, a homogeneous medium and a linear definition of the form

$$\boldsymbol{D}(\boldsymbol{r}, t) = \int \boldsymbol{\varepsilon}(\boldsymbol{r} - \boldsymbol{r}', t - t') \cdot \boldsymbol{E}(\boldsymbol{r}', t')\, \mathbf{d}\boldsymbol{r}' \tag{1.7}$$

and Fourier transform according to $\exp[\mathrm{i}(\boldsymbol{k} \cdot \boldsymbol{r} - \omega t)]$:

$$\boldsymbol{D}(\boldsymbol{k}, \omega) = \boldsymbol{\varepsilon}(\boldsymbol{k}, \omega) \cdot \boldsymbol{E}(\boldsymbol{k}, \omega). \tag{1.8}$$

There are two tensor forms which are invariant against all rotations in space, namely the unit tensor $\boldsymbol{I}$ and the dyadic form $\boldsymbol{kk}$. Thus a tensor describing an isotropic medium is in general a combination of these two forms. It is actually more convenient to introduce the longitudinal and transverse projectors

$$\boldsymbol{L} = \boldsymbol{kk}/k^2, \qquad \boldsymbol{T} = \boldsymbol{I} - \boldsymbol{kk}/k^2. \tag{1.9}$$

Thus in general

$$\boldsymbol{\varepsilon}(\boldsymbol{k}, \omega) = \varepsilon_\text{L}(\boldsymbol{k}, \omega)\boldsymbol{L} + \varepsilon_\text{T}(\boldsymbol{k}, \omega)\boldsymbol{T}. \tag{1.10}$$

This defines the *longitudinal and transverse dielectric functions* ε_L and ε_T.

Fourier transforming (1.4) we introduce the susceptibility tensor

$$4\pi\boldsymbol{\chi}(\boldsymbol{k},\omega)=\boldsymbol{\varepsilon}(\boldsymbol{k},\omega)-\boldsymbol{I}, \tag{1.11}$$

so that $\boldsymbol{P}$ is $\boldsymbol{\chi}\cdot\boldsymbol{E}$ in Fourier transform. Equation (1.11) entails the corresponding definitions of $\chi_L(\boldsymbol{k},\omega)$ and $\chi_T(\boldsymbol{k},\omega)$. If we set up our perturbation theory this will produce precisely χ_L and χ_T, which thus appear as the natural elements in the (L, T) scheme.

Now, if we had a vacuum, from the Fourier transforms of (1.6) we would write

$$\mathrm{i}\boldsymbol{k}\wedge\boldsymbol{H}=-\mathrm{i}(\omega/c)\boldsymbol{E}, \tag{1.12}$$

whereas for a dielectric:

$$\mathrm{i}\boldsymbol{k}\wedge\boldsymbol{H}=-\mathrm{i}(\omega/c)\boldsymbol{\varepsilon}_b\cdot\boldsymbol{E}. \tag{1.13}$$

On introducing free carriers we need to define another linear constitutive relation, namely

$$\boldsymbol{J}_f(\boldsymbol{k},\omega)=\boldsymbol{\sigma}(\boldsymbol{k},\omega)\cdot\boldsymbol{E}(\boldsymbol{k},\omega). \tag{1.14}$$

This defines the *conductivity tensor* $\boldsymbol{\sigma}$. Now in a medium with both bound and free charges:

$$\mathrm{i}\boldsymbol{k}\wedge\boldsymbol{H}=-\mathrm{i}(\omega/c)[\boldsymbol{\varepsilon}_b+\mathrm{i}(4\pi/\omega)\boldsymbol{\sigma}]\cdot\boldsymbol{E}, \tag{1.15}$$

whence the *Appleton–Hartree* (*A.H.*) *equation*,

$$\boldsymbol{\varepsilon}(\boldsymbol{k},\omega)=\boldsymbol{\varepsilon}_b(\boldsymbol{k},\omega)+\mathrm{i}(4\pi/\omega)\boldsymbol{\sigma}(\boldsymbol{k},\omega). \tag{1.16}$$

Again, we have σ_L and σ_T and the two corresponding A.H. equations. It is useful to introduce the electromagnetic response function or Green function in the following manner. Take (1.6) and eliminate $\boldsymbol{H}$ and $\boldsymbol{D}$ from the two curl equations. Then $\boldsymbol{E}$ is linearly related to the external source by a linear differential equation. Solving for $\boldsymbol{E}$ amounts to finding the *resolvent operator* of the problem. In Fourier transform we have

$$\boldsymbol{k}\wedge\boldsymbol{E}=(\omega/c)\boldsymbol{H}, \tag{1.17}$$

and

$$\mathrm{i}\boldsymbol{k}\wedge\boldsymbol{H}=-\mathrm{i}(\omega/c)(\varepsilon_L\boldsymbol{L}+\varepsilon_T\boldsymbol{T})\cdot\boldsymbol{E}+(4\pi/c)\boldsymbol{J}_{ext}. \tag{1.18}$$

Notice that

$$\boldsymbol{k}\wedge(\boldsymbol{k}\wedge\boldsymbol{E})=\boldsymbol{kk}\cdot\boldsymbol{E}-k^2\boldsymbol{E}=-k^2\boldsymbol{T}\cdot\boldsymbol{E}. \tag{1.19}$$

This equation, which holds for any vector, will often be used. Now take the vector product of $\boldsymbol{k}$ and (1.17). Eliminating $\boldsymbol{H}$ between this and (1.18) we have

$$\mathrm{i}\omega[\varepsilon_{\mathrm{L}}\boldsymbol{L}+(\varepsilon_{\mathrm{T}}-c^2k^2/\omega^2)\boldsymbol{T}]\cdot\boldsymbol{E}=4\pi\boldsymbol{J}_{\mathrm{ext}}, \tag{1.20}$$

whence we obtain the resolvent operator,

$$\hat{\boldsymbol{G}}(\boldsymbol{k},\omega)=\mathrm{i}\frac{4\pi}{\omega}\left[-\frac{\boldsymbol{k}\boldsymbol{k}}{k^2\varepsilon_{\mathrm{L}}}+\frac{\omega^2}{c^2}\frac{(k^2\boldsymbol{I}-\boldsymbol{k}\boldsymbol{k})}{k^2(k^2-\varepsilon_{\mathrm{T}}\omega^2/c^2)}\right], \tag{1.21}$$

with longitudinal and transverse parts,

$$\hat{G}_{\mathrm{L}}(\boldsymbol{k},\omega)=-\mathrm{i}\frac{4\pi}{\omega\varepsilon_{\mathrm{L}}},\qquad \hat{G}_{\mathrm{T}}(\boldsymbol{k},\omega)=\frac{\mathrm{i}4\pi\omega}{c^2(k^2-\varepsilon_{\mathrm{T}}\omega^2/c^2)}. \tag{1.22}$$

This is the central concept in this scheme. If we were to seek, say, the plasmon dispersion relation from perturbation theory we would calculate ε_{L} and set it equal to zero, which of course is just the pole of $\hat{G}_{\mathrm{L}}$. Likewise the pole of $\hat{G}_{\mathrm{T}}$ gives the transverse mode, or *polariton*, in which the photon mode is coupled to a medium with transverse dielectric constant ε_{T}. Thus the results of perturbation theory serve to construct $\hat{\boldsymbol{G}}$, which yields the e.m. field,

$$\boldsymbol{E}=\hat{\boldsymbol{G}}\cdot\boldsymbol{J}_{\mathrm{ext}}=(\hat{G}_{\mathrm{L}}\boldsymbol{L}+\hat{G}_{\mathrm{T}}\boldsymbol{T})\cdot\boldsymbol{J}_{\mathrm{ext}};\qquad \boldsymbol{H}=\hat{G}_{\mathrm{T}}(c/\omega)\boldsymbol{k}\wedge\boldsymbol{J}_{\mathrm{ext}}. \tag{1.23}$$

The role of the longitudinal and transverse response functions is to describe the screening of the longitudinal and transverse external field. For example, take the longitudinal part of $\boldsymbol{E}$ in (1.23):

$$\boldsymbol{E}_{\mathrm{L}}(\boldsymbol{k},\omega)=-\mathrm{i}(4\pi/\omega\varepsilon_{\mathrm{L}})\boldsymbol{J}_{\mathrm{L,ext}}(\boldsymbol{k},\omega). \tag{1.24}$$

The external source would create by itself a field which, in the absence of the medium, has a longitudinal component,

$$\boldsymbol{E}_{\mathrm{L,ext}}(\boldsymbol{k},\omega)=-\mathrm{i}(4\pi/\omega)\boldsymbol{J}_{\mathrm{L,ext}}(\boldsymbol{k},\omega). \tag{1.25}$$

This corresponds simply to setting $\varepsilon_{\mathrm{L}}=\varepsilon_{\mathrm{T}}=1$ in (1.22). Thus

$$\boldsymbol{E}_{\mathrm{L}}(\boldsymbol{k},\omega)=(1/\varepsilon_{\mathrm{L}})\boldsymbol{E}_{\mathrm{L,ext}}(\boldsymbol{k},\omega), \tag{1.26}$$

which describes longitudinal screening.

Likewise

$$\boldsymbol{E}_{\mathrm{T}}(\boldsymbol{k},\omega)=\mathrm{i}\frac{4\pi\omega}{c^2(k^2-\varepsilon_{\mathrm{T}}\omega^2/c^2)}\boldsymbol{J}_{\mathrm{T,ext}}(\boldsymbol{k},\omega), \tag{1.27}$$

whereas

$$\boldsymbol{E}_{\mathrm{T,ext}}(\boldsymbol{k},\omega)=\mathrm{i}\frac{4\pi\omega}{c^2(k^2-\omega^2/c^2)}\boldsymbol{J}_{\mathrm{T,ext}}(\boldsymbol{k},\omega), \tag{1.28}$$

whence the screening of a transverse external field is given by

$$\boldsymbol{E}_{\mathrm{T}}(\boldsymbol{k},\omega)=\frac{1}{1+\omega^2(\varepsilon_{\mathrm{T}}-1)/(\omega^2-c^2k^2)}\boldsymbol{E}_{\mathrm{T,ext}}(\boldsymbol{k},\omega), \tag{1.29}$$

or, in terms of the transverse conductivity,

$$\boldsymbol{E}_{\mathrm{T}}(\boldsymbol{k},\omega)=\frac{1}{1+\mathrm{i}4\pi\omega\sigma_{\mathrm{T}}/(\omega^2-c^2k^2)}\boldsymbol{E}_{\mathrm{T,ext}}(\boldsymbol{k},\omega). \tag{1.30}$$

Let us now concentrate on conductors. What does transverse screening mean? Clearly transverse screening must be somehow related to the *diamagnetic response* of the electron gas. But we seem to have started from a system of differential equations in which no magnetic effects have been considered. It is at this stage that we must establish the connection between the (L, T) and the (ε, μ) scheme which can be introduced as follows. Always in Fourier transform, *define* the *dielectric function* $\varepsilon=\varepsilon_{\mathrm{L}}$, the *electrical conductivity* $\sigma=\sigma_{\mathrm{L}}$, and the *electrical susceptibility* $\chi=\chi_{\mathrm{L}}$. In this scheme the total current associated with the motion of the free carriers is divided into two parts, that is

$$\boldsymbol{J}_{\mathrm{f}}=\boldsymbol{J}_{\mathrm{E}}+\boldsymbol{J}_{\mathrm{M}}. \tag{1.31}$$

The first one, $\boldsymbol{J}_{\mathrm{E}}$, is the *electric conduction current*, which from the definition of σ is

$$\boldsymbol{J}_{\mathrm{E}}=\sigma\boldsymbol{E}, \tag{1.32}$$

while $\boldsymbol{J}_{\mathrm{M}}$ is the *magnetic induction current*. According to Ampère's law this is related to the *magnetisation* $\mathfrak{M}$ by

$$\boldsymbol{J}_{\mathrm{M}}=c\nabla\wedge\mathfrak{M}=\mathrm{i}c\boldsymbol{k}\wedge\mathfrak{M}. \tag{1.33}$$

But the total free carrier current of (1.31) is identical to $\boldsymbol{J}_{\mathrm{f}}$ of the (L, T) scheme, in which we would express it as

$$\boldsymbol{J}_{\mathrm{f}}=\sigma_{\mathrm{L}}\boldsymbol{E}_{\mathrm{L}}+\sigma_{\mathrm{T}}\boldsymbol{E}_{\mathrm{T}}. \tag{1.34}$$

Notice the crucial difference with (1.32) in which, by definition, $\boldsymbol{E}$ is the complete electric field $\boldsymbol{E}_{\mathrm{L}}+\boldsymbol{E}_{\mathrm{T}}$. Putting together (1.31)–(1.34), we have

$$c\boldsymbol{k}\wedge\mathfrak{M}=\omega(\chi_{\mathrm{L}}-\chi_{\mathrm{T}})\boldsymbol{E}_{\mathrm{T}}, \tag{1.35}$$

and using (1.19), we obtain

$$\mathfrak{M}=(\omega/ck^2)(\chi_{\mathrm{T}}-\chi_{\mathrm{L}})\boldsymbol{k}\wedge\boldsymbol{E}. \tag{1.36}$$

Now, in the (ε, μ) scheme we still define a *magnetic field* $\mathfrak{H}$, but the fundamental field is really the *magnetic induction* $\mathfrak{B}$, which is given by

$$\mathfrak{B}=\mathfrak{H}+4\pi\mathfrak{M}. \tag{1.37}$$

The *magnetic susceptibility* χ_M is defined in the constitutive equation $\mathfrak{M} = \chi_M \mathfrak{H}$, whence

$$\mathfrak{B} = \mu \mathfrak{H} = (1 + 4\pi\chi_M)\mathfrak{H}. \tag{1.38}$$

The connection with χ_L and χ_T is now obvious by looking at (1.36) and (1.38), from which

$$\boldsymbol{k} \wedge \boldsymbol{E} = \frac{ck^2(\mu - 1)}{4\pi\omega\mu(\chi_T - \chi_L)} \mathfrak{B}. \tag{1.39}$$

We now identify this with $\boldsymbol{k} \wedge \boldsymbol{E}$ of (1.6) in which $\boldsymbol{H}$ of the (L, T) scheme becomes the fundamental field $\mathfrak{B}$ of the (ε, μ) scheme. Hence

$$[\mu(\boldsymbol{k}, \omega) - 1]/\mu(\boldsymbol{k}, \omega) = (\omega^2/c^2k^2)[\varepsilon_T(\boldsymbol{k}, \omega) - \varepsilon_L(\boldsymbol{k}, \omega)]. \tag{1.40}$$

Thus diamagnetism is related to the difference between the longitudinal and transverse response functions.

Notice that this is equivalent to saying that, from (1.39),

$$\nabla \wedge \boldsymbol{E} = -(1/c)\partial\mathfrak{B}/\partial t, \tag{1.41}$$

and then the corresponding divergence equation is

$$\nabla \cdot \mathfrak{B} = 0.$$

Finally, since $\boldsymbol{H}$ becomes $\mathfrak{B}$, we must see how the last of equations (1.6) is re-expressed to be consistent with the new scheme. For this we separate out in $\boldsymbol{D}$ the effects of bound and free charges, using (1.31) for the latter. Thus

$$\nabla \wedge \mathfrak{B} = \frac{1}{c}\frac{\partial \boldsymbol{D}_b}{\partial t} + \frac{4\pi}{c}(\boldsymbol{J}_E + \boldsymbol{J}_M) + \frac{4\pi}{c}\boldsymbol{J}_{ext},$$

which, by taking $\boldsymbol{J}_M$ to the left-hand side (l.h.s.) and using (1.33) and (1.37), becomes

$$\nabla \wedge \mathfrak{H} = \frac{1}{c}\frac{\partial \boldsymbol{D}_b}{\partial t} + \frac{4\pi}{c}\boldsymbol{J}_E + \frac{4\pi}{c}\boldsymbol{J}_{ext}. \tag{1.42}$$

Thus we have two choices. In the (L, T) scheme we define $\boldsymbol{P}$, $\boldsymbol{E}$ and $\boldsymbol{H}$. The scheme is completed by defining $\boldsymbol{D}$ as in (1.4) and the constitutive equation (1.8), equivalent to two scalar definitions. The field equations are then (1.6), and ε_L and ε_T are obtained from perturbation theory. in the (ε, μ) scheme we define the fields $\boldsymbol{P}$, $\boldsymbol{E}$ and $\mathfrak{B}$. We define likewise $\boldsymbol{D}$ as in (1.4) and also explicitly the magnetisation $\mathfrak{M}$ as in (1.37), which is equivalent to defining $\mathfrak{H}$. We now have two scalar constitutive equations, namely $\boldsymbol{D} = \varepsilon\boldsymbol{E}$ – equivalent to (1.32) – and (1.38). The field equations are then

$$\left.\begin{array}{ll}\nabla\cdot\mathfrak{B}=0, & \nabla\wedge\boldsymbol{E}=-\dfrac{1}{c}\dfrac{\partial\mathfrak{B}}{\partial t},\\ \nabla\cdot\boldsymbol{D}=4\pi\sigma_{\text{ext}}, & \nabla\wedge\mathfrak{H}=\dfrac{1}{c}\dfrac{\partial\boldsymbol{D}_{\text{b}}}{\partial t}+\dfrac{4\pi}{c}\boldsymbol{J}_{\text{E}}+\dfrac{4\pi}{c}\boldsymbol{J}_{\text{ext}}.\end{array}\right\}\tag{1.43}$$

The connection between the two schemes is provided by $\varepsilon=\varepsilon_{\text{L}}$ and (1.40). It is also interesting to derive the Green function in the (ε,μ) scheme. This is easily obtained by Fourier transforming the two curls in (1.43) and eliminating $\mathfrak{B}$. This yields

$$\hat{\mathfrak{G}}=\mathrm{i}\frac{4\pi}{\omega\varepsilon}\left[-\frac{\boldsymbol{kk}}{k^2}+\frac{\omega^2}{c^2}\varepsilon\mu\frac{(\boldsymbol{I}-\boldsymbol{kk})}{k^2(k^2-\varepsilon\mu\omega^2/c^2)}\right],\tag{1.44}$$

and now

$$\boldsymbol{E}=\hat{\mathfrak{G}}\cdot\boldsymbol{J}_{\text{ext}},\qquad \boldsymbol{B}=\hat{\mathfrak{G}}_{\text{T}}(c/\omega)\boldsymbol{k}\wedge\boldsymbol{J}_{\text{ext}}.\tag{1.45}$$

The meaning of this analysis is illustrated by the following physical problem. Suppose we apply an external source. In the (L, T) scheme we calculate a magnetic field $\boldsymbol{H}$ which, from (1.22) and (1.23), is

$$\boldsymbol{H}(\boldsymbol{k},\omega)=\mathrm{i}\frac{4\pi}{c}\frac{\boldsymbol{k}\wedge\boldsymbol{J}_{\text{ext}}(\boldsymbol{k},\omega)}{k^2-\varepsilon_{\text{T}}(\boldsymbol{k},\omega)\omega^2/c^2}.\tag{1.46}$$

Now, this is the magnetic field inside the system, and we want to see what this corresponds to in the (ε,μ) scheme. The way to find out is to express ε_{T}, from (1.40), in terms of ε and μ and to transform the denominator of (1.46) into

$$k^2-\frac{\omega^2}{c^2}\varepsilon_{\text{T}}=\frac{1}{\mu}\left(k^2-\frac{\omega^2}{c^2}\varepsilon\mu\right);\tag{1.47}$$

then (1.46) becomes

$$\boldsymbol{H}(\boldsymbol{k},\omega)=\mathrm{i}\frac{4\pi}{c}\frac{\mu}{(k^2-\varepsilon\mu\omega^2/c^2)}\boldsymbol{k}\wedge\boldsymbol{J}_{\text{ext}}(\boldsymbol{k},\omega)=\mathfrak{B}(\boldsymbol{k},\omega),\tag{1.48}$$

as follows immediately from (1.44) and (1.45). Notice that in the magnetostatic limit ($\omega=0$) the magnetic induction $\mathfrak{B}$ still depends on the properties of the medium, whereas dividing (1.48) by μ we have the field $\mathfrak{H}$ which of course does not depend on the properties of the medium. The lesson to be learned from this exercise is that if we set up our perturbation theory to calculate the response functions and use the (L, T) scheme (1.6) to calculate $\boldsymbol{H}$ we are calculating the magnetic field *inside* the medium. Thus, no matter what we do we are always looking at $\mathfrak{B}$ and never seeing $\mathfrak{H}$.

1.2. Bulk and surface diamagnetism

In order to evaluate the response functions we must choose some model. Suppose we take a semiclassical picture of the electron gas. On the basis of a Boltzmann equation formulation one can calculate the longitudinal and transverse conductivity functions and thus obtain the *semiclassical Lindhard formulae* for ε_L and ε_T. We are concerned with a degenerate electron gas with Fermi velocity $\boldsymbol{v}_F$ and *Fermi–Thomas screening constant* k_{FT} related to the plasma frequency ω_p by $k_{FT}^2 = 3\omega_p^2/v_F$. We are not interested now in lifetime effects due to collisions of the charge carriers. We want to concentrate on diamagnetism, not on the problem of electrical conductivity as such. Furthermore, we are interested in the static limit. Thus, for $\omega < v_F k$, the results we are interested in are

$$\varepsilon_L(\boldsymbol{k}, \omega) = 1 + \frac{k_{FT}^2}{k^2}\left(1 - \frac{\omega^2}{v_F^2 k^2}\right), \tag{1.49}$$

and

$$\varepsilon_T(\boldsymbol{k}, \omega) = 1 - \frac{k_{FT}^2}{k^2} + \mathrm{i}\frac{k_{FT}^2 v_F}{4k\omega}\left(1 - \frac{\omega^2}{v_F^2 k^2}\right). \tag{1.50}$$

It is clear that the semiclassical model predicts

$$\lim_{\omega \to 0} \frac{\omega^2}{c^2 k^2}[\varepsilon_T(\boldsymbol{k}, \omega) - \varepsilon_L(\boldsymbol{k}, \omega)] = 0; \tag{1.51}$$

hence, by (1.40), *no classical magnetism*, a well-known result. The *quantum-mechanical Lindhard formulae*, under the same conditions, are rather different. For ε_L we have

$$\varepsilon_L(\boldsymbol{k}, \omega) = 1 + \frac{k_{FT}^2}{k^2}\left(1 + \mathrm{i}\frac{\pi\omega}{2v_F k} - \frac{\omega^2}{v_F^2 k^2} - \frac{k^2}{12k_F^2} + \ldots\right). \tag{1.52}$$

It is clear that now

$$\lim_{\omega \to 0} \frac{\omega^2}{c^2 k^2}\varepsilon_L(\boldsymbol{k}, \omega) = 0, \tag{1.53}$$

thus we need be concerned only with ε_T. This is somewhat more complicated, namely

$$\varepsilon_T(\boldsymbol{k}, \omega) = -3\frac{\omega_p^2}{\omega^2}\left[(1 \pm a^2) - \frac{1}{2a}(1 - a^2)^2 \ln\left|\frac{a+1}{a-1}\right|\right], \tag{1.54}$$

where $a = k/2k_F$. Thus now we have some diamagnetism. For our purposes it is sufficient to look at the two limits of (1.54) for $a \to 0$ and

$a\to\infty$. From (1.40) we find

$$\lim_{a\to 0}\frac{\mu-1}{\mu}=-\frac{\omega_{\rm p}^2}{4c^2k_{\rm F}^2},\qquad \lim_{a\to\infty}\frac{\mu-1}{\mu}=-\frac{\omega_{\rm p}^2}{c^2k^2}. \tag{1.55}$$

We may think of an approximate interpolation formula of the form

$$\frac{\mu(\boldsymbol{k})-1}{\mu(\boldsymbol{k})}=-\frac{k_{\rm p}^2}{4k_{\rm F}^2+k^2}\quad(\omega_{\rm p}=ck_{\rm p}), \tag{1.56}$$

whence

$$\mu(\boldsymbol{k})=\frac{4k_{\rm F}^2+k^2}{4k_{\rm F}^2+k_{\rm p}^2+k^2},\qquad \chi_{\rm M}(\boldsymbol{k})=-\frac{k_{\rm p}^2}{4\pi(4k_{\rm F}^2+k_{\rm p}^2+k^2)}. \tag{1.57}$$

This is only a plausible device to introduce some spatial dispersion into our picture of the static diamagnetic response. For a bulk system this is not too important. Dropping k^2 in the denominator of $\chi_{\rm M}$ we find the usual *Landau diamagnetism*. But we seek a surface problem, in which we want to study the inhomogeneous surface region. For this we shall use the dispersion function of (1.57).

What we really would like to know is the diamagnetic response function of a surface system, meaning by this the bulk material and its surface. This is in general a very difficult problem, but there is a simple model which can be solved fairly easily. Suppose we take our medium bounded by the surface $z=0$ and contained in $z>0$ and we fill up the region $z<0$ with the mirror image of the medium. We are then inventing a hypothetical extended medium which fills up the entire space and gives a false picture of the real problem for $z<0$, where we have a vacuum. But this does not matter. We only need a description of the actual medium in $z>0$, where we want to know the resulting magnetic induction $\mathfrak{B}$ which describes the response to a constant external field $\mathfrak{H}$. Such a field could be thought of as produced by a hypothetical external current whose amplitude we define to be located strictly at $z=0$; namely

$$\boldsymbol{J}_{\rm ext}(\boldsymbol{r})=J_{\rm s}\,{\rm e}^{{\rm i}\kappa x}\delta(z)\boldsymbol{y}^0. \tag{1.58}$$

Notice that our $J_{\rm ext}$ has only y-component and varies with x and z. Its Fourier transform is

$$\boldsymbol{J}_{\rm ext}(\boldsymbol{k})=J_{\rm s}(0,1,0). \tag{1.59}$$

Thus, by choosing $\boldsymbol{k}=(\kappa,0,q)$, $\boldsymbol{J}_{\rm ext}$ is purely transverse and

$$\boldsymbol{k}\wedge\boldsymbol{J}_{\rm ext}=-J_{\rm s}q\boldsymbol{x}^0+J_{\rm s}\kappa\boldsymbol{z}^0. \tag{1.60}$$

We shall actually be interested in the limit $\kappa\to 0$. Thus we shall be concerned only with the x-component of (1.60).

In the static case we have, from (1.44) and (1.45),

$$\frac{c}{\omega}\mathfrak{G}_{\mathrm{T}}=\mathrm{i}\frac{4\pi}{ck^2}\mu. \tag{1.61}$$

Thus $\boldsymbol{J}_{\mathrm{ext}}$ creates a magnetic field with Fourier transform

$$\boldsymbol{\mathfrak{H}}(\boldsymbol{k})=\mathrm{i}\frac{4\pi}{ck^2}\boldsymbol{k}\wedge\boldsymbol{J}_{\mathrm{ext}}(\boldsymbol{k});\qquad \mathfrak{H}_x(q)=-\mathrm{i}\frac{4\pi q}{ck^2}. \tag{1.62}$$

In real space we have

$$\mathfrak{H}_x(z)=\frac{1}{2\pi}\int \mathrm{e}^{\mathrm{i}qz}\mathfrak{H}_x(q)\,\mathrm{d}q. \tag{1.63}$$

The evaluation of this integral depends on the sign of z. For $z>0$ we close the integration contour through the upper half circle at infinity. This yields

$$\mathfrak{H}_x(z)=(2\pi/c)J_{\mathrm{s}}\,\mathrm{e}^{-\kappa z}\rightarrow(2\pi/c)J_{\mathrm{s}}\quad(z>0). \tag{1.64}$$

For $z<0$ we must close through the lower half circle and

$$\mathfrak{H}_x(z)=-(2\pi/c)J_{\mathrm{s}}\,\mathrm{e}^{\kappa z}\rightarrow-(2\pi/c)J_{\mathrm{s}}\quad(z<0). \tag{1.65}$$

Thus the field $\mathfrak{H}$ is antisymmetric, whereas if we evaluate $\boldsymbol{E}_{\mathrm{ext}}$ we find that it is symmetric. The same happens if we evaluate $\boldsymbol{\mathfrak{B}}$ and $\boldsymbol{E}$. But of course this is just what mirror image means. The electric field *vector* is mirrored into its symmetric image, but $\boldsymbol{\mathfrak{B}}$ is a *skew vector*, and its mirror image is precisely antisymmetric. In physical terms $\boldsymbol{\mathfrak{B}}$ induces a rotation of the free carriers whose mirror reflection is shown in Fig. 1. This is the meaning of the transformation law $(\mathfrak{B}_x, \mathfrak{B}_y, \mathfrak{B}_z)\rightarrow(-\mathfrak{B}_x, -\mathfrak{B}_y, \mathfrak{B}_z)$, whereas $(E_x, E_y, E_z)\rightarrow(E_x, E_y, -E_z)$, and this is what happens when we fill up the region $z<0$ with the mirror image of the medium under consideration. We may regard $2\pi J_{\mathrm{s}}/c$ as the amplitude of an external magnetic field H_0 which is going to create a magnetic induction inside the medium. This is easy to obtain from (1.62), which yields

$$\mathfrak{B}_x(q)=-\mathrm{i}\frac{4\pi q}{ck^2}\mu(\boldsymbol{k})J_{\mathrm{s}}=-\mathrm{i}\frac{2\mu(\boldsymbol{k})q}{k^2}H_0. \tag{1.66}$$

It is here that we use (1.57), which yields

$$\mathfrak{B}_x(z)=-\frac{\mathrm{i}H_0}{\pi}\int \mathrm{e}^{\mathrm{i}qz}\frac{(4k_{\mathrm{F}}^2+k^2)q}{k^2(4k_{\mathrm{F}}^2+k_{\mathrm{p}}^2+k^2)}\mathrm{d}q. \tag{1.67}$$

The poles of the integrand, for $z>0$, are at $q=\mathrm{i}\kappa$ and $q=\mathrm{i}q_{\mathrm{s}}$, where we

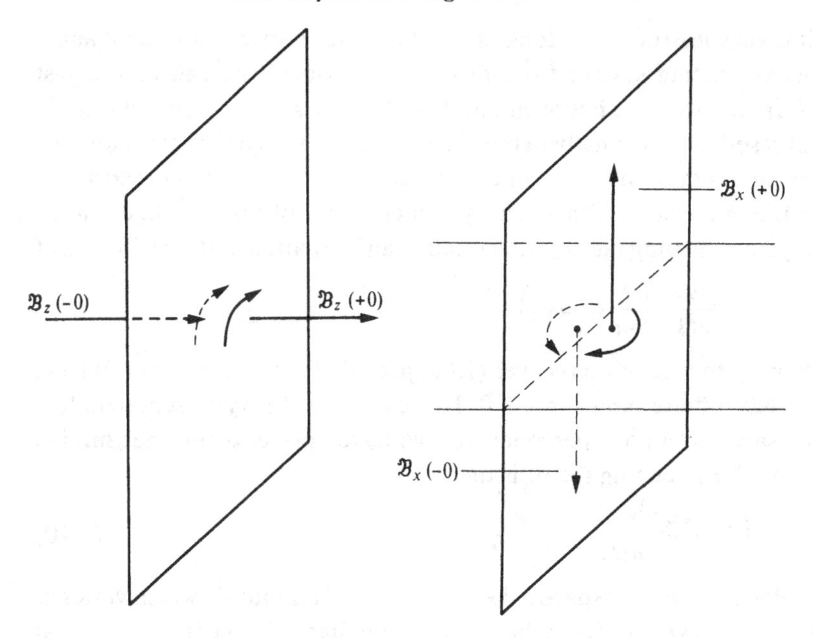

Fig. 1. Mirror image of magnetic induction and induced free-carrier trajectories. The surface is the plane $z=0$.

define $q_s^2=4k_F^2+k_p^2$. Let us also write, from (1.57),

$$\mu_0=4k_F^2/(4k_F^2+k_p^2). \tag{1.68}$$

We then find

$$\mathfrak{B}_x(z)=\mu_0H_0[1+(k_p^2/4k_F^2)\,e^{-q_sz}]. \tag{1.69}$$

At this stage it is convenient to consider the scope and limitations of what we have done. First, we are studying a particular model. When we fill up the region $z<0$ with the mirror image of the medium, we are actually introducing a model of a *specular surface*. Secondly, we shall not take the trouble to solve a complete matching problem, in which one would enquire about real fields on both sides of the surface. Both the question of the model of the surface and the general matching problem will be further investigated in this and the following chapter. For the time being we are only interested in finding by a particular *ad hoc* trick, a physically plausible model of the inhomogeneity associated with surface diamagnetism, and this is what (1.69) gives. In the bulk, $z\to\infty$ and $\mathfrak{B}_x\to\mu_0H_0$, the Landau value. The decay of the inhomogeneous term is governed by the screening constant q_s. In practice k_p is usually negligible compared with k_{FT} and the decay constant is essentially $2k_F$.

It is very instructive to look at the difference between a normal and a superconducting system from this point of view. The calculation just performed to find $\mathfrak{B}_x$ has been carried out in the (ε, μ) scheme but it could be phrased more formally in the (L, T) scheme by saying that we are just working out the consequences of the transverse screening as described in (1.30). For a system with a density n of electrons of charge $\boldsymbol{e}$, in the static limit, the screening factor, from (1.40), can be written as the reciprocal of

$$1+\frac{4\pi e^2}{c^2k^2}\left(\frac{n}{m}+\frac{\omega^2}{e^2}\chi_{\mathrm{T}}\right).$$

Following this track we recover (1.69) provided the system is normal, i.e. $\omega^2\chi_{\mathrm{T}}$ has a finite limit for $\omega\rightarrow 0$. However, the theory of superconductivity says that in a superconductor we have perfect diamagnetism, i.e. $\chi_{\mathrm{T}}=0$. The screening factor is then

$$\left(1+\frac{4\pi e^2 n}{c^2k^2m}\right)^{-1}=\frac{k^2}{k_{\mathrm{p}}^2+k^2}. \tag{1.70}$$

This describes in real space a decay of the *total* amplitude which varies as $\exp(-k_{\mathrm{p}}z)$, a very different behaviour from that reflected in (1.69). What (1.70) says is that in a superconductor the magnetic field cannot penetrate into the bulk beyond a thin surface layer where we have an inhomogeneous amplitude with decay constant k_{p}. After this exercise we shall not explicitly study magnetic effects. In the remainder of this chapter we shall study the general problem of matching electromagnetic fields at surfaces using the (L, T) scheme.

1.3. Electromagnetic matching at surfaces

In principle the matching conditions at some chosen *matching plane* which supposedly separates two media can be stated very simply: the tangential components of $\boldsymbol{E}$ and $\boldsymbol{H}$ must be continuous. We shall see in due course that this does not always suffice to determine the e.m. field inside the media.

For the time being we shall concentrate on the effects of the surface on the matching problem. In order to proceed by gradual stages we shall explicitly use in this discussion the ideal model in which the two media extend, homogeneous and unperturbed with their own bulk properties, up to the matching plane which, in this model, is identified with the surface of (abrupt) separation between the two media. More realistic models, which take account of smooth-surface inhomogeneities, will be discussed later. At this stage we only intend to focus on the distinct features of the problems which arise with matching at a given *matching plane* (see chapter 2), which we shall choose at $z=0$. However, the formalism of the matching methods presently to be described are equally applicable to other models, although the evaluation of the resulting formulae may be more difficult.

What we are going to do now is to set up a systematic method for carrying out the sort of analysis which appeared as an *ad hoc* trick in § 1.2. The idea is that given the two half-media joined at the matching plane we shall imagine each one extended to fill up the other side and we shall reconstitute the real system by appropriately matching the two real halves and the fields therein contained. This implies that we must match correctly the tangential components of $\boldsymbol{E}$ and $\boldsymbol{H}$. For this it proves convenient to introduce the concept of surface impedance, as follows.

We define the *surface impedance tensor* $\boldsymbol{Z}$, so that $\boldsymbol{E}_{\|}(s)$, the electric field parallel to the surface and evaluated *at the surface*, is related to $\boldsymbol{H}(s)$ by

$$\boldsymbol{E}_{\|}(s) = (c/4\pi)\boldsymbol{Z}\cdot[\boldsymbol{n}\wedge\boldsymbol{H}(s)]. \tag{1.71}$$

In line with the program just outlined, we define a surface impedance for each extended medium and we shall use the value of it calculated from each 'real' side. This definition applies separately to the incident, reflected and transmitted waves; and $\boldsymbol{n}$ is the *outer normal* to the surface. Thus $\boldsymbol{n} = (0, 0 \pm 1)$ where $\pm$ holds for a medium contained in $z \lessgtr 0$. Next we specify the polarisation of the e.m. field. Thus, for P-mode, $\boldsymbol{E} = (E_x, 0, E_z)$, $\boldsymbol{H} = (0, H_y, 0)$, and we have

$$Z_{\mathrm{P}}^{\lessgtr} = Z_{xx}^{\lessgtr} = \mp(4\pi/c)[E_x/H_y]_{\mp 0}. \tag{1.72}$$

The sign conventions must be kept in mind to avoid errors in the calculations. For S-mode, $\boldsymbol{E} = (0, E_y, 0)$,

$$\boldsymbol{H} = (H_x, 0, H_z) \quad \text{and} \quad Z_{\mathrm{S}}^{\lessgtr} = Z_{yy}^{\lessgtr} = \pm(4\pi/c)[E_y/H_x]_{\mp 0}. \tag{1.73}$$

Different surface properties related to the response of the surface can be readily evaluated if the surface impedances of the two media joined at the matching plane are known. For example, consider an incident e.m. wave in the P-mode approaching the surface from the left (fig. 2). Then

$$\left.\begin{aligned}
\boldsymbol{E}_{\mathrm{i}} &= (E_{x\mathrm{i}}\boldsymbol{x}^0 + E_{z\mathrm{i}}\boldsymbol{z}^0)\,\mathrm{e}^{\mathrm{i}(\kappa x + qz - \omega t)},\\
\boldsymbol{H}_{\mathrm{i}} &= H_{y\mathrm{i}}\boldsymbol{y}^0\,\mathrm{e}^{\mathrm{i}(\kappa x + qz - \omega t)},\\
\boldsymbol{E}_{\mathrm{r}} &= (E_{x\mathrm{r}}\boldsymbol{x}^0 + E_{z\mathrm{r}}\boldsymbol{z}^0)\,\mathrm{e}^{\mathrm{i}(\kappa x - qz - \omega t)},\\
\boldsymbol{H}_{\mathrm{r}} &= H_{y\mathrm{r}}\boldsymbol{y}^0\,\mathrm{e}^{\mathrm{i}(\kappa x - qz - \omega t)},\\
\boldsymbol{E}_{\mathrm{t}} &= (E_{x\mathrm{t}}\boldsymbol{x}^0 + E_{z\mathrm{t}}\boldsymbol{z}^0)\,\mathrm{e}^{\mathrm{i}(\kappa x + q'z - \omega t)},\\
\boldsymbol{H}_{\mathrm{t}} &= H_{y\mathrm{t}}\boldsymbol{y}^0\,\mathrm{e}^{\mathrm{i}(\kappa x + q'z - \omega t)},
\end{aligned}\right\} \tag{1.74}$$

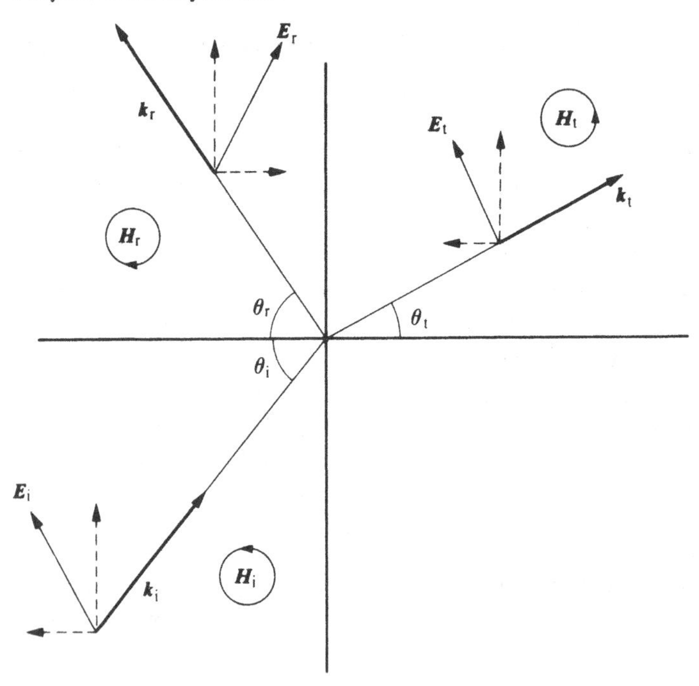

Fig. 2. Incident, reflected and transmitted e.m. field with P-mode geometry.

where the subscripts i, r, and t stand for incident, reflected, and transmitted. Let 1 and 2 indicate the media in $z<0$ and $z>0$, respectively, and notice that $\boldsymbol{H}$ changes sign upon reflection, as $\boldsymbol{k}\wedge\boldsymbol{E}$ changes sign. Then, from (1.72), (1.73) and

$$E_{x\mathrm{i}}+E_{x\mathrm{r}}=E_{x\mathrm{t}}, \qquad H_{y\mathrm{i}}+H_{y\mathrm{r}}=H_{y\mathrm{t}}, \tag{1.75}$$

we find the reflection amplitude

$$\frac{H_{y\mathrm{r}}(-0)}{H_{y\mathrm{i}}(-0)}=\frac{Z_{\mathrm{P}}^{1<}+Z_{\mathrm{P}}^{2>}}{Z_{\mathrm{P}}^{1<}-Z_{\mathrm{P}}^{2>}}. \tag{1.76}$$

For S-mode, in the same way,

$$\frac{E_{y\mathrm{r}}(-0)}{E_{y\mathrm{i}}(-0)}=\frac{Z_{\mathrm{S}}^{2>}-Z_{\mathrm{S}}^{1>}}{Z_{\mathrm{S}}^{2>}+Z_{\mathrm{S}}^{1<}}. \tag{1.77}$$

The denominators in (1.76) and (1.77) are just the secular determinants of (1.75) with $E_{\mathrm{i}}=H_{\mathrm{i}}=0$, which corresponds to a self-sustained eigenmode of the surface system in which there is no *input* and yet there is a finite *output.* In other words, we look at the reflectivity as a surface response function and obtain the surface mode from the pole of the reflectivity, which reads

$$Z_{\mathrm{P}}^{1<} - Z_{\mathrm{P}}^{2>} = 0, \quad \text{or} \quad Z_{\mathrm{S}}^{1<} + Z_{\mathrm{S}}^{2>} = 0. \tag{1.78}$$

Consider the simple case in which we have a vacuum in $z < 0$ and $\boldsymbol{E}_{\mathrm{i}}$ is purely transverse. Then $E_{\mathrm{i}} = H_{\mathrm{i}}$, $E_{x\mathrm{i}} = E_{\mathrm{i}} \cos \theta_{\mathrm{i}}$, and

$$Z_{\mathrm{P}}^{\mathrm{vac}<} = -(4\pi/c) \cos \theta_{\mathrm{i}}. \tag{1.79}$$

We shall see later that this holds even if $\boldsymbol{E}_{\mathrm{i}}$ has some longitudinal component. Thus, the reflectivity and the surface mode dispersion relation (s.m.d.r.) for a medium in $z > 0$ are, in the P-mode,

$$R = \left| \frac{Z_{\mathrm{P}}^{>} - (4\pi/c) \cos \theta_{\mathrm{i}}}{Z_{\mathrm{P}}^{>} + (4\pi/c) \cos \theta_{\mathrm{i}}} \right|^2, \qquad Z_{\mathrm{P}}^{>} + (4\pi/c) \cos \theta_{\mathrm{i}} = 0. \tag{1.80}$$

The form of these results is suitable for reflectivity studies, in which we think of ω and θ_{i} as the natural variables, but in a dispersion relation we look for a relation between ω and κ. From fig. 2, since in a vacuum $\omega = ck$, we have $\kappa = \omega \sin \theta / c$, thus we rewrite the s.m.d.r. more conveniently as

$$Z_{\mathrm{P}}(\kappa, \omega) + (4\pi/c)(1 - c^2\kappa^2/\omega^2)^{1/2} = 0. \tag{1.81}$$

It is practical to keep in mind the relation between surface impedance, reflectivity and s.m.d.r. because then we only need to do the work for one of these, and we have the solution for the other two.

Suppose now that the two media are simply described by frequency dependent dielectric functions ε_1 and ε_2. Then $\sin \theta_{\mathrm{i}} = c\kappa/\omega\varepsilon_1^{1/2}$ and $\sin \theta_{\mathrm{t}} = c\kappa/\omega\varepsilon_2^{1/2}$, whence the s.m.d.r. acquires the well-known and simple form

$$\left(\kappa^2 - \frac{\omega^2}{c^2}\varepsilon_1\right)^{1/2} \Big/ \varepsilon_1 + \left(\kappa^2 - \frac{\omega^2}{c^2}\varepsilon_2\right)^{1/2} \Big/ \varepsilon_2 = 0. \tag{1.82}$$

Before tackling less trivial problems it is convenient to summarise the picture that emerges from this simple model.

Consider the case $\varepsilon_1 = 1$, i.e., a vacuum in $z < 0$. Rearranging (1.82) and taking the square we find another familiar and simple equation, namely

$$(c\kappa/\omega)^2 = \varepsilon/(1 + \varepsilon), \tag{1.83}$$

but in taking the square we may introduce spurious solutions. Remember we are looking for stable surface modes with amplitudes that decay on going into the bulk. For P-mode geometry we are looking for a magnetic field of the form

$$\boldsymbol{H} = H_y(z)\, \mathrm{e}^{\mathrm{i}(\kappa x - \omega t)}. \tag{1.84}$$

Eliminating E from the two curl equations in (1.6) we find

$$\partial^2 H_y/\partial z^2 + [(\omega^2/c^2)\varepsilon - \kappa^2]H_y = 0. \tag{1.85}$$

The oscillatory or exponentially damped behaviour of H_y depends on the sign of the quantity in brackets. The condition for a stable surface mode is

$$\kappa^2 - (\omega^2/c^2)\varepsilon > 0. \tag{1.86}$$

With this criterion we can discard the spurious solutions that may arise on solving (1.83) for ω in terms of κ. The specific form of the bulk or surface modes obtained from the poles of (1.21) and from (1.82) or (1.83) depends on $\varepsilon(\omega)$. For a metal we have

$$\varepsilon = 1 - \omega_p^2/\omega^2 \tag{1.87}$$

and, solving (1.83):

$$\omega_\pm^2 = \tfrac{1}{2}\omega_p^2 + c^2\kappa^2 + \tfrac{1}{2}\sqrt{(\omega_p^4 + 4c^2\kappa^2)}. \tag{1.88}$$

With the criterion (1.86) we discard ω_+ as physically unrealisable. The corresponding picture is shown in fig. 3. For metals, retardation effects are actually unimportant in this context. Letting $c \to \infty$ in (1.82) we have, in this extremely crude picture, the non-dispersive surface plasmon

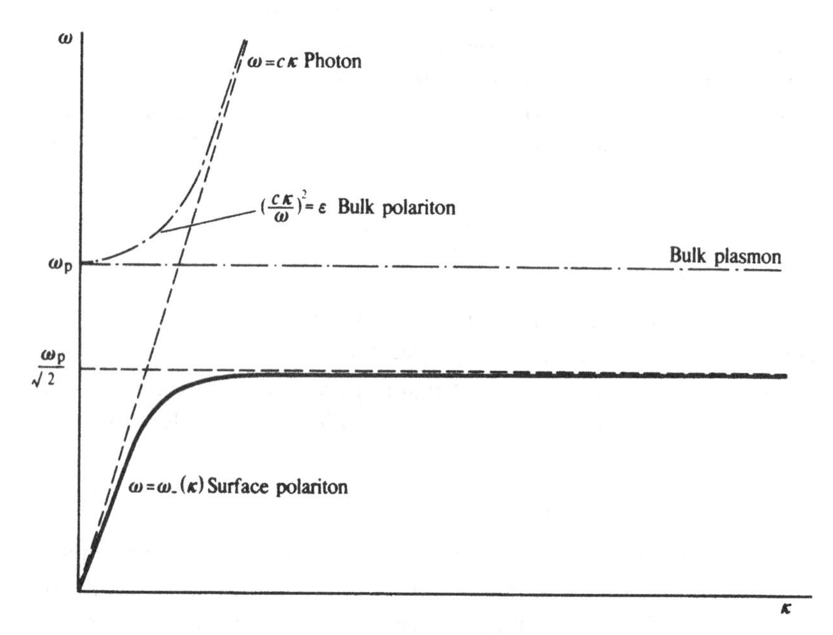

Fig. 3. Dynamical (plasmon) and electrodynamical (polariton) modes for a non-dispersive free-electron metal. Dot-dashed lines: bulk modes. Solid line: surface mode.

frequency ω_s given by

$$\varepsilon_1(\omega_s)+\varepsilon_2(\omega_s)=0, \tag{1.89}$$

and, in particular, for the vacuum–metal case,

$$\omega_s=\omega_p/\sqrt{2}, \tag{1.90}$$

which would correspond to raising to infinity the slope of the photon line in fig. 3. For a dielectric crystal matrix we have

$$\varepsilon=\varepsilon_c[1+\Omega_p^2/(\omega_T^2-\omega^2)], \tag{1.91}$$

with longitudinal bulk frequency $\omega_L^2=\omega_T^2+\Omega_p^2$, static dielectric constant $\varepsilon_0=\varepsilon_c(1+\Omega_p^2/\omega_T^2)$, and optical or high-frequency dielectric constant $\varepsilon_\infty=\varepsilon_c$. The bulk polariton now has two branches (fig. 4)

$$\omega_\pm^2=\frac{1}{2}\left(\frac{c^2\kappa^2}{\varepsilon_c}+\omega_T^2+\Omega_p^2\right)\pm\frac{1}{2}\sqrt{\left(\frac{c^2\kappa^2}{\varepsilon_c}+\omega_T^2+\Omega_p^2\right)^2-\frac{4c^2\kappa^2}{\varepsilon_T}\omega_T^2}\Bigg]. \tag{1.92}$$

The important thing is that there is a gap of forbidden bulk eigenfrequencies. This allows for surface eigenvalues provided (1.86) is

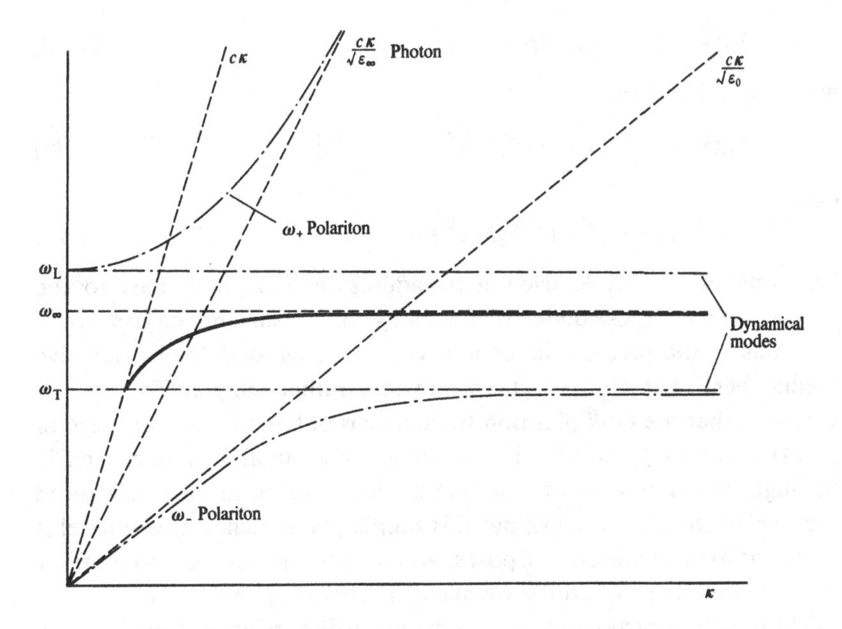

Fig. 4. Dynamical (phonon, exciton) and electrodynamical (polariton) modes for a non-dispersive dielectric. Dot-dashed lines: bulk modes. Solid line: surface mode.

satisfied. With a dielectric function like that of (1.91), starting from $\omega = 0$ we see that this condition is satisfied for $\omega = \omega_T$, when $\varepsilon \to \infty$ and, from (1.83) $\kappa = \omega_T/c$. Here starts the surface polariton branch whose frequency, for $\kappa \to \infty$, tends to

$$\omega_\infty = \sqrt{\{(\varepsilon_0 + 1)/(\varepsilon_\infty + 1)\}}\omega_T, \tag{1.93}$$

i.e. the surface eigenfrequency stays always inside the forbidden gap (fig. 4).

Now, in this extremely crude picture the dielectric functions do not depend on the wavevector, i.e. the model makes no allowance for non-local interactions. Of course we should also include friction or dissipation: the bulk modes need not have an infinite lifetime. But we are not concerned just now with these effects. Let us rather concentrate on the effects of dispersion, still making use of simple models. The moment we introduce some $\boldsymbol{k}$-dependence we can have two different dielectric functions, for longitudinal and transverse responses. For example, in the simplest dispersive model for metals we can take, for a homogeneous system,

$$\varepsilon_L(\boldsymbol{k}, \omega) = 1 + \omega_p^2/(\beta_L^2 k^2 - \omega^2) \tag{1.94}$$

and

$$\varepsilon_T(\boldsymbol{k}, \omega) = 1 + \omega_p^2/(\beta_T^2 k^2 - \omega^2), \tag{1.95}$$

while for a dielectric,

$$\varepsilon_L(\boldsymbol{k}, \omega) = \varepsilon_c[1 + \Omega_p^2/(\beta_L^2 k^2 + \omega_T^2 - \omega^2)], \tag{1.96}$$

and

$$\varepsilon_T(\boldsymbol{k}, \omega) = \varepsilon_c[1 + \Omega_p^2/(\beta_T^2 k^2 + \omega_T^2 - \omega^2)]. \tag{1.97}$$

We could also include friction by adding an imaginary part to the denominators of these dielectric functions, but we shall not be concerned with this in the present discussion. Of course as models of dispersive media these dielectric functions are still rather simple. To say, for example, that the bulk plasmon frequency is obtained, from the zero of (1.94), as $\omega^2 = \omega_p^2 + \beta_L^2 k^2$ is just too much of a simplification, especially for high values of k. All these details are discussed in more advanced theories of the electron gas, but this simple phenomenological model is sufficient for our present purposes, and in fact it is not too bad in many practical instances, provided we avoid short-wave phenomena.

The most conspicuous changes occur when dispersion is introduced in a dielectric. The bulk dynamical modes – e.g. excitons or optical phonons – now become dispersive. The upper bulk polariton branch changes only

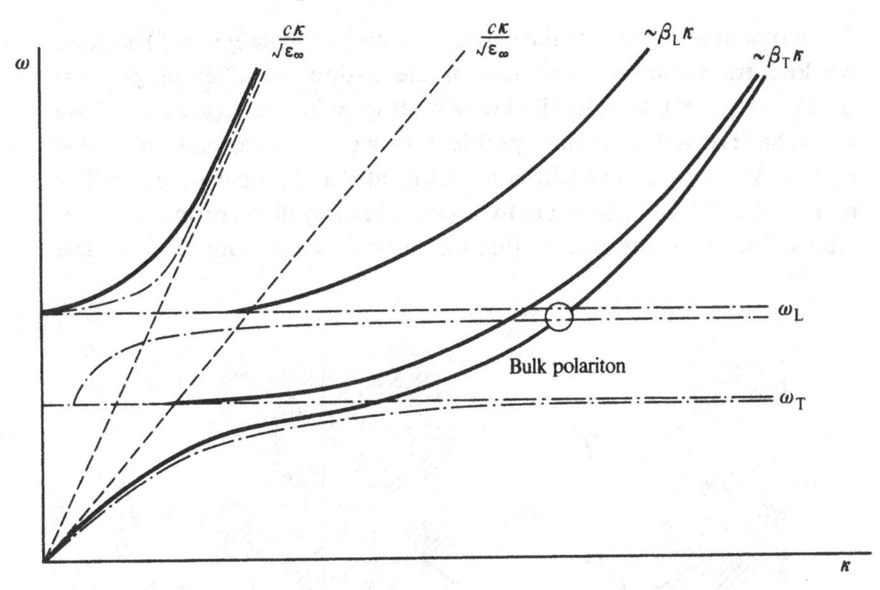

Fig. 5. Bulk and surface modes for dispersive (solid lines) and non-dispersive (dot-dashed lines) dielectrics. ⊙, Possible coupling.

slightly but the lower one (fig. 5) bends up following the transverse dynamical mode, thus crossing the frequency range that was formerly forbidden. This is bound to affect drastically the surface polariton branch. Nothing similar to the surface mode of the non-dispersive case could now be expected to survive in general because this crosses the bulk polariton and there we may have a coupling between bulk and surface polariton modes. Since there are no forbidden bulk frequencies in the dispersive model this coupling should become possible somewhere. However, this is only a possibility; and how do we know the strength of this coupling? In a way we could phrase the situation in this manner: the general matching conditions, as discussed above, suffice to determine the surface solutions of Maxwell's equations for the non-dispersive case, but dispersion introduces a further unknown, namely the admixture of bulk polariton which now goes into the making of the surface mode, and it looks as if we need further information.

There is another way of looking at this problem. Dispersion means non-local interactions in real space. Thus, for example, an electric field $\boldsymbol{E}$ regarded as a stimulus at a point $\boldsymbol{r}'$ produces as response a polarisation $\boldsymbol{P}$ at another point $\boldsymbol{r}$. Indeed $\boldsymbol{P}(\boldsymbol{r})$ will collect the results of all the stimuli $\boldsymbol{E}(\boldsymbol{r}')$:

$$\boldsymbol{P}(\boldsymbol{r}) = \int \chi_s(\boldsymbol{r}, \boldsymbol{r}') \cdot \boldsymbol{E}(\boldsymbol{r}')\, \mathrm{d}\boldsymbol{r}'. \tag{1.98}$$

The moment we introduce the surface, we cannot write $\chi_s(\boldsymbol{r}-\boldsymbol{r}')$ because we lose translational invariance in the z-direction. Let us put $\boldsymbol{r}=(\boldsymbol{\rho}, z)$, $\boldsymbol{r}'=(\boldsymbol{\rho}', z')$. If we really knew what to write for $\chi_s(\boldsymbol{\rho}, z; \boldsymbol{\rho}', z')$ we would have solved the surface problem. However, we can make plausible models. We think of an excitation starting at $\boldsymbol{r}'$ and propagating to $\boldsymbol{r}$. The point is that this can happen in two ways, either by direct propagation, or after reflection at the surface (fig. 6), and each contributes a term. The

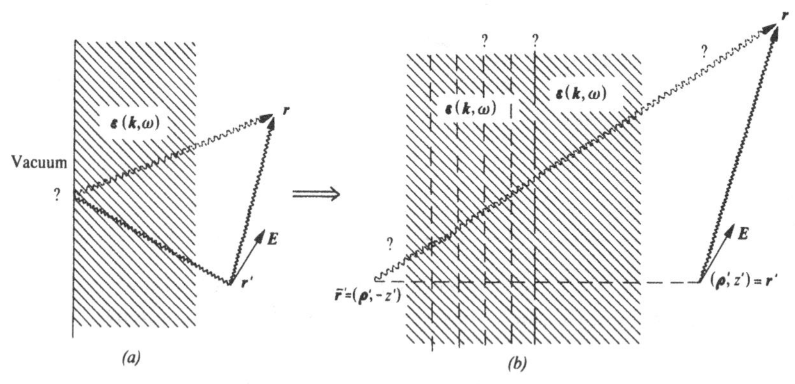

Fig. 6. A semi-infinite medium (a) with a surface, can be simulated by an infinitely extended medium (b), provided the extended field has an adequate behaviour. This can be created by a suitable hypothetical stimulus which in principle can be anywhere except inside the real material.

first of these is clear. The input propagates through bulk material with the 'propagator' $\hat{\boldsymbol{G}}$ of (1.21). This would contribute to (1.98) a term with $\chi(\boldsymbol{\rho}, z; \boldsymbol{\rho}', z')=\chi(\boldsymbol{\rho}-\boldsymbol{\rho}', z-z')$, where χ is the bulk susceptibility. The question now is what to write for the reflected term. *This is where we make the model of surface reflection.* An intuitive way of describing this would be to imagine that we fill up the entire space with bulk material and introduce a fictitious field at the mirror image point $\bar{\boldsymbol{r}}'=(\boldsymbol{\rho}', -z')$. This hypothetical input is then assumed to propagate with $\hat{\boldsymbol{G}}(\boldsymbol{k}, \omega)$ and to produce at $\boldsymbol{r}$ the same polarisation as the reflected term. We are then confronted with two questions, namely (i) we must find some suitable fictitious stimulus such that it creates the real field $\boldsymbol{E}$ at $\boldsymbol{r}'$ and the desired hypothetical field at $\bar{\boldsymbol{r}}'$ and (ii) $\boldsymbol{E}(\bar{\boldsymbol{r}}')$ must be such that it corresponds to our chosen model of surface scattering. With this we can rewrite (1.98) as

$$\boldsymbol{P}(\boldsymbol{r})=\int_{z'>0}[\chi(\boldsymbol{r}, \boldsymbol{r}')\cdot\boldsymbol{E}(\boldsymbol{r}')+\chi(\boldsymbol{r}, \bar{\boldsymbol{r}}')\cdot\boldsymbol{E}(\bar{\boldsymbol{r}}')]\,\mathbf{d}\boldsymbol{r}', \tag{1.99}$$

and now it all depends on the relation between $\boldsymbol{E}(\bar{\boldsymbol{r}}')$, which is hypothetical and serves to define the model, and $\boldsymbol{E}(\boldsymbol{r}')$, which is the real field inside the material.

In this model we are assuming that the unperturbed homogeneous bulk material extends right up to the matching plane. This may be an unwarranted simplification, depending on different physical problems, but we now focus on the effect of surface scattering which by itself is not a trivial difficulty. The way to deal with surface reflection depends on the physical nature of the elementary excitations or quasiparticles being reflected at the surface and this must be discussed by bearing in mind the type of material under consideration.

1.4. Electrodynamics of dielectric surfaces

In a dielectric the quasiparticles undergoing surface reflection are polaritons resulting from the coupling of the e.m. field and dynamical modes like, e.g. excitons or polar optical phonons. Although a phenomenological model can be introduced in a more abstract way, it is intuitively helpful to keep in mind a concrete situation. Let us think of the phonon case. The excitation reaches the bulk atoms, these start vibrating and re-emitting, thus passing on the excitation. Periodically-repeated positions house identical atoms which, under identical constraints, respond in the same way. A lattice eigenmode is thus made up, until the surface is reached. There the atoms are subject to different constraints, their response is different and this is essentially what determines the way the excitation is re-emitted back into the bulk. Let us ignore possible geometric irregularities of the surface, which presumably would affect the short waves. If there is no energy loss, the simplest non-trivial model which seems plausible is to assume that the different vibrational responses of the surface atom causes a re-emission of the wave with a certain phase lag ϕ. This is conveniently described in terms of a parameter $p = e^{i\phi}$. The models corresponding to different values of p are shown in fig. 7. For $p = 1$ we have *specular* scattering; the surface atoms are assumed to respond ideally like the bulk atoms. This situation is simulated by an electric field at $\bar{\boldsymbol{r}}'$ which is just the mirror image of $\boldsymbol{E}(\boldsymbol{r}')$. The surface atoms then behave like bulk atoms receiving from $\bar{\boldsymbol{r}}'$ the excitation due to the mirror image field and passing it on to $\boldsymbol{r}$ without phase lag. The opposite extreme would be that in which the surface atoms are totally constrained. Then $p = -1$, i.e. $\phi = \pi$. The amplitude of the excitation at $z = 0$ is zero, as the incident and reflected waves have exactly opposite phases. The field $\boldsymbol{E}(\bar{\boldsymbol{r}}')$ is then the *antispecular* image of $\boldsymbol{E}(\boldsymbol{r}')$. Fig. 7*d* would correspond to some intermediate value of ϕ, describing a situation

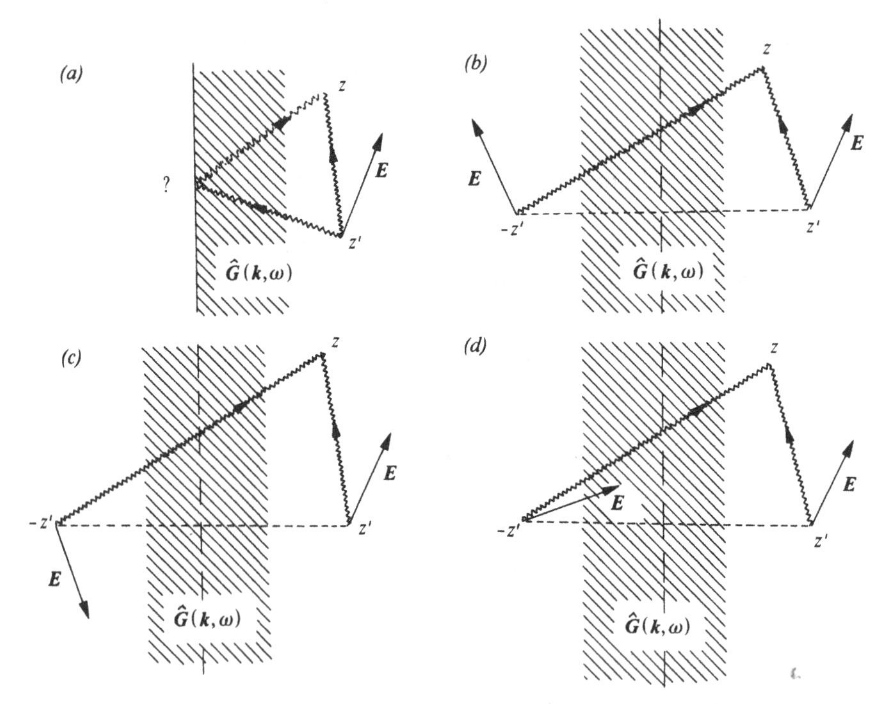

Fig. 7. (*a*) The initial problem. (*b*) The specular model ($\phi = 0$, $p = +1$). (*c*) The antispecular model ($\phi = \pi$, $p = -1$). (*d*) Arbitrary ϕ-model.

in which the surface atoms are somewhat constrained, but not totally. The problem now is to find some appropriate hypothetical external stimulus which creates an electric field with the desired behaviour.

Consider the ideal specular case and P-mode geometry. We can use a trick like the one of § 1.2, but let us write it down a little more formally. Define a hypothetical extended medium M with dielectric tensor $\boldsymbol{\varepsilon}(\boldsymbol{k}, \omega)$, and assume that a fictitious surface current is introduced as

$$\boldsymbol{J}^{\mathrm{M}}_{\mathrm{ext}}(\boldsymbol{r}, t) = J_{\mathrm{s}}\, \mathrm{e}^{\mathrm{i}(\kappa x - \omega t)} \delta(z) \boldsymbol{x}^0. \tag{1.100}$$

This creates an e.m. field which can be obtained at once by using (1.21) and which has mirror symmetry, i.e. $\boldsymbol{E}^{\mathrm{M}}$ behaves symmetrically and $\boldsymbol{H}^{\mathrm{M}}$ behaves antisymmetrically. We take these to be the real fields under consideration in $z \geqslant 0$. Of course they bear no relation to reality in $z < 0$, but we shall only use their values in $z \geqslant 0$. In fact it suffices to know

$$E^{\mathrm{M}}_x(+0) = \frac{J_{\mathrm{s}}}{2\pi} \int \mathrm{e}^{\mathrm{i}q\eta} \hat{G}_{xx}(q)\, \mathrm{d}q. \tag{1.101}$$

Here and henceforth in integrals like this $\eta \to 0$ always, without further specification. Note that the surface projection of $\hat{\boldsymbol{G}}(\boldsymbol{r}, \boldsymbol{r}')$, i.e. $\hat{\boldsymbol{G}}(z = 0,$

$z' = 0$), is just

$$\frac{1}{2\pi}\int e^{\pm iqz}\hat{\boldsymbol{G}}(q)\,dq \equiv \boldsymbol{G}^{(\pm)}. \tag{1.102}$$

This equation *defines* the surface projection $\boldsymbol{G}^{(\pm)}$. Thus

$$E_x^{M}(+0) = G_{xx}^{(+)} J_s. \tag{1.103}$$

Furthermore, bearing in mind that $H_y^M(-0) = -H_y^M(+0)$ and integrating $\nabla \wedge \boldsymbol{H}^M$ from $-\eta$ to $+\eta$, we find

$$J_s = -(c/2\pi) H_y^M(+0). \tag{1.104}$$

The surface impedance follows by eliminating J_s. Actually, from the structure of $\hat{\boldsymbol{G}}$ as a function of $\boldsymbol{k}$ in (1.21), the convergence factor $e^{\pm iq\eta}$ is unnecessary as long as we are only concerned with (x, y) components of $\boldsymbol{G}$. Thus we can omit this factor and the $(\pm)$ superscript in the remainder of this chapter. We then have

$$Z_P^{>} = -2G_{xx} \quad (p = 1). \tag{1.105}$$

Notice that we have obtained a physically meaningful result by introducing into the analysis a fictitious element – i.e. J_s – which is finally eliminated. Thus this element need not even be physically meaningful in itself. This gives us the clue as to how to generate the antispecular model, corresponding to $p = -1$.

Imagine that a magnetic current $\boldsymbol{\Gamma}$ existed. We would then start from the curl equations,

$$\nabla \wedge \boldsymbol{E} = -\frac{1}{c}\frac{\partial \boldsymbol{H}}{\partial t} + \frac{4\pi}{c}\boldsymbol{\Gamma}_{\text{ext}}, \qquad \nabla \wedge \boldsymbol{H} = \frac{1}{c}\frac{\partial \boldsymbol{D}}{\partial t}, \tag{1.106}$$

and Fourier transforming and eliminating $\boldsymbol{E}$ we would find

$$\boldsymbol{H}(\boldsymbol{k}, \omega) = \hat{\boldsymbol{K}}(\boldsymbol{k}, \omega) \cdot \boldsymbol{\Gamma}_{\text{ext}}(\boldsymbol{k}, \omega), \tag{1.107}$$

where

$$\hat{\boldsymbol{K}}(\boldsymbol{k}, \omega) = -i\frac{4\pi\omega}{c^2}\frac{\varepsilon_T \boldsymbol{I} - c^2\boldsymbol{kk}/\omega^2}{k^2 - \omega^2\varepsilon_T/c^2}. \tag{1.108}$$

A world of magnetic currents would be governed by $\varepsilon_T(\boldsymbol{k}, \omega)$ only, but the response has both longitudinal and transverse parts and div $\boldsymbol{H}$ need not vanish. Now we have a neat way to derive the surface impedance for the antispecular model. It suffices to choose as fictitious surface stimulus a transverse magnetic current, i.e.

$$\boldsymbol{\Gamma}_{\text{ext}}^{M}(\boldsymbol{r}, t) = \Gamma_s e^{i(\kappa x - \omega t)}\,\delta(z)\boldsymbol{y}^0. \tag{1.109}$$

In our extended medium we now have a symmetric $\boldsymbol{H}^{\mathrm{M}}$ field and an antisymmetric $\boldsymbol{E}^{\mathrm{M}}$ field. Thus

$$H_y^{\mathrm{M}}(+0)=K_{yy}^{(+)}\Gamma_s, \tag{1.110}$$

while, integrating $\nabla \wedge \boldsymbol{E}^{\mathrm{M}}$, we have

$$\Gamma_s=\frac{c}{2\pi}E_x^{\mathrm{M}}(+0). \tag{1.111}$$

Again the convergence factor and the $(\pm)$ superscript are unnecessary for (x, y) components and we find

$$\frac{1}{Z_{\mathrm{P}}^{>}}=\frac{c^2}{8\pi^2}K_{yy} \quad (p=-1). \tag{1.112}$$

The way to deal with S-mode geometry is identical, with appropriate substitutions, and we find

$$Z_{\mathrm{S}}^{>}=-2G_{yy} \quad (p=1), \tag{1.113}$$

for the specular model while, for the antispecular case,

$$\frac{1}{Z_{\mathrm{S}}^{>}}=\frac{c^2}{8\pi^2}K_{xx} \quad (p=-1). \tag{1.114}$$

In order to evaluate the surface impedance we need only choose the model dielectric functions ε_{L} and ε_{T} and use (1.21) or (1.108). As a simple exercise consider a vacuum, $\varepsilon_{\mathrm{L}}=\varepsilon_{\mathrm{T}}=1$. In this case the interactions are local and surface scattering plays no part. Consider, for example, P-mode geometry; then (1.105) and (1.112) do indeed yield the same result, and this agrees with (1.79). However, we must take care to remember the sign conventions for incoming and outgoing waves. Thus, if we have a medium 1 in $z<0$ and a medium 2 in $z>0$, then the s.m.d.r. is

$$\int \mathrm{d}q|\hat{G}_{xx}^1(q;\kappa,\omega)+\hat{G}_{xx}^2(q;\kappa,\omega)|=0 \tag{1.115}$$

for $p=1$, or

$$\int \mathrm{d}q|\hat{K}_{yy}^1(q;\kappa,\omega)+\hat{K}_{yy}^2(q;\kappa,\omega)|=0 \tag{1.116}$$

for $p=-1$. We can also see now in a more revealing light the condition (1.86) obtained in §-1.3 for the localisation of a surface mode.

Consider the specular case and the free surface of a medium in $z>0$, where the extended field $\boldsymbol{E}^{\mathrm{M}}$ is equal to the real field $\boldsymbol{E}$, and just concentrate on E_x, which can be written as

$$E_x(z) = \frac{2iJ_s}{\omega} \int e^{iqz} \left[\frac{-\kappa^2}{k^2 \varepsilon_L} + \frac{\omega^2}{c^2} \frac{q^2}{k^2(k^2 - \omega^2 \varepsilon_T/c^2)} \right] dq. \tag{1.117}$$

We need not be concerned with J_s. We are looking for the z-dependence, which comes from the poles of the integrand. If we have a pole at some $q = q_n$, then we have a term that varies as $e^{iq_n z}$. Suppose we have a local medium and we choose to close the integration contour through the upper half circle at infinity. It is then easy to see that the net contribution from the pole at $q = +i\kappa$ is zero. We are left with a residue from the pole at

$$q = +i(\kappa^2 - \omega^2/c^2)^{1/2}, \tag{1.118}$$

and this gives a localised surface mode, with exponential decay for $z > 0$, provided the square root is real, which is the condition we found in (1.86). We can see now that the situation will be different and more complicated as soon as we study non-local interactions. The z-dependence of (1.117) will then depend on the structure of $\varepsilon(\boldsymbol{k}, \omega)$, i.e. on the model for the bulk material.

There is also a less superficial complication, the understanding of which is rather more subtle. From now on let us concentrate on the physically more interesting P-polarisation. Consider, say, E_x^M as a function of z. In line with the discussion of § 2.3, we want it to behave in the following way:

$$E_x^M(-z) = pE_x^M(z), \tag{1.119}$$

for arbitrary p. The situations for $p = \pm 1$ are clear but, how do we satisfy this equation for any other value of p? One would think off hand of trying a combination of symmetric and antisymmetric fields, but this cannot work. Such a combination can never diagonalise the group formed by the identity and the inversion, whose basis functions can only be purely symmetric ($p = 1$) or antisymmetric ($p = -1$). Yet there is nothing else we can use as fictitious surface stimuli other than an electric or a magnetic stimulus. Since no combination thereof can result in (1.119), the conclusion is obvious: the problem, for arbitrary p, cannot be solved with a combination of *surface* stimuli *only*, *We must add bulk stimuli.* But remember the hypothetical field is to be identified with the real one inside the material, which we shall take to be in $z > 0$. Thus the fictitious bulk stimuli must be defined so that they vanish for $z > 0$.

We are now ready to formulate the problem in a more articulate way. We shall obtain real physical results for real (model) systems by doing the electrodynamics of pseudosystems with pseudosurfaces. We start from a vacuum in $z < 0$ and define an extended pseudovacuum, labelled V, filling up the entire space and with an e.m. field ($\boldsymbol{E}^{\text{V}}, \boldsymbol{H}^{\text{V}}$) which is equal to the real field ($\boldsymbol{E}, \boldsymbol{H}$) in $z<0$. Likewise we define the extended pseudomedium such that ($\boldsymbol{E}^{\text{M}}, \boldsymbol{H}^{\text{M}}$) is equal to ($\boldsymbol{E}, \boldsymbol{H}$) in $z>0$. The matching of tangential components then reads

$$H_y^{\text{V}}(-0) = H_y^{\text{M}}(+0), \qquad E_x^{\text{V}}(-0) = E_x^{\text{M}}(+0). \tag{1.120}$$

The definition of the extended pseudosystems includes (i) the statement that they fill up the entire space with the corresponding bulk dielectric functions, and (ii) the definition of the fictitious stimuli contained in them. It is this that determines the extended pseudofields, such as (1.119). This guarantees that they describe in their respective domains a physical field which is just (1.120). Consider the extended pseudomedium (M). We know that E_x^{M}, E_z^{M} and H_y^{M} behave in different ways. One concise way of expressing this is to impose the behaviour of the rotational and the irrotational parts of $\boldsymbol{E}^{\text{M}}$; i.e. we require that, by reference to a generic $z>0$

$$\nabla \cdot \boldsymbol{E}^{\text{M}}(-z) = p\nabla \cdot \boldsymbol{E}^{\text{M}}(z), \tag{1.121}$$

while

$$\nabla \wedge \boldsymbol{E}^{\text{M}}(-z) = -p\nabla \wedge \boldsymbol{E}^{\text{M}}(z). \tag{1.122}$$

This is not yet all. We have prescribed the behaviour of derivatives of $\boldsymbol{E}^{\text{M}}$ and now we must add one condition about the field itself at some point. The obvious condition is, for example,

$$E_x^{\text{M}}(-0) = pE_x^{\text{M}}(+0). \tag{1.123}$$

The scheme now is the following. We must define the fictitious stimuli. This will introduce certain parameters; equations (1.121)–(1.123) are the *subsidiary conditions* needed to eliminate these. We do likewise for the extended pseudovacuum (V), and then use all this in the fundamental *matching equations* (1.120). We must make sure that the number of conditions fits the number of unknowns.

We know from the discussion earlier that in general we must introduce as fictitious stimuli a combination of electric and magnetic currents located at the surface and in the physically unrealisable half of the bulk of the extended pseudomedia. Since the vacuum is a non-dispersive medium, we can use for V the simplest stimulus, which is a surface electric current $\boldsymbol{J}^{\text{V}} = (J_{\text{s}}^{\text{V}}, 0, 0)\delta(z)$. This originates a symmetric e.m. field $\boldsymbol{E}^{\text{V}}, \boldsymbol{H}^{\text{V}}$, which we can write down immediately using (1.21). For M, however, we must introduce the electric stimulus with Fourier transform

$$\boldsymbol{J}^{\text{M}} = \boldsymbol{J}_{\text{s}}^{\text{M}}(\kappa) + \boldsymbol{J}_{\text{b}}^{\text{M}}(\kappa, q).$$

For P-mode geometry

$$\boldsymbol{J}_{\mathrm{s}}^{\mathrm{M}}=(J_{\mathrm{s}}^{\mathrm{M}},0,0);\qquad \boldsymbol{J}_{\mathrm{b}}^{\mathrm{M}}=(J_{\mathrm{bx}},0,J_{\mathrm{bz}}). \tag{1.124}$$

Both J_{bx} and J_{bz} are functions of (κ, q) such that their Fourier transform with respect to q is zero for $z>0$. Thus we know so far that, as functions of a complex q, they must be analytic in the upper half plane. We also need the magnetic stimulus specified by

$$\boldsymbol{\Gamma}_{\mathrm{s}}^{\mathrm{M}}=(0,\Gamma_{\mathrm{s}}^{\mathrm{M}},0);\qquad \boldsymbol{\Gamma}_{\mathrm{b}}^{\mathrm{M}}=(0,\Gamma_{\mathrm{b}}^{\mathrm{M}},0), \tag{1.125}$$

where again $\Gamma_{\mathrm{b}}^{\mathrm{M}}(\kappa, q)$, as a function of a complex q, must be analytic in the upper half plane. The complete e.m. field in M will consist of

$$\boldsymbol{E}^{\mathrm{M}}=\boldsymbol{E}^{\mathrm{M}J}+\boldsymbol{E}^{\mathrm{M}\Gamma};\qquad \boldsymbol{H}^{\mathrm{M}}=\boldsymbol{H}^{\mathrm{M}J}+\boldsymbol{H}^{\mathrm{M}}\Gamma. \tag{1.126}$$

These fields can be immediately written down in Fourier transform. From (1.21),

$$\boldsymbol{E}^{\mathrm{M}J}=\left\{-\mathrm{i}\frac{4\pi}{\omega}\frac{\boldsymbol{kk}}{k^2\varepsilon_{\mathrm{L}}}+\mathrm{i}\frac{4\pi\omega}{c^2}\frac{(k^2\boldsymbol{I}-\boldsymbol{kk})}{k^2[k^2-(\omega^2/c^2)\varepsilon_{\mathrm{T}}]}\right\}\cdot\boldsymbol{J}^{\mathrm{M}}, \tag{1.127}$$

and

$$\boldsymbol{H}^{\mathrm{M}J}=\mathrm{i}\frac{4\pi}{c}\frac{\boldsymbol{k}\wedge\boldsymbol{J}^{\mathrm{M}}}{k^2-(\omega^2/c^2)\varepsilon_{\mathrm{T}}}; \tag{1.128}$$

while from (1.108),

$$\boldsymbol{E}^{\mathrm{M}\Gamma}=\mathrm{i}\frac{4\pi}{c}\frac{\boldsymbol{k}\wedge\boldsymbol{\Gamma}^{\mathrm{M}}}{k^2-(\omega^2/c^2)\varepsilon_{\mathrm{T}}}, \tag{1.129}$$

and

$$\boldsymbol{H}^{\mathrm{M}\Gamma}=-\mathrm{i}\frac{4\pi\omega}{c^2}\frac{\varepsilon_{\mathrm{T}}}{k^2-(\omega^2/c^2)\varepsilon_{\mathrm{T}}}\boldsymbol{\Gamma}^{\mathrm{M}}. \tag{1.130}$$

It is actually more convenient to write the equations that will follow in terms of div $\boldsymbol{J}_{\mathrm{b}}^{\mathrm{M}}$ and curl $\boldsymbol{J}_{\mathrm{b}}^{\mathrm{M}}$. Thus we redefine

$$\kappa J_{\mathrm{bx}}^{\mathrm{M}}+qJ_{\mathrm{bz}}^{\mathrm{M}}=f_1(\kappa,q)\kappa J_{\mathrm{s}}^{\mathrm{M}}(\kappa)=f_1(\kappa,q)\mathscr{J}(\kappa), \tag{1.131}$$

and

$$qJ_{\mathrm{bx}}^{\mathrm{M}}-\kappa J_{\mathrm{bz}}^{\mathrm{M}}=f_2(\kappa,q)\kappa J_{\mathrm{s}}^{\mathrm{M}}(\kappa)=f_2(\kappa,q)\mathscr{J}(\kappa). \tag{1.132}$$

Then

$$\boldsymbol{k}\cdot\boldsymbol{J}^{\mathrm{M}}=\mathscr{J}(1+f_1);\qquad \boldsymbol{k}\wedge\boldsymbol{J}^{\mathrm{M}}=\mathscr{J}(f_2+q/\kappa)\boldsymbol{y}^0. \tag{1.133}$$

These expressions are more useful in writing down the subsidiary conditions (1.121) and (1.122). For the only component of Γ_s^M we define

$$\Gamma_s^M(\kappa)+\Gamma_b^M(\kappa, q)=\Gamma_s^M(\kappa)[1+\Gamma_b^M(\kappa, q)/\Gamma_s^M(\kappa)]$$
$$=\Gamma_s^M[1+f_3(\kappa, q)]. \tag{1.134}$$

Now, we have the extended fields in Fourier transform, i.e. in particular as functions of q, while both the subsidiary conditions and the matching equations are expressed in z-space, retaining κ and ω-dependence. Thus (1.120)–(1.123) take the form of equalities between integrals over q, as in (1.101) or (1.117). By introducing the parameters

$$\alpha=\mathrm{i}J^x/\mathcal{J}; \qquad \beta=-\mathrm{i}(\omega/c)\Gamma_s^M/\mathcal{J}, \tag{1.135}$$

the matching equations (1.120) become

$$\alpha\int\frac{\mathrm{e}^{-\mathrm{i}q\eta}q\,\mathrm{d}q}{k^2-\omega^2/c^2}$$
$$=\beta\int\frac{\mathrm{e}^{\mathrm{i}q\eta}\varepsilon_{\mathrm{T}}(1+f_3)}{k^2-(\omega^2/c^2)\varepsilon_{\mathrm{T}}}\,\mathrm{d}q+\mathrm{i}\int\frac{\mathrm{e}^{\mathrm{i}q\eta}(f_2+q/\kappa)}{k^2-(\omega^2/c^2)\varepsilon_{\mathrm{T}}}\,\mathrm{d}q, \tag{1.136}$$

and

$$\alpha\left\{-\kappa^2\int\frac{\mathrm{e}^{\mathrm{i}q\eta}}{k^2}\,\mathrm{d}q+\frac{\omega^2}{c^2}\int\frac{\mathrm{e}^{-\mathrm{i}q\eta}q^2}{k^2(k^2-\omega^2/c^2)}\,\mathrm{d}q\right\}$$
$$=\beta\int\frac{\mathrm{e}^{\mathrm{i}q\eta}q(1+f_3)}{k^2-(\omega^2/c^2)\varepsilon_{\mathrm{T}}}\,\mathrm{d}q+\mathrm{i}\left\{-\kappa\int\frac{\mathrm{e}^{\mathrm{i}q\eta}(1+f_1)}{k^2\varepsilon_{\mathrm{L}}}\right.$$
$$\left.+\frac{\omega^2}{c^2}\int\frac{\mathrm{e}^{\mathrm{i}q\eta}q\left(f_2+\frac{q}{\kappa}\right)}{k^2[k^2-(\omega^2/c^2)\varepsilon_{\mathrm{T}}]}\,\mathrm{d}q\right\}. \tag{1.137}$$

Remember that $\eta\to 0$. The sign of the exponent determines whether the integration contours must be closed through the upper or lower half circle at infinity.

One of the two parameters, say α, can be eliminated at once. We thus obtain

$$\mathrm{i}\kappa\int\frac{\mathrm{e}^{\mathrm{i}q\eta}(1+f_1)}{k^2\varepsilon_{\mathrm{L}}}\,\mathrm{d}q=\beta\int\frac{\mathrm{e}^{\mathrm{i}q\eta}(\mathrm{i}\nu\varepsilon_{\mathrm{T}}+q)(1+f_3)}{k^2-(\omega^2/c^2)\varepsilon_{\mathrm{T}}}\,\mathrm{d}q$$
$$+\int\frac{\mathrm{e}^{\mathrm{i}q\eta}\left(\mathrm{i}\frac{\omega^2}{c^2}\frac{q}{k^2}-\nu\right)\left(f_2+\frac{q}{\kappa}\right)}{k^2-(\omega^2/c^2)\varepsilon_{\mathrm{T}}}\,\mathrm{d}q, \tag{1.138}$$

where

$$\nu=(\kappa^2-\omega^2/c^2)^{1/2}.$$

This is actually the s.m.d.r. in quite general form, but it still contains several unknowns, namely the parameter β and the functions f_1, f_2, f_3 which, as functions of q, we know must be analytic in the upper half plane. At this stage we resort to the subsidiary conditions, using the definitions of (1.133). For example, from (1.121) we have

$$\int e^{-iqz}\frac{(1+f_1)}{\varepsilon_L}\,dq = p\int e^{iqz}\frac{(1+f_1)}{\varepsilon_L}\,dq, \tag{1.139}$$

and likewise for the other conditions. It is clear that the subsidiary equations involve just the unknowns we want to eliminate, but we must sort this out more explicitly. For example, the *functional form* of f_1, f_2 and f_3, on the face of equations like (1.139), obviously depends on the form of ε_L and ε_T. To proceed further we must choose the dielectric functions, i.e. we define here our model for the bulk system to which we shall add a surface.

Typical phenomenological problems of interest may be, for example, the consideration of surface polaritons – i.e. their dispersion and damping – or the structure of the reflectivity as related to various elementary excitations of the system. In these problems one is seldom interested in very short waves. A plausible model, consistent with this semiclassical phenomenological spirit, which is tractable in practice and yet contains the essential long-wave features of a dispersive system, has the following form:

$$\varepsilon_L = \varepsilon_c\left(1+\frac{\Omega_p^2}{\omega_T^2-\omega^2+\beta_L^2k^2}\right); \qquad \varepsilon_T = \varepsilon_c\left(1+\frac{\Omega_p^2}{\omega_T^2-\omega^2+\beta_T^2k^2}\right). \tag{1.140}$$

This describes dispersive bulk dynamical modes – phonons, excitons – according to

$$\omega_L(k)^2 = \omega_L^2+\beta_L^2k^2; \qquad \omega_T(k)^2 = \omega_T^2+\beta_T^2k^2. \tag{1.141}$$

The numerical values of β_L and β_T may perhaps be rather similar, but for a proper physical interpretation of the formal analysis they must be kept distinct. We shall see in due course that their physical roles are different. It is convenient to express ε_L as

$$\varepsilon_L = \varepsilon_c\frac{(q+iQ_L)(q-iQ_L)}{(q+iq_L)(q-iq_L)}, \tag{1.142}$$

where

$$Q_L = \left(\kappa^2+\frac{\omega_T^2+\Omega_p^2-\omega^2}{\beta_L^2}\right)^{1/2}; \qquad q_L = \left(\kappa^2+\frac{\omega_T^2-\omega^2}{\beta_L^2}\right)^{1/2}. \tag{1.143}$$

Similarly we write

$$\varepsilon_{\mathrm{T}}=\varepsilon_{\mathrm{c}}\frac{(q+\mathrm{i}Q_{\mathrm{T}})(q-\mathrm{i}Q_{\mathrm{T}})}{(q+\mathrm{i}q_{\mathrm{T}})(q-\mathrm{i}q_{\mathrm{T}})}, \tag{1.144}$$

with

$$Q_{\mathrm{T}}=\left(\kappa^{2}+\frac{\omega_{\mathrm{T}}^{2}+\Omega_{\mathrm{p}}^{2}-\omega^{2}}{\beta_{\mathrm{T}}^{2}}\right)^{1/2}; \qquad q_{\mathrm{T}}=\left(\kappa^{2}+\frac{\omega_{\mathrm{T}}^{2}-\omega^{2}}{\beta_{\mathrm{T}}^{2}}\right)^{1/2}. \tag{1.145}$$

After this we put

$$\frac{1}{k^{2}-(\omega^{2}/c^{2})\varepsilon_{\mathrm{T}}}=\frac{(q+\mathrm{i}q_{\mathrm{T}})(q-\mathrm{i}q_{\mathrm{T}})}{(q+\mathrm{i}Q_{\mathrm{m}})(q-\mathrm{i}Q_{\mathrm{m}})(q+\mathrm{i}q_{\mathrm{m}})(q-\mathrm{i}q_{\mathrm{m}})}, \tag{1.146}$$

where

$$Q_{\mathrm{m}}^{2}=\frac{1}{2}\left\{\left(\kappa^{2}+q_{\mathrm{T}}^{2}-\varepsilon_{\mathrm{c}}\frac{\omega^{2}}{c^{2}}\right)+\sqrt{\left[\left(\kappa^{2}+q_{\mathrm{T}}^{2}-\varepsilon_{\mathrm{c}}\frac{\omega^{2}}{c^{2}}\right)^{2}-4\left(\kappa^{2}q_{\mathrm{T}}^{2}-\varepsilon_{\mathrm{c}}\frac{\omega^{2}}{c^{2}}Q_{\mathrm{T}}^{2}\right)\right]}\right\}, \tag{1.147}$$

and

$$q_{\mathrm{m}}^{2}=\frac{1}{2}\left\{\left(\kappa^{2}+q_{\mathrm{T}}^{2}-\varepsilon_{\mathrm{c}}\frac{\omega^{2}}{c^{2}}\right)-\sqrt{\left[\left(\kappa^{2}+q_{\mathrm{T}}^{2}-\varepsilon_{\mathrm{c}}\frac{\omega^{2}}{c^{2}}\right)^{2}-4\left(\kappa^{2}q_{\mathrm{T}}^{2}-\varepsilon_{\mathrm{c}}\frac{\omega^{2}}{c^{2}}Q_{\mathrm{T}}^{2}\right)\right]}\right\}. \tag{1.148}$$

We are now ready to see how to deal with integrals like those of (1.139). The point to notice is that we are trying to construct extended fields whose z-dependence is imposed by conditions like (1.139). Now, for the integral on the r.h.s. we must close the integration contour above. Since f_1 is analytic in the upper half plane, the only term will come from the pole at $q=\mathrm{i}Q_{\mathrm{L}}$, from (1.142). Thus, on the l.h.s. – where we must close the integration contours *below* – we cannot allow for any z-dependent term other than $\exp(-Q_{\mathrm{L}}|z|)$, which results from the pole at $q=-\mathrm{i}Q_{\mathrm{L}}$. By looking at (1.142) it is clear that f_1 must have the form

$$f_1=B_0/(q+\mathrm{i}q_{\mathrm{L}}). \tag{1.149}$$

This ensures that no undesired residue is picked out on performing the integration, because the factor $(q+\mathrm{i}q_{\mathrm{L}})$ is cancelled out in $(1+f_1)/\varepsilon_{\mathrm{L}}$. Of course B_0 is still a function of (κ,ω), but this dependence is understood throughout. On performing the integrals of (1.139) we find on both sides a z-dependence that varies as $\exp(-Q_{\mathrm{L}}|z|)$, and equating the coefficients on both sides, we have

$$1+\frac{B_0}{\mathrm{i}(q_{\mathrm{L}}-Q_{\mathrm{L}})}=p\left[1+\frac{B_0}{\mathrm{i}(q_{\mathrm{L}}+Q_{\mathrm{L}})}\right]. \tag{1.150}$$

Notice that we have sorted out the functional form of f_1 and reduced the problem to the introduction of the subsidiary parameter B_0, for which we have the subsidiary equation (1.150).

The detailed study of this condition, and of the form of f_1 which follows from it, is an example of the kind of argument one uses in this analysis. The next subsidiary condition, namely (1.122), is dealt with in much the same manner, although it is a little more involved. The appropriate form of f_2 turns out to be

$$f_2 = \frac{(\kappa^2 + \mathrm{i}q_\mathrm{L}q_\mathrm{T})}{\kappa(\kappa^2 - q_\mathrm{L}^2)}B_0 = \frac{(q - iq_\mathrm{T})}{(\kappa^2 - q_\mathrm{T}^2)}B_1 + \frac{B_1}{(q + \mathrm{i}q_\mathrm{T})}, \tag{1.151}$$

with the only pole at $q = -\mathrm{i}q_\mathrm{T}$; and that of f_3 is

$$f_3 = B_2/(q + \mathrm{i}\kappa), \tag{1.152}$$

with a pole at $q = -\mathrm{i}\kappa$. We have introduced two further subsidiary parameters, namely B_1 and B_2. These are determined from (1.222) by performing the integrations and equating, independently, the coefficients of $\exp(-Q_\mathrm{m}|z|)$ and of $\exp(-q_\mathrm{m}|z|)$. This yields two equations. One is

$$(\kappa^2 - q_\mathrm{m}^2)\Gamma_\mathrm{s}^\mathrm{M}\left\{(1+p) + \left(\frac{1}{\kappa - q_\mathrm{m}} + \frac{p}{\kappa + q_\mathrm{m}}\right)\frac{B_2}{\mathrm{i}}\right\}$$

$$= \frac{\omega}{c}\mathscr{I}\left\{(1+p)B' + \frac{(p-1)\mathrm{i}q_\mathrm{m}}{\kappa}(1 - B'')\right.$$

$$\left. + \left(\frac{1}{q_\mathrm{T} - q_\mathrm{m}} + \frac{p}{q_\mathrm{T} + q_\mathrm{m}}\right)\frac{B_1}{\mathrm{i}}\right\}, \tag{1.153}$$

where

$$B' = -\frac{\kappa}{(q_\mathrm{L}^2 - \kappa^2)}B_0 + \frac{\mathrm{i}q_\mathrm{T}}{(q_\mathrm{T}^2 - \kappa^2)}B_1;$$

$$B'' = \frac{\mathrm{i}q_\mathrm{L}}{(q_\mathrm{L}^2 - \kappa^2)}B_0 + \frac{\kappa}{(q_\mathrm{T}^2 - \kappa^2)}B_1. \tag{1.154}$$

The other equation is

$$(\kappa^2 - Q_\mathrm{m}^2)\Gamma_\mathrm{s}^\mathrm{M}\left\{(1+p) + \left(\frac{1}{\kappa - Q_\mathrm{m}}\,\frac{p}{\kappa + Q_\mathrm{m}}\right)\frac{B_2}{\mathrm{i}}\right\}$$

$$= \frac{\omega}{c}\mathscr{I}\left\{(1+p)B' + \frac{(p-1)\mathrm{i}Q_\mathrm{m}}{\kappa}(1 - B'')\right.$$

$$\left. + \left(\frac{1}{q_\mathrm{T} - Q_\mathrm{m}}\,\frac{p}{q_\mathrm{T} + Q_\mathrm{m}}\right)\frac{B_1}{\mathrm{i}}\right\}. \tag{1.155}$$

This involves the ratio Γ_s^M/J, i.e. the parameter β defined in (1.135) and carried over to the s.m.d.r. (1.138), but we still have the last subsidiary condition (1.123), which yields

$$2(c/\omega)\mathrm{i}\kappa B_2\Gamma_s^M = \mathcal{J}\{[1+f_1(-\mathrm{i}\kappa)]/\mathrm{i}-f_2(-\mathrm{i}\kappa)+\mathrm{i}\}$$
$$+p\mathcal{J}\{[1+f_1(\mathrm{i}\kappa)]/\mathrm{i}+f_2(\mathrm{i}\kappa)+\mathrm{i}\}. \tag{1.156}$$

Here $f_1(\pm\mathrm{i}\kappa)$ means setting $q=\pm\mathrm{i}\kappa$ in (1.149). Let us pause to see where we stand. We have four unknown parameters, namely B_0, B_1, B_2 and β. For these we have the four subsidiary equations, (1.150), (1.153), (1.155) and (1.156). With these we can determine the four unknowns and use B_0 to construct f_1 from (1.149), and B_1 and B_2 to construct f_2 and f_3 from (1.151) and (1.152). Using all of this in (1.138) we obtain the explicit form of the s.m.d.r., which of course depends on the model chosen for ε_L and ε_T.

In this model it turns out that $B_2=0$, a fact which would not hold generally and which greatly simplifies matters. To begin with we have $f_3=0$ in the s.m.d.r. The condition (1.150) for B_0 is unchanged and for the two remaining parameters, i.e. B_1 and β, we have then

$$(p+1)(q_m^2-\kappa^2)\beta = \mathrm{i}\frac{\omega^2}{c^2}\Bigg\{\frac{(p-1)\mathrm{i}q_m}{\kappa}$$
$$+\left[\frac{(p-1)q_Lq_m}{\kappa}-(p+1)\kappa\right]\frac{B_0}{q_L^2-\kappa^2}$$
$$+\frac{\mathrm{i}(\kappa^2-q_m^2)[(1+p)q_T+(1-p)q_m]}{(q_T^2-\kappa^2)(q_T^2-q_m^2)}B_1\Bigg\}, \tag{1.157}$$

and

$$(p+1)(Q_m^2-\kappa^2)\beta = \mathrm{i}\frac{\omega^2}{c^2}\Bigg\{\frac{(p-1)\mathrm{i}Q_m}{\kappa}$$
$$+\left[\frac{(p-1)q_LQ_m}{\kappa}-(p+1)\kappa\right]\frac{B_0}{q_L^2-\kappa^2}$$
$$+\frac{i(\kappa^2-Q_m^2)[(1+p)q_T+(1-p)Q_m]}{(q_T^2-\kappa^2)(q_T^2-Q_m^2)}B_1\Bigg\}. \tag{1.158}$$

Thus we have solved the problem in terms of p. Incidentally, the situation for $p=-1$ cannot be taken over literally from here. We must go back a step to before (1.135) because in this case $\mathcal{J}=0$. Alternatively, we can simply use the explicit solution found for this case in (1.112) or (1.116).

It is interesting to consider one question related to the general discussion of § 1.3, where surface impedance, reflectivity, and s.m.d.r. were discussed together. Suppose we have found a s.m.d.r. of the form $g(\kappa, \omega) = 0$. We cannot from here extract unambiguously the formula for the surface impedance because our function $g(\kappa, \omega)$ need not be identical to the l.h.s. of (1.81). It might contain some non-vanishing factor multiplying all terms of it. This ambiguity can be sorted out by taking a particular case, the simplest one being $p = 1$. The above equations are then greatly simplified. In fact B_0, B_1 and β all vanish, as can be seen from (1.150), (1.157) and (1.158). This was of course obvious on physical grounds, if we remember the role of the fictitious stimuli represented by the functions f_1, f_2 and f_3. Only a surface electric current need be used in this case, as we saw earlier on, from (1.100)–(1.105). Identifying the solution there obtained for the specular model, with the one we have now setting $p = 1$, we obtain the non-vanishing factor, which turns out to be $4\kappa/\omega$. Thus we have two physically meaningful ways of expressing the solution we have obtained. One is to say that using B_0, B_1, β and the form of f_1, f_2 and f_3 in (1.138) we have in explicit form the s.m.d.r. for arbitrary p. The other one is to say that we obtain the explicit formula for the surface impedance by using the same information in the equation

$$\mathrm{i}\pi\frac{\omega}{c}\left(1-\frac{c^2\kappa^2}{\omega^2}\right)^{1/2} - \mathrm{i}\frac{\omega}{4}Z_{\mathrm{P}}$$

$$= \int\left\{\frac{\kappa^2(1+f_1)}{\kappa^2}\left(\frac{1}{\varepsilon_{\mathrm{L}}}+1\right) + \frac{\mathrm{i}\kappa\beta(\mathrm{i}\nu\varepsilon_{\mathrm{T}}+q)}{[k^2-(\omega^2/c^2)\varepsilon_{\mathrm{T}}]} + \frac{[\mathrm{i}(\omega^2/c^2)q-\nu](f_2+q/\kappa)\mathrm{i}\kappa}{[k^2-(\omega^2/c^2)\varepsilon_{\mathrm{T}}]} - \frac{\kappa^2(1+f_1)}{k^2}\right\} \mathrm{e}^{\mathrm{i}q\eta}\,\mathrm{d}q, \tag{1.159}$$

which results from inserting the non-vanishing factor $4\kappa/\omega$ in its correct place.

As it stands, the scheme is now ready for use as a practical device to calculate the s.m.d.r., or to study the reflectivity or any other desired element in terms of the parameter p. But apart from specific applications it is interesting to see the physical picture contained in this formal solution. The central issue, discussed in § 1.3, is that in a dispersive system the surface modes can couple to bulk modes. Now we can see how this comes about. It suffices to look at the magnetic field in real space, i.e.

$$\boldsymbol{H}(\boldsymbol{r}) = \frac{1}{(2\pi)^3}\int \mathrm{e}^{\mathrm{i}\boldsymbol{k}\cdot\boldsymbol{r}}\boldsymbol{H}(\boldsymbol{k})\,\mathrm{d}\boldsymbol{k}. \tag{1.160}$$

The integrand is the sum of (1.128) and (1.130). On performing the integrations we pick out the residues at two poles. One is the bulk polariton $c^2k^2 = \omega\varepsilon_T$. The other one, i.e. $\varepsilon_T^{-1} = 0$, is also a bulk transverse mode, but a purely dynamical one – a transverse phonon or exciton – uncoupled to the e.m. field. In the quasistatic limit, this is the only mode that survives. Although this limit may not be very useful in practical applications to dielectrics, it is considerably simpler to work out and is sufficient to give the physical picture we require. So let us concentrate on the quasistatic limit.

If the fields of the surface system have an admixture of bulk modes, this provides a damping mechanism for the surface modes. Thus we look for damping in the s.m.d.r. In view of the above discussion we expect that this will depend on the transverse dispersive coefficient β_T. Now, in the quasistatic limit we have $q_m^2 \to q_T^2$, $\kappa^2 - Q_m^2 \to \varepsilon\omega^2/c^2$, where ε is the non-dispersive dielectric function of (1.91). The above analysis yields the s.m.d.r.,

$$(1+\varepsilon)\left\{\left[1+\frac{B_0}{\mathrm{i}(q_L+\kappa)}\right]\frac{1}{\varepsilon}-\beta\right\}$$

$$=\frac{\Omega_p^2}{(\omega^2-\omega_T^2)}\left\{\frac{1}{\varepsilon}\left[1+\frac{B_0}{\mathrm{i}(Q_L+q_L)}\right]\frac{1}{Q_L}+\frac{\varepsilon_c\beta}{q_T}\right\}\kappa. \tag{1.161}$$

Here B_0 is still given by (1.150), while

$$\beta=\frac{1}{\varepsilon(1+p)}\left[(p-1)+\mathrm{i}B_0\left(\frac{1}{q_L-\kappa}-\frac{p}{q_L+\kappa}\right)\right].$$

For our purposes it suffices to consider the long-wave limit $\kappa \to 0$. Defining

$$\omega_0^2 = \omega_T^2 + [\varepsilon_c/(1+\varepsilon_c)]\Omega_p^2,$$

we find, from (1.161),

$$\omega=\omega_0+\frac{(1+p)(\varepsilon_s-\varepsilon_c)^{1/2}}{2(1+\varepsilon_c)^{1/2}}\frac{\omega_T}{\omega_0}\left\{\frac{\varepsilon_c^{1/2}(1+p)\beta_L}{[(1-p)^2+\varepsilon_c(1+p)^2]}\right.$$

$$\left.-\mathrm{i}(1-p)\left[\frac{\beta_L}{(1-p)^2+\varepsilon_c(1+p)^2}+\frac{\beta_T}{2(1+p)}\right]\right\}\kappa \tag{1.162}$$

Now we have found what we were looking for. The damping, which reflects the strength of the coupling to the bulk mode, *depends on the model of surface scattering*, as was intuitively obvious. In this case it depends very simply – through the factor $(1-p)$ – but very crucially: the surface mode is stable for the specular surface model. This explains the appearance of a term in β_T, but (1.162) contains also a damping term in β_L, which is equally affected by $(1-p)$ but does not correspond to a surface–bulk coupling. In order to see the meaning of this term we have to wait until we prove a general theorem in § 1.8.

Our entire picture depends on the phenomenological parameter p. Although the solution has been obtained as a function of p, irrespective of its value or of how one obtains it, we anticipated on intuitive grounds that p would be of the form $\exp(i\phi)$. It is interesting to see what this means physically. Let us look at energy conservation. To the standard electromagnetic terms of Poynting's theorem we must add another term which can only be evaluated from a model. If we think of a system of oscillators which respond to the electric field, and call $\boldsymbol{P}_d$ the part of the polarisation which is directly associated with the dynamical modes – phonons or excitons – of the oscillators, then the rate at which $\boldsymbol{E}$ is doing work on the system of oscillators is $\boldsymbol{E}\cdot\dot{\boldsymbol{P}}_d$. Thus Poynting's theorem reads

$$\tfrac{1}{4}\varepsilon_c(\boldsymbol{E}\cdot\dot{\boldsymbol{E}}+\boldsymbol{H}\cdot\dot{\boldsymbol{H}})+\nabla\cdot\boldsymbol{S}_E+\boldsymbol{E}\cdot\dot{\boldsymbol{P}}_d=0. \qquad (1.163)$$

Here $\boldsymbol{S}_E$ is the ordinary Poynting vector of the e.m. field. The last term in general cannot be expressed in terms of phenomenological coefficients, except for simple cases. In general we must evaluate it from a given model. In our case this is already defined in (1.140). To simplify the discussion let us take $\beta_L=\beta_T=\beta_I$. Then the dielectric function of (1.140) corresponds to an equation of motion of the form

$$(1/NM)\boldsymbol{E}=\ddot{\boldsymbol{P}}_d+\omega_T^2\boldsymbol{P}_d-\beta_I^2\nabla^2\boldsymbol{P}_d, \qquad (1.164)$$

where N and M are the density and effective mass of the oscillators. Moreover, the dielectric function in real space, by Fourier transforming (1.140), is

$$\varepsilon(z,z')=\varepsilon_c\delta(z-z')+i(\varepsilon_c\Omega_p^2/2F\beta_I^2)\exp(iF|z-z'|), \qquad (1.165)$$

where

$$F^2=(\omega^2-\omega_T^2)/\beta_I^2-\kappa^2.$$

For the frequency range of interest, $\omega^2>\omega_T^2+\beta_I^2\kappa^2$, F is real. So far we have the model of the bulk material. For the effect of the surface we have assumed that we can write, for $z\geqslant 0$,

$$\boldsymbol{D}(z)=\varepsilon_c\boldsymbol{E}(z)$$
$$+i\frac{\varepsilon_c\Omega_p^2}{2F\beta_I^2}\int_0^\infty[\exp(iF|z-z'|)+p\exp iF(z+z')\boldsymbol{\alpha}]\cdot\boldsymbol{E}(z')\,dz', \qquad (1.166)$$

where the tensor

$$\boldsymbol{\alpha} = \begin{bmatrix} 1 & 0 & 0 \\ 0 & 0 & 0 \\ 0 & 0 & -1 \end{bmatrix}$$

is introduced to abbreviate the notation. Equation (1.166) follows directly from our model of surface scattering and from (1.165). The term in Ω_p^2 would vanish in the absence of the oscillators. Thus for $\boldsymbol{P}_d$ we have, for $z \geqslant 0$,

$$4\pi \boldsymbol{P}_d(z) = i\frac{\varepsilon_c \Omega_p^2}{2F\beta_I^2} \int_0^\infty [\exp iF|z-z'| + p \exp iF(z+z')\boldsymbol{\alpha}] \cdot \boldsymbol{E}(z')\, dz'. \tag{1.167}$$

The problem is simply this: our model is completely defined with (1.164) and (1.167). We want to study (1.163) for this model.

We must express (1.163) rather more carefully, since this is a real energy balance and we use in our formalism complex quantities which vary as $\exp(-i\omega t)$. Thus any product of the form $\boldsymbol{A} \cdot \boldsymbol{B}$ must be understood as $\frac{1}{4}(\boldsymbol{A} + \boldsymbol{A}^*) \cdot (\boldsymbol{B} + \boldsymbol{B}^*)$. Let us now take the time average of (1.163) over one period $2\pi/\omega$. It is easy to see that the first term vanishes because the integrand varies as $\sin 2\omega t$. We are left with

$$\tfrac{1}{4}\langle(\boldsymbol{E} + \boldsymbol{E}^*) \cdot (\dot{\boldsymbol{P}}_d + \dot{\boldsymbol{P}}_d^*)\rangle + \nabla \cdot \langle \boldsymbol{S}_E \rangle = 0. \tag{1.168}$$

Let us study the first term. The products $\boldsymbol{E} \cdot \dot{\boldsymbol{P}}_d$ and $\boldsymbol{E}^* \cdot \dot{\boldsymbol{P}}_d^*$ integrate out to zero. We only need the sum of cross products like $\boldsymbol{E} \cdot \dot{\boldsymbol{P}}_d^*$. Using (1.164) we have

$$\tfrac{1}{4}\langle(\boldsymbol{E} + \boldsymbol{E}^*) \cdot (\dot{\boldsymbol{P}}_d + \dot{\boldsymbol{P}}_d^*)\rangle = -\tfrac{1}{4}NM\beta_I^2 \langle \dot{\boldsymbol{P}}_d^* \cdot \nabla^2 \boldsymbol{P}_d + \boldsymbol{P}_d \cdot \nabla^2 \boldsymbol{P}_d^* \rangle. \tag{1.169}$$

It is convenient to write this scalar quantity as the divergence of a vector which, by considering (1.168), we see would assume for the dynamical modes of the oscillators the same role as $\boldsymbol{S}_E$ for the e.m. field. For this we define the vector components

$$S_{di} = -\frac{NM\beta_I^2}{4}\left[\left(\frac{\partial \boldsymbol{P}_d}{\partial x_i} + \frac{\partial \boldsymbol{P}_d^*}{\partial x_i}\right) \cdot (\dot{\boldsymbol{P}}_d + \dot{\boldsymbol{P}}_d^*)\right] \quad (x_i = x, z).$$

On taking the time average,

$$\langle S_{di} \rangle = -\frac{NM\beta_I^2}{4}\left\langle \frac{\partial \boldsymbol{P}_d}{\partial x_i} \cdot \dot{\boldsymbol{P}}_d^* + \frac{\partial \boldsymbol{P}_d^*}{\partial x_i} \cdot \dot{\boldsymbol{P}}_d \right\rangle, \tag{1.170}$$

whence

$$\nabla\cdot\langle \boldsymbol{S}_{\rm d}\rangle = -\frac{NM\beta_{\rm I}^2}{4}\Big\langle \dot{\boldsymbol{P}}_{\rm d}^*\cdot\nabla^2\boldsymbol{P}_{\rm d}+\dot{\boldsymbol{P}}_{\rm d}\cdot\nabla^2\boldsymbol{P}_{\rm d}^*\Big\rangle$$

$$-\frac{NM\beta_{\rm I}^2}{4}\Big\langle\sum_{x_i}\Big[\frac{\partial \boldsymbol{P}_{\rm d}}{\partial x_i}\cdot\frac{\partial \dot{\boldsymbol{P}}_{\rm d}^*}{\partial x_i}+\frac{\partial \boldsymbol{P}_{\rm d}^*}{\partial x_i}\cdot\frac{\partial \dot{\boldsymbol{P}}}{\partial x_i}\Big]\Big\rangle.$$

The last time average is again zero and, comparing with (1.169), we have

$$\tfrac{1}{4}\langle(\boldsymbol{E}+\boldsymbol{E}^*)\cdot(\dot{\boldsymbol{P}}_{\rm d}+\dot{\boldsymbol{P}}_{\rm d}^*)\rangle=\nabla\cdot\langle\boldsymbol{S}_{\rm d}\rangle,$$

which used in (1.168) yields

$$\nabla\cdot\langle\boldsymbol{S}_{\rm d}+\boldsymbol{S}_{\rm E}\rangle=0, \tag{1.171}$$

which holds for all $z\neq 0$. Integrating this with respect to z between $-\eta$ and $+\eta$, letting $\eta\to 0$, and using the fact that $\boldsymbol{S}_{\rm E}$ is continuous across the surface we have

$$\langle S_{{\rm d}z}(+0)\rangle-\langle S_{{\rm d}z}(-0)\rangle=0.$$

But $S_{{\rm d}z}$ is identically zero for all $z\leqslant 0$, outside the polarisable medium. From this and (1.170) we have

$$\langle S_{{\rm d}z}(+0)\rangle=-\frac{{\rm i}\omega NM\beta_{\rm I}^2}{4}\Big\langle\frac{\partial\boldsymbol{P}_{\rm d}}{\partial z}\cdot\boldsymbol{P}_{\rm d}^*-\frac{\partial\boldsymbol{P}_{\rm d}^*}{\partial z}\cdot\boldsymbol{P}_{\rm d}\Big\rangle=0. \tag{1.172}$$

This is a consequence of our model for the bulk – through the equation of motion – and of energy conservation. Now we use our model for the surface, i.e. (1.167), to find an expression which relates $\boldsymbol{P}_{\rm d}$ to its normal derivative and which is useful in evaluating (1.172). First we have, from (1.167),

$$4\pi P_{{\rm d}x}(+0)={\rm i}\frac{\varepsilon_{\rm c}\Omega_{\rm p}^2}{2F\beta_{\rm I}^2}\int_0^\infty(1+p)\exp({\rm i}Fz')E_x(z')\,{\rm d}z',$$

while – indicating the normal derivative by a prime –

$$4\pi P'_{{\rm d}x}(+0)={\rm i}\frac{\varepsilon_{\rm c}\Omega_{\rm p}^2}{2F\beta_{\rm I}^2}\int_0^\infty {\rm i}F(p-1)\exp({\rm i}Fz')E_x(z')\,{\rm d}z'.$$

Hence

$$P_{{\rm d}x}(+0)+\frac{\rm i}{F}\frac{(p+1)}{(p-1)}P'_{{\rm d}x}(+0)=0. \tag{1.173}$$

Likewise we find

$$P_{{\rm d}z}(+0)+\frac{\rm i}{F}\frac{(p-1)}{(p+1)}P'_{{\rm d}z}(+0)=0. \tag{1.174}$$

This simplifies the evaluation of (1.172), where we can write $P'_d(+0)$ and $P'^*_d(+0)$ from (1.173), (1.174) and their complex conjugates. Doing this we find that, apart from non-vanishing factors, the condition for energy conservation at the surface is

$$\left\langle |P_{dx}|^2 \mathrm{Re}\left[F\frac{(p-1)}{(p+1)}\right] + |P_{dz}|^2 \mathrm{Re}\left[F\frac{(p+1)}{(p-1)}\right]\right\rangle = 0,$$

i.e.

$$\mathrm{Re}\left[F\frac{(p-1)}{(p+1)}\right] = 0 = \mathrm{Re}\left[F\frac{(p+1)}{(p-1)}\right]. \tag{1.175}$$

Since we have assumed a non-dissipative bulk (we are just concentrating on the effects of the surface), for the frequencies of physical interest, F is real. Hence the above condition implies that $(p-1)/(p+1)$ must be pure imaginary, which requires p to be of the form $\exp(\mathrm{i}\phi)$, as we had anticipated on intuitive grounds. The sign of ϕ, varying between 0 and π, is determined by the sign convention for the exponentials used to represent the fields and is fixed by imposing standard causality requirements.

One final comment is in order. We have formally solved the problem as a function of p. Thus, in this solution p could be anything and the above formulae, up to (1.162), are equally valid. We have argued that the form $\exp(\mathrm{i}\phi)$ is physically consistent with energy conservation, but other forms of p are not at all inconceivable. A situation with a different value of p, i.e. not of the form $\exp(\mathrm{i}\phi)$, would correspond to some energy source or sink at the surface and this is conceivable, provided some external interaction supplies the appropriate energy sink or source, as the case may be. The question of solving formally and exactly the complete matching problem for given p is a separate issue, which we have solved in this section.

1.5. Charge carriers in surface systems

Before studying various problems of electrodynamics of conducting surfaces we shall investigate some questions related to the presence of free charge carriers and the effects of the surface on them. We shall concentrate on fairly simple models, aiming at obtaining a broad picture and raising a wide range of questions. Suppose we write, in the relaxation time approximation for bulk collisions, the following Boltzmann equation for the distribution function $f(\boldsymbol{r}, \boldsymbol{p}, t)$:

$$(\partial/\partial t + \dot{\boldsymbol{r}} \cdot \nabla_{\boldsymbol{r}} + \dot{\boldsymbol{p}} \cdot \nabla_{\boldsymbol{p}})f = -(f - f_0)/\tau. \tag{1.176}$$

Let $\boldsymbol{r}(t), \boldsymbol{p}(t)$ describe the free trajectories under the influence of *all* fields acting on the carriers, but otherwise undisturbed by *any* collisions. We introduce the concept of *local* time such that at local time t the collisionless trajectory of a carrier has taken it to the six-dimensional phase space point $\boldsymbol{r}=\boldsymbol{r}(t)$, $\boldsymbol{p}=\boldsymbol{p}(t)$. We then rewrite (1.176) as

$$\left(\frac{\mathrm{d}}{\mathrm{d}t}+\frac{1}{\tau}\right)f_t=\left(\frac{1}{\tau}f_0\right)_t. \tag{1.177}$$

In this formalism everything is a function of local time t, even for time-independent problems. For example, τ may be a function of the kinetic energy of the carrier, which depends on $\boldsymbol{p}(t)$.

Suppose a carrier has been introduced at some previous local time t'. This is represented by a unit δ-function pulse $\delta(t-t')$ and then the probability of finding the carrier at present local time t is the Green function $\hat{G}(t, t')$ obeying

$$\left(\frac{\mathrm{d}}{\mathrm{d}t}+\frac{1}{\tau}\right)\hat{G}(t,t')=\delta(t-t'). \tag{1.178}$$

This casts the differential operator in a concise simple form, and now we must specify boundary conditions. If we find the appropriate Green function, then the solution of (1.177) is, formally,

$$f_t=\int_{-\infty}^{\infty}\hat{G}(t,t')(f_0/\tau)_{t'}\,\mathrm{d}t. \tag{1.179}$$

All we have to do is to interpret this explicitly in the notation of (1.176). Suppose, for example, we have an infinite medium with no surface. The only condition on $\hat{G}(t, t')$ is that it must be causal. Consider $t'<t_0<t_1$; then, from (1.178),

$$\hat{G}(t_1,t')=\hat{G}(t_0,t')\exp\left[-\int_{t_0}^{t_1}\frac{\mathrm{d}t''}{\tau(t'')}\right]. \tag{1.180}$$

Also, integrating (1.178) from $t=t'-\eta$ to $t=t'+\eta$ and letting $\eta\to 0$,

$$\hat{G}(t'+0,t')=1. \tag{1.181}$$

Thus, taking $t_1=t$, $t_0=t'+0$, we have

$$\hat{G}(t,t')=\theta(t-t')\exp\left[-\int_{t'}^{t}\frac{\mathrm{d}t''}{\tau(t'')}\right]. \tag{1.182}$$

The unit step function $\theta(t-t')$ is there to ensure causality. With this formula we could study nonlinear transport, but we shall not consider this here. Rather, we rewrite (1.176) and (1.177) in a way which is more suitable for subsequent linearisation and also for studying possible inhomogeneities. First we put

$$f = f_0 + f_1. \tag{1.183}$$

Thus f_1 measures the departure from equilibrium and is the part of f which carries net currents. Consider $\dot{\boldsymbol{p}}$ as the sum of $\dot{\boldsymbol{p}}_\mathrm{i}$, due to the effect of the fields associated with a possible static inhomogeneity, and $\dot{\boldsymbol{p}}_\mathrm{f}$, due to all other fields acting on the charge carrier. This could be, for example, an external electric field, or the field associated with a plasmon in the gas of carriers. The main thing is that all possible effects of static inhomogeneities are included in $\dot{\boldsymbol{p}}_\mathrm{i}$. Then we rewrite (1.176) as

$$\begin{aligned}(\partial/\partial t) + \dot{\boldsymbol{r}} \cdot \nabla_{\boldsymbol{r}} + \dot{\boldsymbol{p}} \cdot \nabla_{\boldsymbol{p}} + 1/\tau)f_1 \\ = -(\partial/\partial t + \dot{\boldsymbol{r}} \cdot \nabla_{\boldsymbol{r}} + \dot{\boldsymbol{p}}_\mathrm{i} \cdot \nabla_{\boldsymbol{p}})f_0 - \dot{\boldsymbol{p}}_\mathrm{f} \cdot \nabla f_0.\end{aligned} \tag{1.184}$$

So far this is mere rewriting. On the r.h.s. we separate $\dot{\boldsymbol{p}}_\mathrm{i}$ and $\dot{\boldsymbol{p}}_\mathrm{f}$ because at equilibrium,

$$(\partial/\partial t + \dot{\boldsymbol{r}} \cdot \nabla_{\boldsymbol{r}} + \dot{\boldsymbol{p}}_\mathrm{i} \cdot \nabla_{\boldsymbol{p}})f_0 = 0, \tag{1.185}$$

even if f_0 is the inhomogeneous distribution corresponding to a static inhomogeneity like a space charge surface layer. Thus (1.177) is equivalent to

$$\left(\frac{\mathrm{d}}{\mathrm{d}t} + \frac{1}{\tau}\right) f_{1t} = -\dot{\boldsymbol{p}}_\mathrm{f} \cdot \nabla_{\boldsymbol{p}} f_0. \tag{1.186}$$

Furthermore,

$$\dot{\boldsymbol{p}}_\mathrm{f} \cdot \nabla_{\boldsymbol{p}} f_0 = e\boldsymbol{v}_{\boldsymbol{p}} \cdot \boldsymbol{E}\,\partial f_0/\partial \mathscr{E}_{\boldsymbol{p}}, \tag{1.187}$$

where $\boldsymbol{v}_{\boldsymbol{p}}$ and $\mathscr{E}_{\boldsymbol{p}}$ are the velocity and energy of an electron in a state $\boldsymbol{p}$. (Let us assume the carriers are electrons. For holes we would simply use $-\boldsymbol{e}$.) Thus we start our analysis from

$$\left(\frac{\mathrm{d}}{\mathrm{d}t} + \frac{1}{\tau}\right) f_{1t} = -\left(e\frac{\partial f_0}{\partial \mathscr{E}_{\boldsymbol{p}}} \boldsymbol{v}_{\boldsymbol{p}} \cdot \boldsymbol{E}\right)_t. \tag{1.188}$$

Notice a very important point. The differential operators of (1.177) and (1.189) are identical. However, the functions f and f_1 may satisfy different boundary conditions and therefore the Green function for (1.189) *need not be the same as* (1.182). It happens to be the same for an infinitely extended medium, where causality is the only condition to satisfy, but this will not hold in general, for example for an arbitrary model of surface scattering, as will be seen presently. Thus, we shall still write the solution in the form

$$f_{1t} = -e \int_{-\infty}^{\infty} \hat{G}(t, t') \left(\frac{\partial f_0}{\partial \mathscr{E}_p} \boldsymbol{v}_p \cdot \boldsymbol{E} \right)_{t'} \mathrm{d}t', \qquad (1.189)$$

but we must use the $\hat{G}$ appropriate to each case. For a bulk system it is immediate to derive from here all the standard textbook formulae of transport theory.

Now introduce the surface. This has two kinds of effects. (i) It may produce an inhomogeneous space charge layer. Typically the scale of this corresponds to the screening length of the system. The effect of the inhomogeneous fields has already been discussed and is contained in the l.h.s. of (1.188), hence it will also be contained in the Green function. (ii) The carriers reaching the surface undergo extra scattering. It is precisely the effect of surface scattering that we must now include as a new boundary condition. This will consequently modify $\hat{G}$. For example, irrespective of the details of the surface scattering, the boundary conditions to be imposed on f and f_1 will in general be different, and so will the corresponding Green functions.

Since we shall be interested in linear transport, it is convenient to see how much (1.189) can be simplified. We bear in mind a situation in which the collisionless trajectories may go through an inhomogeneous region. Thus, while all effects due to the field $\boldsymbol{E}$ are negligible as higher order terms, we must be rather careful about the effects of the inhomogeneous fields. This will not affect $\partial f_0/\partial \mathscr{E}_p$ because the total energy of the carrier, if the effect of $\boldsymbol{E}$ is neglected, is constant along the trajectory and this derivative is equal to the derivative with respect to the total energy. Thus we can take this term outside the integrand. This cannot be done in general with $\boldsymbol{v}_p$, which changes if the trajectory goes through an inhomogeneity. We have not yet found $\hat{G}(t, t')$ for a general case. However, it is obvious that it will always contain the exponential appearing in (1.182). Can we simplify this exponential? This merits a separate discussion.

Everything hinges on the fact that τ depends in general on the energy of the carrier. If this changes because of the acceleration due to the effect of $\boldsymbol{E}$, such a correction is again a higher-order term and can be neglected in the linear approximation, but what happens if the trajectory crosses an inhomogeneity? The total energy of the carrier – if we neglect the effect of $\boldsymbol{E}$ – remains constant, but *the kinetic energy changes.* So the problem is reduced to the following: is τ a function of the total or of the kinetic energy of the carrier? We would intuitively expect that the scattering probability is determined by the kinetic energy, since this ultimately determines the wavelength associated with the state of motion of the carrier. This can be put on a formal basis if the inhomogeneity can be treated in the W.K.B. (Wentzel–Kramers–Brillouin) approximation, which ultimately is the very justification of the semiclassical picture we are using here. What one does in the W.K.B. approximation is actually to introduce the local wavelength or momentum, i.e. kinetic energy as a locally valid quantum number, to label the states. But our only concern is with the implications of this on the evaluation of f_{1t} by means of (1.189). If a carrier during its path from t' to t goes through an inhomogeneity, then its kinetic energy changes and the exponential of (1.182) cannot be simplified. This could be an important factor in the theory of surface galvanomagnetic effects, since the different transport coefficients in the presence of a magnetic field are rather sensitive to the dispersive properties of τ, i.e. to its energy dependence. However, we shall not study these problems in detail.

Having made this point, we start from the following picture: the exponential appearing in $\hat{G}(t, t')$ is simplified to $\exp[-(t-t')/\tau]$. This means (a) we linearise in $\boldsymbol{E}$, and (b) we neglect effects due to the dispersive properties of τ. The factor $\partial f_0/\partial \mathscr{E}_{\boldsymbol{p}}$, as explained above, is taken outside the integrand. We concentrate on the effects of surface scattering. Consider first the ideal specular case. A carrier starting at any past local time t' and reaching the surface, is by assumption reflected specularly (fig. 8) so that it reaches local time t as if it had been travelling through an infinitely extended medium with mirror symmetry. The Green function is

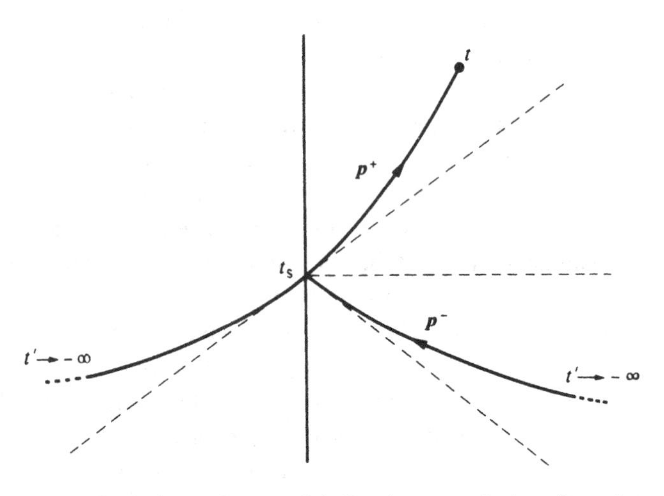

Fig. 8. Specular surface model. Carriers specularly reflected reach local time t as if they came from $t' \to -\infty$ after travelling through the symmetrically extended medium.

formally identical to (1.182) of the bulk problem. As explained above, we shall use

$$\hat{G}_s(t, t') = \theta(t-t') \exp\left[\frac{-(t-t')}{\tau}\right], \tag{1.190}$$

labelling with s for specular. Notice what this means. The non-equilibrium part of the distribution collected at present local time is

$$f_{1t} = -e\frac{\partial f_0}{\partial \mathscr{E}_p}\int_{-\infty}^{t} \exp\left[\frac{-(t-t')}{\tau}\right](\boldsymbol{v}_p \cdot \boldsymbol{E})_{t'}\, dt', \tag{1.191}$$

formally as in a bulk system. The difference may lie in the collisionless trajectory. Suppose $\boldsymbol{E}$ is constant in space and time and suppose we make a model in which our half medium is homogeneous right up to the surface, i.e. we concentrate exclusively on the effects of surface scattering, as in § 1.4. Then the specular model gives

$$f_{1t} = -e\frac{\partial f_0}{\partial \mathscr{E}_p}\boldsymbol{v}_p \cdot \boldsymbol{E}\int_{-\infty}^{t} \exp\left[\frac{-(t-t')}{\tau}\right] dt', \tag{1.192}$$

exactly as in a bulk system. This simplification is also valid in the presence of an inhomogeneous surface layer if $\boldsymbol{E}$ is parallel to the surface, as is usually the case, for example, in a surface mobility experiment. In this argument the steady-state distribution need not be independent of $(\boldsymbol{r}, t)$. If $\boldsymbol{E}(\boldsymbol{r}', t)$ is of the form $\boldsymbol{E}(\boldsymbol{k}, \omega) \exp \mathrm{i}\,(\boldsymbol{k} \cdot \boldsymbol{r}' - \omega t')$, we put it back inside the integrand writing $\boldsymbol{r}' = \boldsymbol{r} + \boldsymbol{v}_p(t-t')$, and using the identity

$$\boldsymbol{E}(\boldsymbol{r}', t') = \boldsymbol{E}(\boldsymbol{k}, \omega) \exp \mathrm{i}\,(\boldsymbol{k} \cdot \boldsymbol{r} - \omega t) \times \exp[\mathrm{i}(\boldsymbol{k} \cdot \boldsymbol{v}_p - \omega)(t'-t)], \tag{1.193}$$

we perform the integration over t' and obtain the standard bulk result which leads to the conductivity tensor,

$$\boldsymbol{\sigma}(\boldsymbol{k}, \omega) = -\frac{2e^2}{h^3}\int \frac{\partial f_0}{\partial \mathscr{E}_p}\frac{\boldsymbol{v}_p\boldsymbol{v}_p}{\tau^{-1} - \mathrm{i}\omega + \mathrm{i}\boldsymbol{k} \cdot \boldsymbol{v}_p}\, d\boldsymbol{p}. \tag{1.194}$$

From this we can obtain the dielectric tensor in the classical model, which we used to obtain the low-frequency limits of § 1.2, (1.49) and (1.50).

As stressed above, these formulae would not be valid, even for a specular surface model, if $\boldsymbol{E}$ had a component perpendicular to the surface and we took account of inhomogeneous space charge layers. These will be discussed in § 1.7. For the time being we shall ignore inhomogeneities and concentrate on the effects of surface scattering. We now face harder questions: how do we describe the scattering of carriers at the surface? This can be a very complicated problem indeed. One can attempt to produce a microscopic model, but the question is whether this can be used as a practical device to calculate, say, the conductivity in a surface system. In general this is rather difficult. It is probably fair to say that one can seldom make real progress beyond the stage of making several simplifying assumptions which, for practical purposes, reduce the model to the introduction of a phenomenological specularity parameter, p. At the very most one could perhaps produce a p which depends on the state of the incident particle, or at least on the angle of incidence. However, while this might help to improve some quantitative estimate, for our purposes it makes little difference. It is perhaps wiser to give up altogether any attempt at relating p to an actual model of scattering probability and to introduce it as a purely phenomenological parameter which measures the fraction of carriers which are specularly reflected. This is very different from the reflection coefficient introduced in § 1.4 for dielectrics. The phase here has no physical meaning. A polarisation vector can point, say, upwards or downwards, and this depends on the phase of the elementary excitation carrying this polarisation. In a conductor the quasiparticles of interest are charge carriers; all that matters is to calculate how much charge is reflected specularly. Thus p has a statistical nature and will be a real number between 0 and 1, if we stick consistently to the above interpretation of this parameter. This is actually an old-standing model, first introduced by J. C. Maxwell. Yet we have to investigate further questions which are far from trivial. Suppose we accept this phenomenological model. We are given a certain value of p. What does this imply in terms of boundary conditions for our problem? And, assuming we have written down the appropriate boundary conditions, how do we then find the correct solution?

Let us go back to (1.189). Suppose we have the extreme case $p=0$: diffuse surface scattering. Take a given carrier with local time t, i.e. one for which at time t, $\boldsymbol{r}=\boldsymbol{r}(t)$ and $\boldsymbol{p}=\boldsymbol{p}(t)$. Now consider all past local times t' corresponding to all past trajectories and let t' run backwards from t. The trajectories which do not touch the surface contribute to the population f_{1t} formally as in a bulk system, although the surface inhomogeneity may affect part of the past trajectory. But suppose we have a path which at some past time $t'=t_s$ touches the surface. In our model, with $p=0$, we can only collect at t the carriers coming from $t'\geqslant t_s$. Those having started at an earlier time are by assumption scattered diffusely and we do not collect them at t. This can be described in terms of the following Green function:

$$\hat{G}_d=\exp\left[\frac{-(t-t')}{\tau}\right]\theta(t-t')\theta(t'-t_s). \tag{1.195}$$

The problem is now reduced to evaluating t_s. This depends on z and on each of the trajectories. For some of them t_s may recede to $-\infty$, but the diffuse Green function of (1.195) formally incorporates the provision to cut off a past trajectory where necessary. This is a very simple argument which we shall soon criticise, but let us provisionally accept it and see where this leads us. For f_{1t} we have now

$$f_{1t} = -e\frac{\partial f_0}{\partial \mathscr{E}_p}\int_{t_s}^{t} \exp\left[\frac{-(t-t')}{\tau}\right](\boldsymbol{v}_p \cdot \boldsymbol{E})_{t'}\,\mathrm{d}t'. \tag{1.196}$$

Thus, in particular, if we evaluate f_{1t} at the surface – on the inside face, of course – we find that $f_{1s} = 0$, i.e. at $z = +0$ the distribution is just the equilibrium distribution. For arbitrary p we define an effective Green function

$$\begin{aligned}\hat{G}_p &= (1-p)\hat{G}_d + p\hat{G}_s \\ &= [(1-p)\theta(t'-t_s) + p]\theta(t-t')\exp\left[\frac{-(t-t')}{\tau}\right],\end{aligned} \tag{1.197}$$

and then

$$\begin{aligned}f_{1t} &= pe\left(-\frac{\partial f_0}{\partial \mathscr{E}_p}\right)\int_{-\infty}^{t_s} \exp\left[\frac{-(t-t')}{\tau}\right](\boldsymbol{v}_p \cdot \boldsymbol{E})_{t'}\,\mathrm{d}t' \\ &\quad + e\left(-\frac{\partial f_0}{\partial \mathscr{E}_p}\right)\int_{t_s}^{t} \exp\left[\frac{-(t-t')}{\tau}\right](\boldsymbol{v}_p \cdot \boldsymbol{E})_{t'}\,\mathrm{d}t'.\end{aligned} \tag{1.198}$$

Let f^+ indicate a distribution of carriers with momentum pointing inwards ($p_z > 0$), and f^- one of carriers with $p_z < 0$, incident on the surface. The solution given in (1.198) satisfies, at the surface $t = t_s$, the condition

$$f_{1s}^+ = pf_{1s}^-, \tag{1.199}$$

i.e. the Green function of (1.197) solves (1.188) subject to the boundary condition (1.199). If this is correct then the problem is reduced to taking proper account of the past trajectories and evaluating the above formulae. For example, for a field that varies as $\exp \mathrm{i}(\boldsymbol{k} \cdot \boldsymbol{r} - \omega t)$, neglecting inhomogeneity effects, we simply write the time difference as $t - t' = (z - z')/v_z$. By setting $t' = t_s$ and $z' = 0$, this yields t_s. The integral over t' is then rewritten as

$$\int \frac{\mathrm{d}z'}{v_z}\exp\left[\frac{-(z-z')}{\tau v_z}\right]\boldsymbol{E}(r', t'),$$

where the unspecified integration limit corresponds to the combination of

(1.198), and this leads to the non-local conductivity,

$$\boldsymbol{\sigma}(z, z'; \omega) = \frac{2e^2}{h^3} \int \mathrm{d}\boldsymbol{p} \left(-\frac{\partial f_0}{\partial \mathscr{E}_{\boldsymbol{p}}} \right) \frac{1}{v_z} \boldsymbol{v}_{\boldsymbol{p}} \boldsymbol{v}_{\boldsymbol{p}} \exp\left[\frac{-(z - z')(\tau^{-1} - \mathrm{i}\omega)}{v_z} \right]. \quad (1.200)$$

The simple picture we have built up is equivalent to the one usually found in standard treatments of various problems in surface transport. However, it has a major drawback. This is most clearly seen in the extreme case, $p = 0$. We say that carriers, having started at $t' < t_s$, do not reach t because they are diffusely scattered. This is in itself correct, but we still have to answer the question, where do they go? Charge must be conserved, and this is not really what (1.199) says. If $p = 0$, then with an incident distribution f_{1s}, carrying a net current into the surface, the net charge flow back into the bulk is zero since $f_{1s}^{+} = 0$. This does not matter if we are only concerned with electric fields and currents parallel to the surface. Then the surface scattering acts simply to slow down the forward drift momentum as, say, impurity scattering does in ordinary bulk transport; there is no distribution carrying a net charge flow in the direction perpendicular to the surface. But this picture is incorrect in general if there are currents in the z-direction. This can happen even with $E_z = 0$: for example, in the presence of a static magnetic field with suitable geometry. In the electrodynamics of conducting surfaces this situation will arise when we study reflectivity, surface impedance or surface plasmons with the physically interesting P-mode geometry, in which case $E_z \neq 0$.

1.6. Electrodynamics of conducting surfaces

We are now in a position to discuss conducting surfaces. In terms of a simple phenomenological model of surface scattering based on the parameter p, conductors differ from dielectric in two important respects, namely (i) p is a real number between 0 and 1, and (ii) with conductors we must face the problem of charge conservation. This means that, although we found in § 1.4 a general solution as a function of p, so that formally p could be anything – even a real number in (0, 1) – this solution *cannot* be used for conductors if fields and currents perpendicular to the surface are involved. We must solve the problem again, this time conserving J_z. The analysis is simpler if it is carried out in quasistatic limit. Since this is an approximation of practical interest in the case of metals, we shall give the explicit argument for P-mode geometry in the quasistatic limit, indicating at the end how the analysis can be extended to include retardation.

The method of approach is the same as we used in § 1.4. We concentrate again on the effects of surface scattering, ignoring inhomogeneities for the time being. We take a half infinite medium with bulk dielectric functions $\varepsilon_L(\boldsymbol{k}, \omega)$ and $\varepsilon_T(\boldsymbol{k}, \omega)$, contained in $z > 0$, with a vacuum in $z < 0$, and define the extended pseudomedium and extended pseudovacuum with appropriate fictitious stimuli. The problem now is to find these stimuli, and this is related to charge conservation. If we are to avoid an artificial charge sink at the surface, the boundary condition on the distribution function cannot be simply (1.199). We must have something reading

$$f_{1s}^{+} = pf_{1s}^{-} + \delta f_{1s}. \tag{1.201}$$

The question is, what is δf_{1s}? We know that (i) it must vanish for $p = 1$, (ii) for any $p \neq 1$ it must give a flow of charge perpendicular to the surface such that it cancels out the incoming flow, but (iii) its form is otherwise not unique. The form of δf_1 is to some extent arbitrary. Choosing it amounts to choosing a model of surface scattering which goes beyond the model used in § 1.5, because now we explicitly provide for charge conservation. This condition will serve to fix some adjustable parameter contained in our choice for δf_{1s}.

A simple way to ensure charge conservation is to concentrate just on the net current J_z returning into the bulk, without worrying about details of the back scattered distribution. This can be generated by a suitably adjusted fictitious electric field perpendicular to the surface. Its role is only to create a source of charge *at the surface* to cancel out the physically unrealisable sink we would otherwise have. But like all fictitious stimuli it must vanish inside the real medium. Thus this field must be of the form $E_{+}\delta(z)\boldsymbol{z}^0$. (It is understood that E_{+} actually means $E_{+}(\boldsymbol{\kappa}, \omega) \exp \mathrm{i}\,(\boldsymbol{\kappa} \cdot \boldsymbol{\rho} - \omega t)$.) With the use of (1.189) this extra field produces an additional contribution to the distribution given by

$$-eE_{+}(\kappa, \omega)\frac{\partial f_0}{\partial \mathscr{E}_{\boldsymbol{p}}} v_z \exp \mathrm{i}\,(\boldsymbol{\kappa} \cdot \boldsymbol{\rho} - \omega t) \propto \frac{\partial f_0}{\partial \mathscr{E}_{\boldsymbol{p}}} \cos \theta \exp \mathrm{i}\,(\boldsymbol{\kappa} \cdot \boldsymbol{p} - \omega t), \tag{1.202}$$

where θ is the angle between $\boldsymbol{v_p}$ and the normal to the surface. Thus we assume a boundary condition of the form (1.201) with

$$\delta f_{1t} \propto (\partial f_0 / \partial \mathscr{E}_{\boldsymbol{p}}) \cos \theta. \tag{1.203}$$

This amounts to a model of surface scattering with some anisotropy, giving greater weight to the direction perpendicular to the surface. We expect this to be also weighted by some factor which vanishes for $p = 1$, when all the distribution is specularly reflected.

Having seen the meaning of our model in terms of a Boltzmann equation approach, let us now return to the method used in § 1.4 and discuss the nature of the fictitious stimuli that we shall use in the quasistatic limit. First consider a symmetric stimulus, corresponding to a specular surface model. Remember the wavevector is always of the form $(\kappa, 0, q)$ and q is ultimately integrated away. We define the extended pseudomedium with a fictitious surface charge $\sigma^{\mathrm{M}}(\kappa, \omega)\delta(z)$, i.e. in Fourier transform, $\sigma^{\mathrm{M}}(\kappa, \omega) = \sigma^{\mathrm{M}}$, satisfying the divergence equation,

$$\mathrm{i}\boldsymbol{k} \cdot \boldsymbol{D}^{\mathrm{M}} = 4\pi\sigma^{\mathrm{M}}. \tag{1.204}$$

The scalar potential in the extended pseudomedium (M) is then

$$\phi^{\mathrm{M}} = (4\pi/k^2\varepsilon_{\mathrm{L}})\sigma^{\mathrm{M}}, \tag{1.205}$$

and the field in M has mirror symmetry. Likewise for the extended pseudovacuum (V),

$$\phi^{\mathrm{V}} = (4\pi/k^2)\sigma^{\mathrm{V}}; \qquad \mathrm{i}\boldsymbol{k} \cdot \boldsymbol{D}^{\mathrm{V}} = 4\pi\sigma^{\mathrm{V}}. \tag{1.206}$$

We have two parameters. The artificial charge source at the surface is not needed in this case because $p = 1$. Now, in the real system both D_z and ϕ must be continuous across the surface, and remember that the real fields are equal to $\boldsymbol{D}^{\mathrm{M}}$, ϕ^{M} in $z > 0$, and to $\boldsymbol{D}^{\mathrm{V}}$, ϕ^{V} in $z < 0$. Thus

$$D_z^{\mathrm{V}}(-0) = D_z^{\mathrm{M}}(+0); \qquad \phi^{\mathrm{V}}(-0) = \phi^{\mathrm{M}}(+0). \tag{1.207}$$

Fourier transforming (1.204)–(1.206) into real space we find from (1.207) that $\sigma^{\mathrm{V}} = -\sigma^{\mathrm{M}}$ and the s.m.d.r. is

$$1 + \frac{\kappa}{\pi}\int \frac{\mathrm{d}q}{k^2\varepsilon_{\mathrm{L}}} = 0. \tag{1.208}$$

This is of course the quasistatic limit of (1.115) or of (1.138) when $p = 1$.

But now consider a hypothetical antisymmetric model, with $p = -1$. By itself this would be a physically unrealisable model for a conductor, but we are studying the nature of the fictitious stimuli which, appropriately combined, will give a physical model for $0 \leqslant p \leqslant 1$. Carrying over the arguments of § 1.4, we start from

$$\nabla \cdot \boldsymbol{D}^{\mathrm{M}} = 0, \qquad \nabla \wedge \boldsymbol{E}^{\mathrm{M}} = \boldsymbol{\Gamma}^{\mathrm{M}}(\boldsymbol{\kappa}, \omega)\delta(z), \tag{1.209}$$

again using a fictitious magnetic stimulus which we shall take in the y-direction. This does not contain anything incompatible with the quasistatic limit. Inside the real system $\nabla \wedge \boldsymbol{E} = 0$ everywhere. In this model $\boldsymbol{D}^{\mathrm{M}}$ is purely transverse and therefore we can always find a vector

$\boldsymbol{\psi}^{\mathrm{M}}$ such that

$$\nabla \cdot \boldsymbol{\psi}^{\mathrm{M}} = 0; \qquad \boldsymbol{D}^{\mathrm{M}} = \nabla \wedge \boldsymbol{\psi}^{\mathrm{M}}.$$

For the same reason $\boldsymbol{D}^{\mathrm{M}} = \varepsilon_{\mathrm{T}} \boldsymbol{E}^{\mathrm{M}}$. We want to find $\boldsymbol{\psi}^{\mathrm{M}}$, which in this geometry is a vector in the y-direction. From this and (1.209) we have

$$\nabla \wedge \boldsymbol{E}^{\mathrm{M}} = (1/\varepsilon_{\mathrm{T}}) \nabla \wedge \boldsymbol{D}^{\mathrm{M}} = -(1/\varepsilon_{\mathrm{T}}) \nabla^2 \boldsymbol{\psi}^{\mathrm{M}} = \boldsymbol{\Gamma}^{\mathrm{M}},$$

i.e. in Fourier transform,

$$\psi^{\mathrm{M}} = (\varepsilon_{\mathrm{T}}/k^2) \Gamma^{\mathrm{M}},$$

whence

$$D_x^{\mathrm{M}} = -\mathrm{i}(q\varepsilon_{\mathrm{T}}/k^2)\Gamma^{\mathrm{M}}; \qquad D_z^{\mathrm{M}} = \mathrm{i}(\kappa\varepsilon_{\mathrm{T}}/k^2)\Gamma^{\mathrm{M}}. \tag{1.210}$$

By setting $\varepsilon_{\mathrm{T}} = 1$, the same equations give $\boldsymbol{D}^{\mathrm{V}}$. Now we express the boundary conditions

$$E_x^{\mathrm{V}}(-0) = E_x(-0) = E_x(+0) = E_x^{\mathrm{M}}(+0), \tag{1.211}$$

and

$$D_x^{\mathrm{V}}(-0) = D_z(-0) = D_z(+0) = D_z^{\mathrm{M}}(+0), \tag{1.212}$$

starting as usual from the Fourier transforms of the extended fields, which we have just derived, and transforming back into real space. From (1.211) we find $\Gamma^{\mathrm{V}} = -\Gamma^{\mathrm{M}}$ and using this in (1.212) we obtain the s.m.d.r.

$$1 + \frac{\kappa}{\pi} \int \frac{\varepsilon_{\mathrm{T}}}{k^2} \, \mathrm{d}q = 0. \tag{1.213}$$

The point of this exercise was to see that this is indeed the quasistatic limit $c \to \infty$ of (1.116), as one can verify. Therefore, *this model does not conserve* J_z, as can be easily verified by working out the current from the extended fields just constructed; we then find $J_z(+0) \neq 0$. The lesson in this is that with only the combination of symmetric and antisymmetric stimuli corresponding to the quasistatic limit of those used in § 1.4 we could never conserve J_z, no matter how we combine them. We need to invent some other kind of antisymmetric stimulus and it must be such that it drives charge in the z-direction, perpendicular to the surface. This can be achieved with a surface dipole, proportional not to $\delta(z)$ but to its derivative. Thus we now start from

$$\nabla \cdot \boldsymbol{D}^{\mathrm{M}} = 4\pi \mathscr{D}^{\mathrm{M}} \delta'(z) = \mathscr{D}_{\mathrm{M}} \delta'(z); \qquad \nabla \wedge \boldsymbol{E}^{\mathrm{M}} = 0, \tag{1.214}$$

which produces an antisymmetric scalar potential with Fourier transform

$$\phi^{\mathrm{M}}(\boldsymbol{\kappa}, \omega) = \mathrm{i}(q/\varepsilon_{\mathrm{L}} k^2) \mathscr{D}_{\mathrm{M}}.$$

Given ϕ^{M} we can derive $\boldsymbol{E}^{\mathrm{M}}$ and, finally,

$$D_z^{\mathrm{M}} = -\mathrm{i}q\varepsilon_{\mathrm{L}}\phi^{\mathrm{M}} = (q^2/k^2)\mathscr{D}^{\mathrm{M}}. \tag{1.215}$$

For the extended pseudovacuum,

$$\phi^{\mathrm{V}} = \mathrm{i}(q/k^2)\mathscr{D}_{\mathrm{V}}; \qquad D_z^{\mathrm{V}} = (q^2/k^2)\mathscr{D}_{\mathrm{V}}.$$

We have again two parameters and two equations. From the continuity of D_z in the real system, $\mathscr{D}_{\mathrm{V}} = \mathscr{D}_{\mathrm{M}}$, which, when used in the equation expressing the continuity of ϕ, yields yet another s.m.d.r., namely,

$$\pi\mathrm{i} + \int \frac{\exp(\mathrm{i}q\eta)q\,\mathrm{d}q}{k^2\varepsilon_{\mathrm{L}}} = 0. \tag{1.216}$$

Notice that both (1.213) and (1.216) are formally valid solutions for different conceivable antisymmetric models, although one is governed by ε_{T} and the other one by ε_{L}. Both are equally invalid physically, as neither one conserves J_z, but now we can see something interesting in the behaviour of the field of (1.215). This can be rewritten as

$$D_z^{\mathrm{M}} = \frac{k^2 - \kappa^2}{k^2}\mathscr{D}_{\mathrm{M}} = \mathscr{D}_{\mathrm{M}} - \frac{\kappa^2}{k^2}\mathscr{D}_{\mathrm{M}}. \tag{1.217}$$

Thus D_z^{M} contains a constant term in q-space. But this is the Fourier transform of a δ-function term in z-space. Such a term produces a contribution to $J_z(+0)$, as discussed above, and thus with suitably adjusted strength it can be used to achieve the desired condition, namely that the total $J_z(+0)$ must vanish.

It is now clear that for a general solution – when $0 \leqslant p \leqslant 1$ – we shall need a combination of the symmetric stimulus and of the two antisymmetric ones just discussed. We could proceed directly from the formulae just given in (1.204), (1.209), and (1.214), but it is rather cumbersome to have to treat the two antisymmetric stimuli on different footings. However, this can be circumvented by means of the following formal device. Notice that even for (1.209), $\nabla \wedge E^{\mathrm{M}}$ is zero everywhere except at $z = 0$. Thus for all $z \neq 0$ this E^{M} can also be derived from a scalar potential. Or, equivalently, we can always find a potential ϕ' such that $E' = -\nabla\phi'$ is equal to E^{M} of (1.209) *everywhere except at* $z = 0$. These two fields therefore differ only in a δ-function term, i.e. their Fourier transforms are equal except for a term independent of q. Thus, since $E'_x = -\mathrm{i}\kappa\phi'$ and $E'_z = -\mathrm{i}q\phi'$, this will have the same behaviour as E^{M} of (1.209) if ϕ' is chosen to be of the form $\phi' = a(\kappa)q/k^2$. Then

$$E'_x = -\mathrm{i}\frac{q}{k^2}\kappa a; \qquad E'_z = -\mathrm{i}\frac{q^2 a}{k^2} = -\mathrm{i}a + \mathrm{i}a\frac{\kappa^2}{k^2}. \tag{1.218}$$

Thus in practical terms we can proceed as follows: we introduce the fictitious field E', making the identification $\kappa a = \Gamma^{\mathrm{M}}$ and, when transforming back into real space, *we first remove the q-independent term* in (1.218). We then guarantee that this introduces an extended field equivalent to $\boldsymbol{E}^{\mathrm{M}}$ of (1.209). Then our complete extended pseudofield derives from a scalar potential whose Fourier transform is

$$\phi^{\mathrm{M}} = \mathrm{i}\frac{4\pi}{k^2}qa + \frac{4\pi}{k^2\varepsilon_{\mathrm{L}}}\sigma^{\mathrm{M}}(1+f) + \mathrm{i}\frac{q\mathscr{D}_{\mathrm{M}}}{k^2\varepsilon_{\mathrm{L}}}. \tag{1.219}$$

Notice the role of the different terms. The first one is the field we have just discussed. Note that the dielectric response functions of the medium do not appear in this term. Its role is only to introduce the functional dependence on q needed to create an antisymmetric field. And we must always, when evaluating the contribution of this term to the fields in real space, make the substitution

$$aq^2/k^2 \to -a\kappa^2/k^2 \tag{1.220}$$

before integrating over q. Next we have the symmetric term and, as we saw in § 1.4, we must in general add a volume term which is zero for $z>0$. Thus $f = f(\kappa, q)$, as a function of q, must be analytic in the upper half plane. The last term is also antisymmetric and describes the response of the system to a surface dipole. Its role is to introduce the charge source needed to ensure that $J_z(+0) = 0$.

As usual, since the vacuum is a local medium, for the extended pseudovacuum we can use any stimulus we choose. The simplest choice is to write

$$\phi^{\mathrm{V}} = (4\pi/k^2)\sigma^{\mathrm{V}}. \tag{1.221}$$

As in § 1.4, we write down two kinds of equations. The *matching equations* express the continuity of ϕ in the real system, i.e. from (1.220) and (1.221),

$$\sigma^{\mathrm{V}}\int\frac{\mathrm{d}q}{k^2} = -a\kappa\int\frac{\mathrm{d}q}{k^2} + \sigma^{\mathrm{M}}\int\frac{(1+f)}{k^2\varepsilon_{\mathrm{L}}}\,\mathrm{d}q + \mathrm{i}\mathscr{D}^{\mathrm{M}}\int\frac{\exp(\mathrm{i}q\eta)q}{k^2\varepsilon_{\mathrm{L}}}\,\mathrm{d}q, \tag{1.222}$$

and the continuity of D_z in the real system. Here we must remember to use (1.220); this yields

$$\sigma^{\mathrm{V}}\int\frac{\mathrm{d}q}{k^2} = a\kappa\int\frac{\varepsilon_{\mathrm{T}}}{k^2}\,\mathrm{d}q - \sigma^{\mathrm{M}}\int\frac{(1+f)}{k^2}\,\mathrm{d}q - \mathrm{i}\mathscr{D}^{\mathrm{M}}\int\frac{\exp(\mathrm{i}q\eta)q}{k^2}\,\mathrm{d}q. \tag{1.223}$$

One parameter can be eliminated straight away. We introduce

$$x_1 = a_\kappa/\sigma^{\mathrm{M}}; \qquad x_2 = \mathscr{D}^{\mathrm{M}}\kappa/\sigma^{\mathrm{M}}; \tag{1.224}$$

and, from (1.222) and (1.223) we have the general form of the s.m.d.r.

$$\int \frac{(1+f)}{k^2}\left(1+\frac{1}{\varepsilon_L}\right) dq - x_1 \int \frac{(1+\varepsilon_T)}{k^2} dq = -i\frac{x_2}{\kappa} \int \frac{\exp(iq\eta)q}{k^2}\left(1+\frac{1}{\varepsilon_L}\right) dq. \tag{1.225}$$

The term in x_2 is the new source term, which would not be present in the quasistatic limit of the results found in § 1.4. This equation still contains the unknown parameters x_1 and x_2, and the function f, about which we know only that it must be analytic in the upper half complex q-plane. Its form is determined by the model used for the dielectric functions. We shall choose the same model as in § 1.4, described in equations (1.142)–(1.145) putting $\varepsilon_c = 1$, $\omega_T = 0$ everywhere. This introduces a third unknown parameter for f. The second kind of equations includes those expressing the subsidiary conditions. These are: (i) The behaviour imposed on $\phi^M(z)$, i.e. $\phi^M(-z) = p\phi^M(z)$ for $z > 0$, or, equivalently,

$$\int \exp(-iq\eta)\phi^M(q)\, dq = p \int \exp(iq\eta)\phi^M(q)\, dq. \tag{1.226}$$

This describes the model of surface scattering in terms of the parameter p. (ii) Charge conservation. This means that at $z = +0$ the incoming net flux must be exactly compensated by the reflected net flux. Thus the total current $J_z(+0)$ must vanish or, equivalently,

$$\int \exp(iq\eta)[E_z^M(q) - D_z^M(a)]\, dq = 0. \tag{1.227}$$

In evaluating (1.226) from (1.219) we have contributions from the residues at $\kappa = \pm iq$ and $\kappa = \pm iQ_L$, which result in terms that vary as $\exp(-\kappa|z|)$ and $\exp(-Q_L|z|)$. Again, for the same reasons as in § 1.4, $f(\kappa, q)$ must have the form (1.149), only now B_0 is different. Now, imposing the condition (1.226) on the two exponentials separately we have

$$x_1 = \left[(p-1) + iB_0\left(\frac{1}{q_L - \kappa} - \frac{p}{p_L + \kappa}\right)\right]\frac{1}{\varepsilon(1+p)} - \frac{1}{\varepsilon}x_2, \tag{1.228}$$

and

$$x_2 = \frac{\kappa}{Q_L(1+p)}\left[(p-1) + iB_0\left(\frac{1}{q_L - Q_L} - \frac{p}{q_L + Q_L}\right)\right], \tag{1.229}$$

where $\varepsilon = \varepsilon(k = 0, \omega) = 1 - \omega_p^2/\omega^2$.

From (1.219) we must derive E_z and D_z for use in (1.226), recalling (1.220) where it pertains. This yields the third desired subsidiary condition:

$$x_1\left[(\varepsilon-1)+\frac{\omega_p^2}{\omega^2}\frac{\kappa}{q_T}\right]+\frac{1}{\varepsilon}\left\{(1-\varepsilon)\left[1+\frac{B_0}{i(\kappa+q_L)}\right]\right.$$
$$\left.-\frac{\omega_p^2}{\omega^2}\left[1+\frac{B_0}{i(Q_L+q_L)}\right]\right\}+\frac{x_2}{\varepsilon}\left[(\varepsilon-1)+\frac{\omega_p^2}{\omega^2}\frac{Q_L}{\kappa}\right]=0. \quad (1.230)$$

On the other hand, for this model of dielectric functions the s.m.d.r. (1.225) becomes

$$(1+\varepsilon)\left[1-\varepsilon x_1+\frac{B_0}{i(\kappa+q_L)}\right]$$
$$=2\varepsilon x_2+(1-\varepsilon)\left\{\left[1+\frac{B_0}{i(q_L+Q_L)}\right]\frac{1}{L}+\frac{\varepsilon}{q_T}x_1\right\}\kappa. \quad (1.231)$$

The three parameters contained in this dispersion relation are found from equations (1.228)–(1.230). It is very easy to see that x_1, x_2 and B_0 are all proportional to $(1-p)$. In the specular case ($p=1$) this leads us back to (1.208). Otherwise this is significantly different from the quasistatic limit of the results found in § 1.4.

It is interesting to look at the long-wave limit. In fact there has been a great deal of experimental and theoretical work on long-wave surface plasmons in metals. The limiting values needed are

$$Q_L\to\omega_p/\beta_L\sqrt{2};\qquad q_L\to-i\omega_p/\beta_L\sqrt{2};\qquad q_T\to-i\omega_p/\beta_T\sqrt{2}$$

and also

$$x_1\to\frac{(1-p)}{(1+p)};\qquad B_0\to 0;\qquad x_2\to\frac{(1-p)\beta_T\sqrt{2}}{(1+p)}\kappa.$$

These yield the long-wave dispersion relation

$$\omega=\frac{\omega_p}{\sqrt{2}}\left\{1+\frac{1}{\sqrt{2}}\left[\beta_L-i\frac{(1-p)\beta_T}{2}\right]\frac{\kappa}{\omega_p}\right\}. \quad (1.232)$$

It suffices to find a model of the form (1.142) – with $\varepsilon_c=1$, $\omega_T=0$ – and to evaluate the dispersive coefficients β_L and β_T. For example, we can start from the complete *Lindhard formulae* and proceed in a manner similar to that used in § 1.2. Taking v_Fk fixed and working out the limits $\omega\ll v_Fk$ and $\omega\gg v_Fk$ we can obtain interpolation formulae of the desired form, with

$$\beta_L=v_F\sqrt{\tfrac{3}{5}};\qquad \beta_T=v_F\sqrt{\tfrac{1}{3}}. \quad (1.233)$$

These yield, for this model,

$$\omega = \frac{\omega_p}{\sqrt{2}}\left[1 + \frac{2\sqrt{3} - i(1-p)}{4\sqrt{5}} - \frac{\sqrt{2}\, v_F \kappa}{\omega_p}\right]. \qquad (1.234)$$

Notice that the dispersion – the real part of the coefficient of κ – does not depend on the kind of surface scattering. This must not be taken too literally. It is merely due to the use of oversimplified dielectric functions in which, by discarding the detailed structure of $\varepsilon_L(k, \omega)$, we introduce a gross picture which ignores details of the charge distribution. If we used a more elaborate dielectric function to evaluate (1.228)–(1.231) we would find some differences. However, the result (1.232) suggests that we should expect these differences to be fairly small, i.e. the effects of such details to be comparatively unimportant, and indeed this turns out to be true in practice. Still, all this would give a very poor description of the long-wave dispersion because we are using a semiclassical approximation which neglects quantum mechanical interference effects between the incident and reflected wavefunctions. It is well known that this can affect the dispersion coefficient considerably, bringing it into closer agreement with experiment. This will be discussed in chapter 5. However, the real purpose of our picture is to emphasise the role of surface scattering and this is most conspicuous in the damping coefficient. This is the most important feature of our results. Notice that the imaginary part in (1.232), as in (1.162), contains the factor $(1-p)$. This indicates that the damping of the surface modes is primarily associated with non-specular scattering. Again, a dielectric function with more structure could give some residual effect, like the *Landau damping* in metals for example, but this is negligible compared with the main term which comes from surface scattering, and is given to a good approximation by the result (1.232). A fuller discussion of the meaning of these results will be given in § 1.8.

So far we have considered the quasistatic limit. For the sake of completeness we shall outline the application of the same analysis to conducting surfaces taking account of both charge conservation and retardation. The simplest thing is to start from § 1.4 and add the extra (dipole) term which we know from the above discussion is necessary to ensure that $J_z(+0) = 0$. We define the extended pseudomedium (M) with the following fictitious stimuli. (i) Electric current $\boldsymbol{J}^M$ as in (1.124). This has both a surface and a volume part. (ii) Magnetic current $\boldsymbol{\Gamma}^M$ as in (1.125), again in general with surface and volume parts. (iii) A surface dipole term of the form $\mathscr{D}^M\delta'(z)$. This can be alternatively described as a 'dipolar current' $\boldsymbol{J}^d$ in the x-direction, related to $\mathscr{D}^M$ by the continuity equation $\kappa J^d = i\omega q \mathscr{D}^M$.

Now, $\boldsymbol{J}^{\mathrm{M}}$ creates the fields $\boldsymbol{E}^{\mathrm{M}J}$, $\boldsymbol{H}^{\mathrm{M}J}$ as in (1.127), (1.128). $\boldsymbol{\Gamma}^{\mathrm{M}}$ creates the fields $\boldsymbol{E}^{\mathrm{M}\Gamma}$, $\boldsymbol{H}^{\mathrm{M}\Gamma}$ as in (1.129), (1.130). The dipole source creates an extra field which obeys the divergence equation,

$$\nabla \cdot \boldsymbol{D}^{\mathrm{Md}} = 4\pi \mathscr{D}^{\mathrm{M}} \delta'(z). \tag{1.235}$$

The extra terms due to this source and needed for our analysis are, in Fourier transform,

$$E_x^{\mathrm{Md}} = 4\pi \frac{\kappa q}{k^2 \varepsilon_{\mathrm{L}}} \mathscr{D}^{\mathrm{M}} - 4\pi \frac{\omega^2}{c^2} \frac{q^3}{\kappa k^2 [k^2 - (\omega^2/c^2)\varepsilon_{\mathrm{T}}]} \mathscr{D}^{\mathrm{M}}, \tag{1.236}$$

and

$$H_y^{\mathrm{Md}} = -4\pi \frac{\omega}{c} \frac{q^2}{\kappa [k^2 - (\omega^2/c^2)\varepsilon_{\mathrm{T}}]} \mathscr{D}^{\mathrm{M}}. \tag{1.237}$$

We can now construct the complete field $\boldsymbol{E}^{\mathrm{M}}$, $\boldsymbol{H}^{\mathrm{M}}$, and likewise for the extended pseudovacuum (V), for which we simply use a surface current $J^{\mathrm{V}}\delta(z)$ in the x-direction. The routine is now the same. The two matching equations resulting from the continuity of E_x and H_y in the real system can be combined to yield the general form of the s.m.d.r.:

$$\begin{aligned}
&\mathrm{i}\kappa \int \frac{(1+f_1)}{k^2 \varepsilon_{\mathrm{L}}} \exp(\mathrm{i}q\eta)\, \mathrm{d}q \\
&= \int \frac{\left(\mathrm{i}\dfrac{\omega^2}{c^2}\dfrac{q}{k^2} - \nu\right)\left(f_2 + \dfrac{q}{\kappa}\right) \exp(\mathrm{i}q\eta)}{k^2 - (\omega^2/c^2)\varepsilon_{\mathrm{T}}}\, \mathrm{d}q \\
&\quad + \beta \int \frac{(q + \mathrm{i}\nu\varepsilon_{\mathrm{T}})(1+f_3)}{k^2 - (\omega^2/c^2)\varepsilon_{\mathrm{T}}} \exp(\mathrm{i}q\eta)\, \mathrm{d}q \\
&\quad + \mathscr{D}' \int \left\{ \frac{\kappa q}{k^2 \varepsilon_{\mathrm{L}}} - \frac{q^2}{\kappa} \frac{\left(\mathrm{i}\nu + \dfrac{\omega^2}{c^2}\dfrac{q}{k^2}\right)}{k^2 - (\omega^2/c^2)\varepsilon_{\mathrm{T}}} \right\} \mathrm{d}q.
\end{aligned} \tag{1.238}$$

Compare with (1.138); the extra term on the r.h.s. is due to the dipole, whose strength is here measured by $\mathscr{D}' = \omega \mathscr{D}^{\mathrm{M}}/\mathscr{J}$. The other terms are identical. The functions f_1, f_2, f_3 have the same meaning and forms as in § 1.4. Thus our dispersion relation now contains five parameters, namely β, B_0, B_1, B_2 and $\mathscr{D}'$. We need five equations. The first four come about as in § 1.4, i.e. one from (1.121), two from (1.122), and one from (1.123). The last one, as in § 1.4, for this model of dielectric functions yields $B_2 = 0$. This simplifies the other subsidiary equations. Thus at this stage we have the following equations: from (1.121),

$$(1-p)-\mathrm{i}B_0\left[\frac{1}{q_\mathrm{L}-Q_\mathrm{L}}-\frac{p}{q_\mathrm{L}-Q_\mathrm{L}}\right]+\mathscr{D}'Q_\mathrm{L}(1+p)=0. \tag{1.239}$$

Compare with (1.150). From the two independent exponentials arising from (1.122) we find

$$\beta(q_\mathrm{m}^2-\kappa^2)(p+1)$$
$$=\mathrm{i}\frac{\omega^2}{c^2}\left\{(p-1)\frac{\mathrm{i}q_\mathrm{m}}{\kappa}\left[(p-1)\frac{q_\mathrm{m}q_\mathrm{L}}{\kappa}-(p+1)\kappa\right]\frac{B_0}{(q_\mathrm{L}^2-\kappa^2)}\right.$$
$$\left.+\mathrm{i}\frac{(\kappa^2-q_\mathrm{m}^2)}{(q_\mathrm{T}^2-\kappa^2)}\left[\frac{p}{q_\mathrm{T}+q_\mathrm{m}}+\frac{1}{q_\mathrm{T}-q_\mathrm{m}}\right]B_1+(p+1)\frac{\omega^2}{c^2}\frac{q_\mathrm{m}^2}{\kappa}\mathscr{D}'\right\} \tag{1.240}$$

(compare with (1.153)), and

$$\beta^2(Q_\mathrm{m}^2-\kappa^2)(p+1)$$
$$=\mathrm{i}\frac{\omega^2}{c^2}\left\{(p-1)\frac{\mathrm{i}Q_\mathrm{m}}{\kappa}\left[(p-1)\frac{Q_\mathrm{m}q_\mathrm{L}}{\kappa}-(p+1)\kappa\right]\frac{B_0}{(q_\mathrm{L}^2-\kappa^2)}\right.$$
$$\left.+\mathrm{i}\frac{(\kappa^2-Q_\mathrm{m}^2)}{(q_\mathrm{T}^2-\kappa^2)}\left[\frac{p}{q_\mathrm{T}+Q_\mathrm{m}}+\frac{1}{q_\mathrm{T}-Q_\mathrm{m}}\right]B_1+(p+1)\frac{\omega^2}{c^2}\frac{Q_\mathrm{m}^2}{\kappa}\mathscr{D}'\right\}. \tag{1.241}$$

Compare with (1.155).

As mentioned above, the condition (1.123) is the one that yields $B_2=0$. And now we add the condition $J_z(+0)=0$ in the form (1.227). This yields

$$\frac{B_1}{(q_\mathrm{T}^2-\kappa^2)}=\mathrm{i}\frac{q_\mathrm{L}}{\kappa}\frac{B_0}{(\kappa^2-q_\mathrm{L}^2)}. \tag{1.242}$$

We now have four equations for the four unknowns β, B_0, B_1, $\mathscr{D}'$. With this we can evaluate the s.m.d.r. (1.238) as a function of p. With some algebra we can verify that this correctly reproduces the two limiting situations already discussed; i.e. in the quasistatic limit $c\to\infty$ we recover the results derived above for conductors in the quasistatic limit, while if we eliminate the dipole source – i.e. set $\mathscr{D}'=0$ and eliminate (1.242) – we recover all the results of § 1.4.

1.7. Surface electrodynamics with inhomogeneous surface layers

In the previous sections the emphasis has been on the effects of surface scattering. We now study the effects of inhomogeneities in the surface region. Most of the basic theory presented here is equally applicable to any kind of conductor, as will be seen in § 1.8 which will be mainly concerned with metals. However, in order to give a more intuitive appeal to the formal analysis through its application to physical systems, in this section we shall mostly be concerned with semiconductors. Typically the inhomogeneity is associated with some given surface charge Q_s. This may be due to charge trapped in surface states, to charge transfer chemisorption, or to the application of an external field. We are also supposed to know all the standard information about the bulk material, i.e. impurity doping, Fermi level position, effective masses and the like. The screening of Q_s is described by an appropriate Poisson equation whose two boundary conditions are the following. (i) At the surface the normal derivative of the electrostatic potential ϕ is $\phi_s' = 4\pi Q_s$. (ii) In the bulk ϕ tends to a fixed value ϕ_b, corresponding to charge neutrality. The parameters of interest are the *surface potential* $\phi_s - \phi_b$ and the *surface excess* of carriers ΔN. For example, for an n-type material with bulk electron concentration $n_b = n$,

$$\Delta N = \int_0^\infty [n(z) - n]\,\mathrm{d}z.$$

The solution of the Poisson equation yields a sort of associated screening length (of the Debye type),

$$L_D = \left(\frac{\varepsilon_c k_B T}{8\pi e^2 n}\right)^{1/2},$$

where ε_c is the dielectric constant of the crystal matrix and k_B is the Boltzmann constant. A typical order of magnitude for L_D can be a few hundred angstroms, while experimentally one is often interested in the infrared region, corresponding to much longer wavelengths. Thus we shall study the long-wave limit $\lambda \gg L_D$.

We assume that an external source, located somewhere, produces an e.m. field normally incident on the surface. There and everywhere inside the medium, $\boldsymbol{J}_{ext} = 0$. We start from (1.6), eliminate H_y from the two curl equations, and obtain

$$\mathrm{d}^2E_x/\mathrm{d}z^2 = -(\omega^2/c^2)\varepsilon_{xx}(\omega, z)E_x. \tag{1.243}$$

We write this as an ordinary differential equation. The x-dependence is irrelevant. Either it does not appear explicitly or else it is negligible compared with the z-dependence because of the long-wave assumption. Furthermore, we introduce an inhomogeneous local dielectric function,

$$\varepsilon_{xx}(\omega, z) = \varepsilon_c + \mathrm{i}(4\pi/\omega)\sigma_{xx}(\omega, z), \tag{1.244}$$

because infrared wavelengths are also longer than typical values of the carrier mean free path $\bar{\lambda}$. Furthermore, we describe the crystal matrix in terms of a simple dielectric constant – which can be a function of ω if we want to include, say, *Reststrahlen* effects in an ionic semiconductor – because the energies of the infrared are also smaller than the gap energies for interband electronic excitations in the crystal. The inhomogeneous conductivity can be evaluated from the model outlined in § 1.5. In (1.198) the electric field has now only x-component, v_x is not affected by the inhomogeneity, and we obtain

$$f_{1t} = -e\frac{\partial f_0}{\partial \mathscr{E}_p}\frac{v_x}{\tau^{-1} - \mathrm{i}\omega}\{1-(1-p)\exp[-(t-t_s)(\tau^{-1}-\mathrm{i}\omega)]\}. \qquad (1.245)$$

The z-dependence is contained in f_0 and in $(t-t_s)$ which not only depends on z – the distance from the surface – but also on the potential profile $\phi(z')$ that the carrier must have crossed on going from the surface to z. This formula can be used to evaluate $\sigma_{xx}(\omega, z)$ in the local approximation, including effects of surface scattering, whence we have (1.244) to be used in (1.243). The standard approach to this kind of problem consists in solving the differential equation (1.243) subject to matching at the surface and to the condition that in the bulk we have a field of the form

$$E_x^{\mathrm{b}} = C_E \exp[\mathrm{i}(\omega/c)\varepsilon_{\mathrm{b}}^{1/2} z], \qquad (1.246)$$

which obeys (1.243) with $\varepsilon_{xx} = \varepsilon_{\mathrm{b}} = \varepsilon_{\mathrm{c}}(1-\omega_{\mathrm{p}}^2/\omega^2)$, where ω_{p} is the plasma frequency of the free carrier gas. In practice this happens for $z > L_{\mathrm{D}}$. The factor $\exp(-\mathrm{i}\omega t)$ is understood. Put

$$\varepsilon_{xx}(z) = \varepsilon_{\mathrm{b}} + \Delta\varepsilon(z); \qquad E_x(z) = E_x^{\mathrm{b}} + \Delta E_x(z). \qquad (1.247)$$

Substituting in (1.243), we have

$$\frac{\mathrm{d}^2 \Delta E_x}{\mathrm{d}z^2} + \left(\frac{\omega}{c}\right)^2 \varepsilon_{xx}\Delta E_x = -\left(\frac{\omega}{c}\right)^2 C_E \Delta\varepsilon \exp\left(\mathrm{i}\frac{\omega}{c}\varepsilon_{\mathrm{b}}^{1/2} z\right). \qquad (1.248)$$

Notice that ΔE_x grows from 0 for $z \gtrsim L_{\mathrm{D}}$ to its maximum value $\Delta E_x(+0) \approx L_{\mathrm{D}}^2\, \mathrm{d}^2\Delta E_x/\mathrm{d}z^2$. Thus the second term on the l.h.s. is of order $(qL_{\mathrm{D}})^2$ compared with the first one and we can neglect it. Then (1.248) can be directly integrated. In fact for the matching problem we only need

$$\Delta E_x(+0) = C_E\left(\frac{\omega}{c}\right)^2 \int_0^\infty \mathrm{d}z' \int_{z'}^\infty \mathrm{d}z''\, \Delta\varepsilon(z''). \qquad (1.249)$$

Likewise we put

$$H_{yt} = H_y^{\mathrm{b}} + \Delta H_y(z); \qquad H_y^{\mathrm{b}} = \varepsilon_{\mathrm{b}}^{1/2} E_x^{\mathrm{b}}. \qquad (1.250)$$

Using (1.249) and Maxwell's equations we find

$$\Delta H_{yt}(+0) = -\mathrm{i}C_E\left(\frac{\omega}{c}\right)\int_0^\infty \Delta\varepsilon(z)\,\mathrm{d}z. \tag{1.251}$$

In fact (1.249) itself is again a second order term compared with E_x^{b}. Neglecting it we have, from (1.246), (1.250) and (1.251),

$$\left.\begin{aligned} &E_x(+0) \approx C_E; \\ &H_y(+0) \approx \varepsilon_{\mathrm{b}}^{1/2} C_E - \mathrm{i}C_E\frac{\omega}{c}\int_0^\infty \Delta\varepsilon\,(z)\,\mathrm{d}z; \end{aligned}\right\} \tag{1.252a}$$

while outside

$$E_x(-0) = E_{x\mathrm{i}} + E_{x\mathrm{r}}; \qquad H_y(-0) = E_{x\mathrm{i}} - E_{x\mathrm{r}}. \tag{1.252b}$$

Matching (1.252*a*) to (1.252*b*), we obtain

$$E_{x\mathrm{i}} + E_{x\mathrm{r}} = C_E$$

$$E_{x\mathrm{i}} - E_{x\mathrm{r}} = C_E\left[\varepsilon_{\mathrm{b}}^{1/2} - \mathrm{i}\frac{\omega}{c}\int_0^\infty \Delta\varepsilon\,(z)\mathrm{d}z\right],$$

whence the reflectivity

$$R = \left|\frac{E_{x\mathrm{r}}}{E_{x\mathrm{i}}}\right|^2 = \left|\frac{1-\varepsilon_{\mathrm{b}}^{1/2} + \mathrm{i}(\omega/c)\displaystyle\int_0^\infty \Delta\varepsilon\,(z)\,\mathrm{d}x}{1+\varepsilon_{\mathrm{b}}^{1/2} - \mathrm{i}(\omega/c)\displaystyle\int_0^\infty \Delta\varepsilon\,(z)\,\mathrm{d}z}\right|^2. \tag{1.253}$$

The effects of the inhomogeneity and of surface scattering are here shown explicitly. When this term is zero we recover the standard *Drude formula*. Note that the extra terms contains the imaginary factor. This interchanges the roles of the real and imaginary parts of $\Delta\varepsilon$, and this can affect significantly the details of the lineshape, thus acquiring relevance in phenomenological applications.

The above argument assumes that the damping of the e.m. wave across the distance L_{D} is practically negligible. This is not unreasonable for the situation studied here. Knowing k_{p} and L_{D} we find $k_{\mathrm{p}}L_{\mathrm{D}} \approx \bar{v}/c$, where $\bar{v}$ is some average thermal velocity. Thus $k_{\mathrm{p}}L_{\mathrm{D}} \ll 1$, which means negligible damping. Even in situations in which damping is not negligible this approach would still work provided the z-dependence associated with the inhomogeneity does not increase too rapidly. The thing to do then is to approach the problem in a W.K.B. manner, trying a solution of the form $E_x \sim \exp[\mathrm{i}qz + \psi(z)]$ for (1.243). However, this would not be a very practical procedure for other problems that we shall study presently.

Hence we now develop a different approach which explicitly assumes negligible damping across the inhomogeneity but is more practical in other situations, like oblique incidence with arbitrary angle θ, when both E_x and E_z are involved for P-mode geometry; or thin-film problems, when we would have to match at two surfaces. This method can be used without requiring a smooth inhomogeneity subject to a W.K.B. approximation. Further interesting applications will be seen in §§ 1.8 and 1.9.

Imagine a sort of *black box* (fig. 9) containing the inhomogeneity. The thickness L is some convenient length such that we are already practically

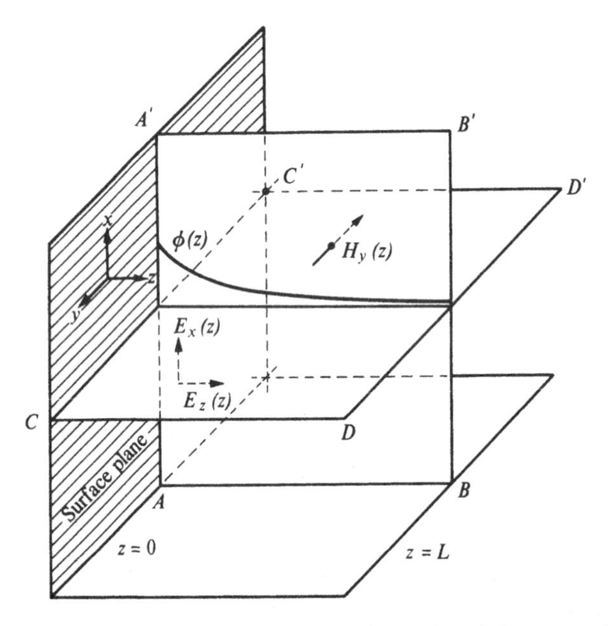

Fig. 9. The black box containing the surface inhomogeneity. The fields correspond to P-mode geometry.

in the homogeneous bulk, but $qL \ll 1$. Let us formulate the problem in general for oblique incidence in the P-mode geometry. We can still treat J_x locally, because of the long-wave assumption, but J_z differs in two respects. First, the analysis of § 1.5 could not be used without violating charge conservation, as explained in § 1.6. Since the emphasis here is on the effect of the inhomogeneity we shall take $p = 1$ for the treatment of J_z. Secondly we have the problem that J_z changes from zero at $z = +0$ to some definite value at a distance of the order of the carrier mean free path $\bar{\lambda}$. Actually $\bar{\lambda}/L_{\mathrm{D}} \approx \omega_{\mathrm{p}}\tau$, and $\omega_{\mathrm{p}}\tau$ is typically at least approximately 1. Thus non-local effects cannot be neglected when treating J_z. The position is then $L_{\mathrm{D}} \lesssim \bar{\lambda} \lesssim L \ll q^{-1}$. Thus for the x-direction we shall use $\sigma_{xx}(\omega, z)$

or $\varepsilon_{xx}(\omega, z)$, while for the z-direction we go back to § 1.5, put $p = 1$ and construct a non-local conductivity $\sigma_{zz}(z, z'; \omega)$ from (1.200). Then we use the non-local dielectric function

$$\varepsilon_{zz}(z, z'; \omega) = \varepsilon_c(z - z') + (4\pi i/\omega)\sigma_{zz}(z, z'; \omega).$$

The local term is written ε_c for short, but it could also have an ω-dependence if the crystal matrix is somewhat ionic, rather than strictly covalent.

The idea of the black-box method is that in the long-wave limit all that matters really is the integrated effect of the inhomogeneity and this should be sufficiently well described, without too many details, in terms of global parameters only, like the surface excess ΔN. To think only of the penetration length of the radiation, which is typically several orders of magnitude larger than the effective thickness of the space charge layer, would be misleading. It may well be that ΔN is comparable to the total number of carriers with which the light interacts in a tube of unit cross-section, and length equal to the penetration length. Then an apparently very thin surface layer can have appreciable effects, which are also easy to interpret in terms of e.m. matching. The surface excess ΔN implies an *excess surface current* – positive or negative – which is bound to affect the matching of the tangential component of $\boldsymbol{H}$, as is obvious from (1.6). Now, in order to calculate the reflectivity we need three linear homogeneous relations between the amplitudes E_i, E_r, E_t. These can be obtained by using the contours of fig. 9 in the following way.

Use the contour $ABB'A$ to calculate the line integral of $\boldsymbol{E}$; use the Stokes theorem and (1.6). Thus

$$\oint \boldsymbol{E} \cdot d\boldsymbol{l} = \iint (\nabla \wedge \boldsymbol{E}) \cdot d\boldsymbol{s} = i\frac{\omega}{c} \iint \boldsymbol{H} \cdot d\boldsymbol{s}. \tag{1.254a}$$

Then write down likewise the line integral of $\boldsymbol{H}$:

$$\oint \boldsymbol{H} \cdot d\boldsymbol{l} = \iint (\nabla \wedge \boldsymbol{H}) \cdot d\boldsymbol{s} = -i\frac{\omega}{c} \iint \boldsymbol{E} \cdot d\boldsymbol{s}. \tag{1.254b}$$

It suffices to evaluate the integrals. We define the bulk amplitude $E^b = E_t\, e^{iq'L}$, and integrate in the x-direction from any arbitrary origin O to any length x. Then, remembering that the x-dependence varies as $e^{i\kappa x}$, we have

$$\oint \boldsymbol{E} \cdot d\boldsymbol{l} = E_x^b \frac{(e^{i\kappa x} - 1)}{i\kappa} - (E_{xi} + E_{xr}) \frac{(e^{i\kappa x} - 1)}{i\kappa} + \int_0^L E_z(z)(1 - e^{i\kappa x})\, dz, \tag{1.255}$$

where we use $E_z(0,z) \approx E_z(x,z) = E_z(z)$. In the bulk we know that

$$H_y^b(x) = \varepsilon_b^{1/2} E^b \, e^{i\kappa x}. \tag{1.256}$$

Thus by again assuming this amplitude to have negligible variation from 0 to L, (1.254a), (1.255) and (1.256) yield

$$E_x^b \frac{(e^{i\kappa x}-1)}{i\kappa} - (E_{xi}+E_{xr}) \frac{(e^{i\kappa x}-1)}{i\kappa} + \int_0^L E_z(z)(1-e^{i\kappa x})\, dz = i\frac{\omega}{c}\varepsilon_b^{1/2} L E^b \frac{(e^{i\kappa x}-1)}{i\kappa}. \tag{1.257}$$

Proceeding likewise with (1.254b), we obtain

$$(E_i - E_r - \varepsilon_b^{1/2} E^b)\, e^{i\kappa x} = -i\frac{\omega}{c} E^b \left(1 - \frac{\sin^2\theta}{\varepsilon_b}\right) e^{i\kappa x} \int_0^L \varepsilon_{xx}(x,z)\, dz. \tag{1.258}$$

We now have the two desired homogeneous equations. We shall consider the x-dependence of ε_{xx} indicated in (1.258) later; for the present problem we can take $\varepsilon_{xx}(z)$, independent of x. We also leave the exponential factors for later reference, and again for this problem we eliminate them from (1.257) and (1.258). Introducing appropriate sine and cosine factors where necessary we have the following homogeneous equations expressed in terms of field amplitudes:

$$\begin{aligned} &E^b\left(1 - \frac{\sin^2\theta}{\varepsilon_b}\right) - (E_i + E_r)\cos\theta - i\frac{\omega}{c}\sin\theta \int_0^L E_z(z)\, dz = i\frac{\omega}{c}\varepsilon_b^{1/2} L E^b \\ &E_i - E_r - \varepsilon_b^{1/2} E^b = -i\frac{\omega}{c} E^b \left(1 - \frac{\sin^2\theta}{\varepsilon_b}\right) \int_0^L \varepsilon_{xx}(z)\, dz. \end{aligned} \tag{1.259}$$

The details of the inhomogeneity are reflected in the two integrals appearing here. For $\varepsilon_{xx}(z)$ we have (1.256) and (1.200), as discussed above.

Consider first normal incidence. Then $E_z = 0$, $\cos\theta = 1$, $\sin\theta = 0$, and equations (1.259) are considerably simplified. It is convenient to define a *surface conductivity* σ_s by

$$L\sigma_s = \int_0^L \sigma_{xx}(z)\, dz. \tag{1.260}$$

Then

$$\int_0^\infty \varepsilon_{xx}(z)\, dz = L\left(\varepsilon_c + i\frac{4\pi\sigma_s}{\omega}\right). \tag{1.261}$$

It is also convenient to put $\sigma_s = \sigma_b + \Delta\sigma$. The measure of the 'strength' of the inhomogeneity, so to speak, is actually $\Delta\sigma$, which apart from surface mobility effects, is essentially a measure of ΔN. Using all this in (1.259) and eliminating E^{b} we obtain the reflectivity formula

$$R = \left|\frac{(1-\varepsilon_{\mathrm{b}}^{1/2})(1-iqL\varepsilon_{\mathrm{b}}^{1/2})-(4\pi/c)L\Delta\sigma}{(1+\varepsilon_{\mathrm{b}}^{1/2})(1-iqL\varepsilon_{\mathrm{b}}^{1/2})+(4\pi/c)L\Delta\sigma}\right|^2. \tag{1.262}$$

Notice the different ways in which L appears here. The factor $(1-iqL\varepsilon_{\mathrm{b}}^{1/2})$ can be taken equal to unity because it is multiplying the leading term in the numerator and in the denominator. Furthermore, $qL \ll 1$, whereas nothing can be said about the term $L\Delta\sigma$ until we know the strength of the inhomogeneity. Also, the integral yielding $\Delta\sigma$ can be indifferently extended either to L or to ∞. We thus recover (1.253). It is interesting to see the form of the results predicted by this formula. The main point is the shape of the curve of R against ω, or λ^{-1}. The most conspicuous feature of the Drude formula is that it has a minimum. The *reflectivity minimum* is in fact the object of active experimental research, and empirical fitting in this range is used to obtain information about parameters like effective masses or carrier relaxation times. But we have remarked that the term in $\Delta\sigma$ interchanges the roles of the real and imaginary parts of $\Delta\varepsilon$ and this may affect the shape of the curve. Since the process of obtaining practical information by empirical fitting is so sensitive to the details of the curve in the range of the reflectivity minimum, the corrections due to $\Delta\sigma$ can be of practical importance. This is illustrated in fig. 10, where R is calculated from (1.262) for a model of

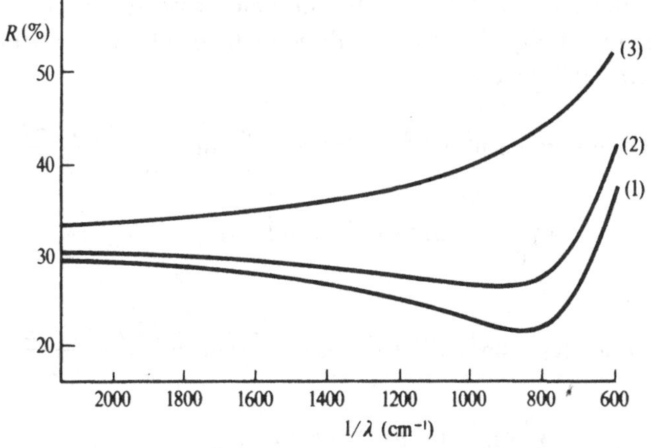

Fig. 10. Reflectivity against reciprocal wavelength calculated from (1.262) for a model of n-Si. (1), (2), (3): Different samples. Numerical data in table 1. After F. Flores, F. García-Moliner & G. Navascués, *Surface Sci.* **24**, 61 (1971).

n-Si. The numerical parameters (table 1) correspond to some actual experimental situations. Curve (1) is simply the Drude formula. It corresponds to a sample with no surface layer. For sample (2) the details of the reflectivity minimum are already considerably changed, and for sample (3) the minimum *disappears.* This feature could never be explained by a Drude formula even at the price of inconceivable fudging. It is easy to understand why $\Delta\sigma$ has such a drastic effect. The value for sample (3) corresponds to $\Delta N \sim 10^{14}$ carriers per cm^2, and this is also for this sample the order of magnitude of the total number of carriers contained in a tube of cross-section equal to 1 cm and length of the order of the penetration length. Thus, an apparently very thin surface layer can have considerable effects.

TABLE 1. *Numerical values of the parameters used for fig. 10*

	(1)	(2)	(3)	Units
T	295	373	573	K
ε_c	12.73	12.73	12.73	
$(\omega_p\tau)^2$	2.8	2.6	1.3	
$L\,\Delta\sigma$	0	2380	6890	μmho
n/m^*	7.90	7.34	7.56	10^{46} cm^{-1} g^{-1}

Now consider the more general case of oblique incidence. Then we must keep the term in $E_z(z)$ in (1.259). It is convenient to introduce explicitly the difference $E_z(z)-E_z^{\mathrm{b}}$, in which case the integral can be extended to infinity. Proceeding directly from (1.259) we then find $R=|\mathrm{Num}/\mathrm{Den}|^2$, where

$$\mathrm{Num}=(\varepsilon_{\mathrm{b}}-\sin^2\theta)^{1/2}-\varepsilon_{\mathrm{b}}\cos\theta+\mathrm{i}\frac{\omega}{c}\sin^2\theta\int_0^\infty \frac{E_z(z)-E_z^{\mathrm{b}}}{E_z^{\mathrm{b}}}\,\mathrm{d}z$$
$$+\mathrm{i}\frac{\omega}{c}(\varepsilon_{\mathrm{b}}-\sin^2\theta)^{1/2}\cos\theta\int_0^\infty[\varepsilon_{xx}(z)-\varepsilon_{\mathrm{b}}]\,\mathrm{d}z \qquad (1.263)$$

and

$$\mathrm{Den}=(\varepsilon_{\mathrm{b}}-\sin^2\theta)^{1/2}+\varepsilon_{\mathrm{b}}\cos\theta+\mathrm{i}\frac{\omega}{c}\sin^2\theta\int_0^\infty \frac{E_z(z)-E_z^{\mathrm{b}}}{E_z^{\mathrm{b}}}\,\mathrm{d}z$$
$$-\mathrm{i}\frac{\omega}{c}(\varepsilon_{\mathrm{b}}-\sin^2\theta)^{1/2}\cos\theta\int_0^\infty[\varepsilon_{xx}(z)-\varepsilon_{\mathrm{b}}]\,\mathrm{d}z. \qquad (1.264)$$

The hard part of the problem consists in evaluating $E_z(z)$ and this depends on the model one uses, but we shall leave such details for practical applications. It is also instructive to exploit the relation between reflectivity and s.m.d.r. discussed in § 1.3. First, for the sake of completeness we give here the result of using the same approach for S-mode geometry. In this case we find

$$R=\left|\frac{\cos\theta-(\varepsilon_b-\sin^2\theta)^{1/2}+\mathrm{i}(\omega/c)\int_0^\infty[\varepsilon_{xx}(z)-\varepsilon_b]\,\mathrm{d}z}{\cos\theta+(\varepsilon_b-\sin^2\theta)^{1/2}-\mathrm{i}(\omega/c)\int_0^\infty[\varepsilon_{xx}(z)-\varepsilon_b]\,\mathrm{d}z}\right|^2. \tag{1.265}$$

Now we can write down the s.m.d.r. including the effects of the surface layer by setting the denominator equal to zero. The variable of interest is now κ, not θ. Thus we put $\sin\theta=c\kappa/\omega$ and likewise for $\cos\theta$.

Consider the physically more interesting P-mode geometry. The s.m.d.r. is then

$$\begin{aligned}\left(\varepsilon_b-\frac{c^2\kappa^2}{\omega^2}\right)^{1/2}&+\varepsilon_b\left(1-\frac{c^2\kappa^2}{\omega^2}\right)^{1/2}\\&=\mathrm{i}\frac{\omega}{c}\left(\frac{c\kappa}{\omega}\right)^2\int_0^\infty\frac{E_z(z)-E_z^b}{E_z^b}\,\mathrm{d}z\\&\quad+\mathrm{i}\frac{\omega}{c}\left(\varepsilon_b-\frac{c^2\kappa^2}{\omega^2}\right)^{1/2}\left(1-\frac{c^2\kappa^2}{\omega^2}\right)^{1/2}\int_0^\infty[\varepsilon_{xx}(z)-\varepsilon_b]\,\mathrm{d}z=0.\end{aligned} \tag{1.266}$$

In the absence of surface-layer effects the r.h.s. vanishes and we recover (1.82). As with the reflectivity, the effect of the surface layer depends ultimately on ΔN. To illustrate this fig. 11 shows the s.m.d.r. for three different values of the dimensionless parameter $\varepsilon_c\omega_p\Delta N/nc$. The calculation corresponds to a sample of n-InSb. Curve (1) is for the case in which there is no boundary layer. Curves (2) and (3) correspond to boundary layers which, for a bulk concentration $n=7\times10^{18}\ \mathrm{cm}^{-3}$, have surface excesses ΔN of order $10^{14}\ \mathrm{cm}^{-2}$, and this again is the order of magnitude of the total number of carriers that the e.m. field interacts with along its full penetration length. The range of the dispersion curve plotted in fig. 11 corresponds to an experimentally accessible region, and the dimensionless plot suggests that the effects of the surface layer can be quite measurable.

The general conclusion is that in the analysis of optical and magneto-optical properties of semiconductor surfaces one may expect to encounter appreciable effects of the space charge boundary layer. The elementary theory outlined can be used as a basis to reformulate the theory of these phenomena. An external static magnetic field could also be added at the price of extra complication, but the method of approach is the same.

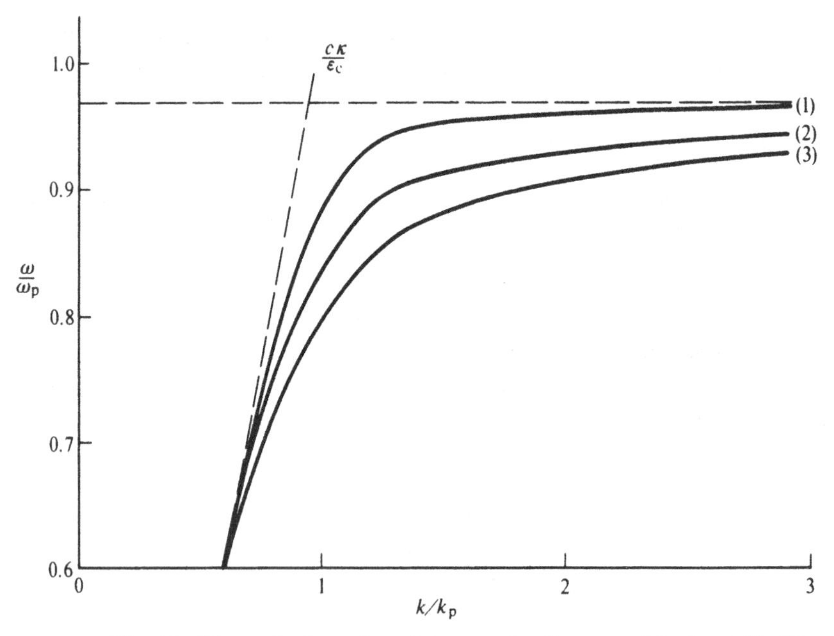

Fig. 11. Dimensionless plot of the s.m.d.r. for a model of n-InSb. The parameter $\varepsilon_c\omega_p\Delta N/nc$ is equal to 0 for (1), 1 for (2) and 2 for (3). After F. Flores & G. Navascués, *Surface Sci.* **34**, 773 (1973).

1.8. The quasistatic limit

Until now, whenever it has been necessary to resort to explicit models, we have mostly used a semiclassical picture. However, many of the basic arguments involve only macroscopic concepts and basic relations, like Maxwell's equations. In view of this, it is important to ascertain to what extent one can derive general results which do not depend on any particular model. We shall carry out such an exercise in the quasistatic limit, where the analysis is considerably simpler. Similar arguments could be put forward to include retardation effects if necessary.

Consider again a black box (fig. 12) containing the inhomogeneity in the z-direction and having essentially irrelevant lengths in the x and y-directions. When using actual models it is necessary to choose the origin $z=0$ somewhere. In fact the problem of where to put this plane is often a tricky one, as we shall see later. For the time being it suffices to think that we shall eventually take $z=0$ somewhere in the region of the surface inhomogeneity and/or of the surface excitation. Our black box is sufficiently large to contain all this, so that A $(z_A<0)$ is definitely outside – say, in the vacuum – and B $(z_B>0)$ is already in the homogeneous bulk material. The inhomogeneity could be, for example, an

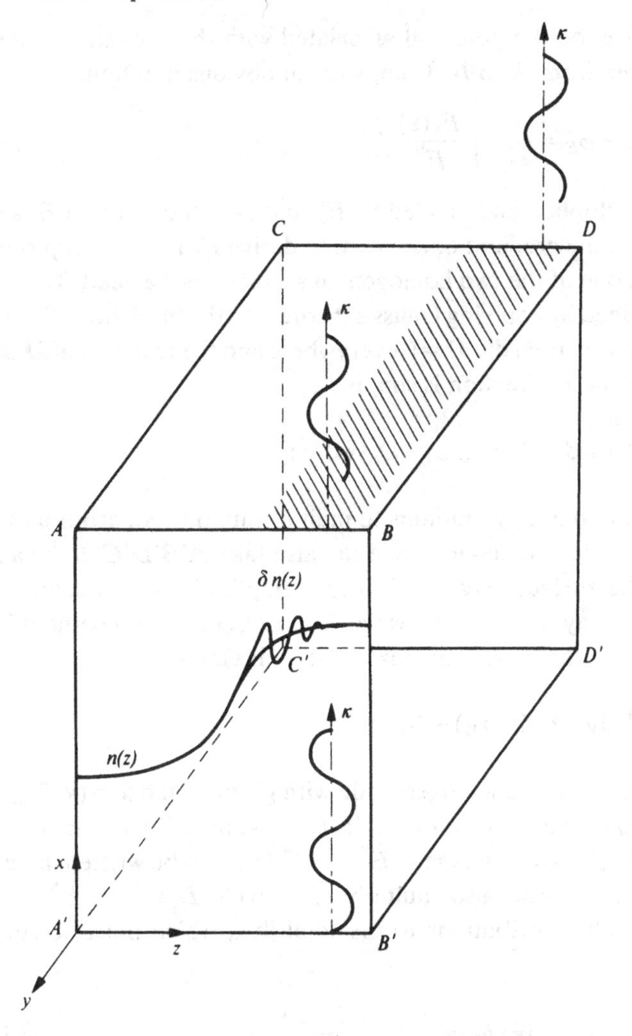

Fig. 12. The black box method in the quasistatic limit. A fluctuating density $\delta n(z)$ is superimposed on a static inhomogeneous density $n(z)$ and travels in the x-direction with wavevector κ. The length $AB = L$.

electron density profile growing from zero outside to some constant bulk value inside. We then look for a surface eigenmode in which some fluctuation $\delta n(z)$, whose amplitude is negligible outside the black box, propagates as $\exp \mathrm{i}\,(\kappa x - \omega t)$ – this exponential factor will be understood throughout – and decays as $\exp(-\kappa|z|)$. Notice that now we are using the black-box argument for arbitrary wavelength. This analysis is strongly reminiscent of the previous one (§ 1.7) but we are really in a different situation.

Now, let ϕ be the potential associated with the excitation. Integrate $E_z = -\partial\phi/\partial z$ from A to B. Then, with an obvious notation,

$$\phi_A - \phi_B = \kappa\phi_B \int \frac{E_z(z)}{E_z^{\mathrm{b}}}\,\mathrm{d}z. \tag{1.267}$$

We have multiplied and divided by E_z^{b} and used the fact that B is in the bulk, so that the gradient of ϕ – used to derive E_z from it – is just $-\kappa\phi_B$. This yields one of the two homogeneous equations we need. The second one is obtained by applying Gauss's theorem to the black box. Since there is no external charge, div $\boldsymbol{D} = 0$ everywhere and the total flux of $\boldsymbol{D}$ is zero. The flux in the x-direction is simply

$$\iint \mathrm{d}y\,\mathrm{d}z\,[D_x(x_A, z) - D_x(x_{A'}, z)].$$

The integration over y amounts to multiplication by an arbitrary factor, which we can take equal to 1. We can also take $A'B'D'C'$ to be a nodal plane of the surface wave, with a real amplitude that is actually proportional to, say, a sinusoidal wave. Then $D_x(x_{A'}, z) = 0$ without loss of generality. Furthermore, the flux in the z-direction is

$$\iint \mathrm{d}y\,\mathrm{d}z\,[D_z(z_B) - D_z(z_A)].$$

Now, we are studying an eigenmode with given ω and $\boldsymbol{k} = (\kappa, 0, q = \mathrm{i}\kappa)$. Thus for this wave, in the bulk, $D_z(z_b) = \varepsilon_{\mathrm{L}}(\kappa, \mathrm{i}\kappa)E_z^{\mathrm{b}}$, while at A we have simply $D_z(z_A) = E_z(z_A)$. Again, E_z^{b} and $E_z(z_A)$ can be written in terms of ϕ_B and ϕ_A. We can also multiply and divide $D_x(z)$ by $E_x^{\mathrm{b}} = -\mathrm{i}\kappa\phi_B$. Adding up all contributions to the total flux, which must be zero, we obtain

$$\phi_A + \varepsilon_{\mathrm{L}}(\kappa, \mathrm{i}\kappa)\phi_B = -\kappa\phi_B \int \frac{D_x(z)}{E_x^{\mathrm{b}}}\,\mathrm{d}z. \tag{1.268}$$

The secular equation for (1.267) and (1.268) yields the s.m.d.r.,

$$1 + \varepsilon_{\mathrm{L}}(\kappa, \mathrm{i}\kappa) + \kappa\left\{\int \frac{E_z(z)}{E_z^{\mathrm{b}}}\,\mathrm{d}z + \int \frac{D_x(z)}{E_x^{\mathrm{b}}}\,\mathrm{d}z\right\} = 0. \tag{1.269}$$

In the quasistatic limit this equation is exact, valid for arbitrary κ and for any model. The details of the model come into the evaluation of the integrals appearing in this formula, and this may require a very elaborate treatment of the surface inhomogeneity, to start with, besides a sophisticated analysis to obtain selfconsistently the field of the surface eigenmode.

To illustrate this point let us transform (1.269) in a suitable manner for the study of the long-wave limit. In this case $\kappa \to 0$ and $\varepsilon_L(\kappa, i\kappa) \to \varepsilon_b$, which is a function of ω only. In the numerator of the first integral add and subtract E_z^b, in the numerator of the second integral add and subtract D_x^b, use the fact that now ($\kappa \to 0$) $D_x^b = \varepsilon_b E_x^b$ and divide by $(1+\kappa L)$. To first order in κ this yields

$$(1+\varepsilon_b)+\kappa\left\{\int \frac{E_z(z)-E_z^b}{E_z^b}\,dz+\int \frac{D_x(z)-D_x^b}{E_x^b}\,dz\right\}=0. \tag{1.270}$$

The advantage of this rewriting is that the integrands vanish outside the range of integration. Now, imagine a model for which the second integral is zero: more will be said about this term presently. The fluctuating electric field component is related to the polarisation associated with the charge density fluctuation. For example, for an electron gas, $\partial E_z/\partial z = -4\pi e\delta n(z)$, because E_x is uniform across the inhomogeneity and div $\boldsymbol{E} = \partial E_z/\partial z$ only. Using this and performing a partial integration we have

$$(1+\varepsilon_b)+\kappa(1-\varepsilon_b)\frac{\int z\delta n(z)\,dz}{\int \delta n(z)\,dz}=0. \tag{1.271}$$

Furthermore, in the long-wave limit we set $\varepsilon_b=-1$ in the term multiplying κ, and $\varepsilon_b = 1-\omega^2/\omega_p^2$ in $(1+\varepsilon_b)$. Finally,

$$\omega^2=\frac{\omega_p^2}{2}\left[1+\kappa\frac{\int z\delta n(z)\,dz}{\int \delta n(z)\,dz}\right]. \tag{1.272}$$

The integrals can effectively be extended to $\pm\infty$. This formula is valid in the long-wave limit for any model. It shows explicitly that the coefficient of κ – an object of experimental interest in the case of metals – is given by something which is itself a macroscopic concept, namely, the dipole moment of the charge density fluctuation, but which can be the object of a very sophisticated microscopic calculation.

Although this argument focuses on one of the terms it is important to go back to (1.269), which contains *two* terms. The first one, as we have just seen, is related to charge density fluctuations in the direction perpendicular to the surface. For a metal it reflects the details of the model used to describe the inhomogeneous electron gas in the surface region. For example, if we use dielectric functions with sufficient structure we can pick out the *Landau damping*, which from the discussion in § 1.6 we expect to be comparatively unimportant; or we can try to include quantum mechanical interference effects, or a finite potential barrier height at the surface, or a non-abrupt potential model. Some of these details may be necessary for a reliable quantitative estimate of, say, the dispersion of long-wave surface plasmons, but the formula (1.269) holds irrespective of the particular models used. Furthermore, there is the second term, containing $D_x(z)$, i.e. the *current fluctuation* – either free of polarisation current – in the direction parallel to the surface. The eigenmode under study is an excitation which, besides entailing charge density fluctuations $\delta n(z)$, also carries a current density fluctuation in the x-direction, and this can be strongly affected by surface scattering. The amplitude of this eigenmode is localised near the surface in such a way that the trajectories of the quasiparticles – e.g. charge carriers – go up and down the x-direction hitting the surface periodically; and it is there that surface scattering can act to disrupt rather efficiently the organised collective surface mode. Moreover, in the evaluation of $D_x(z)$ we may in general expect both the longitudinal and the transverse dielectric functions to play a part.

We can now complete the interpretation of the results found in §§ 1.4 and 1.6. Consider first (1.162). We found that the damping coefficient contained not only β_T, but also β_L, and this could not be attributed to surface–bulk coupling. It is clear now that this is a contribution from the current density fluctuation term, which therefore vanishes if the surface is specular. Of course, part of the term in β_T can also be a part of the current fluctuation term. Thus, altogether this term can be as important as the coupling to bulk eigenmodes. Now consider (1.232). Here there is no coupling to bulk eigenmodes, unless we really study short waves seriously, in which case we should have to use far more elaborate dielectric functions. Also, our simple dielectric functions do not contain anything like *Landau damping.* All the damping in this case comes from the said effect of surface scattering on the current density fluctuations, and we have seen that this is by far the largest contribution.

We can also throw further light on the problem of the conservation of J_z discussed in § 1.6. Suppose we did the wrong calculation, i.e. instead of starting from (1.219), which contains the source term needed to cancel out the physically unrealisable charge sink, we omit this term. Then the dispersion relation still has the form (1.231) but now $x_2 = 0$, and x_1 and B_0 are different. This leads to the following long-wave s.m.d.r.:

$$\frac{\omega\sqrt{2}}{\omega_p} = 1 + \frac{1}{2\sqrt{2}}\left\{\frac{(1+p)^2}{(1+p^2)}\beta_L - i\left[\frac{(1-p^2)}{(1+p^2)}\beta_L + (1-p)\beta_T\right]\right\}\frac{\kappa}{\omega_p}. \qquad (1.273)$$

This becomes (1.232) *only* if $p = 1$. Otherwise this result is rather different. Let us focus on the real part. In practice it turns out that the slope predicted by the semiclassical model is too large and positive. A more sophisticated treatment including interference effects produces an appreciable lowering of the slope of ω/κ and this is regarded as an improvement. If we used (1.273) uncritically we would seem to obtain almost as good results but our apparent success would be quite false. It is clear what is happening in (1.273). The physically unrealisable charge sink at the surface produces an incorrect picture of the fluctuating charge density profile and predicts the wrong value for its dipole moment. For $p = 0$ the slope of (1.273) is reduced to one half of its correct value. The lesson to be learned from this is that the question of charge conservation must be taken seriously when fields and currents with components perpendicular to the surface are involved.

1.9. Non-planar surfaces

We have seen so far that the theory of reflectivity is intimately related to the theory of s.m.d.r., but we have not actually developed a *dynamical theory* of the coupling between, say, incident photons and surface plasmons. In fact the surface excitations we have described so far could never couple to an incident photon because it is impossible to conserve energy and momentum simultaneously. This is illustrated in fig. 13*a*. Notice that, for a given frequency, the momentum (of the incident photon) parallel to the surface is too small for matching to be possible. Although a detailed study of this problem would take us too far away from our main theme, it is instructive to see how the physical idea behind the methods used to circumvent this difficulty are related to the body of theoretical principles we are concerned with.

There are essentially three main ways to solve the problem. One consists in increasing the momentum of the the incident photons by first passing them through a prism. The situation (fig. 13*b*) corresponds to total reflection on the inner side of the prism. But total reflection is a macroscopic notion. Actually the field amplitude is evanescent beyond the surface at which the e.m. wave is reflected. The trick consists in putting the surface of the sample sufficiently close to the prism, so that we do have some amplitude in the surface region of the sample. But this is the amplitude of the field which has been passed through the prism, i.e. of incident photons with larger momentum. This method is usually called *frustrated* or *attenuated total reflection*. We have now three media, namely, the glass of the prism, the air gap, and the medium under study. The analysis is an exercise in the use of the general matching principles and techniques we have discussed. It does not really involve new ideas and we shall not delve any further into it.

Another way, since photons are so difficult, is not to use photons; or, rather, to use them in a very clever way. The theoretical basis is also a

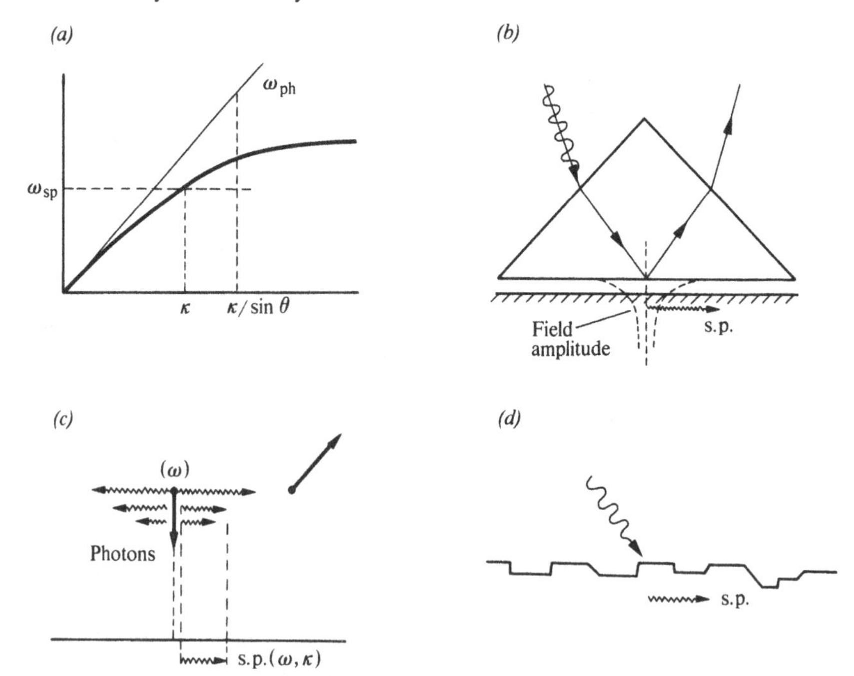

Fig. 13. It is impossible to conserve energy and momentum simultaneously unless (*a*) the photon momentum is increased, (*b*) by passing it through a prism. (*c*) An incident electron – a moving charge – is a source of photons which can excite surface plasmons (s.p.). (*d*) For non-planar surfaces an incident photon can also excite a surface plasmon, as κ (photon) and κ (plasmon) need not be equal.

direct application of the theory given in § 1.1. Consider (fig. 13*c*) an electron approaching the surface with velocity $\boldsymbol{v}$. This amounts to a current $J_{\text{ext}}(\boldsymbol{r}, t) = e\boldsymbol{v}\delta(\boldsymbol{r} - \boldsymbol{v}t)$, which can be rewritten as

$$\boldsymbol{J}_{\text{ext}}(\boldsymbol{r}, t) = \frac{e\boldsymbol{v}}{(2\pi)^4}\int e^{i(\boldsymbol{k}\cdot\boldsymbol{r}-\omega t)}\delta(\omega - \boldsymbol{k}\cdot\boldsymbol{v})\,\mathbf{d}\boldsymbol{k}\,d\omega.$$

That is to say, the Fourier transform of $\boldsymbol{J}_{\text{ext}}(\boldsymbol{r}, t)$ is

$$\boldsymbol{J}(\boldsymbol{k}, \omega) = 2\pi e\boldsymbol{v}\delta(\omega - \boldsymbol{k}\cdot\boldsymbol{v}). \tag{1.274}$$

Now, from (1.21), this creates a field with Fourier transform

$$\boldsymbol{E}(\boldsymbol{k}, \omega) = i\frac{e8\pi^2}{\omega}\left[-\frac{\boldsymbol{k}\boldsymbol{k}\cdot\boldsymbol{v}}{k^2} + \frac{\omega^2}{c^2}\frac{\left(\boldsymbol{v} - \dfrac{\boldsymbol{k}\boldsymbol{k}\cdot\boldsymbol{v}}{k^2}\right)}{\left(k^2 - \dfrac{\omega^2}{c^2}\right)}\right]\delta(\omega - \boldsymbol{k}\cdot\boldsymbol{v}). \tag{1.275}$$

This means that the moving electron is a white source of photons. For given ω these photons can have wavevectors $\boldsymbol{k}$ according to the following conditions. The projections of $\boldsymbol{k}$ parallel to $\boldsymbol{v}$ must be equal to ω/v, but the normal component – parallel to the surface – is not restricted to any particular value. Thus, for given ω we can always find a photon emitted by the moving source with a κ equal to the κ of the corresponding surface mode. The surface mode can then be excited while the electron is reflected losing energy $\hbar\omega$ and momentum $\hbar\kappa$.

There is yet another way to circumvent the difficulty. The problem, as we have seen, arises from the impossibility of matching energy and momentum simultaneously. But momentum parallel to the surface must be conserved only if the surface is plane, i.e. if we have translational symmetry parallel to the surface. Suppose we have a rough surface. Then this requirement disappears and there is no objection to the coupling between photons and surface modes. This poses new theoretical questions. We have discussed matching at surfaces taking account of surface scattering of the quasiparticles reaching it from *inside* but, although a model to produce some non-specular scattering of, say, electrons might involve something like a surface roughness met by the incident electrons, we have always assumed that the incoming e.m. field meets a plane surface. The situation we have now is different. The incident e.m. wave meets a surface whose average position may be planar, but which has some height $h(\boldsymbol{\rho})$ with respect to this average position, $\boldsymbol{\rho}$ being a vector (x, y) in the average surface plane. We have not studied e.m. matching at such non-planar surfaces, and this requires more theory.

We actually consider something like roughness on an atomic scale, or at any rate such that the amplitude of the height function h is much smaller than the wavelengths of interest. Thus h will be treated like a perturbation and therefore it suffices to consider one single Fourier component. A stochastically rough surface can always be described in terms of standard statistical functions obtainable from the Fourier transform of h. An equivalent situation is that in which the planarity of a surface initially prepared to be as nearly ideal as possible, is deliberately destroyed by producing regularly spaced grooves. This is actually an experimental technique used in practice to study the s.m.d.r. The position then is the following: we need further matching theory, and the prototype of surface we need to study is reduced to a model with a periodicity which we can take in the x-direction, with regular spacing d. We shall describe this by the height function

$$z_s = h(x) = \sum_m h_m\, e^{ig_m x}. \tag{1.276}$$

Here g_m is the reciprocal lattice vector associated with the periodicity d:

$$g_m = mg = 2\pi m/d, \tag{1.277}$$

and h_m is treated as a small perturbation.

The main point in this problem is that κ is not uniquely defined, i.e. it is only defined *modulo* a reciprocal lattice vector g_m. Consider an incident e.m. wave as in fig. 2. The component of the reflected and transmitted wavevectors parallel to the surface is now $\kappa_n = \kappa + g_n$. We have a whole spectrum of new 'channels', the case $n = 0$ being the principal channel, corresponding to specular reflection of the incident e.m. wave. Thus for the reflected waves we have

$$k_\mathrm{r}^2 = k_\mathrm{i}^2 = \kappa_n^2 + q_n^2, \tag{1.278}$$

i.e. the perpendicular component also changes. For the transmitted waves we need again a description of the medium starting from the surface inwards. In this discussion we shall focus on the effect on non-planarity and use the simplest model for the rest. We assume the medium to be simply described by the local bulk dielectric function $\varepsilon_\mathrm{b}(\omega)$. The analysis could be extended, at the price of greater complication, to more elaborate models, be we shall only discuss the simplest case needed to bear out the effects of a non-planar surface. Then for the transmitted waves we have

$$k_\mathrm{t}^2 = \kappa_n^2 + q_n'^2 = (\omega^2/c^2)\varepsilon_\mathrm{b}. \tag{1.279}$$

For P-mode geometry the incident fields are still expressed as in (1.74), but now the rest must be written in the form

$$\left.\begin{aligned} \boldsymbol{E}_\mathrm{r} &= \sum_n (E_{x\mathrm{r}n}\boldsymbol{x}_0 + E_{z\mathrm{r}n}\boldsymbol{z}_0)\,\mathrm{e}^{\mathrm{i}(\kappa_n x + q_n z)}, \\ \boldsymbol{E}_\mathrm{t} &= \sum_n (E_{x\mathrm{t}n}\boldsymbol{x}_0 + E_{z\mathrm{t}n}\boldsymbol{z}_0)\,\mathrm{e}^{\mathrm{i}(\kappa_n x + q_n' z)}. \end{aligned}\right\} \tag{1.280}$$

The time factor is implicit in these expressions and similarly $\boldsymbol{H}_\mathrm{r}$, $\boldsymbol{H}_\mathrm{t}$ are obtainable from them. The problem is to find the amplitudes reflected or transmitted in all the channels as functions of the perturbations h_m.

The principles applying to the matching are always the same. We could say for example that the tangential components of $\boldsymbol{E}$ and $\boldsymbol{H}$ must be continuous. But the implementation of this program with this geometry is rather cumbersome as the orientation of the tangent plane changes along the surface. Fortunately there is no need to apply the method literally. We can use the black-box method as in § 1.7. This yields the desired linear homogeneous equations. The integration contours are oriented as in fig. 9

but now we can, without loss of generality, choose the origin of z as shown in fig. 14 and also choose the origin of x so that $h = 0$ for $x = 0$. The integrations over x are then extended from 0 to a generic arbitrary x. On evaluating (1.253) we must now use the expression for the fields as in (1.280). We shall find an integral of $E_z(x, z)$ from $z = 0$ to L. We can define a local inhomogeneous dielectric function $\varepsilon(x, z)$ which is equal to

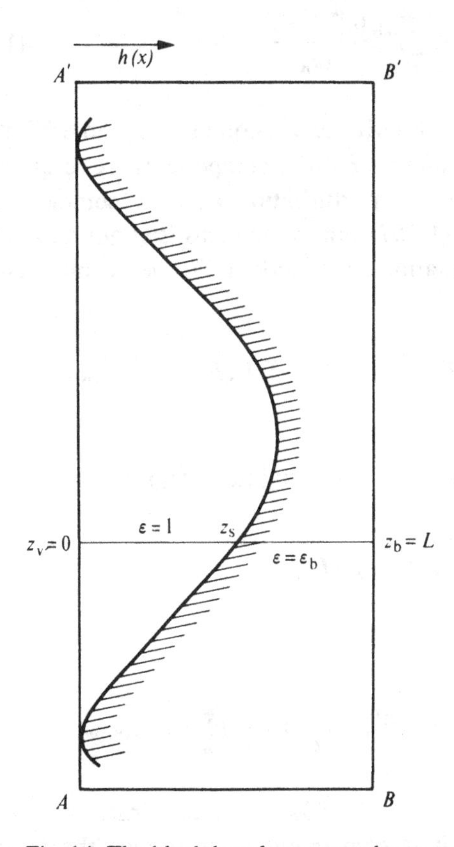

Fig. 14. The black box for a non-planar surface.

1 for $z < h(x)$ and equal to ε_b for $z > h(x)$. Since D_z is continuous, we can use the equality $\varepsilon(x, z)E_z(x, z) = \varepsilon_\mathrm{b}E_z(x, L) = \varepsilon_\mathrm{b}E_z^\mathrm{b}(x)$ to express $E_z(x, z)$ as $\varepsilon_\mathrm{b}E_z^\mathrm{b}(x)/\varepsilon(x, z)$. Then (fig. 14)

$$\int_0^L E_z(x, z)\,\mathrm{d}z = \int_0^\infty \frac{\varepsilon_\mathrm{b}}{\varepsilon(x, z)} E_z^\mathrm{b}(x)\,\mathrm{d}z$$

$$= [(\varepsilon_\mathrm{b} - 1)h(x) + L]\sum_n E_z^\mathrm{b}\,\mathrm{e}^{\mathrm{i}\kappa_n x}, \tag{1.281}$$

and (1.253) yields

$$\sum_n E^{\mathrm{b}}_{xn}\frac{(\mathrm{e}^{\mathrm{i}\kappa_n x}-1)}{\mathrm{i}\kappa_n}-E_{x\mathrm{i}}\frac{(\mathrm{e}^{\mathrm{i}\kappa x}-1)}{\mathrm{i}\kappa}-\sum_n E_{x\mathrm{r}n}\frac{(\mathrm{e}^{\mathrm{i}\kappa_n x}-1)}{\mathrm{i}\kappa_n}$$

$$+L\sum_n E^{\mathrm{b}}_{zn}(1-\mathrm{e}^{\mathrm{i}\kappa_n x})-(\varepsilon_{\mathrm{b}}-1)h(x)\sum_n E^{\mathrm{b}}_{zn}\mathrm{e}^{\mathrm{i}\kappa_n x}$$

$$=\mathrm{i}\frac{\omega}{c}\varepsilon_{\mathrm{b}}^{1/2}L\sum_n E^{\mathrm{b}}_{n}\frac{(\mathrm{e}^{\mathrm{i}\kappa_n x}-1)}{\mathrm{i}\kappa_n}. \tag{1.282}$$

Notice that $h(x)$ has its own Fourier expansion (1.276) and $\mathrm{e}^{\mathrm{i}g_m x}\,\mathrm{e}^{\mathrm{i}\kappa_n x}=\mathrm{e}^{\mathrm{i}\kappa_{(n+m)}x}$. It is here that the different Fourier components are coupled with amplitudes proportional to the perturbation h_m. The common factor which we eliminated from (1.257) cannot be factorised out now. This is the crucial difference. Equating independent Fourier components of (1.282) we obtain

$$E^{\mathrm{b}}_{xn}-E_{x\mathrm{i}}\delta_{n0}-E_{x\mathrm{r}n}-\mathrm{i}L\kappa_n E^{\mathrm{b}}_{zn}-\mathrm{i}(\varepsilon_{\mathrm{b}}-1)\kappa_n\sum_m h_m E^{\mathrm{b}}_{z(n-m)}=\mathrm{i}(\omega/c)\varepsilon_{\mathrm{b}}LE^{\mathrm{b}}_{n}. \tag{1.283}$$

Proceeding similarly with (1.254) and using the integral

$$\int_0^z \varepsilon(x,z)\,\mathrm{d}z=(1-\varepsilon_{\mathrm{b}})h(x)+L\varepsilon_{\mathrm{b}}, \tag{1.284}$$

we obtain

$$E_{\mathrm{i}}\delta_{n0}-E_{\mathrm{r}n}-\varepsilon_{\mathrm{b}}^{1/2}E^{\mathrm{b}}_{n}=-\mathrm{i}\frac{\omega}{c}K\varepsilon_{\mathrm{b}}E^{\mathrm{b}}_{xn}-\mathrm{i}\frac{\omega}{c}(1-\varepsilon_{\mathrm{b}})\sum_m h_m E^{\mathrm{b}}_{x(n-m)}. \tag{1.285}$$

The matching equations are now (1.283) and (1.285) instead of (1.259).

For practical purposes it is convenient to introduce the following notation: let θ_n/θ'_n be the values of $\theta_{\mathrm{r}}/\theta_{\mathrm{t}}$ for the different channels. We define $s_n=\sin\theta_n$, $c_n=\cos\theta_n$, $s'_n=\sin\theta'_n$, $c'_n=\cos\theta'_n$. Then

$$s_n=s_0\frac{\kappa_n}{\kappa};\qquad c_n=(1-s_n)^{1/2};\qquad s'_n=\frac{s_n}{\varepsilon_{\mathrm{b}}^{1/2}};\qquad c'_n=\left(1-\frac{s_n^2}{\varepsilon_{\mathrm{b}}}\right)^{1/2}, \tag{1.286}$$

and we write the field components as $E_{x\mathrm{r}n}=E_{\mathrm{r}n}c_n$ etc. We also rearrange (1.283) and (1.286) so that the incident amplitude is regarded as an input in terms of which we seek $E_{\mathrm{r}n}$ and E^{b}_{n}. Thus we start our analysis from the two equations

$$\left.\begin{aligned}&-c_nE_{rn}+\{c'_n+\mathrm{i}[\kappa_n s'_n-(\omega/c)\varepsilon_b^{1/2}]L\}E_n^b\\&+\mathrm{i}(\varepsilon_b-1)\kappa_n\sum_m h_m s'_{(n-m)}E^b_{(n-m)}=c_0\delta_{n0}E_i;\\&E_{rn}+[\varepsilon_b^{1/2}-\mathrm{i}(\omega/c)\varepsilon_b c'_n]E_n^b\\&+\mathrm{i}(\varepsilon_b-1)\frac{\omega}{c}\sum_m h_m c'_{(n-m)}E^b_{t(n-m)}=\delta_{n0}E_i.\end{aligned}\right\}\tag{1.287}$$

The use of angle or momentum variables depends on whether we study reflectivity or the s.m.d.r. For the analysis of reflectivity it is convenient to put $\kappa=s_0\omega/c$, $\kappa_n=s_n\omega/c$ everywhere, and to define

$$\left.\begin{aligned}&C'_n=c'_n+\mathrm{i}(\omega/c)[(s_n s'_n-\varepsilon_b^{1/2})L+(\varepsilon_b-1)s_n s'_0 h_0];\\&N_n=\varepsilon_b^{1/2}+\mathrm{i}(\omega/c)[(\varepsilon_b-1)h_0c'_n-\varepsilon_b Lc'_n].\end{aligned}\right\}\tag{1.288}$$

Then (1.287) becomes

$$\begin{aligned}&C-c_nE_{rn}+C'_nE_n^b+\mathrm{i}(\varepsilon_b-1)\frac{\omega}{c}s_n\sum_{m\neq 0}h_m s'_{(n-m)}E^b_{(n-m)}=c_0\delta_{n0}E_i;\\&E_{rn}+N_nE_n^b+\mathrm{i}(\varepsilon_b-1)\frac{\omega}{c}\sum_{m\neq 0}h_m c'_{(n-m)}E^b_{(n-m)}=\delta_{n0}E_i.\end{aligned}\tag{1.289}$$

Notice that with our choice of origin for z (fig. 14), $h_0\neq 0$ is the average depth of the non-planar surface.

The mixing of Fourier components reflected in (1.283)–(1.288) would in theory generate an infinite set of coupled linear equations, as in the nearly-free-electron model of a band structure, but also as there, one makes approximations in practice. The essence of the results obtained from the analysis is sufficiently illustrated by considering the effect of just one Fourier component, say h_1. In order to have a real height function we must also include h_{-1}. Then we have four unknowns, namely, E_{r0}, E_{r1}, E_0^b, E_1^b, and four equations which are

$$\left.\begin{aligned}&-c_0E_{r0}+C'_0E_0^b+\mathrm{i}(\varepsilon_b-1)(\omega/c)s_0h_{-1}s'_1E_1^b=c_0E_i;\\&E_{r0}+N_0E_0^b+\mathrm{i}(\varepsilon_b-1)(\omega/c)h_{-1}c'_1E_1^b=E_i;\\&-c_1E_{r1}+C'_1E_1^b+\mathrm{i}(\varepsilon_b-1)(\omega/c)s_1h_1s'_0E_0^b=0;\\&E_{r1}+N_1E_1^b+\mathrm{i}(\varepsilon_b-1)(\omega/c)h_1c'_0E_0^b=0.\end{aligned}\right\}\tag{1.290}$$

The solution of this system is given by

$$E_{r0} = [(C_0' - c_0 N_0)(C_1' + c_1 N_1) + (\varepsilon_b - 1)^2 (\omega/c)^2 (s_0 s_1' - c_0 c_1')(s_1 s_0' + c_1 c_0')|h_1|^2] E_i/\Delta;$$

$$E_{r1} = i2c_0(\varepsilon_b - 1)(\omega/c)(s_1 s_0' N_1 - c_0' C_1') h_1 E_i/\Delta; \quad (1.291)$$

$$E_0^b = 2c_0(C_1' + c_1 N_1) E_i/\Delta;$$

$$E_1^b = -i2c_0(\varepsilon_b - 1)(\omega/c)(s_1 s_0' + c_1 c_0') h_1 E_i/\Delta,$$

where Δ (proportional to) the determinant of the coefficients of the unknowns in (1.290) is

$$\Delta = (C_0' + c_0 N_0)(C_1' + c_1 N_1) + (\varepsilon_b - 1)^2 (\omega/c)^2 (s_0 s_1' + c_0 c_1')(s_1 s_0' + c_1 c_0')|h_1|^2. \quad (1.292)$$

The various reflection and transmission coefficients can be evaluated from (1.291) and the s.m.d.r. from the vanishing of Δ.

The solution here obtained is in general a rather complicated function of ω, κ, and d, and besides contains the perturbation characterised by the two parameters h_0 and h_1, but it takes rather simple and meaningful forms in some special ranges. Notice that a drastic change takes place in the main reflectivity amplitude E_{r0}/E_i, depending on whether we are close to or removed from a situation in which $C_1' + c_1 N_1$ vanishes. What does this mean? Suppose we go back to (1.289) and take $h_m = 0$, $L = 0$, i.e. we flatten out the surface. Then, formally, for $n = 0$,

$$\left.\begin{aligned} -c_0 E_{r0} + C_0' E_0^b &= c_0 E_i; \\ E_{r0} + N_0 E_0^b &= 0, \end{aligned}\right\} \quad (1.293)$$

while for every $n \neq 0$

$$\left.\begin{aligned} -c_n E_{rn} + C_n' E_n^b &= 0; \\ E_{rn} + N_n E_n^b &= 0. \end{aligned}\right\} \quad (1.294)$$

We must also evaluate (1.288) for all n with the condition $h_n = 0, L = 0$, in which case $N_n = \varepsilon_b^{1/2}$ and $C_n' = c_n'$. It is easy to see that (1.293) yields simply the standard Drude formula for oblique incidence on a medium with local dielectric function ε_b. Define

$$\nu_b = [\kappa^2 - (\omega^2/c^2)\varepsilon_b]^{1/2}. \quad (1.295)$$

Under conditions in which there are localised surface modes, ν_b is real and we find

$$R=\left|\frac{c_0'-c_0\varepsilon_b^{1/2}}{c_0'+c_0\varepsilon_b^{1/2}}\right|^2=\left|\frac{\varepsilon_b\cos\theta-i(c/\omega)\nu_b}{\varepsilon_b\cos\theta+i(c/\omega)\nu_b}\right|^2=1, \tag{1.296}$$

i.e. we have total reflection in this range. Also, in (1.294) E_{rn} and E_n^b must vanish unless $C_n'+c_nN_n$ vanishes, i.e. unless

$$1/\nu_{vn}+\varepsilon_b/\nu_{bn}=0, \tag{1.297}$$

where ν_{bn} is the value of ν_b for $\kappa=\kappa_n$ and ν_{vn} is the same in the vacuum, where $\varepsilon=1$. This is nothing but (1.82), i.e. the s.m.d.r. for each particular value of $\kappa=\kappa_n$. We have simply hit upon the *unperturbed* eigenmodes, as in an empty lattice test in the nearly-free-electron theory of a crystal band structure. In fact equations (1.294) are superfluous in this case. They do not convey any more information than is contained in (1.293) or (1.296), where it suffices to set the denominator equal to zero in order to obtain the standard unperturbed s.m.d.r. But this argument shows the meaning of the vanishing of $C_1'+c_1N_1$ in (1.291) and (1.292): we are simply 'near resonance', i.e. close to a surface eigenmode for the particular value $\kappa=\kappa_1$, although there is a difference between the unperturbed (flat) and the perturbed (non planar) s.m.d.r. So far we are treating the different Fourier components h_n as decoupled. For a more accurate picture of the surface eigenmodes of the non-planar surface it is necessary to introduce some coupling. We shall presently see how this modifies the s.m.d.r. but, before doing this, let us study the situation described by the solution (1.291), (1.292).

First suppose we are 'near resonance', which corresponds to, say, the point ω_1 of fig. 15. Just at this point $C_1'+c_1N_1$ vanishes. Using this condition and sticking consistently to the same order of approximation we find

$$E_{r1}\approx 2i\left(\frac{c}{\omega}\right)\frac{\varepsilon_b^{1/2}h_1}{(\varepsilon_b-1)(s_0s_1'+c_0c_1')|h_1|^2}E_i;$$

$$E_1^b\approx -2i\left(\frac{c}{\omega}\right)\frac{c_0h_1}{(\varepsilon_b-1)(s_0s_1'+c_0c_1')|h_1|^2}E_i.$$

Thus the amplitudes scattered – by either reflection or transmission – into channel κ_1 become very large. However, this does not necessarily mean that the energy flux is propagated in the directions of E_{r1} and E_1^b. We must look at the corresponding spatial dependences, which vary as $\exp(iq_1z)$ and $\exp(iq_1'z)$, respectively. But we have just seen that when the s.m.d.r. is satisfied, ν_{vn} and ν_{bn} – for $n=1$ in this case – are real, i.e. q_1 and q_1' are imaginary, and E_{r1} and E_1^b decay as $\exp(-\nu_{v1}z)$ and $\exp(\nu_{b1}z)$. Thus the

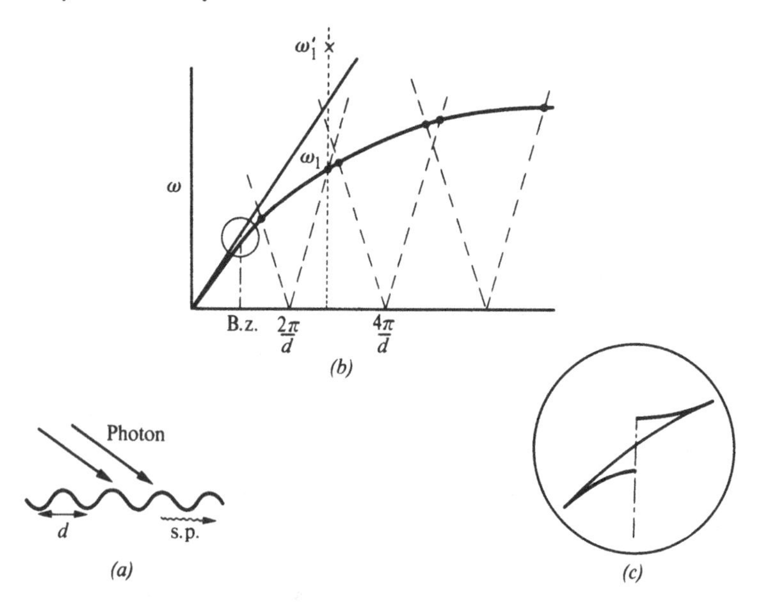

Fig. 15. (a) Photons incident on a non-planar surface can excite surface plasmons (s.p.). (b) The points show the frequencies for which, for given angle of incidence, the condition $\kappa = \omega \sin \theta/c + 2\pi n/d$ is satisfied. (c) A closer look at the Brillouin zone edge shows the appearance of a gap of width proportional to $1/d$.

large values of the scattered amplitudes signal the excitation of the surface mode, whose amplitude is localised near $z = 0$, and simply propagates parallel to the surface. Of course, in a complete dynamical theory one should describe the transient process through which the surface mode is excited, but we shall not do this here. We are simply investigating the dispersion relation for stationary surface modes. Under these conditions the only way in which energy could be dissipated through the surface mode is if we provide some model for its damping. This can be done, as we have discussed earlier in this chapter, and we can also easily add some appropriate imaginary part to the dielectric function, to describe a dissipative conductor. But as long as we have no damping or dissipation in our model, if energy is neither reflected nor transmitted in the channel κ_1, nor dissipated in the corresponding surface mode, then it must all be reflected in the main (specular) channel. Indeed, under these circumstances (1.291) and (1.292) yield

$$R_0 = \left|\frac{E_{r0}}{E_i}\right|^2 \approx \left|\frac{s_0 s_1' - c_0 c_1'}{s_0 s_1' + c_0 c_1'}\right|^2 = \left|\frac{s_0 s_1' - \mathrm{i} c_0 \dfrac{c}{\omega \varepsilon_b} \nu_{b1}}{s_0 s_1' + \mathrm{i} c_0 \dfrac{c}{\omega \varepsilon_b} \nu_{b1}}\right|^2 = 1. \tag{1.298}$$

A more detailed study shows that R_0 is strongly peaked at the point ω_1 (fig. 15) and decays rather sharply as ω, for fixed κ_1, moves away from ω_1, as is typical of all resonances.

This takes us to the opposite situations, 'away from resonance': for example, to a point like ω_1' (fig. 15) for which propagation in a vacuum with wavevector $(\kappa_1, 0, q_1)$ is possible. We now expect that the specular reflectivity will decrease, as some amplitude will be scattered into channel κ_1. This can be followed up by studying the reflected amplitude E_{r1} from (1.291) and (1.292) again. Treating the perturbation h_1 to lowest order we find

$$E_{r1} \approx 2i\left(\frac{\omega}{c}\right)\frac{(\varepsilon_b-1)c_0(s_1s_0'\varepsilon_b^{1/2}-c_0'c_1')h_1}{(c_0'+c_0\varepsilon_b^{1/2})(c_1'+c_1\varepsilon_b^{1/2})}E_i. \tag{1.299}$$

For physical purposes we are actually interested in $|E_{r1}/E_i|^2$ and this requires some care. Consider, for example, the simple case of normal incidence. Then $\kappa = 0$ (but $\kappa_1 \neq 0$) and

$$\frac{E_{r1}}{E_i} \approx -i\frac{2(\varepsilon_b-1)}{(1+\varepsilon_b^{1/2})c}\frac{\left(1-\dfrac{c^2\kappa_1^2}{\omega^2\varepsilon_b}\right)^{1/2}}{\left[\left(1-\dfrac{c^2\kappa_1^2}{\omega^2\varepsilon_b}\right)^{1/2}+\varepsilon_b^{1/2}\left(1-\dfrac{c^2\kappa_1^2}{\omega^2}\right)^{1/2}\right]}h_1. \tag{1.300}$$

We must now separate real and imaginary parts, remembering that $\kappa_1 = 2\pi/d$. A not uncommon value of d in actual experiments may be of the order of 10^{-4} cm, so that the condition $1 > c^2\kappa_1^2/\omega^2$ may be satisfied for frequencies above $10^{15}\ \mathrm{s}^{-1}$. If we assume further that at these frequencies $\omega > \omega_p$, i.e. we study frequencies for which no e.m. waves can propagate in the medium, then $\varepsilon_b < 0$ and its square root is imaginary. In this case we find

$$R_1 = \left[\frac{E_{r1}}{E_i}\right]^2 \approx 4\frac{\omega^2}{c^2}\frac{[\kappa_1^2-(\omega^2/c^2)\varepsilon_b]}{[\kappa_1^2(1+\varepsilon_b)-(\omega^2/c^2)\varepsilon_b]}|h_1|^2.$$

This formula only holds under the above conditions; otherwise we must go back to (1.300) or, for oblique incidence, to (1.299). Notice that even (1.298) and (1.299) describe only a crude approximation, as they hold only very close to or very removed from 'resonance'. For a more complete description of the reflectivity we must go back to (1.291) and (1.292). But these are also the result of treating the different Fourier components h_m as decoupled from each other. For a more accurate analysis we must go back to (1.298), where the Fourier components are mixed. The detailed algebra can be left for explicit applications, but we can clearly anticipate the sort of picture which emerges from this. In higher approximations we shall find, in the reflected amplitudes in the channel κ_m, coupling terms proportional to $h_m h_n$. The physical meaning of this is related to indirect scattering into κ_m via resonance with the surface eigenmode for κ_n.

It is also interesting to see how the unperturbed surface eigenmode spectrum is modified by the perturbation even to a low-order approximation. For this it is sufficient to set (1.292) equal to zero. The argument applies equally to every Fourier component h_n. The main object we must study is $C'_n + c_n N_n$ which, from (1.288), is conveniently expressed as

$$(c'_n + c_n \varepsilon_b^{1/2}) + i(\omega/c)[(s_n s'_n - c_n c'_n \varepsilon_b) - \varepsilon_b^{1/2}]L \\ + i(\omega/c)(\varepsilon_b - 1)[s_n s'_n + c_n c'_n]h_0,$$

which is to be evaluated to lowest order. Thus, in the square brackets we assume the unperturbed s.m.d.r. holds in the form

$$(\varepsilon_b - s_n^2)^{1/2} = -\varepsilon_b(1 - \varepsilon_n^2)^{1/2},$$

whence $s_n^2 = \varepsilon_b/(1+\varepsilon_b)$. After some algebra we find that the coefficients of L and h_0 then vanish identically, whence to leading order

$$C'_n + c_n N_n \approx i(c/\omega\varepsilon_b^{1/2})(\nu_{bn} + \varepsilon_b \nu_{vn}),$$

and the s.m.d.r. becomes

$$-(c^2/\omega^2\varepsilon_b)(\nu_b + \varepsilon_b\nu_v)(\nu_{bn} + \varepsilon_b\nu_{vn}) \\ + (\omega^2/c^2)(\varepsilon_b - 1)^2(s_0 s'_n + c_0 c'_n)(s_n s'_0 + c_n c'_0)|h_n|^2 = 0. \tag{1.301}$$

We are studying the effects of a periodic perturbation of a plane surface. The intuitive idea behind this study is that since the momentum parallel to the surface is now only defined *modulo* any reciprocal lattice vector, photons and surface modes can couple as shown in fig. 15. The result of the formal analysis actually conveys some more information, as can be easily seen from (1.301). Since we have introduced a periodicity with its own scheme of Brillouin zones, we expect the effects of the perturbation to be strongest near the edges of the Brillouin zones. Consider, for example, the case $n = 1$ in (1.301) and take $\kappa = -\frac{1}{2}g_1$. Then $\kappa_1 = \frac{1}{2}g_1 = -\kappa$, $c_1 = c_0$, $c'_1 = c'_0$, $s_1 = -s_0$ and the product $(\nu_b + \varepsilon_b\nu_v)(\nu_{bn} + \varepsilon_b\nu_{vn})$ is simply the square of $(\nu_b + \varepsilon_b\nu_v)$. Under these conditions – and for $n = 1$ – (1.301) can be rewritten, after some algebra, as

$$\frac{1}{\nu_v} + \frac{\varepsilon_b}{\nu_b} = \left(\kappa^2 - \frac{\omega^2}{c^2}\right)^{-1/2} + \varepsilon_b\left(\kappa^2 - \frac{\omega^2}{c^2}\varepsilon_b\right)^{-1/2} = \pm 2(\varepsilon_b - 1)|h_1|. \tag{1.302}$$

This means that a gap opens up at the Brillouin zone edge (fig. 15), of width proportional to $|h_1|$. What we have here is the strict analogue of a nearly-free-electron band structure to lowest order.

2

The method of surface Green-function matching

2.1. Formal theory of surface Green-function matching

In chapter 1 we solved a few problems by seemingly *ad hoc* methods in which we defined fictitious extended systems, introduced hypothetical – sometimes non-existent or even physically impossible – stimuli and, in short, obtained physically valid results for a real system by solving the problem for a *pseudosystem*. In this chapter we shall build up a systematic formalism for 'pseudising' surface problems.

Let us stress from the start that 'pseudism' in itself is substantially different from model making. This is obvious in the case of a pseudopotential formulation of the electronic theory of band structure in solids. A *pseudohamiltonian* can be constructed so that, while it is known to give the wrong answer for part of the spectrum – say, below the valence band – it is guaranteed to give in principle the correct answer for the part of the spectrum we are interested in, e.g. the valence band and above. The practical matter of introducing simplifications – i.e. ultimately, models – comes later, at the stage of feeding into the calculation something manageable and hopefully reasonable. But one can actually build up a systematic formal method for generating different pseudopotentials from first principles and the pseudohamiltonians thus constructed answer – correctly or incorrectly – all the questions, above or below the valence band. A model is substantially different in that we know from the start that it will give only an approximately valid answer to only part of the questions, and it may even leave other questions totally unanswered; but the idea is that we shall hopefully use all the information yielded by our model, whereas with a pseudohamiltonian we start out knowing that we must on principle discard a well-defined part of the information therein contained.

The formalism developed in this chapter is entirely an exercise in pseudism carried out in real space. Of course one always introduces models in the end because, no matter which formalism one uses, the practical questions involved in obtaining explicit results come up sooner or later, but we stress that the formal theory presently to be developed does not depend on any particular model. The idea is to define a *matching surface* separating two domains of the real system, which acts like a purely geometrical probe giving information about the state of affairs at the interface. We shall study in detail the case of a plane surface, but it should be obvious that all the basic formulae hold for a surface of arbitrary geometry. This will be further discussed in § 2.5.

Now, consider a *surface system*. This can be a 'half medium' with a free surface, or something like a metal–semiconductor junction or a solid–liquid system with an interface. Define the matching plane at $z=0$ separating domain $1(z<0)$ from domain $2(z>0)$. let $m=1,2$ indicate the two domains. Everything so far defined belongs to the real system. For example, the definition of a matching plane is not incompatible with a smooth transition region from 1 to 2, if we know how to describe such an inhomogeneity. Take the real material on each side m and *define* an *extended pseudomedium* M_m such that it is identical to the real system on the side m of the surface and is arbitrary on the other side. These extended pseudomedia in principle have their own Green functions – electronic, electromagnetic, elastic, etc. – which we shall call $\hat{\boldsymbol{G}}_m$. These can be scalars or, in general, tensors. As repeatedly stressed, $\hat{\boldsymbol{G}}_m$ need not correspond to, e.g. the Green function of an infinite, ideal, perfect crystal. The question of knowing $\hat{\boldsymbol{G}}_m$ is a practical one. Our formal problem here is the following: taking $\hat{\boldsymbol{G}}_1$ and $\hat{\boldsymbol{G}}_2$ as given, we want to find $\hat{\boldsymbol{G}}_s$, the Green function of the (real) surface system.

We shall use $(i,j)=(1,2,3)$ to indicate the spatial components, with $(x_1,x_2,x_3)=\boldsymbol{r}=(\boldsymbol{\rho},z)$ and $(k_1,k_2,k_3)=(\boldsymbol{\kappa},q)$, and introduce two-dimensional objects as follows. Given a three-dimensional object like $\hat{\boldsymbol{G}}$, with matrix elements $\langle\boldsymbol{r}|\hat{\boldsymbol{G}}|\boldsymbol{r}'\rangle$, we define the corresponding two-dimensional one

$$\langle\boldsymbol{\rho}|\boldsymbol{G}|\boldsymbol{\rho}'\rangle\equiv\langle\boldsymbol{r}=\boldsymbol{\rho}|\hat{\boldsymbol{G}}|\boldsymbol{r}'=\boldsymbol{\rho}'\rangle. \tag{2.1}$$

However, not only do we set $z=z'=0$ in $\hat{\boldsymbol{G}}$; we also *define* $\boldsymbol{G}$ *with an inverse only in the two-dimensional space* of $\boldsymbol{\rho}$. Thus we distinguish between the three-dimensional and two-dimensional units

$$\hat{\mathscr{I}}=\int \mathrm{d}^3r|\boldsymbol{r}\rangle\langle\boldsymbol{r}|;\qquad \mathscr{I}=\int \mathrm{d}^2\rho|\boldsymbol{\rho}\rangle\langle\boldsymbol{\rho}|. \tag{2.2}$$

We can also look upon $\mathscr{I}$ as the *surface projector*. Then, given, say, $\hat{\boldsymbol{A}}$, $\hat{\boldsymbol{B}}$ and $\boldsymbol{A}$, $\boldsymbol{B}$ we must carefully define different products. We shall use the convention that $\hat{\boldsymbol{A}}\cdot\hat{\boldsymbol{B}}$ indicates simply the ordinary tensor index contraction, but $\boldsymbol{A}\cdot\boldsymbol{B}$ indicates *also* the two-dimensional scalar product in the following sense:

$$(\boldsymbol{A}\cdot\boldsymbol{B})_{ij}=\sum_r\int \mathrm{d}^2\rho A_{ir}|\boldsymbol{\rho}\rangle\langle\boldsymbol{\rho}|B_{rj}. \tag{2.3}$$

The meaning of the inverse $\boldsymbol{A}^{-1}$ is then that

$$\boldsymbol{A}^{-1} \cdot \boldsymbol{A} = \boldsymbol{I}\mathcal{I}. \tag{2.4}$$

This is not a trivial equality. It is very important to realise that, while $\boldsymbol{A}$ is directly related to $\hat{\boldsymbol{A}}$ by (2.1), such a relation does not hold between the inverses. Physically if $\hat{G}$ – a scalar here – is, for example, the resolvent of a Hamiltonian, *G is not, and does not have the properties of*, a resolvent. We shall also often encounter expressions in which both tensor index contraction and scalar product in the two-dimensional $\boldsymbol{\rho}$-space involve three-dimensional objects. We shall indicate this by inserting $\mathcal{I}$ since, from (2.2),

$$\hat{\boldsymbol{A}} \cdot \mathcal{I}\hat{\boldsymbol{B}} = \hat{\boldsymbol{A}}\mathcal{I} \cdot \hat{\boldsymbol{B}} = \int \mathrm{d}^2\rho \hat{\boldsymbol{A}}|\boldsymbol{\rho}\rangle \cdot \langle\boldsymbol{\rho}|\hat{\boldsymbol{B}}. \tag{2.5}$$

The explicit insertion of the surface projector $\mathcal{I}$ is unnecessary, by convention, whenever the two factors are two-dimensional and also if one of them only is three-dimensional. For example, in $\hat{\boldsymbol{A}} \cdot \boldsymbol{B}$, since $\boldsymbol{B}$ by definition only exists in two dimensions, we have automatically:

$$\hat{\boldsymbol{A}} \cdot \boldsymbol{B} = \int \mathrm{d}^2\rho \hat{\boldsymbol{A}}|\boldsymbol{\rho}\rangle \cdot \langle\boldsymbol{\rho}|\boldsymbol{B} = \hat{\boldsymbol{A}}\mathcal{I} \cdot \boldsymbol{B}. \tag{2.6}$$

The normal derivative, i.e. $\partial/\partial z$ for a plane surface, will be indicated by a prime. But $\hat{\boldsymbol{G}}$ is a function of two arguments $(\boldsymbol{r}, \boldsymbol{r}')$ or, with $(\boldsymbol{\rho}, \boldsymbol{\rho}')$ fixed, of (z, z'). Thus $'\hat{\boldsymbol{G}}$ and $\hat{\boldsymbol{G}}'$ will correspond to differentiating with respect to the first or second argument. However, care must be taken when taking the surface projections because the derivatives may be discontinuous. In practice $\hat{\boldsymbol{G}}$ is usually the Green function corresponding to a *second*-order differential equation and it is a general mathematical principle that the *first* derivatives are then discontinuous. In chapter 1 we saw an example of a *first*-order differential equation and then it was the Green function itself that was discontinuous. In this case we define

$$\left.\begin{aligned} {}'\boldsymbol{G}^{(\pm)}\langle\ \rangle \lim_{z'\to\pm 0} \left|\frac{\partial\langle z|\hat{\boldsymbol{G}}|z'\rangle}{\partial z}\right|_{z=0}; \\ {}^{(\pm)}\boldsymbol{G}'\langle\ \rangle \lim_{z\to\pm 0} \left|\frac{\partial\langle z|\hat{\boldsymbol{G}}|z'\rangle}{\partial z'}\right|_{z'=0}. \end{aligned}\right\} \tag{2.7}$$

We are now ready to write down a formal theory of matching in terms of Green functions. The ultimate goal of this analysis is to construct the complete $\hat{\boldsymbol{G}}_{\mathrm{s}}$, so that it suffices to find two kinds of matrix elements between points lying either on the same side – e.g. $\boldsymbol{r}_2$, $\boldsymbol{r}_2'$ – or on opposite sides – e.g. $\boldsymbol{r}_1$, $\boldsymbol{r}_2'$ – of the surfaces. The corresponding matrix elements will be briefly indicated as $\langle 2|\hat{\boldsymbol{G}}|2'\rangle$ and $\langle 1|\hat{\boldsymbol{G}}|2'\rangle$. We can regard the Green function as the response function which yields some 'field' created at $\boldsymbol{r}$ as output or response corresponding to an input or stimulus at $\boldsymbol{r}'$. Let $\boldsymbol{\delta}|2'\rangle$ be a standard unit input at $\boldsymbol{r}'=\boldsymbol{r}_2'$. The general form of $\langle 2|\hat{\boldsymbol{G}}_s|2'\rangle$ can be written as

$$\langle 2|\hat{\boldsymbol{G}}_s\cdot\boldsymbol{\delta}|2'\rangle=\langle 2|\hat{\boldsymbol{G}}_2\cdot\boldsymbol{\delta}|2'\rangle+\langle 2|\hat{\boldsymbol{G}}_2\cdot\boldsymbol{N}_2\cdot\hat{\boldsymbol{G}}_2\cdot\boldsymbol{\delta}|2'\rangle. \tag{2.8}$$

The physical meaning of this is represented in fig. 16. For example, the last term means that the standard unit input $\boldsymbol{\delta}|2'\rangle$ propagates through

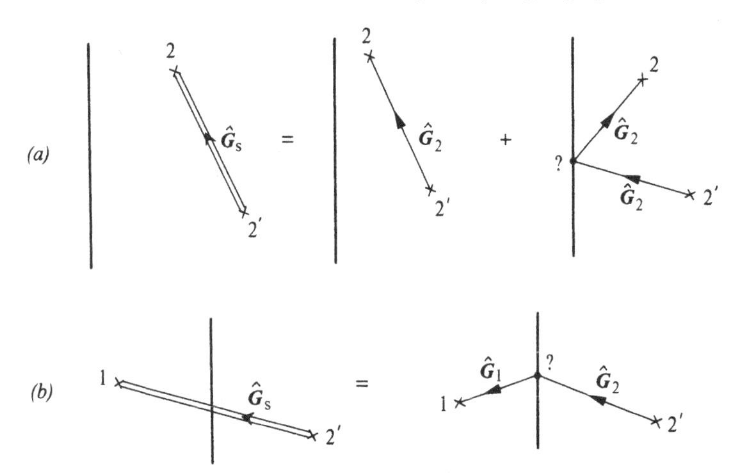

Fig. 16. (*a*) The form of a matrix element of $\hat{\boldsymbol{G}}_s$ between points on the same side of the surface. (*b*) The same for the two points on different sides.

domain 2 with 'propagator' $\hat{\boldsymbol{G}}_2$, hits the surface where something happens that we still do not know and which we represent by the two-dimensional object $\boldsymbol{N}_2$ – and then propagates again with propagator $\hat{\boldsymbol{G}}_2$ until it reaches the point $\boldsymbol{r}=\boldsymbol{r}_2$. Similarly for the cross matrix element,

$$\langle 1|\hat{\boldsymbol{G}}_s\cdot\boldsymbol{\delta}|2'\rangle=\langle 1|\hat{\boldsymbol{G}}_1\cdot\boldsymbol{N}_1\cdot\hat{\boldsymbol{G}}_2\cdot\boldsymbol{\delta}|2'\rangle. \tag{2.9}$$

We keep formally the standard unit input $\boldsymbol{\delta}$ because if this is a vector – e.g. a force or an electric field – then the response at $\boldsymbol{r}$ is also a vector – e.g. a displacement or a polarization – and we shall want to match vector fields in terms of tensor Green functions. In the end $\boldsymbol{\delta}$ is always eliminated from the physically interesting results.

The problem would be solved if we knew $\boldsymbol{N}_1$ and $\boldsymbol{N}_2$. For the sake of simplicity it is convenient to introduce equivalent unknowns $\hat{\boldsymbol{F}}_m$ defined as

$$\boldsymbol{N}_m\cdot\hat{\boldsymbol{G}}_2\cdot\boldsymbol{\delta}|2'\rangle=\mathscr{I}\hat{\boldsymbol{F}}_m|2'\rangle \quad (m=1,2), \tag{2.10}$$

Notice that $\hat{\boldsymbol{F}}_m$ (a vector) depends on the point where the input acts and $\mathscr{I}\hat{\boldsymbol{F}}_m$ has matrix elements between surface and bulk. We then start from

$$\langle 2|\hat{\boldsymbol{G}}_s \cdot \boldsymbol{\delta}|2'\rangle = \langle 2|\hat{\boldsymbol{G}}_2 \cdot \boldsymbol{\delta}|2'\rangle + \langle 2|\hat{\boldsymbol{G}}_2 \cdot \mathscr{I}\hat{\boldsymbol{F}}_2|2'\rangle; \tag{2.11a}$$

$$\langle 1|\hat{\boldsymbol{G}}_s \cdot \boldsymbol{\delta}|2'\rangle = \qquad \langle 1|\hat{\boldsymbol{G}}_1 \cdot \mathscr{I}\hat{\boldsymbol{F}}_1|2'\rangle. \tag{2.11b}$$

The problem is to find $\hat{\boldsymbol{F}}_1$ and $\hat{\boldsymbol{F}}_2$, which are to be obtained from the matching conditions. These are usually of two kinds, namely, (i) some fields must be continuous across the matching surface, and (ii) some linear combination of normal derivatives must also be continuous: for example, the electronic wavefunction and its derivative, or the tangential components of the electric field $\boldsymbol{E}$ and of $\boldsymbol{H} \propto \text{curl}\, \boldsymbol{E}$. The first condition is easy to express. It suffices to take the surface projection of (2.11) from the left, i.e. to take $2 \to \boldsymbol{\rho}$, $1 \to \boldsymbol{\rho}$. Then the first matching condition reads

$$\mathscr{I}\hat{\boldsymbol{G}}_2 \cdot \boldsymbol{\delta}|2'\rangle + \boldsymbol{G}_2 \cdot \hat{\boldsymbol{F}}_2|2'\rangle = \boldsymbol{G}_1 \cdot \hat{\boldsymbol{F}}_1|2'\rangle. \tag{2.12}$$

Notice, in accordance with the above remarks about notation, that the surface projectors appearing in (2.11) in $\hat{\boldsymbol{G}}_m\mathscr{I} \cdot \hat{\boldsymbol{F}}_m$ are now unnecessary.

The second matching condition requires closer scrutiny. Although we are developing a formal analysis it is better as this stage to appeal to physical intuition by considering some concrete physical systems. For example, take an isotropic elastic solid with Lamé coefficients λ, μ, and mass density ρ; an external force $\boldsymbol{f}$ produces a displacement field $\boldsymbol{u}$ obeying the equation of motion

$$-\mu\nabla^2 u_i - (\lambda+\mu)\nabla_i(\nabla \cdot \boldsymbol{u}) + \rho\ddot{u}_i = f_i. \tag{2.13}$$

Here and henceforth ∇_i means the ith component of ∇, i.e. $\partial/\partial x_i$. The Green function of the bulk material satisfies the equation

$$-\mu\nabla^2\hat{G}_{ij} - (\lambda+\mu)\nabla_i\nabla_r\hat{G}_{rj} + \rho\ddot{\hat{G}}_{ij} = \delta_{ij}\delta(\boldsymbol{r}-\boldsymbol{r}')\delta(t-t'). \tag{2.14}$$

(Summation over repeated indices is understood whenever the repeated index does not take a specific value 1, 2, or 3.) We are explicitly concerned with the z-dependence. Thus we take Fourier transforms as in chapter 1, leaving the dependence on $(\boldsymbol{\kappa}, \omega)$ understood where this is not explicitly shown. Now define the surface increment

$$\Delta' G_{ij} = {}'G_{ij}^{(+)} - {}'G_{ij}^{(-)}. \tag{2.15}$$

Integrating (2.14) from $z = -\eta$ to $z = +\eta$, letting $\eta \to 0$ and *then* using (2.7), we find

$$\mu\Delta' G_{ij} + (\lambda+\mu)\delta_{i3}\Delta' G_{3j} = \delta_{ij}\mathscr{I}. \tag{2.16}$$

The linear form appearing here is directly related to another object of physical interest. The *stress* tensor $\boldsymbol{\tau}$ is defined by

$$\tau_{ij} = \mu(\nabla_i u_j + \nabla_j u_i) + \lambda\delta_{ij}\nabla \cdot \boldsymbol{u}, \tag{2.17}$$

and the second matching condition requires the continuity of $\boldsymbol{n} \cdot \boldsymbol{\tau}$, i.e. of τ_{3j}, since $n_i = \delta_{i3}$. Now, define the linear form

$$\hat{A}_{ij} = \mu n_r(\nabla_r \hat{G}_{ij} + \nabla_i \hat{G}_{rj}) + \lambda \nabla_p \hat{G}_{pj}\delta_{ir} n_r, \tag{2.18}$$

which is naturally suggested when $\boldsymbol{u}$ is thought of as given by the response function $\hat{\boldsymbol{G}}$ times some stimulus or, more formally, by comparing the equations of motion, (2.13) and (2.14). For practical purposes it is convenient to rewrite this in the form

$$\hat{A}_{ij} = \mu'\hat{G}_{ij} + (\lambda + \mu)\delta_{i3}'\hat{G}_{3j} + [\mu(1-\delta_{i3})\nabla_i \hat{G}_{3j} + \lambda(1-\delta_{p3})\nabla_p \hat{G}_{pj}\delta_{i3}]. \tag{2.19}$$

Integrating across the surface from $-\eta$ to $+\eta$ the terms in square brackets – which are continuous – integrate out to zero while, from (2.16), we have

$$\Delta \boldsymbol{A} = \boldsymbol{I}\mathcal{I}, \tag{2.20}$$

where $\boldsymbol{I}$ is the diagonal unit tensor as in chapter 1. It is instructive to consider the physical meaning of this property. The equation of motion says that the external driving force must be in dynamical equilibrium with the sum of the inertial and elastic forces. The ith component of the latter can be expressed in the standard way as the divergence of the stress tensor. For the purposes of this argument we can ignore the inertial term, which is continuous across the surface. Consider an equation of the form

$$\sum_j \partial\tau_{ij}/\partial x_j = f_i - \rho\ddot{u}_i.$$

When the external stimulus is a unit δ-function input the integral of this from $-\eta$ to $+\eta$ is equal to unity. This is the physical meaning of (2.20). We shall see presently that this property of the surface increment of $\hat{\boldsymbol{A}}$ is of central importance. Moreover, the second matching condition, i.e. the continuity of τ_{3j}, can be expressed in terms of the surface projections of the linear forms $\hat{\boldsymbol{A}}_1$ and $\hat{\boldsymbol{A}}_2$ of the two extended media in the following way. The displacement $\boldsymbol{u}$ in the actual surface system can be expressed in terms of $\hat{\boldsymbol{G}}_s$ by taking the expression for $\hat{\boldsymbol{G}}_s$ alternatively from (2.11a) and (2.11b) and taking appropriate surface projections. Then (compare (2.19) and (2.17)) the condition $\boldsymbol{n} \cdot \boldsymbol{\tau}^{(2)} = \boldsymbol{n} \cdot \boldsymbol{\tau}^{(1)}$ reads

$$\mathcal{I}\hat{\boldsymbol{A}}_2 \cdot \boldsymbol{\delta}|2'\rangle + \boldsymbol{A}_2^{(-)} \cdot \hat{\boldsymbol{F}}_2|2'\rangle = \boldsymbol{A}_1^{(+)} \cdot \hat{\boldsymbol{F}}_1|2'\rangle. \tag{2.21}$$

The signs of the surface projections $\boldsymbol{A}_m$ are written down explicitly and consistently with the definition (2.7) because these linear forms involve normal derivatives, as explained. This process must be carried out rather carefully to avoid mistakes. The sign on $\boldsymbol{A}$, or on $'\boldsymbol{G}$, is actually determined by the sign of the difference between the two arguments of the Green functions involved.

Consider another important physical example. Electromagnetic waves were discussed at length in chapter 1. The equation of motion was expressed in Fourier transform in (1.20) but it is convenient to express it here in space variables, i.e.

$$-\nabla^2 E_i + \nabla_i \nabla_r E_r - \frac{\omega^2}{c^2} \int \mathrm{d}\boldsymbol{r}'' \varepsilon_{ir}(\boldsymbol{r} - \boldsymbol{r}'') E_r(\boldsymbol{r}'') = \mathrm{i} \frac{4\pi\omega}{c^2} J_i. \qquad (2.22)$$

Hence the corresponding Green function satisfies the equation

$$\nabla^2 \hat{G}_{ij}(\boldsymbol{r}, \boldsymbol{r}') - \nabla_i \nabla_r \hat{G}_{rj}(\boldsymbol{r}, \boldsymbol{r}') + \frac{\omega^2}{c^2} \int \mathrm{d}\boldsymbol{r}'' \varepsilon_{ir}(\boldsymbol{r}, \boldsymbol{r}'') \hat{G}_{rj}(\boldsymbol{r}'', \boldsymbol{r}') = -\mathrm{i} \frac{4\pi\omega}{c^2} \delta_{ij} \delta(\boldsymbol{r} - \boldsymbol{r}'). \qquad (2.23)$$

The first matching condition requires the continuity of the tangential components of $\boldsymbol{E}$. This yields an equation identical in form to (2.12), except that the tensors involved have only (x, y) components. The second matching condition requires the continuity of the (x, y) components of $\boldsymbol{H}$, i.e. of curl $\boldsymbol{E}$. Expressed in terms of Green functions these two components are of the form $-'\hat{G}_{ij} + \nabla_i \hat{G}_{3j}$ $(i, j = 1, 2)$. It is intuitively obvious that, apart from scalar constant factors, this will give the form of the linear combination of derivatives, $\hat{\boldsymbol{A}}$, needed for this case. Indeed, it is easy to see, by a similar argument, that for this case,

$$\hat{A}_{ij} = (\mathrm{i}\, c^2/4\pi\omega)[-'\hat{G}_{ij} + \mathrm{i}\, k_i \hat{G}_{3j}]. \qquad (2.24)$$

Remember that now $(i, j) = (1, 2)$. Since ∇_i does not involve any normal derivative, in Fourier transform this introduces simply the factor $\mathrm{i}\, k_i$. Integrating (2.23) from $-\eta$ to $+\eta$ we find that this linear form again satisfies the central property (2.20) with the only difference that $\mathscr{I}$ this time is the 2×2 unit matrix. Similarly we can write down the second matching condition exactly in the form (2.21). The physical interpretation of (2.20) this time is similar to that of the elastic case. From Maxwell's equations, since D_z is continuous across the surface, looking at curl $\boldsymbol{H}$ in (1.6), we see that a unit δ-function input – in this case a unit surface current – produces a unit jump in the tangential projection of $\boldsymbol{H}$ on the surface.

A physically very important and algebraically very simple example is that of electronic states. Here the field under consideration is the electronic wavefunction. The Green function is a scalar. The first matching condition is identical to (2.12) but without vector and tensor indices, i.e.

$$\mathscr{I}\hat{G}_2|2'\rangle + G_2\hat{F}_2|2'\rangle = G_1\hat{F}_1|2'\rangle. \tag{2.25}$$

The second matching condition is simpler: we just require the continuity of the normal derivative of the wavefunction. Thus the linear form $\hat{A}$ for this problem is simply $-'\hat{G}$ (for the change of sign see § 2.4) and the second matching condition reads

$$\mathscr{I}'\hat{G}_2|2'\rangle + 'G_2^{(-)}\hat{F}_2|2'\rangle = 'G_1^{(+)}F_1|2'\rangle. \tag{2.26}$$

The relative simplicity of this example can be used to clear up some technical points worth discussing. First the signs in (2.26). In this case we have 'something' like $\langle \boldsymbol{\rho}''|\hat{F}_2|2'\rangle$ coming out of the surface and propagating to $\langle 2|$, yielding the term

$$\int \mathrm{d}^2\rho''\langle 2|\hat{G}_2|\rho''\rangle\langle \rho''|\hat{F}_2|2'\rangle = \langle 2|\hat{G}_2\mathscr{I}\hat{F}_2|2'\rangle,$$

The important point to notice is that $\mathscr{I}\hat{F}_2$ has its first argument at the surface. Thus, when taking normal derivatives we have to study the term

$$\partial\langle \boldsymbol{\rho}, z_2|\hat{G}_2|\boldsymbol{\rho}''\rangle/\partial z_2,$$

in which the second argument is already at the surface (i.e. $z''=0$), *before* taking $\partial/\partial z_2$. Then the sign of $z''-z_2$ is always negative and this yields $'G_2^{(-)}$. The other sign is similarly explained. The second and more important point to notice is that (2.26) *does not* contradict the principle that the derivatives of the Green functions present discontinuities. What we are saying is that either (2.11*a*) or (2.11*b*) can be used to calculate $'\hat{G}_s\rangle$, with the second argument still in the bulk (fig. 17).

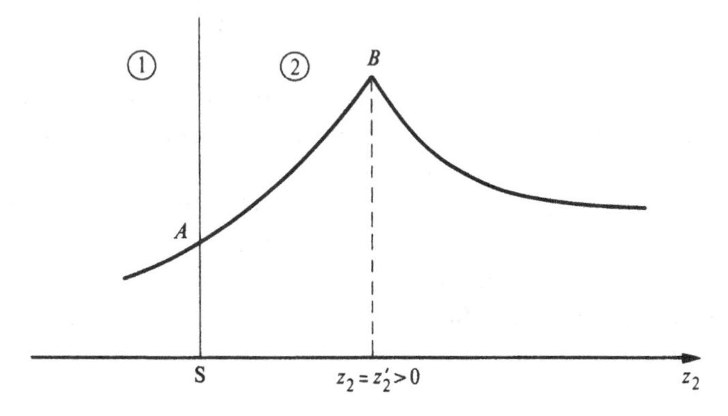

Fig. 17. The discontinuity of the derivative of a Green function appears at B, as a function of z_2, for fixed z_2'. But the slope at A is continuous, whether it is calculated from the left or from the right. S, surface.

2.2. Solution of the matching problem

Let us now return to the general formulation. The position is as follows: the form of $\hat{\boldsymbol{G}}_s$ is given in (2.11*a, b*). The problem amounts to finding the two unknowns, $\hat{\boldsymbol{F}}_1$ and $\hat{\boldsymbol{F}}_2$, for which we have the two matching conditions, (2.12) and (2.21). For each physical problem there is a suitable linear form $\hat{\boldsymbol{A}}$ – to be found from the physics of the problem each time – which has two central properties, namely, (i) it satisfies (2.20), and (ii) the second matching condition can be expressed in terms of $\hat{\boldsymbol{A}}_1$ and $\hat{\boldsymbol{A}}_2$ as in (2.21). In order to solve the problem we shall proceed in three stages.

2.2.1. The secular equation for surface eigenmodes

This corresponds to the situation in which there is no external stimulus. The inhomogeneous equations, (2.12) and (2.21) become the homogeneous equations,

$$\left.\begin{aligned} \boldsymbol{G}_2 \cdot \boldsymbol{F}_2 - \boldsymbol{G}_1 \cdot \boldsymbol{F}_1 = 0; \\ \boldsymbol{A}_2^{(-)} \cdot \boldsymbol{F}_2 - \boldsymbol{A}_1^{(+)} \cdot \boldsymbol{F}_1 = 0. \end{aligned}\right\} \tag{2.27}$$

These result from setting the external input term equal to zero and taking $z_2' \to +0$. In this case $\boldsymbol{F}_1$ and $\boldsymbol{F}_2$ have the nature of auxiliary parameters which are to be eliminated from (2.27) in order to yield a secular equation. To simplify the algebra it is convenient to transform from $\boldsymbol{\rho}$ to $\boldsymbol{\kappa}$. The two-dimensional scalar product involves a convolution in $\boldsymbol{\rho}$-space and its Fourier transform is simply the ordinary algebraic product, supplemented of course by tensor index contractions. In this representation we simply handle the objects appearing in (2.27) as matrices that are functions of $(\boldsymbol{\kappa}, \omega)$. By the ordinary rules of matrix multiplication it is easy to eliminate $\boldsymbol{F}_1$ and $\boldsymbol{F}_2$, obtaining the secular equation

$$\det [\boldsymbol{A}_2^{(-)} \cdot \boldsymbol{G}_2^{-1} - \boldsymbol{A}_1^{(+)} \cdot \boldsymbol{G}_1^{-1}] = 0. \tag{2.28}$$

This is one of the central results of the formal analysis, yielding the dispersion relation between ω and $\boldsymbol{\kappa}$ for the surface eigenmodes. The meaning of the determinant must be interpreted in each physical situation, as will be seen in the applications.

2.2.2. The surface projection of $\hat{\boldsymbol{G}}_s$

Everything we are doing here can be done in terms of 'wavefunctions', where this term is understood in a broad sense; e.g. it could be a displacement field. The reason it is useful to do the matching in terms of Green functions is that these contain all the relevant physical information for the model they describe. We shall see that $\boldsymbol{G}_s$, the surface projection of $\hat{\boldsymbol{G}}_s$, has a central role in the physical analysis of the problem. We now want to obtain the formula for $\boldsymbol{G}_s$, so we must go back to the inhomogeneous equations (2.12) and (2.21) and again take $z_2' \to +0$. This yields

$$\boldsymbol{G}_2 \cdot \boldsymbol{F}_2 - \boldsymbol{G}_1 \cdot \boldsymbol{F}_1 = -\boldsymbol{G}_2 \cdot \boldsymbol{\delta}\mathcal{I}; \tag{2.29a}$$

$$\boldsymbol{A}_2^{(-)} \cdot \boldsymbol{F}_2 - \boldsymbol{A}_1^{(+)} \cdot \boldsymbol{F}_1 = -\boldsymbol{A}_2^{(+)} \cdot \boldsymbol{\delta}\mathcal{I}. \tag{2.29b}$$

Multiplying (2.29a) by $\boldsymbol{A}_2^{(-)} \cdot \boldsymbol{G}_2^{-1}$ and subtracting from (2.29b), we have

$$(\boldsymbol{A}_2^{(-)} \cdot \boldsymbol{G}_2^{-1} - \boldsymbol{A}_1^{(+)} \cdot G_1^{-1}) \cdot \boldsymbol{G}_1 \cdot \boldsymbol{F}_1 = (\boldsymbol{A}_2^{(-)} - \boldsymbol{A}_2^{(+)}) \cdot \boldsymbol{\delta}\mathcal{I}. \tag{2.30}$$

It is at this stage that the property (2.20) plays a central part. Furthermore $\boldsymbol{G}_1 \cdot \boldsymbol{F}_1$, if we take the surface projection of (2.11b) from both sides, is precisely $\boldsymbol{G}_s \cdot \boldsymbol{\delta}$, which allows us to eliminate $\boldsymbol{\delta}$ from both sides of (2.30), since the equality holds for *any* vector $\boldsymbol{\delta}$. This yields the desired formula in the form

$$\boldsymbol{G}_s^{-1} = \boldsymbol{A}_1^{(+)} \cdot \boldsymbol{G}_1^{-1} - \boldsymbol{A}_2^{(-)} \cdot \boldsymbol{G}_2^{-1}, \tag{2.31}$$

another central result which, incidentally, proves (compare with (2.28)) that the secular equation is

$$\det \boldsymbol{G}_s^{-1} = 0. \tag{2.32}$$

This result is not surprising in the least. It embodies in a concise way the general principle that the spectrum of a system is contained in the singularities of its Green function. However, we are talking about surface eigenstates and it is not immediately obvious that the secular determinant should be precisely that of the reciprocal of the surface projection of the Green function of the entire system. For example, we have stressed that the surface projection itself, $\boldsymbol{G}_s$, *is not and does not have the properties of* a resolvent operator. The theorem of (2.32) must actually be regarded as a general one for surface systems, and one which in fact holds even if (2.31) does not, a situation which will be studied in § 2.8. Furthermore, suppose we know that (2.32) holds generally before obtaining (2.28) and (2.31). (Indeed this can be proved (§ 2.4) by a direct argument.) Even then, having derived only (2.28) we would not be entitled to conclude that $\boldsymbol{G}_s^{-1}$ is given by (2.31) because when deriving a secular equation we have no way of knowing whether it contains – or lacks – possible non-vanishing factors. This is not only a matter of constant factors, as in (1.159) or in the discussion of § 1.4. The reason it is important to get this point straight is that these factors may themselves be functions of $(\omega, \boldsymbol{\kappa})$, as can easily happen if one derives a secular equation by some direct physical argument, rather than following literally the formal analysis leading up to

(2.28). In this case one could obtain erroneous results for the density of states (§ 2.3) of a surface system.

2.2.3. The full Green function of a surface system

The ultimate goal is to obtain $\hat{\boldsymbol{G}}_s$ fully and explicitly, for which it suffices to know the two typical matrix elements (2.11*a*, *b*). We shall solve this problem *without using the matching conditions*. For this we go back to (2.11*a*) and take $z_2=0$. Thus

$$\langle\boldsymbol{\rho}|\hat{\boldsymbol{G}}_s\cdot\boldsymbol{\delta}|2'\rangle=\langle\boldsymbol{\rho}|\hat{\boldsymbol{G}}_2\cdot\boldsymbol{\delta}|2'\rangle+\langle\boldsymbol{\rho}|\hat{\boldsymbol{G}}_2\cdot\mathscr{I}\hat{\boldsymbol{F}}_2|2'\rangle. \tag{2.33}$$

Next we take, also in (2.11*a*), $z_2'=0$, i.e.,

$$\langle 2|\hat{\boldsymbol{G}}_s\cdot\boldsymbol{\delta}|\boldsymbol{\rho}'\rangle=\langle 2|\hat{\boldsymbol{G}}_2\cdot\boldsymbol{\delta}|\boldsymbol{\rho}'\rangle+\langle 2|\hat{\boldsymbol{G}}_2\cdot\mathscr{I}\mathrm{F}_2|\boldsymbol{\rho}'\rangle. \tag{2.34}$$

Furthermore, in (2.11*a*) again we take both $z_2=0$ and $z_2'=0$, whence

$$\boldsymbol{F}_2=\boldsymbol{G}_2^{-1}\cdot(\boldsymbol{G}_s-\boldsymbol{G}_2)\cdot\boldsymbol{\delta}\mathscr{I}, \tag{2.35}$$

which we insert in (2.34). We have again an equality that holds for any arbitrary vector $\boldsymbol{\delta}$, and which can therefore be eliminated. This yields

$$\langle 2|\hat{\boldsymbol{G}}_s|\boldsymbol{\rho}'\rangle=\langle 2|\hat{\boldsymbol{G}}_2|\boldsymbol{\rho}'\rangle+\langle 2|\hat{\boldsymbol{G}}_2\cdot\boldsymbol{G}_2^{-1}\cdot(\boldsymbol{G}_s-\boldsymbol{G}_2)|\boldsymbol{\rho}'\rangle, \tag{2.36}$$

from which we have

$$\langle\boldsymbol{\rho}|\hat{\boldsymbol{G}}_s|2'\rangle=\langle\boldsymbol{\rho}|\hat{\boldsymbol{G}}_2|2'\rangle+\langle\boldsymbol{\rho}|(\boldsymbol{G}_s-\boldsymbol{G}_2)\cdot\boldsymbol{G}_2^{-1}\cdot\hat{\boldsymbol{G}}_2|2'\rangle. \tag{2.37}$$

Comparing (2.37) and (2.33) we obtain the unknown term

$$\mathscr{I}\hat{\boldsymbol{F}}_2|2'\rangle=\boldsymbol{G}_2^{-1}\cdot(\boldsymbol{G}_s-\boldsymbol{G}_2)\cdot\boldsymbol{G}_2^{-1}\cdot\hat{\boldsymbol{G}}_2\cdot\boldsymbol{\delta}|2'\rangle. \tag{2.38}$$

This is to be used in (2.11*a*). On doing this we obtain again an equation from which $\boldsymbol{\delta}$ can be eliminated, and finally:

$$\langle 2|\hat{\boldsymbol{G}}_s|2'\rangle=\langle 2|\hat{\boldsymbol{G}}_2|2'\rangle+\langle 2|\hat{\boldsymbol{G}}_2\cdot\boldsymbol{G}_2^{-1}\cdot(\boldsymbol{G}_s-\boldsymbol{G}_2)\cdot\boldsymbol{G}_2^{-1}\cdot\hat{\boldsymbol{G}}_2|2'\rangle. \tag{2.39}$$

We can likewise start from (2.11*b*) and following a similar argument obtain the second formula

$$\langle 1|\hat{\boldsymbol{G}}_s|2'\rangle=\langle 1|\hat{\boldsymbol{G}}_1\cdot\boldsymbol{G}_1^{-1}\cdot\boldsymbol{G}_s\cdot\boldsymbol{G}_2^{-1}\cdot\hat{\boldsymbol{G}}_2|2'\rangle. \tag{2.40}$$

We stress that the matching conditions have not been invoked to derive (2.39) and (2.40). These two equations express a general theorem about the form of the matrix elements of $\hat{\boldsymbol{G}}_s$ and hold *irrespective of the formula for the surface projection* $\boldsymbol{G}_s$.

Let us summarise the position. The form of the matrix elements of $\hat{\boldsymbol{G}}_s$ is given by (2.39) and (2.40), which are absolutely general. These formulae involve the surface projection $\boldsymbol{G}_s$. The formula for $\boldsymbol{G}_s$ depends on the matching conditions. When these amount to continuity of the field and of linear combinations of normal derivatives thereof, then $\boldsymbol{G}_s$ is given by (2.31). In any case a general property of $\boldsymbol{G}_s$ is that the secular equation for surface eigenmodes is (2.32) which, for the matching conditions studied here, takes the form (2.28). Physically interesting situations in which the matching conditions are significantly different will be studied in §§ 2.8, 2.9.

Finally, it is interesting to see how the analysis is simplified when the matching plane is a mirror image symmetry plane for the extended media. The easiest example to imagine is perfect unchanged bulk material abruptly terminated at the matching plane. Then $\hat{\boldsymbol{G}}_m$ is the Green function of a perfect bulk system. However, the formal theory is not restricted to this model. A conceivable situation is described in fig. 18. The practical problem then is how to construct the extended Green functions $\hat{\boldsymbol{G}}_m$ by perturbing the Green functions of perfect bulk media, but this does not concern the formal analysis. The real simplification

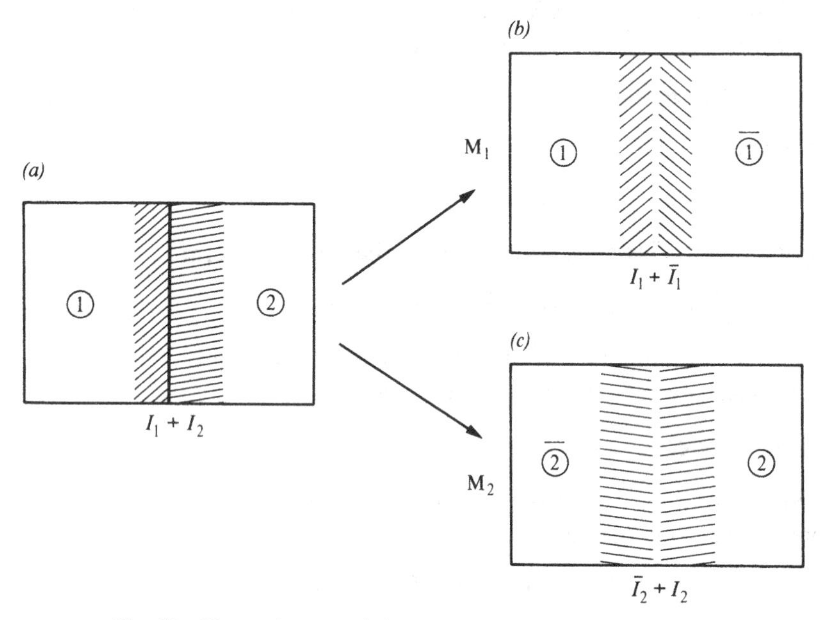

Fig. 18. The real system (*a*) may contain inhomogeneities on both sides of the matching plane. The definition of the extended systems M_1 and M_2 is arbitrary beyond the matching plane. In particular M_1 could be defined as shown in (*b*) by filling up the other side of space with the mirror image of the real material on side 1. This would include the mirror image $\bar{I}_1$ of the inhomogeneity I_1. The extended system M_1 would then be a perfect bulk medium perturbed by the inhomogeneity $I_1 + \bar{I}_1$. Likewise (*c*) shows the situation for M_2.

concerns the normal derivatives. With symmetry the derivatives taken from both sides ($z \to \pm 0$) are equal in magnitude and opposite in sign. This implies, from (2.20),

$$\boldsymbol{A}^{(+)} = -\boldsymbol{A}^{(-)} = \tfrac{1}{2}\boldsymbol{I}\mathcal{I}. \tag{2.41}$$

Then (2.31) takes a much simpler form, namely,

$$\boldsymbol{G}_s^{-1} = \tfrac{1}{2}(\boldsymbol{G}_1^{-1} + \boldsymbol{G}_2^{-1}) = \tfrac{1}{2}\boldsymbol{G}_1^{-1}(\boldsymbol{G}_1 + \boldsymbol{G}_2)\boldsymbol{G}_2^{-1}. \tag{2.42}$$

By separating out non-vanishing factors the secular equations for surface eigenstates can then be written in the considerably simpler form,

$$\det[\boldsymbol{G}_1 + \boldsymbol{G}_2] = 0. \tag{2.43}$$

This illustrates the comments made after (2.32). Indeed, for the particular case of symmetry it would be easy to derive (2.43) directly by an *ad hoc* argument and this would certainly yield the correct surface state eigenvalues, but (2.43) could not be used to surmise the correct form of $\boldsymbol{G}_s^{-1}$. Here we would be missing the non-vanishing factors the importance of which will be seen in § 2.3.

2.3. The density of states of a surface system

Besides obtaining the secular equation surface eigenmodes, we have strived in § 2.2 to obtain the full $\hat{\boldsymbol{G}}_s$ because it contains all the physical information of interest. Let us now exploit this knowledge by studying one of the most important physical questions concerning a system. We want to know the density of states – or modes – of a surface system: all of it, bulk and surface effects included.

First of all we must specify the variable in which a density of states is defined. For electronic states this variable is the energy. For elastic waves the natural variable is ω^2. Since we must write something down explicitly, we shall perform the analysis for a density of states – or modes – in ω^2. It should be obvious, however, that the argument is quite general and can be easily adapted to any other variable. We shall also omit the explicit specification of the vanishingly small imaginary part which must be added to the variable. The sign of this depends on the sign convention for $\exp(\pm i\omega t)$. In our case ω^2 must be understood as $\omega^2 - i\eta$ when $\eta \to 0$. With this understanding the total density of modes of a surface system will be evaluated from the general formula

$$N_t(\omega^2) = -(1/\pi)\,\mathrm{Tr}\,\mathrm{Im}\,\hat{\boldsymbol{G}}_s(\omega^2). \tag{2.44}$$

In order to evaluate this formula we shall need (2.39) and the identical equality holding on side 1. With the convention that both points stay always on the same side of the surface we have from (2.39), for $m = 1, 2$

$$\begin{aligned} \boldsymbol{G}_s^{-1} \cdot \hat{\boldsymbol{G}}_s &= \boldsymbol{G}_s^{-1} \cdot \hat{\boldsymbol{G}}_m + \boldsymbol{G}_s^{-1} \cdot \boldsymbol{G}_m \cdot \boldsymbol{G}_m^{-1} \cdot \boldsymbol{G}_s \cdot \boldsymbol{G}_m^{-1} \cdot \hat{\boldsymbol{G}}_m \\ &\quad - \boldsymbol{G}_s^{-1} \cdot \boldsymbol{G}_m \cdot \boldsymbol{G}_m^{-1} \cdot \boldsymbol{G}_m \cdot \boldsymbol{G}_m^{-1} \cdot \hat{\boldsymbol{G}}_m \\ &= \boldsymbol{G}_s^{-1} \cdot \hat{\boldsymbol{G}}_m + \boldsymbol{G}_m^{-1} \cdot \hat{\boldsymbol{G}}_m - \boldsymbol{G}_s^{-1} \cdot \hat{\boldsymbol{G}}_m = \boldsymbol{G}_m^{-1} \cdot \hat{\boldsymbol{G}}_m. \end{aligned} \tag{2.45}$$

It is convenient to introduce the projectors $P_m (m = 1, 2)$, which project onto side (1, 2). Then (2.45) is written more precisely as

$$\boldsymbol{G}_s^{-1} \cdot \hat{\boldsymbol{G}}_s P_m = \boldsymbol{G}_m^{-1} \cdot \hat{\boldsymbol{G}}_m P_m. \tag{2.46}$$

The l.h.s. of (2.39) is likewise written for either side as $P_m \hat{\boldsymbol{G}}_s P_m$. By using (2.46) this becomes

$$\begin{aligned} P_m \hat{\boldsymbol{G}}_s P_m = P_m \hat{\boldsymbol{G}}_m P_m + P_m \hat{\boldsymbol{G}}_s \cdot \boldsymbol{G}_s^{-1} \cdot \hat{\boldsymbol{G}}_s P_m \\ - P_m \hat{\boldsymbol{G}}_m \cdot \boldsymbol{G}_m^{-1} \cdot \hat{\boldsymbol{G}}_m P_m, \end{aligned} \tag{2.47}$$

whence

$$P_m (\hat{\boldsymbol{G}}_s - \hat{\boldsymbol{G}}_s \cdot \boldsymbol{G}_s^{-1} \cdot \hat{\boldsymbol{G}}_s) P_m = P_m (\hat{\boldsymbol{G}}_m - \hat{\boldsymbol{G}}_m \cdot \boldsymbol{G}_m^{-1} \cdot \hat{\boldsymbol{G}}_m) P_m. \tag{2.48}$$

This equality is very useful for our purposes. We want to evaluate Tr $\hat{\boldsymbol{G}}_s$ in the full three-dimensional system and this involves integrating the diagonal matrix element over domains 1 and 2. But the $\hat{\boldsymbol{G}}_s$ involved in (2.48) is *the same* for $m = 1, 2$. Thus summing the two identical equalities and taking traces we have

$$\mathrm{Tr}\, \hat{\boldsymbol{G}}_s = \mathrm{Tr}\, \hat{\boldsymbol{G}}_s \cdot \boldsymbol{G}_s^{-1} \cdot \hat{\boldsymbol{G}}_s + \mathrm{Tr} \sum_m P_m (\hat{\boldsymbol{G}}_m - \hat{\boldsymbol{G}}_m \cdot \boldsymbol{G}_m^{-1} \cdot \hat{\boldsymbol{G}}_m) P_m. \tag{2.49}$$

The first term on the r.h.s. can be transformed in an interesting manner. In concise notation, we are discussing Green functions satisfying an equation of motion of the general form

$$(\hat{\boldsymbol{M}} - \omega^2 \boldsymbol{I}) \cdot \hat{\boldsymbol{G}} = \boldsymbol{I}.$$

It is easy to see from here that

$$\hat{\boldsymbol{G}} \cdot \hat{\boldsymbol{G}} = \mathrm{d}\hat{\boldsymbol{G}}/\mathrm{d}\omega^2 \tag{2.50}$$

and, by using the cyclic invariance of the trace,

$$\mathrm{Tr}\, \hat{\boldsymbol{G}} \cdot \boldsymbol{G}^{-1} \cdot \hat{\boldsymbol{G}} = \mathrm{Tr}\, \boldsymbol{G}^{-1} \cdot \mathrm{d}\hat{\boldsymbol{G}}/\mathrm{d}\omega^2. \tag{2.51}$$

Notice that, since $\boldsymbol{G}^{-1}$ is two-dimensional and the trace implies taking diagonal elements, the r.h.s. of (2.51) selects only the surface matrix elements of the last factor. Indicating the two-dimensional integration over $\boldsymbol{\rho}$ by the symbol tr we have, for any Green function,

$$\mathrm{Tr}\, \boldsymbol{G}^{-1} \cdot \mathrm{d}\hat{\boldsymbol{G}}/\mathrm{d}\omega^2 = \mathrm{tr}\, \boldsymbol{G}^{-1} \cdot \mathrm{d}\boldsymbol{G}/\mathrm{d}\omega^2.$$

Hence

$$\begin{aligned}\mathrm{Im}\,\mathrm{Tr}\, \hat{\boldsymbol{G}} \cdot \boldsymbol{G}^{-1} \cdot \hat{\boldsymbol{G}} &= \mathrm{Im}\,\mathrm{tr}\, \boldsymbol{G}^{-1} \cdot \mathrm{d}\boldsymbol{G}/\mathrm{d}\omega^2 \\ &= \mathrm{Im}\,\{\mathrm{d}/\mathrm{d}\omega^2\,(\ln \det \boldsymbol{G}^{-1})\}. \qquad (2.52)\end{aligned}$$

In particular we can apply these equalities to $\hat{\boldsymbol{G}}_\mathrm{s}$ in the first term on the r.h.s. of (2.49). We then obtain the formula for the total density of modes *per unit volume*†

$$N_\mathrm{t} = -(1/\pi)\,\mathrm{Im}\,\mathrm{Tr} \sum_{m=1,2} P_m(\hat{\boldsymbol{G}}_m - \hat{\boldsymbol{G}}_m \cdot \boldsymbol{G}_m^{-1} \cdot \hat{\boldsymbol{G}}_m)P_m + \frac{1}{\pi}\frac{\mathrm{d}}{\mathrm{d}\omega^2} \arg \det \boldsymbol{G}_\mathrm{s}^{-1}. \qquad (2.53)$$

This is as far as one can go in giving a general formula for N_t. Let us now interpret this result in physical terms.

Suppose we take each extended medium m and cut it through the matching plane introducing there an infinitely repulsive *hard wall*. We just have material m on side m and a vacuum on the other side and we do not allow for matching solutions. Then the cross matrix elements of $\hat{\boldsymbol{G}}_\mathrm{s}^\infty$, the Green function of this surface system, vanish. By (2.40) this means that $\boldsymbol{G}_\mathrm{s}^\infty = 0$, a fact which is also obvious if we think of the spectral representation of $\langle z|\hat{\boldsymbol{G}}_\mathrm{s}^\infty|z'\rangle$ in terms of its own eigenfunctions: with an infinitely repulsive hard wall the eigenfunctions of $\hat{\boldsymbol{G}}_\mathrm{s}^\infty$ vanish for $z = z' = 0$. Therefore, by (2.39),

$$\hat{\boldsymbol{G}}_\mathrm{s}^\infty \equiv P_m \hat{\boldsymbol{G}}_\mathrm{s}^\infty P_m = P_m(\hat{\boldsymbol{G}}_m - \hat{\boldsymbol{G}}_m \cdot \boldsymbol{G}_m^{-1} \cdot \hat{\boldsymbol{G}}_m)P_m. \qquad (2.54)$$

Notice that, although we do not have matching solutions with the hard wall, we *do have surface effects*. If the space were filled with bulk material m up to the surface, without accounting for any surface effects at all, then the Green function of this hypothetical system would be identical to G_m, satisfying the same boundary conditions at infinity, but it would only exist in the domain of P_m. This Green function indeed would be simply $P_m\hat{\boldsymbol{G}}_m P_m$, which is the first term on the r.h.s. in (2.54). The next term, containing $\boldsymbol{G}_m^{-1}$, represents the surface effects of the hard wall, analogous to the hard core terms in scattering theory. Notice that it does not correspond to surface eigenstates proper, but it does describe an effect localised near the surface because it contains matrix elements between

† For applications in practice see (2.118).

surface and bulk points, which decay as these are further away from the surface. This term describes the way in which the hard wall by itself would distort the spectrum of the bulk system because, even without allowing for matching solutions, the wavefunctions of the system would have to adapt themselves to the new situation of having to vanish at the surface. Thus, if we define

$$N^{\infty} = -\frac{1}{\pi} \operatorname{Tr} \operatorname{Im} \sum_m P_m(\hat{\boldsymbol{G}}_m - \hat{\boldsymbol{G}}_m \cdot \boldsymbol{G}_m^{-1} \cdot \hat{\boldsymbol{G}}_m)P_m, \qquad (2.55)$$

then we can rewrite (2.53) in a more transparent way as

$$N_t = N^{\infty} + \frac{1}{\pi} \frac{\mathrm{d}}{\mathrm{d}\omega^2} \arg \det \boldsymbol{G}_s^{-1}. \qquad (2.56)$$

What we are doing is building up the actual surface system in two stages. First we cut the media with a hard wall, which keeps both domains separate. Then we lower the infinite barrier and allow for physical coupling, i.e. matching, between the two sides. The contribution of the matching solutions is then contained in the second term. Indeed, $\det \boldsymbol{G}_s^{-1}$ is a function of ω^2 and $\boldsymbol{\kappa}$ which vanishes whenever ω^2 and $\boldsymbol{\kappa}$ satisfy the dispersion relation; that is to say, whenever we reach a surface eigenmode the argument of this determinant changes by π and its derivative contributes a δ-function of strength 1 to the density of modes, as shown in (2.56). This approach to the actual surface system in two stages will be further used in chapter 3.

The formula for the density of modes takes a particularly simple form if the matching surface is a plane of mirror image symmetry of the extended media. In this case

$$\operatorname{Tr} P_m\hat{\boldsymbol{G}}_m \cdot \boldsymbol{G}_m^{-1} \cdot \hat{\boldsymbol{G}}_m P_m = \tfrac{1}{2} \operatorname{Tr} \hat{\boldsymbol{G}}_m \cdot \boldsymbol{G}_m^{-1} \cdot \hat{\boldsymbol{G}}_m,$$

and applying (2.52) to each $\hat{\boldsymbol{G}}_m$, N_t simplifies to†

$$N_t = \frac{1}{2} \sum_m \left[N_m + \frac{1}{\pi} \frac{\mathrm{d}}{\mathrm{d}\omega^2} \arg \det \boldsymbol{G}_m \right] + \frac{1}{\pi} \frac{\mathrm{d}}{\mathrm{d}\omega^2} \arg \det \boldsymbol{G}_s^{-1}. \qquad (2.57)$$

Here N_m is the density of states for the entire extended medium m. Clearly

$$\operatorname{Tr} P_m\hat{\boldsymbol{G}}_m P_m = \tfrac{1}{2} \operatorname{Tr} \hat{\boldsymbol{G}}_m,$$

and this term yields simply the density of states we would have by taking only one half of material m without any surface effects, matching or hard

† For applications in practice see (2.118).

wall. We can define the *surface density of states* N_s as

$$N_s = N_t - \tfrac{1}{2}\sum_m N_m. \tag{2.58}$$

This applies to either (2.57) or (2.56) (recall (2.52)). It includes all surface effects and must not be confused with the *density of surface states* – i.e. new matching solutions – which is given by the last term of (2.56) or (2.57). We shall see several practical applications of these formulae.

2.4. The meaning of the formal theory for electronic states

Given a Hamiltonian operator H one usually defines the resolvent operator $\hat{G}$ by

$$\hat{G}(\omega) = (\omega - H)^{-1}. \tag{2.59}$$

Here ω is the complex energy variable $E + i\eta$ and the physical results are to be understood in the limit $\eta \to 0$. The representation – in $\boldsymbol{r}$ or $\boldsymbol{k}$-space – of this operator yields the Green function for the theory of electronic states. If this definition is used, then some of the signs given so far must be changed (remember (2.50)). The first change occurs in the Schrödinger-like equation satisfied by $\hat{G}(\boldsymbol{r}, \boldsymbol{r}')$. It is easy to see, by integrating this, for $z' = 0$ fixed, between $z = -\eta$ and $z = +\eta$, and letting $\eta \to 0$, that

$$'G^{(+)} - 'G^{(-)} = -\mathscr{I}, \tag{2.60}$$

and this is where the sign is changed with respect to (2.20). This leads to

$$G_s^{-1} = 'G_2^{(-)}G_2^{-1} - 'G_1^{(+)}G_1^{-1}, \tag{2.61}$$

instead of (2.31). The secular equations (2.28), or (2.43) for the case of symmetry, are the same – the sign is irrelevant here – and they must be understood in the following sense. The Green functions and their normal derivatives can be evaluated in any chosen representation; suppose we choose the momentum representation. Then $\hat{G}$ is a function of $\boldsymbol{k} = (\boldsymbol{\kappa}, q)$ and of ω, which for the secular equation is equal to the real eigenvalue we are seeking. The surface projections are evaluated from the formulae

$$\left.\begin{aligned} G(\boldsymbol{\kappa}, \omega) = G &= \frac{1}{2\pi}\int_{-\infty}^{\infty} e^{iq\eta}\hat{G}(\boldsymbol{\kappa}, \omega; q)\, dq; \\ 'G^{(\pm)} &= \frac{1}{2\pi}\int e^{\mp iq\eta} iq\hat{G}(\boldsymbol{\kappa}, q; \omega)\, dq. \end{aligned}\right\} \tag{2.62}$$

Then $k_3 = q$ is integrated away and we are left with functions of $(\boldsymbol{\kappa}, \omega)$. Now, in the momentum representation, if the crystal symmetry in the two-dimensional space $\boldsymbol{\rho}$ is conserved, then we must use a crystal momentum, defined up to two-dimensional reciprocal lattice vectors γ_n.

Thus the precise meaning of *det* in this case is that the secular equation is

$$\det\left[\langle \boldsymbol{\kappa}+\boldsymbol{\gamma}_n|G_s^{-1}|\boldsymbol{\kappa}+\boldsymbol{\gamma}_{n'}\rangle\right]=0. \tag{2.63}$$

The rows and columns are labelled by the reciprocal lattice vectors coupled by the symmetry in the surface plane, as in the momentum representation of a three-dimensional crystal band structure in which the surface state eigenvalues are functions of $\boldsymbol{\kappa}$.

For the density of states we observe that it follows from (2.59) that now

$$\hat{G}^2=-\mathrm{d}\hat{G}/\mathrm{d}\omega, \tag{2.64}$$

instead of (2.50). Then $N_t(E)$ is given by

$$N_t=N^{\infty}-\frac{1}{\pi}\frac{\mathrm{d}}{\mathrm{d}E}\arg\det G_s^{-1}. \tag{2.65}$$

We stress that here the determinant is a function of $E+\mathrm{i}\eta$ in the limit $\eta\to 0$. For the case of symmetry†:

$$N_t=\frac{1}{2}\sum_m\left[N_m-\frac{1}{\pi}\frac{\mathrm{d}}{\mathrm{d}E}\arg\det G_m\right]-\frac{1}{\pi}\frac{\mathrm{d}}{\mathrm{d}E}\arg\det G_s^{-1}. \tag{2.66}$$

The formula for N is the same as (2.55) with scalar Green functions, because this follows from the general forms (2.39) and (2.40) which, as stressed before, represent a general theorem which does not depend on the structure of the equation of motion.

In our discussion so far we have stressed that the Green function yields the surface state eigenvalues and the density of states of the entire system, including the *surface density of states.* This term denotes all effects due to the surface, including the density of states due to new – matching – solutions, which can be of different types (chapter 4) and the distortion of the bulk continuum. The part of the surface density of states specifically due to new matching solutions will be denoted as *density of surface states.* Other spectral functions are directly related to this. For example, the *local density of states* $N_t(\boldsymbol{r}, E)$ is obtained by applying to $\hat{G}_s$ the general formula

$$N_t(\boldsymbol{r}, E)=-(1/\pi)\,\mathrm{Im}\,\langle\boldsymbol{r}|\hat{G}_s|\boldsymbol{r}\rangle, \tag{2.67}$$

for which we only need to know the diagonal matrix elements. At each energy level this formula gives a local description of the density of states which shows the role of the hard wall and matching terms much as we have seen in the total integrated density of states. In fact it is from the integral of (2.67) over $\boldsymbol{r}$ that we obtain $N_t(E)$, while the integral over E

† For applications in practice see (2.118).

yields the particle density as a function of $\boldsymbol{r}$. For all these problems the starting points are the form (2.39) of $\langle \boldsymbol{r}|\hat{\boldsymbol{G}}_s|\boldsymbol{r}\rangle$ and the formula (2.61) for G_s. But the Green function still contains more physical information. An interesting example concerns the scattering amplitudes, which can be discussed in much the same way.

Imagine, for example, a crystal on the left (domain 1) and a vacuum on the right (domain 2) and consider an incoming eigenstate $|\psi_{\boldsymbol{k}}\rangle$ of M_2 – e.g. a plane-wave state in this case – incident on the surface. By the same line of argument the wavefunction in domain 2 will consist of the incident and the reflected waves, and will therefore be of the form

$$\phi(\boldsymbol{r}_2) = \langle 2|\phi\rangle = \langle 2|\psi_{\boldsymbol{k}}\rangle + \langle 2|\hat{G}_2 \mathscr{I} F_2\rangle, \tag{2.68}$$

while the transmitted wavefunction will have the form

$$\phi(\boldsymbol{r}_1) = \langle 1|\phi\rangle = \langle 1|\hat{G}_1 \mathscr{I} F_1\rangle. \tag{2.69}$$

Our unknowns are $\mathscr{I}F_1\rangle$ and $\mathscr{I}F_2\rangle$, which we find by matching the wavefunction and its normal derivative. We then have two matching equations of the form

$$\mathscr{I}\psi_{\boldsymbol{k}} + G_2 F_2 = G_1 F_1; \tag{2.70a}$$

$$\mathscr{I}\psi'_{\boldsymbol{k}} + {}'G_2^{(-)} F_2 = {}'G_1^{(+)} F_1. \tag{2.70b}$$

We can eliminate F_1 for instance by multiplying (2.70a) from the left by ${}'G_1^{(+)} G_1^{-1}$ and subtracting. This yields

$$({}'G_2^{(-)} G_2^{-1} - {}'G_1^{(+)} G_1^{-1}) G_2 F_2 + \mathscr{I}\psi'_{\boldsymbol{k}} - {}'G_1^{(+)} G_1^{-1} \psi_{\boldsymbol{k}} = 0.$$

Now we use (2.60), whence

$$G_2 F_2 = G_s({}'G_1^{(+)} G_1^{-1} \cdot \psi_{\boldsymbol{k}} - \mathscr{I}\psi'_{\boldsymbol{k}}). \tag{2.71}$$

We want to use this in (2.68) to form the complete wavefunction ϕ and we would like to get rid of $\mathscr{I}\psi'_{\boldsymbol{k}}$ in (2.71). This can be accomplished by means of the following formal device. Imagine that we reconstitute the entire medium M_2 by matching it onto itself. The same formal analysis would hold throughout with $\hat{G}_1 = \hat{G}_2$ everywhere. But then the reflected term would be identically zero. Hence

$${}'G_2^{(+)} G_2^{-1} \psi_{\boldsymbol{k}} = \mathscr{I}\psi'_{\boldsymbol{k}}, \tag{2.72}$$

and by using this in (2.71),

$$G_2 F_2 = G_s({}'G_1^{(+)} G_1^{-1} - {}'G_2^{(+)} G_2^{-1}) \psi_{\boldsymbol{k}},$$

which, by applying (2.60) to ${}'G_2$, becomes

$$G_2 F_2 = G_s({}'G_1^{(+)} G_1^{-1} - {}'G_2^{(-)} G_2^{-1} + G_2^{-1}) \psi_{\boldsymbol{k}},$$

and, by using (2.61) again,

$$G_2F_2=(G_s-G_2)G_2^{-1}\psi_{\mathbf{k}}. \tag{2.73}$$

We have found the desired scattered term, but we must now specify the correct labels. The scattered wavefunction ϕ has three quantum numbers, namely, the energy E and the two components of $\boldsymbol{\kappa}$. With the dependence on E understood we write, for scattered wavefunction at $z=+0$:

$$\phi_{\boldsymbol{\kappa}}(+0)=\psi_{\boldsymbol{\kappa}}(+0)+(G_s-G_2)G_2^{-1}\psi_{\boldsymbol{\kappa}}(+0).$$

This explains the physical meaning of the form of the 'diagonal' matrix element (2.39). The probability amplitude for scattering from an incoming state $(\boldsymbol{\kappa}, E)$ to an outgoing state $(\boldsymbol{\kappa}', E)$ is

$$\langle\boldsymbol{\kappa}'|\mathscr{R}|\boldsymbol{\kappa}\rangle=\langle\boldsymbol{\kappa}'|(G_s-G_2)G_2^{-1}|\boldsymbol{\kappa}\rangle, \tag{2.74}$$

where $|\boldsymbol{\kappa}\rangle$ means a plane-wave state with energy E and momentum $\boldsymbol{\kappa}$ parallel to the surface. This gives the general matrix element of the *reflection operator* $\mathscr{R}$. Of course, the two-dimensional crystal symmetry of the surface will give non-vanishing matrix elements only when $\boldsymbol{\kappa}'$ differs from $\boldsymbol{\kappa}$ by a two-dimensional reciprocal lattice vector, as in scattering problems in three dimensions.

The transmission probability can be studied in much the same manner. We could be interested, for example, in studying a *Bloch state* $|\psi_{\mathbf{k}}\rangle$ coming from the left (domain 1). The form of the scattered wavefunction is then

$$\langle 2|\phi\rangle=\langle 2|\hat{G}_2\mathscr{I}\hat{F}_2\rangle \tag{2.75}$$

on the right, and

$$\langle 1|\phi\rangle=\langle 1|\psi_{\mathbf{k}}\rangle+\langle 1|\hat{G}_1\mathscr{I}\hat{F}_1\rangle \tag{2.76}$$

on the left. Again, matching the wavefunction and its normal derivative we have

$$\mathscr{I}\psi_{\mathbf{k}}+G_1\hat{F}_1=G_2\hat{F}_2$$

$$\mathscr{I}\psi'_{\mathbf{k}}+{}'G_1^{(+)}\hat{F}_1={}'G_2^{(-)}\hat{F}_2.$$

F_1 can be eliminated by following an identical procedure and using (2.61) again. This yields

$$G_s^{-1}G_1\hat{F}_1=\mathscr{I}\psi'_{\mathbf{k}}-{}'G_2^{(-)}G_2^{-1}\psi_{\mathbf{k}},$$

where $\mathscr{I}\psi'_{\mathbf{k}}$ can be eliminated by the same argument as before; i.e. reconstituting M_1 by matching it onto itself so that $G_2=G_1$, we have

$${}'G_1^{(-)}G_1^{-1}\psi_{\mathbf{k}}=\mathscr{I}\psi'_{\mathbf{k}}, \tag{2.77}$$

whence

$$G_2\hat{F}_2 = G_s G_1^{-1}\psi_{\mathbf{k}}.$$

Thus the *transmission operator* is

$$\mathcal{T} = G_s G_1^{-1}, \tag{2.78}$$

which explains the physical meaning of the form of the cross matrix element (2.40).

This analysis yields, incidentally, another interesting formula. The same formal trick we used to derive (2.72) and (2.77) could be applied to reconstitute hypothetically the surface system itself. But there is a subtlety concerning signs which must be analysed with care. It is intuitively obvious that the factors on the l.h.s. of (2.72) and (2.77) have the nature of logarithmic derivative operators. Take (2.72) for instance; notice that the (+) superscript comes from $'G_1^{(+)}$ in (2.71), which means that the derivative is calculated from the left (side 1). If we want to use (2.72) in order to obtain the correct formula for the logarithmic derivative from the right (side 2) we must change the sign. Thus the correct interpretation of (2.72) is that the logarithmic derivative operator from the right is

$$\mathcal{L}^{(+)} = {}'G_2^{(-)} G_2^{-1}. \tag{2.79}$$

Similarly the logarithmic derivative operator from the left is

$$\mathcal{L}^{(-)} = {}'G_1^{(+)} G_1^{-1}. \tag{2.80}$$

Then G_s^{-1} in (2.61) acquires a very suggestive form, namely,

$$G_s^{-1} = \mathcal{L}^{(+)} - \mathcal{L}^{(-)}, \tag{2.81}$$

and the secular equation says, literally, that the logarithmic derivative is continuous, i.e.

$$\det\left[\mathcal{L}^{(+)} - \mathcal{L}^{(-)}\right] = 0, \tag{2.82}$$

as of course it must be, since the whole surface Green-function matching argument is simply a formal device for matching the wavefunction and its normal derivative. We can now answer a question which has remained untouched throughout the formal development. We know $\hat{G}_s$ is a resolvent, but what is its surface projection G_s? It would be erroneous to think of G_s as anything like a two-dimensional true Green function – i.e. a proper resolvent. The meaning of G_s is very neatly expressed by (2.81).

2.5. The shape of the matching surface

So far we have been using a plane matching surface whenever the shape of this surface has been referred to explicitly. However, the formal theory is quite general and applicable to matching surfaces of arbitrary shape, differentiation with respect to z being replaced by the normal derivative, with appropriate geometry, sign conventions, etc. For example, we might be interested in studying an impurity. We might then define a sphere surrounding the impurity as the matching surface, with the convention that 1 means the inside and 2 means the outside of the sphere. The positive normal is then the outer normal, the normal derivative is just the radial derivative, and sign conventions follow by consistency.

The practical difficulty when choosing a matching surface or arbitrary shape is to choose an adequate representation in which the secular determinant takes a tractable form. Of course this is always determined by the symmetry of the problem. For a plane surface we have seen that the two-dimensional crystal symmetry suggests the use of crystal momenta, which gives the secular determinant the form of (2.63). With spherical symmetry the obvious thing to do is to use the angular momentum representation. As an example of application of the formalism we shall discuss a lattice of *muffin-tins*. The principle that the shape of the matching surface is arbitrary is here demonstrated in an extreme form. Not only do we have a non-planar – in this case a spherical – matching surface, but we have an infinite crystalline repetition of spheres, and proper matching must be ensured at everyone of them. Our matching surface is multiply connected. Moreover, we must combine two representations: one with angular momentum quantum numbers on *one* sphere, and another one with crystal momenta to describe the periodic lattice of muffin-tins. Let us see how to go about this.

The most intuitive way to see how to approach the problem consists in recalling the meaning of a Green function $\hat{G}(\boldsymbol{r}, \boldsymbol{r}')$ as the response at $\boldsymbol{r}$ to a unit input at $\boldsymbol{r}'$. Let this Green function be $\hat{G}_0(\boldsymbol{r}, \boldsymbol{r}')$, the vacuum or free electron Green function, to start with. In order to generate Bloch waves, i.e. linear combinations of plane waves having the Bloch property, it proves useful to introduce the *structural Greenian*

$$\hat{G}_{\boldsymbol{k}}(\omega)=\frac{(2\pi)^3}{\Omega}\sum_n \frac{|\boldsymbol{k}+\boldsymbol{g}_n\rangle\langle \boldsymbol{k}+\boldsymbol{g}_n|}{\omega-|\boldsymbol{k}+\boldsymbol{g}_n|^2}. \tag{2.83}$$

Here Ω is the volume of the unit cell. The state vector $|\boldsymbol{k}+\boldsymbol{g}_n\rangle$ is a plane-wave state with wavevector $\boldsymbol{k}+\boldsymbol{g}_n$ and in this argument we take $\hbar^2/2m=1$ to simplify the notation. This Greenian clearly has the Bloch property we desire since the matrix element is

$$\hat{G}_{\boldsymbol{k}}(\boldsymbol{r}, \boldsymbol{r}')=\langle \boldsymbol{r}|\hat{G}_{\boldsymbol{k}}(\omega)|\boldsymbol{r}'\rangle=\frac{1}{\Omega}\sum_n \frac{\exp[\mathrm{i}(\boldsymbol{k}+\boldsymbol{g}_n)\cdot(\boldsymbol{r}-\boldsymbol{r}')]}{\omega-|\boldsymbol{k}+\boldsymbol{g}_n|^2}, \tag{2.84}$$

and it is therefore obvious that upon a translation by a lattice vector $\boldsymbol{R}_n$, it transforms as

$$\hat{G}_{\boldsymbol{k}}(\boldsymbol{r}+\boldsymbol{R}_n, \boldsymbol{r}') = \mathrm{e}^{\mathrm{i}\boldsymbol{k}\cdot\boldsymbol{R}_n}\hat{G}_{\boldsymbol{k}}(\boldsymbol{r}, \boldsymbol{r}'). \tag{2.85}$$

This is thus the appropriate structural Green function to generate a Bloch wave with wavevector $\boldsymbol{k}$; which is why it is labelled $\boldsymbol{k}$. Now, by using standard closure relations applicable to lattice sums, (2.84) is easily rewritten as

$$\hat{G}_{\boldsymbol{k}}(\boldsymbol{r}, \boldsymbol{r}') = \sum_n \hat{G}_0(\boldsymbol{r}, \boldsymbol{r}'+\boldsymbol{R}_n)\,\mathrm{e}^{\mathrm{i}\boldsymbol{k}\cdot\boldsymbol{R}_n}, \tag{2.86}$$

which is most intuitive. $\hat{G}_{\boldsymbol{k}}(\boldsymbol{r}, \boldsymbol{r}')$ gives, at $\boldsymbol{r}$, the response to a periodic array of stimuli $\mathrm{e}^{\mathrm{i}\boldsymbol{k}\cdot\boldsymbol{R}_n}$ located at all points $\boldsymbol{r}'+\boldsymbol{R}_n$, including $\boldsymbol{R}_n = 0$. That is to say, suppose we put our input at a generic point $\boldsymbol{r}'$: the use of the structural Green function $\hat{G}_{\boldsymbol{k}}$ automatically creates a lattice of points $\boldsymbol{r}'+\boldsymbol{R}_n$, puts on each one of them a unit input with appropriate phase factor, and collects at $\boldsymbol{r}$ the response to all these stimuli (fig. 19*a*). This is just the device we need to turn the multiply-periodic matching problem around for the lattice of muffin-tins (fig. 19*b*). Instead of defining an input at $\boldsymbol{r}'$ and studying the response at all periodically repeated points on all matching spheres, the crystal periodicity is introduced by studying the matching on just *one* sphere, but using the structural Green function (2.86), which neatly separates out the structural aspects of the problem.

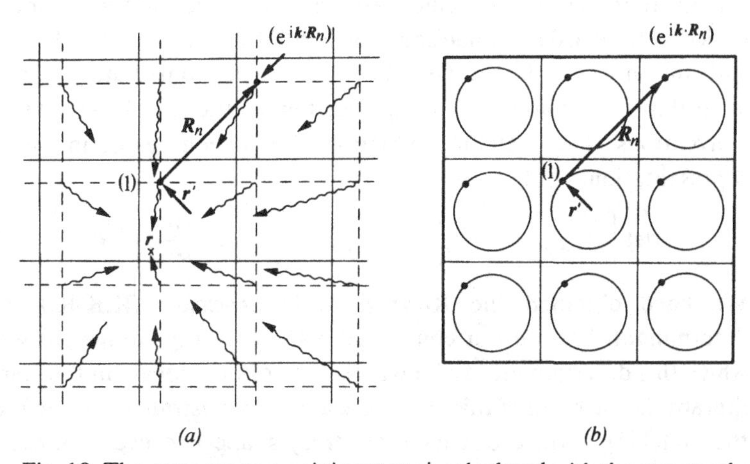

Fig. 19. The response to a unit input at $\boldsymbol{r}'$, calculated with the structural Green function $\hat{G}_{\boldsymbol{k}}$ is equivalent to the response to a periodic array of stimuli with phase factors $\mathrm{e}^{\mathrm{i}\boldsymbol{k}\cdot\boldsymbol{R}_n}$, calculated with the free-electron Green function. (*a*) Matching on one *muffin-tin* sphere and using $\hat{G}_{\boldsymbol{k}}$ is equivalent to multiply periodic matching (*b*) on a crystalline array of *muffin-tin* spheres.

So we now concentrate on one matching sphere, using the angular momentum representation. One quantum number is the energy $E = k^2$ of the matching solution we want. The other two are (l, m), which will sometimes be indicated by L. We know from (2.80) that, taking the Lth matrix element,

$$\left[{}'G_1^{(+)}G_1^{-1}\right]_{L,L'} = (R_l'/R_l)\,\delta_{ll'}\delta_{mm'}. \qquad (2.87)$$

Here R_l is the radial wavefunction which must be found by solving the appropriate Schrödinger equation inside the muffin-tin sphere. The prime on functions – not on the labels l, m, L – indicates normal derivative, and the argument of the radial wavefunctions and Bessel functions that will presently appear, as well as of their derivatives, is (kr), evaluated at $r = r_0$, the muffin-tin radius, as in all well-known textbook formulae. This is the term which contains the scattering properties of the potential inside the muffin-tin sphere.

The structural properties are contained in the representation of (2.79), because now $\hat{G}_2$ is $\hat{G}_k$ of (2.86). Thus all we have to do is to use the standard angular momentum representation of the plane waves and of the free electron Green function. As is well known this yields

$$(G_k)_{L,L'} = A_{L,L'}j_lj_{l'} + k\delta_{ll'}\delta_{mm'}j_ln_l, \qquad (2.88)$$

which involves the Bessel functions and the *structure constants* $A_{L,L'}$ – actually functions of kr_0 – which are evaluated once and for all for a given lattice by standard formulae common in solid state textbooks. With (2.88) and the normal derivative ${}'G^{(-)}_{k,LL'}$ we can immediately form the logarithmic derivative from outside, taking the usual care with sign conventions. The matching equation (2.82) then reads, in the (l, m) representation,

$$\det\left[A_{L,L'}j_l\left(j_{l'} - j_{l'}'\frac{R_{l'}}{R_{l'}'}\right) + \mathrm{k}\delta_{ll'}\delta_{mm'}\left(n_{l'} - n_{l'}'\frac{R_{l'}}{R_{l'}'}\right)\right] = 0 \qquad (2.89)$$

We have obtained the *Korringa–Kohn–Rostoker* (K.K.R.) secular determinant. We are not concerned with the many equivalent ways in which this determinant can be written and often appears in the standard literature. The point of this exercise was to demonstrate the principle that the matching surface can have arbitrary shape and even connectivity. Incidentally, this reproduces the K.K.R. method in a rather concise and intuitive way, while illustrating the meaning of the logarithmic derivative operators (2.79) and (2.80).

Another example of physical interest, in which the matching surface is not just a plane, is the *two-surface* problem: as with, for example, a thin film or a sandwich of some material between two different media. This can be tackled in different ways. Let the three media be denoted 1, 2, 3. One could for instance, construct the Green function $\hat{G}_{12}$ of the system formed by matching 1 and 2 and then use $\hat{G}_{12}$ and $\hat{G}_3$ to form the complete Green function $\hat{G}_s = \hat{G}_{123}$. One could also go back to (2.8) and (2.9) and reformulate the problem using the same approach but for a more complicated situation. We would have to write down the general form of $\langle 1|\hat{G}_s|3'\rangle$, $\langle 2|\hat{G}_s|3'\rangle$, $\langle 3|\hat{G}_s|3'\rangle$ and $\langle 2|\hat{G}_s|2'\rangle$ introducing suitable unknowns to represent all possible reflections and transmissions involved. There is no essential difficulty in doing all this if necessary, but it would amount to a mere exercise with a more complicated algebra without emphasising any new or interesting feature. We shall instead discuss the problem in another way which stresses the use of surfaces with unconventional shapes, more in the spirit in which we have discussed the K.K.R. method.

Our problem is defined as follows (fig. 20). We have a material labelled l on the left, a material labelled f in the middle – the film – and a material labelled r on the right. We *define* the matching surface to consist of the two faces of the film, with positive outward normal as shown in fig. 20,

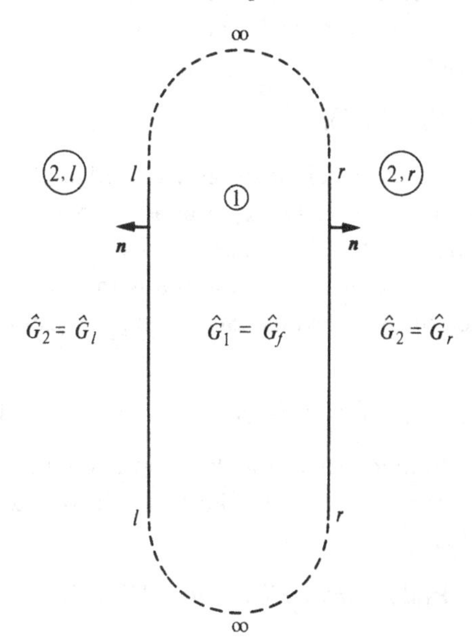

Fig. 20. Geometry for the two-surface problem. The material inside domain 1 is bounded by *one* single surface, consisting of the *two* planes, oriented according to the outer normals as shown, joined at infinity. Domain 2 then consists of two disconnected domains, l and r, containing different materials.

joined by the lines at infinity. Notice that we are defining *one* matching surface, which obviously is not a plane. Domain 1 is now the film and domain 2 consists of *two disconnected* subdomains, l and r. It proves convenient to reformulate the problem in a way which is entirely equivalent to the analysis of §§ 2.1, 2.2 and which, incidentally, stresses the equivalence of Green-function matching and wavefunction matching. To this effect let us forget for a moment about the film problem and consider an identity which can be easily checked. Let $\psi(z)$ be a function which satisfies appropriate conditions – actually satisfied by the square integrable wavefunctons acceptable as physical solutions in quantum mechanics. Let $\theta_{\pm}$ be step function $\theta(\pm z)$. Using symbolic calculus one can prove that

$$\frac{\mathrm{d}^2}{\mathrm{d}z^2}(\theta_-\psi) = \theta_-\frac{\mathrm{d}^2\psi}{\mathrm{d}z^2} - \delta(z)\phi'(0) - \delta'_-(z)\phi(0). \tag{2.90}$$

The minus subscript to δ' means that the function to be eventually multiplied by this term and integrated must be differentiated from the left ($z<0$). A similar identity can be proved which involves θ_+. Now consider a three-dimensional surface problem. Let θ_- be the P_1 projector and θ_+ the P_2 projector and let $\mathrm{d}/\mathrm{d}z$ be the normal derivative operator ∇_n. Then (2.90) becomes, more generally,

$$\nabla_n^2 P_m - P_m \nabla_n^2 = \pm(\mathscr{I}\nabla_n + \nabla_n\mathscr{I}), \tag{2.91}$$

i.e. the + sign holds when $m=1$. The technicalities involved in (2.90) and (2.91) are not too important. Let us take (2.91) as an identity given from symbolic calculus. This can be used to relate in an interesting manner the values of ψ and ψ' at the matching surface, in the following way.

Consider an actual surface problem, whose Schrödinger equation can be identically expressed as

$$H_s|\psi\rangle = (H_0 + P_1 V_1 + P_2 V_2)|\psi\rangle = E|\psi\rangle. \tag{2.92}$$

Here ψ is the matching wavefunction we are seeking. The point to notice is that H_0 contains the operator ∇_n^2. Now take the P_1 projection and use (2.91) with $m=1$. This yields

$$P_1 H_s|\psi\rangle = (H_0 + V_1)P_1|\psi\rangle - (\mathscr{I}\nabla_n + \nabla_n\mathscr{I})|\psi\rangle = EP_1|\psi\rangle,$$

whence

$$P_1|\psi\rangle = -\hat{G}_1(\mathscr{I}\nabla + \nabla\mathscr{I})|\psi\rangle, \tag{2.93}$$

which relates the wavefunction everywhere in domain 1 to $\psi(-0)$ and $\psi'(-0)$. In fact, since ψ is the actual wavefunction, both ψ and ψ' must be continuous at the matching surface and it is not necessary to specify signs. It will suffice to write ψ_s and ψ_s'. For the P_2 we can obtain in the same way

$$P_2|\psi\rangle = \hat{G}_2(\mathscr{I}\nabla + \nabla\mathscr{I}|\psi\rangle. \tag{2.94}$$

Now let $\boldsymbol{r}_1$ in (2.93) and $\boldsymbol{r}_2$ in (2.94) tend to the matching surface; this yields

$$\psi_s = -G_1\psi_s' + {}^{(-)}G_1'\psi_s, \tag{2.95a}$$

$$\psi_s = G_2\psi_s' - {}^{(+)}G_2'\psi_s. \tag{2.95b}$$

The definition of ${}^{(\pm)}G'$ is analogous to that of ${}'G^{\pm}$ (2.7). All that matters is the sign of $(z'-z)$, where z' is always the second argument of the Green function. These two equations can also be used to find the matching secular equation. By rearranging terms and using the standard routine this yields

$$\det\left[G_1^{-1}(\mathscr{I} - {}^{(-)}G_1') + G_2^{-1}(\mathscr{I} + {}^{(+)}G_2')\right] = 0. \tag{2.96}$$

This looks rather different from the secular equation which results from equating det G_s^{-1} to zero by taking G_s^{-1} from (2.61), but in fact is the same. It suffices to use the identity

$${}'G^{(\pm)}G^{-1} = G^{-1}\,{}^{(\pm)}G', \tag{2.97}$$

with which ${}^{(-)}G_1'$ and ${}^{(+)}G_2'$ can be expressed in terms of ${}'G_1^{(-)}$ and ${}'G_2^{(+)}$. By then using (2.60) it is immediate to see that both secular equations are identical. In fact the formal argument leading to (2.95*a*, *b*) can be omitted altogether if one is interested only in the surface projections of ψ and ψ'. Indeed, it is easy to see, by using (2.97) and (2.60), that (2.95*a*) is identical to (2.80), and (2.95*b*) is identical to (2.79). Thus, we have two options, namely, either to start from (2.95*a*, *b*) and end up with (2.96), or else to start from (2.79, 2.80) and end up with (2.61). Similarly, the argument presently to be developed for the film problem can be replaced by a similar argument starting from (2.79), (2.80), and this leads to a different – but of course equivalent – form of the secular determinant for the film. The formal argument presented here is more complete, as it gives the wavefunction at every point in the volume in terms of ψ_s and ψ_s', thus connecting with the standard mathematical formulation for boundary value problems.

Let us now go back to the film problem as defined above (fig. 20). We shall derive the corresponding secular equation using (2.95*a*, *b*), where ψ_s and ψ_s' mean the corresponding values at the surface. All we have to do is to interpret this scheme as adapted to our problem. Take for instance

(2.95a). This can be written down explicitly for a generic point on either face of the surface. For this it is convenient to define the face projectors $\mathcal{I}_l$ and $\mathcal{I}_r$ such that the entire two-dimensional surface unit or projector is $\mathcal{I} = \mathcal{I}_r + \mathcal{I}_l$. Then a term like $G_1\psi'_s$, i.e. $G_f\psi'_s$, must be interpreted in the following way:

$$G_f\psi'_s = G_f\mathcal{I}\psi'_s = G_f(J_l + J_r)\psi'_s = G_f|l\rangle\psi'_l + G_f|r\rangle\psi'_r.$$

Now take (2.95a) and specify it for a generic point on the left face. This yields

$$\psi_l = -\langle l|G_f|l\rangle\psi'_l - \langle l|G_f|r\rangle\psi'_r + \langle l|^{(-)}G'_f|l\rangle\psi_l + \langle l|^{(-)}G'_f|r\rangle\psi_r. \tag{2.98}$$

Similarly, for a point on the right face,

$$\psi_r = -\langle r|G_f|l\rangle\psi'_l - \langle r|G_f|r\rangle\psi'_r + \langle r|^{(-)}G'_f|l\rangle\psi_l + \langle r|^{(-)}G'_f|r\rangle\psi_r. \tag{2.99}$$

Next we use (2.95b) starting, say, from the subdomain (2, l). This is disconnected from (2, r), so that G_l has only the matrix elements $\langle l|G_l|l\rangle$. Then:

$$\psi_l = \langle l|G_l|l\rangle\psi'_l - \langle l|^{(+)}G'_l|l\rangle\psi_l. \tag{2.100}$$

Similarly, starting from the subdomain (2, r):

$$\psi_r = \langle r|G_r|r\rangle\psi'_r - \langle r|^{(+)}G'_r|r\rangle\psi_r. \tag{2.101}$$

We have a system of four linear homogeneous equations for the four unknowns ψ_l, ψ_r, ψ'_l, ψ'_r. The secular determinant is

$$\begin{vmatrix} \langle l|\mathcal{I} + {}^{(+)}G'_l|l\rangle & \langle l|G_l|l\rangle & 0 & 0 \\ \langle l|\mathcal{I} - {}^{(-)}G'_f|l\rangle & \langle l|G_f|l\rangle & \langle l|\mathcal{I} - {}^{(-)}G'_f|r\rangle & \langle l|G_f|r\rangle \\ \langle r|\mathcal{I} - {}^{(-)}G'_f|l\rangle & \langle r|G_f|1\rangle & \langle r|\mathcal{I} - {}^{(-)}G'_f|r\rangle & \langle r|G_f|r\rangle \\ 0 & 0 & \langle r|\mathcal{I} + {}^{(+)}G'_r|r\rangle & \langle r|G_r|r\rangle \end{vmatrix} \tag{2.102}$$

Notice the presence of the cross terms between left and right faces. As the thickness of the film increases, these cross matrix elements decrease in the manner typical of Green functions, and this decouples the two faces. In this limit (2.102) factorises out into two secular determinants. It is easy to see that this reproduces the one-surface secular equation (2.96) for each face independently.

2.6. Electromagnetic surface Green-function matching

Electromagnetic matching was discussed at length in chapter 1. We shall only discuss it very briefly here, just to demonstrate how this can be encompassed within the method of surface Green-function matching. Let the two media have e.m. Green functions as in (1.21). The corresponding linear form $\hat{\boldsymbol{A}}$ has been discussed in § 2.1. We recall that in this case only the (x, y) components are involved in the matching. Now, if the dielectric functions depend on the modulus of $\boldsymbol{k}$ one can prove that (i) the surface projection $\boldsymbol{G}$ is diagonal, (ii) $\boldsymbol{A}^{(\pm)}$ is also diagonal, and (iii) $\boldsymbol{A}^{(+)} = -\boldsymbol{A}^{(-)}$. Then, by (2.20),

$$\boldsymbol{A}^{(+)} = -\boldsymbol{A}^{(-)} = \tfrac{1}{2}\mathcal{I}.$$

In this case the equation (2.31) simplifies to

$$\boldsymbol{G}_{\mathrm{s}}^{-1} = \tfrac{1}{2}(\boldsymbol{G}_1^{-1} + \boldsymbol{G}_2^{-1}). \tag{2.103}$$

Therefore, the secular equation for the surface modes is

$$\det[\boldsymbol{G}_1 + \boldsymbol{G}_2] = 0. \tag{2.104}$$

But we have said that $\boldsymbol{G}_1$ and $\boldsymbol{G}_2$ are diagonal, so (2.104) is factorised into two equations, namely,

$$G_{1,xx} + G_{2,xx} = 0, \tag{2.105}$$

and

$$G_{1,yy} + G_{2,yy} = 0. \tag{2.106}$$

What does this mean? The answer is provided by the physical interpretation of the formalism of §§ 2.1, 2.2 applied to this case. The fictitious surface stimuli of (2.27) now have the nature of surface currents $\boldsymbol{J}_1^{\mathrm{s}}$ and $\boldsymbol{J}_2^{\mathrm{s}}$. The media on both sides of the surface respond to this with their Green functions $\hat{\boldsymbol{G}}_1$, $\hat{\boldsymbol{G}}_2$. In each extended medium the resulting electric field, $\hat{\boldsymbol{G}}_m \cdot \boldsymbol{J}_m^{\mathrm{s}}$ is symmetric about the surface plane and we are matching the field according to

$$\boldsymbol{G}_1 \cdot \boldsymbol{J}_1^{\mathrm{s}} = \boldsymbol{G}_2 \cdot \boldsymbol{J}_2^{\mathrm{s}}.$$

But the magnetic fields in the extended media are antisymmetric, i.e. curl $\boldsymbol{E}_m$ is antisymmetric. It is easy to see that this implies $\boldsymbol{J}_1^{\mathrm{s}} = -\boldsymbol{J}_2^{\mathrm{s}}$, whence (2.104). So far this would be a mere repetition of the argument leading to (2.104), but now we can go to a step farther and interpret the factorisation of (2.104). Suppose we choose $\boldsymbol{J}^{\mathrm{s}}$ in the x-direction. Then the extended electric fields have x- and z-components, given by $G_{xx}J_x^{\mathrm{s}}$ and $G_{zx}J_x^{\mathrm{s}}$, which does not appear in the matching equations. Thus (2.105) is the s.m.d.r. for P-mode geometry. If, on the other hand, we choose J^{s} to have only y-component, then the extended electric fields have only y-component, $G_{yy}J_y^{\mathrm{s}}$. Thus (2.106) is the s.m.d.r. for S-mode geometry. The analysis in terms of surface Green-function matching yields in a concise way, in compact tensor form, a secular equation which contains both polarisations.

It is important to notice that $\hat{G}$ (1.21) is precisely the response function to an electric – symmetric – stimulus. Thus *the analysis we have just sketched applies only to the specular surface model.* It is obvious how to describe the antispecular surface model. The formal analysis is identical, but this time we say that the media have e.m. Green functions $\hat{\boldsymbol{K}}_1$ and $\hat{\boldsymbol{K}}_2$ as in (1.108). This implies automatically that we are introducing a magnetic – antisymmetric – surface current, as discussed in § 1.4. Then, with an argument identical to the one leading to (2.24) we find that the appropriate linear differential form for this problem is

$$\hat{A}_{ij} = -\mathrm{i}(c^2/2\pi\omega\varepsilon_\mathrm{T})(-'\hat{K}_{ij} + \mathrm{i}k_i K_{3j}) \quad (i, j = x, y), \tag{2.107}$$

and this has the same properties as (2.24). The secular equation is then

$$\det[\boldsymbol{K}_1 + \boldsymbol{K}_2] = 0,$$

which is again factorised in two equations. By a similar analysis it is easy to see that

$$K_{1,yy} + K_{2,yy} = 0$$

is now the s.m.d.r. for P-mode geometry, while

$$\mathrm{K}_{1,xx} + \mathrm{K}_{2,xx} = 0$$

is the s.m.d.r. for S-mode geometry.

The more general case, in which the surface is neither totally specular nor totally antispecular is rather more involved and has been discussed at length. The above exercise was simply meant to demonstrate the relationship with the formal method developed in this chapter. The point is that while the arguments of chapter 1 lead only to the evaluation of surface modes, or reflectivity, or surface impedance – three equivalent problems, as we saw in § 1.6 – here we have a more complete analysis which shows how to construct the entire three-dimensional Green function of the surface system. This can provide a basis for studying, for example, problems associated with the surface density of modes and related sum rule problems.

2.7. Elastic surface waves in solids

We now turn to a different type of problem, namely, elastic waves at the interface between two media, of which one in particular could be the vacuum. In order to demonstrate the use of this formalism in a practical application involving tensor Green functions we shall take a simple model, namely, an isotropic elastic solid with a surface, and we shall study it thoroughly, making systematic use of all the formal machinery built up in §§ 2.1, 2.2, and 2.3. The Green function in this case satisfies the equation of motion (2.14). Fourier transforming from $(\boldsymbol{r}, t)$ to $(\boldsymbol{k}, \omega)$ it is easy to see that $\hat{\boldsymbol{G}}$ must be a combination of the unit tensor $\boldsymbol{I}$ and the dyad $\boldsymbol{kk}$; this yields

$$\hat{G}(k,\omega)=\frac{1}{(\mu k^2-\rho\omega^2)}\left[\boldsymbol{I}-\frac{(\lambda+\mu)\boldsymbol{kk}}{(\lambda+2\mu)k^2-\rho\omega^2}\right]. \tag{2.108}$$

It is convenient to express the Lamé coefficients λ, μ in terms of l and t, the velocities of the longitudinal and transverse waves, respectively, related to λ, μ by

$$\rho l^2=\lambda+2\mu, \qquad \rho t^2=\mu. \tag{2.109}$$

We shall study surface waves in the x-direction on the surface $z=0$ of an isotropic elastic solid contained in $z>0$. The wavevector will be of the form $\boldsymbol{k}=(\boldsymbol{\kappa}, 0, q)$ and the surface projection $\boldsymbol{G}$ is evaluated by the formulae (2.62) applied to tensor Green functions. This yields

$$\boldsymbol{G}=\frac{\mathscr{I}}{\rho}\begin{vmatrix}\dfrac{(\kappa^2-b_tb_l)}{2\omega^2 b_l} & 0 & 0\\ 0 & \dfrac{1}{2b_t t^2} & 0\\ 0 & 0 & \dfrac{(\kappa^2-b_tb_l)}{2\omega^2 b_t}\end{vmatrix}, \tag{2.110}$$

where

$$b_l=(\kappa^2-\omega^2/l^2)^{1/2}; \qquad b_t=(\kappa^2-\omega^2/t^2)^{1/2}.$$

The surface derivative is likewise evaluated, yielding

$$'\boldsymbol{G}^{(\pm)}=\frac{\mathscr{I}}{\rho}\begin{vmatrix}\pm\dfrac{1}{2t^2} & 0 & \mathrm{i}\dfrac{\kappa(b_t-b_l)}{2\omega^2}\\ 0 & \pm\dfrac{1}{2t^2} & 0\\ \mathrm{i}\dfrac{\kappa(b_t-b_l)}{2\omega^2} & 0 & \pm\dfrac{1}{2l^2}\end{vmatrix}. \tag{2.111}$$

Notice that

$$\Delta' \boldsymbol{G} = {}'G^{(+)} - {}'G^{(-)} = \mathscr{I} \begin{vmatrix} \dfrac{1}{\mu} & 0 & 0 \\ 0 & \dfrac{1}{\mu} & 0 \\ 0 & 0 & \dfrac{1}{\lambda + 2\mu} \end{vmatrix}. \tag{2.112}$$

Compare with (2.60) for the Schrödinger equation. Since the evaluation of the integrals appearing in this kind of analysis requires some care in the handling of residues with consistent sign conventions, it is convenient in practice to have occasional checks on the results whenever possible. Here it suffices to compare with (2.16), which follows quite generally from a direct integration of the equation of motion. By setting $i, j = 1, 2, 3$ we obtain nine equalities which express just (2.112).

Knowing (2.111) and (2.110) – which has a trivial inversion – we obtain the surface projection of the linear form of interest for this problem, i.e. the sum of the first two terms on the r.h.s. of (2.19). Omitting the rest – which is continuous across the surface – we find

$$\boldsymbol{A}^{(\pm)} = \mathscr{I} \begin{vmatrix} \pm\frac{1}{2} & 0 & \dfrac{-\mathrm{i}\kappa[\omega^2 + 2t^2(b_t b_l - \kappa^2)]}{2\omega^2 b_t} \\ 0 & \pm\frac{1}{2} & 0 \\ \dfrac{\mathrm{i}\kappa[\omega^2 + 2t^2(b_t b_l - \kappa^2)]}{2\omega^2 b_l} & 0 & \pm\frac{1}{2} \end{vmatrix}, \tag{2.113}$$

which obviously satisfies (2.20).

With this we evaluate (2.31), obtaining

$$\boldsymbol{G}_{\mathrm{s}}^{-1} = \rho \mathscr{I} \begin{vmatrix} \dfrac{\omega^2 b_l}{(\kappa^2 - b_t b_l)} & 0 & \dfrac{\mathrm{i}\kappa[\omega^2 + 2t^2(b_t b_l - \kappa^2)]}{(\kappa^2 - b_t b_l)} \\ 0 & t^2 b_t & 0 \\ \dfrac{-\mathrm{i}\kappa[\omega^2 + 2t^2(b_t b_l - \kappa^2)]}{(\kappa^2 - b_t b_l)} & 0 & \dfrac{\omega^2 b_t}{(\kappa^2 - b_t b_l)} \end{vmatrix}. \tag{2.114}$$

We are now ready to study the effects of the surface on the vibrational spectrum of an elastic solid.

First we use the secular equation for surface modes. From (2.114) we have

$$\det G_s^{-1} = \frac{\rho^3 t^6 b_t}{(\kappa^2 - b_t b_l)} \left[4\kappa^2 b_t b_l - \left(2\kappa^2 - \frac{\omega^2}{t^2}\right)^2 \right]. \tag{2.115}$$

The dispersion relation is obtained by setting this equal to zero. By omitting the non-vanishing factors in front, the secular equation is simply

$$4\kappa^2\left(\kappa^2 - \frac{\omega^2}{t^2}\right)^{1/2}\left(\kappa^2 - \frac{\omega^2}{l^2}\right)^{1/2} - \left(2\kappa^2 - \frac{\omega^2}{t^2}\right)^2 = 0. \tag{2.116}$$

This is the well-known dispersion relation for surface *Rayleigh waves*. The solution is $\omega = \xi t \kappa$, where ξ is a number rather close to 1, but actually $\xi < 1$ always.

Now, the non-vanishing factors in (2.115) are not all irrelevant. We can certainly discard them in seeking just the s.m.d.r. But suppose we want to study the density of modes. Then b_t and $(\kappa^2 - b_t b_l)$ are both functions of ω whose argument, for fixed κ, can vary as ω and therefore (remember (2.56)) these terms can contribute to the density of modes. This demonstrates the general remarks made in § 2.2.2. Actually in this case the formula (2.57) is applicable. Thus we need det $\boldsymbol{G}$. From (2.110) this is

$$\det \boldsymbol{G} = (\kappa^2 - b_t b_l)^2 / 8\rho^3 \omega^4 t^2 b_t^2 b_l. \tag{2.117}$$

In writing down the explicit result for the density of modes it is necessary to be rather precise. A formula like (2.53) or (2.57) actually gives the density of modes *for fixed κ and as a function of ω^2*. Furthermore, this density of modes is defined *per unit volume of the system* and if we keep track of all the steps leading to (2.53) or (2.57) we find that the surface terms are actually multiplied by a factor area/volume: this is just an example of another general theorem known as *Lederman's theorem*. The factors A and V were implicitly taken equal to unity in the discussion of § 2.3. In order to be precise, the formula we must use here is

$$N_{\kappa,t}(\omega^2) = \tfrac{1}{2} N_{\kappa,\mathrm{b}}(\omega^2) + \frac{A}{V}\left[\frac{1}{2\pi}\frac{\mathrm{d}}{\mathrm{d}\omega^2}\arg\det \boldsymbol{G} + \frac{1}{\pi}\frac{\mathrm{d}}{\mathrm{d}\omega^2}\arg\det \boldsymbol{G}_s^{-1}\right], \tag{2.118}$$

where $\boldsymbol{G}$ and $\boldsymbol{G}_s$ are functions of ω^2 and this is $\lim(\omega^2 + \mathrm{i}\eta)$ as $\eta \to 0$. The volume factor is usually absorbed in the definition of density per unit volume. The results derived from the surface term then give a surface contribution proportional to A. The factor in square brackets gives the surface contribution *per unit surface area* which we shall denote as $N_{\kappa,s}(\omega^2)$, and it is in this form that we shall henceforth present the results.

We then change the variable in which a density of modes is defined, from ω^2 to ω, and what we are ultimately interested in is the complete density for all κ as a function of ω only. The formula is simply

$$N_{\kappa,s}(\omega) = 2\pi N_{\kappa,s}(\omega^2); \qquad N_s(\omega) = \int \frac{d^2\kappa}{(2\pi)^2} N_{\kappa,s}(\omega). \qquad (2.119)$$

Now, for a given κ we have the *Rayleigh mode* $\xi t\kappa$ and also a bulk mode $t\kappa$. Notice that this is the *transverse threshold*: above $\omega = t\kappa$ we have (fig. 21) a continuum of transverse bulk modes whose wavevector component normal to the surface starts growing from zero. Likewise, $\omega = l\kappa$ is the *longitudinal threshold*, above which we have the continuum of bulk longitudinal modes. In chapters 3 and 4 we shall see that important contributions to the density of states arise from the *band edges*, which are – for the problem of electronic states – the analogue of the bulk threshold in figure 21. There are two important differences. First, the band edge effects we shall see in chapters 3 and 4, together with the changes in the rest of the spectrum, must comply with overall sum rules

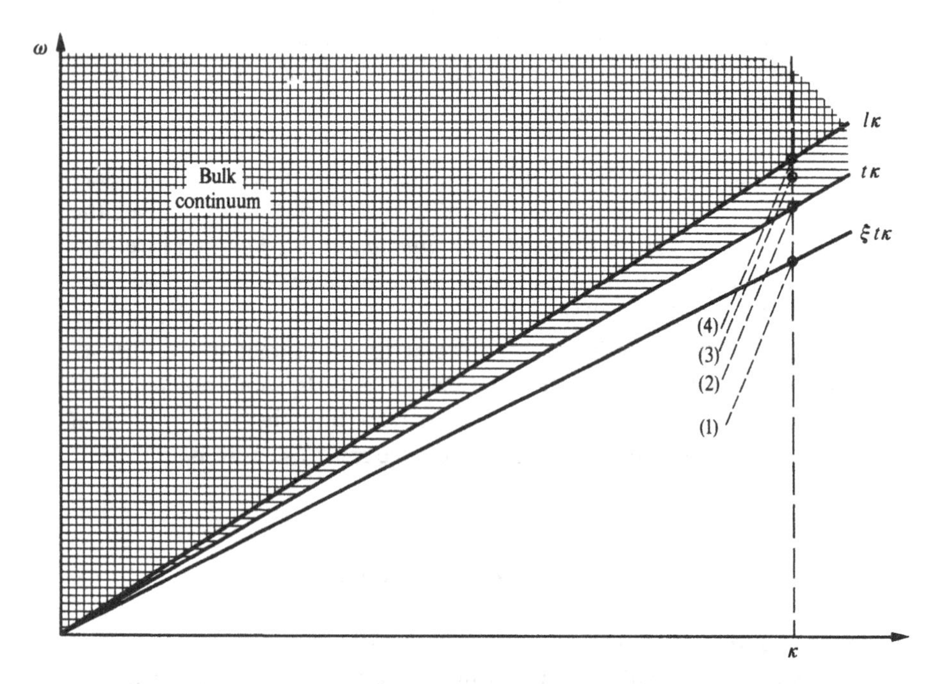

Fig. 21. For a given κ the contributions to the surface density of elastic modes come from (1) the Rayleigh pole; (2) the transverse threshold; (3) the continuous range between the two bulk thresholds; (4) the longitudinal threshold. Horizontal lines: continuum of bulk transverse modes. Vertical lines: continuum of bulk longitudinal modes.

which express the conservation of the total number of states. This is just an example of a general theorem in perturbation theory, *Levinson's theorem*, which holds for electronic states whether they are perturbed by, say, the potential of an impurity or by the potential barrier at the surface. The formal proof of this theorem is based on the analytic properties of a phase function from which the density of states can be evaluated. For example, for the electronic state problem, when (2.66) holds the phase function in question is

$$\Phi(\omega) = -\tfrac{1}{2}\sum_m \arg\det G_m(\omega) - \arg\det G_s^{-1}(\omega), \qquad (2.120)$$

where ω is the complex energy variable. In particular the behaviour of this function at infinity is of crucial importance for the proof of the theorem, which depends on $\Phi(\omega)$ vanishing sufficiently fast as $|\omega| \to \infty$. If one studies the corresponding phase function for the problem of elastic waves it turns out that it fails to fulfill this requirement and Levinson's theorem does not hold. We shall not get involved in such technical details here, but we shall come back to this question presently in more physical terms. The second difference is that there are two separate band edges or thresholds in the problem of elastic waves, and we shall see that this range has a distinct contribution.

The various contributions to the density of modes are rather neatly picked out with the Green function formulae. We proceed in the following way: first we concentrate on the *surface density of modes* for given $\boldsymbol{\kappa}$, i.e. on $N_{s,\boldsymbol{\kappa}}(\omega)$ defined as in (2.119), by using (2.118) for $N_{s,\boldsymbol{\kappa}}(\omega)$. Keeping $\boldsymbol{\kappa}$ fixed we let ω increase from zero (fig. 21) and follow the behaviour of the ω-dependent terms in (2.115) and (2.117), evaluating the contributions to $N_{s,\boldsymbol{\kappa}}(\omega)$ which arise for different values or ranges of ω. We then integrate these over $\boldsymbol{\kappa}$ as in (2.119) and express the final results as contributions to $N_s(\omega)$. The terms we find arise as follows.

Nothing happens until we reach (fig. 21) the *Rayleigh mode* (1). Here $\det \boldsymbol{G}_s^{-1}$ changes sign, as it goes through zero according to (2.116), and therefore its argument changes by π. This yields the first contribution,

$$N_s(1) = \omega/2\pi\xi^2 t^2.$$

The two arguments of interest stay constant after this, until we reach the *transverse threshold* (2). Notice that here b_t, defined in (2.110), goes through zero changing from real to imaginary. But we must add the contributions of (2.115) and (2.117) and it turns out that they cancel out exactly. Thus

$$N_s(2) = 0.$$

This shows the importance of keeping track of all the terms. Next we get into the continuous range between thresholds. Here we must watch the behaviour of the terms containing b_t and b_l simultaneously. We find

$$N_s(3) = \frac{(3l^4 + t^4 - 2l^2t^2)\omega}{8\pi l^2 t^2 (l^2 - t^2)} - \frac{\omega}{2\pi \xi^2 t^2},$$

a part of which, curiously, is going to cancel out exactly the contribution of the *Rayleigh mode*. The last contribution comes from the *longitudinal threshold*, where b_l goes through zero. This yields

$$N_s(4) = -\omega / 8\pi l^2.$$

Adding up all these contributions we have

$$N_s(\omega) = \frac{(3l^4 + 2t^4 - 3l^2t^2)\omega}{8\pi l^2 t^2 (l^2 - t^2)}. \tag{2.121}$$

This is a central result. It includes all the effects of the surface – per unit surface area, as explained – which contribute to the new value of the density of modes.

It might be in order at this stage to consider the meaning of what we are doing. We are describing the vibrational properties of a solid crystal in the long-wave limit, as if it were an – isotropic – elastic medium. This is the *Debye model*, in which we introduce a maximum cutoff frequency by imposing the physical requirement that the total number of degrees of freedom, for a crystal with $N_V = N/V$ atoms per unit volume, must be $3N_V$. This is the condition which yields ω_D. Then we introduce the surface and we find that *Levinson's theorem* does not hold. But of course we do have a condition on the change in the density of modes or, rather, on the new value of this density. This comes from the physical requirement, which applies equally to the surface system, that the total number of degrees of freedom must still be $3N_V$, except that they are now distributed differently between bulk and surface modes. In doing this with the new *total* density of states we must be careful to put the factor A/V in its proper place. This yields the new (maximum) cutoff value, which now applies to the entire system:

$$\omega_M = \omega_D \left[1 - \frac{A}{V} \frac{(6\pi^2)^{2/3} B c_s^2}{144\pi N_V^{1/3}} \right], \tag{2.122}$$

where

$$\frac{3}{c_s^3} = \frac{1}{l^3} + \frac{2}{t^3}; \qquad B = \frac{(3l^4 + 2t^4 - 3l^2t^2)}{l^2 t^2 (l^2 - t^2)}. \tag{2.123}$$

Not surprisingly, the change in ω_D contains the factor A/V. This change is negative (remember $l > t$) because the net change in the density of modes (2.121) is positive; thus with a larger density the same total number of degrees of freedom is reached sooner, i.e. with a smaller cutoff frequency.

Formulae (2.121) and (2.122) can constitute the basis of the two-dimensional surface version of a *Debye model.* If we know the density of modes of a system we can calculate the vibrational entropy by using standard formulae of statistical thermodynamics. Therefore we can obtain the *surface entropy*, which includes all surface effects. We find

$$S_s = \frac{Bk_B\omega_D^3}{16\pi}\left[\frac{3}{Y^2}\int_0^Y \frac{x^2}{e^x - 1}\,dx - \frac{Y}{e^Y - 1}\right], \tag{2.124}$$

where $Y = \hbar\omega_D/k_BT = \Theta_D/T$, while the zero-point energy is

$$E_s(0) = -B\hbar\omega_D^3/96\pi. \tag{2.125}$$

With this we can find all the standard thermodynamical functions of interest.

The picture which emerges from this very simple basis is quite reasonable. Let us discuss it keeping in mind what we would expect to be the main phenomenological features. At low temperatures, of course, the specific heat in both cases should tend to zero. It is easy to see, by differentiating (2.124) with respect to T and studying the behaviour of the result for $Y \gg 1$, that the low T surface specific heat, per unit area, is given by

$$C_s = 3\pi\zeta(3)\frac{(3l^4 + 2t^4 - 3l^2t^2)}{l^2t^2(l^2 - t^2)}\frac{k_B^3T^2}{h^2}, \tag{2.126}$$

a well-known formula. The main feature of this result is that C_s tends to zero as T^2, which of course one must expect on dimensional grounds, for the same reason that the specific heat of a bulk crystal tends to zero as T^3. This just demonstrates that (2.121) does indeed describe a spectrum of two-dimensional modes – including the distortions of the bulk localised near the surface – much as the distinct surface modes themselves do, as emphasised in § 2.3. In the opposite limit, at high temperatures, we expect the *surface* – vibrational – *energy* to tend to zero: when all the oscillators are excited it cannot make any difference whether they behave as bulk or surface degrees of freedom. They all contribute equally to the internal energy of the system. Indeed, the internal energy derived – in the limit $Y \ll 1$ – from (2.124) by standard thermodynamical arguments is exactly cancelled out by the zero point term (2.125) and we find $E_s = 0$ for $Y \to 0$. This implies that the surface specific heat should have the qualitative behaviour described in fig. 22, and this is also the behaviour one finds

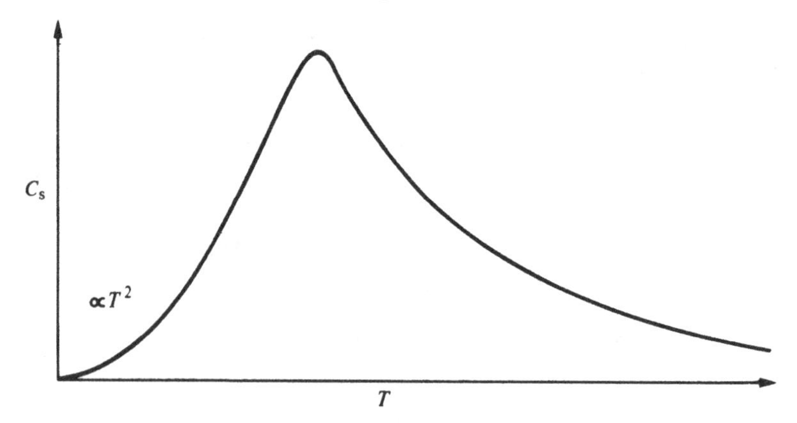

Fig. 22. Expected qualitative behaviour of the – constant volume – surface specific heat as a function of temperature.

if one starts from (2.124). Now, the high-temperature internal energy cannot change when the surface effects are introduced, for the reason just stated, but *the entropy can*, because as we have seen, the same total number of degrees of freedom is accommodated within a redistribution in the new density of states. Thus we expect a non-vanishing surface entropy for $T \gg \Theta_D$. Indeed, in this limit we find from (2.124):

$$S_s \approx (B\omega_D^2 k_B/32\pi)[1-0.042(\Theta_D/T)^2-\ldots]. \tag{2.127}$$

Thus the qualitative picture which emerges from this simple model is quite acceptable – that is, just for the class of thermodynamical problems for which the *Debye model* is reasonable for bulk systems. In practice it turns out that even quantitative estimates based on (2.124)–(2.127) are reasonable. Having seen with this brief digression into surface thermodynamics how the method of surface Green-function matching can be used directly for practical purposes, let us return to the dynamical problem.

The physical meaning of the matching equations can be seen intuitively in the following way: the secular equation gives the eigenvalues corresponding to the set of homogeneous equations

$$\sum_j \boldsymbol{G}_s^{-1}{}_{ij} F_j = 0. \tag{2.128}$$

The F_js are the fictitious surface stimuli which, in this case, have the physical nature of a force. Thus, if we know the F_js we can write down the actual displacement inside the material, which is simply the response of a medium with elastic Green function $\hat{\boldsymbol{G}}$ to the surface stimuli F_j. This gives the eigenvalues

$$u(z;\kappa;\omega)=\frac{1}{2\pi}\int dq\, e^{iqz}\hat{G}(q,\kappa;\omega)\cdot F(\kappa,\omega), \tag{2.129}$$

which are thus explicitly given except for a normalisation factor.

Now, it is clear from (2.114) that $F_2=0$, because b_t can only be zero if $\omega=t\kappa$, which is not the case for the Rayleigh wave. Hence $u_2=0$, i.e. the displacement associated with this surface mode is polarised in the sagittal plane. The actual form of the z-dependence of $u_1(z)$ and $u_3(z)$ comes from the q-dependence of G (2.108). Since F is independent of q we can rewrite (2.128) as

$$u_i(z;\kappa;\omega)=\sum_j \hat{G}_{ij}(z;\kappa;\omega)F_j(\kappa,\omega), \tag{2.130}$$

and

$$\hat{G}_{ij}(z;\kappa;\omega)=\frac{1}{2\pi}\int e^{iqz}\hat{G}_{ij}(q,\kappa;\omega)\, dq. \tag{2.131}$$

We find

$$\left.\begin{aligned}
\hat{G}_{11}&=\frac{1}{2\omega^2}\left[\frac{\kappa^2}{b_l}e^{-b_lz}-b_t\, e^{-b_tz}\right];\\
\hat{G}_{31}=\hat{G}_{13}&=\frac{i\kappa}{2\omega^2}\left[e^{-b_lz}-e^{-b_tz}\right];\\
\hat{G}_{33}&=\frac{1}{2\omega^2}\left|-b_l\, e^{-b_lz}+\frac{\kappa^2}{b_t}e^{-b_tz}\right|.
\end{aligned}\right\} \tag{2.132}$$

Notice that the column vectors $(\hat{G}_{11}, \hat{G}_{13})$ and $(\hat{G}_{31}, \hat{G}_{33})$ are neatly separated into a longitudinal part (associated with e^{-b_lz}) which has zero curl, and a transverse part (associated with e^{-b_tz}) with zero divergence. Thus the linear combinations (2.130) give the wave amplitudes decomposed into their longitudinal and transverse parts. In order to construct the actual form of the amplitudes u_1 and u_3 we use (2.128) to determine the ratio F_3/F_1 and remember that all this is for a surface mode for which (2.116) holds. Putting $\omega=\xi t\kappa$ and using (2.116) we find, up to some amplitude factor u_0,

$$\left.\begin{aligned}
u_1&=-i\kappa u_0[e^{-b_lz}-(1-\tfrac{1}{2}\xi^2)\, e^{-b_tz}]e^{i(\kappa x-\omega t)};\\
u_3&=b_lu_0\,[e^{-b_lz}-(1-\tfrac{1}{2}\,\xi^2)^{-1}\, e^{-b_tz}]e^{i(\kappa x-\omega t)},
\end{aligned}\right. \tag{2.133}$$

which is the well-known textbook result describing an elliptical motion whose shape changes with depth. Since (2.116) has no real root $\xi>1$, the condition $\omega < t\kappa < l\kappa$ is always satisfied for a Rayleigh mode. Hence b_l and b_t are both real and (2.133) describes a stationary wave whose amplitude is localised near the surface.

We now know the s.m.d.r., the form of the surface wave amplitudes, the density of modes of the surface system, and $\boldsymbol{G}_s$. Finally, the complete three-dimensional Green function $\hat{\boldsymbol{G}}_s$ of the surface system can be found by evaluating (2.39) and (2.40), for which we have all the necessary information. It is only a matter of standard algebra which will be omitted here.

The application of the method to other problems of interest can be carried out in the same manner. For example, suppose we have a medium with coefficients labelled 1 on side $z<0$, and 2 on side $z>0$. The definitions of b_{l1}, b_{l2}, etc. are the obvious extensions of the definitions of b_l, b_t. Repeating the above steps we find

$$\det \boldsymbol{G}_s^{-1} = \frac{(\rho_2 t_2^2 b_{t2} + \rho_1 t_1^2 b_{t1})}{(\kappa^2 - b_{t1}b_{l1})(\kappa^2 - b_{t2}b_{l2})}$$
$$\times\{\omega^4[(\rho_1-\rho_2)^2\kappa^2 - (\rho_1 b_{t2} + \rho_2 b_{t1})(\rho_1 b_{l2} + \rho_2 b_{l1})]$$
$$+4\omega^2\kappa^2(\mu_1-\mu_2)[\rho_1 b_{t2}b_{l2} - \rho_2 b_{t1}b_{l1} - \rho_1\kappa^2 + \rho_2\kappa^2]$$
$$+4\kappa^2(\mu_1-\mu_2)^2[(b_{t2}b_{l2}-\kappa^2)(b_{t1}b_{l1}-\kappa^2)]\}. \qquad (2.134)$$

From this we obtain at once the dispersion relation for *Stoneley waves*, which are the surface modes of this interfacial problem. Moreover, the same analysis which produced (2.121) yields the surface density of modes for this system, which is now, by again omitting the factor A/V,

$$N_s(\omega) = \frac{\omega}{8\pi}\left[\frac{2U_1}{U_2} - \frac{(1-r_d)(r_a-1)}{(r_a+1)}\right], \qquad (2.135)$$

where

$$r_a = \frac{\mu_1}{\mu_2}; \qquad r_b = \frac{t_2^2}{l_2^2}; \qquad r_c = \frac{t_1^2}{l_1^2}; \qquad r_d = \frac{t_2^2}{t_1^2};$$

$$U_1 = r_a^2[-(1-r_c)(1+r_b^2) + r_d(1+r_b)(r_c^2+r_c+1)]$$
$$-r_a[r_c(2r_b^2 - r_b + 1) + r_d r_b(2r_c^2 - r_c + 1)]$$
$$+[(1+r_c)(r_b^2 - r_b + 1) - r_d(1-r_b)(1+r_c^2)];$$
$$U_2 = r_a^2(1+r_b)(1-r_c) + 2r_a(1+r_b r_c) + (1+r_c)(1-r_b).$$

All these examples demonstrate the central role of $\boldsymbol{G}_s$. The whole idea of the method developed in this chapter is to obtain basic formulae which indicate in a concise way how some physical results can be found without having to construct the entire three-dimensional Green function of the surface system. An example of central importance is the density of modes, from which other physical properties follow. For example, for the low-temperature specific heat the practical rule is simply to replace ω by the factor $\zeta(3)6k_B^3T^2/\hbar^2$. This economy of algebra can be a practical advantage as soon as we go beyond the simple Rayleigh model. A case in point is anisotropic crystals. The situation there can be rather complicated because it turns out that, for a given crystal face, the analytical nature of the surface modes may depend on the direction of propagation. One can also find resonant modes, for example, and this pattern changes with the orientation of the crystal face. It is easy to imagine that these are just the features which are characteristically associated with Green functions and their analytic properties. For example, knowing $\boldsymbol{G}_s$ we have different options at our disposal. One is to study the resulting secular equation. In general we may find different types of solutions, e.g. real and complex. It helps a great deal in visualising the meaning of these solutions if we can look at the actual form of the surface eigenwaves, and this can be obtained by the same argument that leads from (2.128) to (2.133). Alternatively, a study of the phase function (2.120) reveals directly the existence of stationary states, resonances, and distortions of the continuum.

2.8. Matching with discontinuities

In § 2.2 we have insisted that several important results expressed in terms of $\boldsymbol{G}_s$ are generally valid irrespective of the formula for $\boldsymbol{G}_s$. These are: (i) the expression of the secular equation (2.32), (ii) the form of the matrix elements of $\hat{\boldsymbol{G}}_s$ (2.39, 2.40), (iii) the formula for the density of states (2.53) (or (2.57)), and for electronic states, (iv) the formulae for the reflection (2.74) and transmission (2.78) coefficients. The question of finding the explicit formula for $\boldsymbol{G}_s$ depends on the boundary conditions. So far we have studied the situations in which the fields in question, and the linear combinations of normal derivatives of them are continuous across the surface. We shall now investigate the situation with discontinuities.

The most common situation is that in which the fields are continuous and the discontinuities affect their normal derivatives. Let $\boldsymbol{M}$ be a tensor which measures this discontinuity. Since we are only interested in the surface projection $\boldsymbol{G}_s$, it suffices to look at the matching conditions obtained by taking the surface projection from both sides. Under these circumstances the first matching condition is still (2.29*a*) but the second one is now

$$\boldsymbol{A}_2^{(+)} \cdot \boldsymbol{\delta \mathscr{I}} + \boldsymbol{A}_2^{(-)} \cdot \boldsymbol{F}_2 - \boldsymbol{A}_1^{(+)} \cdot \boldsymbol{F}_1 = \boldsymbol{n} \cdot \boldsymbol{M}, \tag{2.136}$$

instead of (2.29b). So we now multiply (2.29a) from the left by $\boldsymbol{A}_2^{(-)} \cdot \boldsymbol{G}_2^{-1}$ and subtract from (2.136). This yields

$$(\boldsymbol{A}_1^{(+)} \cdot \boldsymbol{G}_1^{-1} - \boldsymbol{A}_2^{(-)} \cdot \boldsymbol{G}_2^{-1}) \cdot \boldsymbol{G}_s \cdot \boldsymbol{\delta \mathscr{I}} = \boldsymbol{I} \cdot \boldsymbol{\delta \mathscr{I}} - \boldsymbol{n} \cdot \boldsymbol{M}. \tag{2.137}$$

Here we have used (2.20) and the surface projection of (2.11b).

Here again it is useful to keep in mind the physical interpretation of the formalism. When trying to obtain the form of the Rayleigh wave we looked at the field as the response of an extended medium to a surface stimulus (2.129). But we can equally consider the field as the response *of the surface system* – i.e. of the system with Green function $\hat{\boldsymbol{G}}_s$ – to the standard unit input. Thus, if we express the field in this form and take surface projections, then the discontinuity on the r.h.s. of (2.137) takes the form $\boldsymbol{m}_s \cdot \boldsymbol{G}_s \cdot \boldsymbol{\delta}$. This amounts to a redefinition of the surface discontinuity in terms of $\boldsymbol{m}_s$ instead of $\boldsymbol{M}$. The explicit form of these tensors must come from the physics of the problem in each case, as we shall soon see. We then have

$$(\boldsymbol{A}_1^{(+)} \cdot \boldsymbol{G}_1^{-1} - \boldsymbol{A}_2^{(-)} \cdot \boldsymbol{G}_2^{-1}) \cdot \boldsymbol{G}_s = \mathscr{I}\boldsymbol{I} - \boldsymbol{m}_s \cdot \boldsymbol{G}_s. \tag{2.138}$$

Thus we have found the desired new formula, namely,

$$\boldsymbol{G}_s^{-1} = \boldsymbol{A}_1^{(+)} \cdot \boldsymbol{G}_1^{-1} - \boldsymbol{A}_2^{(-)} \cdot \boldsymbol{G}_2^{-1} + \boldsymbol{m}_s. \tag{2.139}$$

Compare with (2.31). The problem is reduced to finding $\boldsymbol{m}_s$.

In order to see how this works, consider a very simple example: the free surface of an incompressible fluid with surface tension γ. In this case we can define a velocity potential Φ satisfying the differential equation

$$\nabla^2 \Phi = 0.$$

The Green function – a scalar here – is simply

$$\hat{G}(k, \omega) = 1/(\kappa^2 + q^2),$$

whence

$$G = 1/2\kappa, \qquad {}'G^{(+)} = \tfrac{1}{2}.$$

Assuming the fluid is contained in $z > 0$ and neglecting gravity effects, we see that the secular equation, from (2.32) and (2.139) is

$$\kappa = -m_s. \tag{2.140}$$

Now we must see what this means.

A surface tension γ implies a discontinuity in the normal stress, given by *Laplace's formula* $\Delta p = \gamma / R_x$, for a wave propagating in the x-direction and carrying a ripple with radius of curvature R_x. This acts as a normal restoring force which must equilibrate the inertial force associated with the fluid motion in the direction perpendicular to the surface. It is an easy matter to see that, in terms of the velocity potential, this condition amounts to the equation found in all textbooks (for small amplitudes):

$$\gamma \frac{\partial}{\partial z} \frac{\partial^2 \Phi}{\partial x^2}\bigg|_{z=0} = \rho \frac{\partial^2 \Phi}{\partial t^2}\bigg|_{z=0}$$

It is from here that we can find our m_s. It suffices to look at the field Φ as G_s – multiplied by the standard unit input which in this case is trivially divided away – evaluated at the surface. The equation holds inside the medium ($z > 0$), while outside we have a vacuum. This yields the desired discontinuity in the normal derivative:

$$-\gamma \kappa^2 \Delta' G_s = \rho \omega^2 G_s. \tag{2.141}$$

In this simple scalar case the discontinuity M introduced above reduces to the scalar quantity $\Delta' G_s$; hence m_s is given by

$$m_s = -\rho \omega^2 / \gamma \kappa^2. \tag{2.142}$$

Using this in (2.140) we obtain the well-known dispersion relation for capillary waves. We could equally have included gravity effects, as will be seen in § 2.10, but our purpose here was only to see with a very simple example how the formalism works. We shall now move on to less trivial problems.

2.9. Elastic surface waves in solids with surface stresses

We have just seen that the formalism for matching with discontinuity conditions can be used to study the effect of surface tension in fluids. Let us pause to discuss the physics of surface tension and related concepts. If we have a surface system with a given total internal energy, and subtract the energy of the homogeneous unperturbed bulk material assumed hypothetically to occupy the same region of space, the difference is what we define as surface energy. The same process defines, for any other thermodynamic potential, the corresponding surface amount of, say, free energy, etc. We know from standard thermodynamics that the Kramers potential (Ω), which is the Helmholtz free energy minus the Gibbs potential, is just pV, the 'mechanical energy' part of the total (internal) energy, so to speak. In the same way the surface value Ω_s is the mechanical term associated with the surface effects. For example, for a fluid,

$$\Omega_s = \gamma A. \tag{2.143}$$

This means that the energy needed to create new area δA at constant temperature and chemical potentials is $\gamma\delta A$. This looks just like a typical *force × displacement* formula. As a matter of fact in a fluid, unless the deformation is too fast, the fluid molecules or atoms are normally capable of diffusing sufficiently quickly to restore local values of the intensive properties – e.g. density or chemical potential – in the surface region when a surface deformation is produced. Thus the energy needed to increase the area by a deformation δA, at constant intensive properties, is also equal to $\gamma\delta A$, hence the term *surface tension*.

The situation is different in a solid. Here (2.143) still holds, and it is proved that Ω_s is the energy needed to create, ideally (i.e. reversibly) and at constant temperature and chemical potentials, the surface area A. The ideal materialisation of this concept would correspond to a process of reversible cleavage, whence the name *reversible cleavage work*, per unit area, for γ, which is also sometimes called *specific surface work*. What this means is that γ itself, for a solid, is a function of the state of deformation of the surface. That is to say, if a surface deformation δA is produced, then the mechanical energy involved is

$$\delta(\gamma A) = \gamma\delta A + A\delta\gamma = (\gamma + \partial\gamma/\partial \ln A)\delta A, \tag{2.144}$$

which can be written in the form, force × displacement, if we define the *surface stress*

$$\tau = \gamma + \partial\gamma/\partial \ln A. \tag{2.145}$$

This is known as *Shuttleworth's equation*. Thus the position is the following: the energy of creation of new area, per unit area, is γ, and the energy of deformation is $\tau\delta A$, i.e. τ per unit area. All these processes are under given thermodynamical conditions, namely, constant temperature and chemical potentials, but we shall omit explicit reference to this in the thermodynamic derivatives presently to be given.

The point of this digression was to emphasise the difference between the various concepts involved in this field of surface physics. It is important to know exactly what is being calculated and what is being measured in a given experiment. For example, in a theoretical calculation one usually obtains the surface energy, while the experiments may measure, say, γ. It is not difficult to prove that the surface entropy is

$$S_s = -\mathrm{d}\gamma/\mathrm{d}T,$$

and hence the surface energy is

$$E_s = \gamma - T\,d\gamma/dT. \qquad (2.146)$$

Thus the comparison between theory and experiment must be made with some care. Most data for γ are obtained at high temperatures and we must work out a correct way of extrapolating to $T=0$. In the simple model of § 2.7, γ, even down to temperatures close to room temperature, is actually linear in T, and the same is true in most simple lattice dynamical models used so far to calculate the surface entropy, which is essentially all vibrational. This is at least an encouraging indication that perhaps estimating E_s by a rough linear extrapolation from high-temperature data gives not a bad result, compared with the value of E_s obtained in a calculation based on an electronic theory of a solid surface.

However, our main concern here is to emphasise that if we want to study the effects of surface stresses on elastic surface waves in solids, we must use the surface stress τ and we must remember (2.145) in order to express the boundary conditions correctly. Let us stick to the convention that our medium is contained in the region $z=x_3>0$. We thus count the curvature associated with a surface wave as positive when, for small amplitudes, $\partial^2 u_3/\partial x_1^2>0$. We assume the wave travels in the x_1-direction. Then the hydrostatic pressure inside is larger than the pressure outside, but we must remember to use τ for the expression of Laplace's formula, which now reads

$$p^{(2)} = p^{(1)} + \frac{\tau}{R_1} = p^{(1)} + \tau\frac{\partial^2 u_3}{\partial x_1^2}. \qquad (2.147)$$

This means a discontinuity in $\tau_{33} = -p$ equal to

$$\Delta\tau_{33} = \tau_{33}^{(2)} - \tau_{33}^{(1)} = -\tau\partial^2 u_3/\partial x_1^2 = \tau\kappa^2 u_3(0). \qquad (2.148)$$

This, with τ replaced by γ, is just what we had for the fluid surface. But now we have an extra condition. According to (2.144), the surface dilation $\delta \ln A$ accompanying the wave involves an extra work. Let us define the following moduli of *surface elasticity*:

$$\Delta_s = \partial\tau/\partial \ln A; \qquad \delta_s = \partial\gamma/\partial \ln A. \qquad (2.149)$$

Notice that, from (2.145), δ_s is just the difference $\tau-\gamma$, which is what makes a solid different from a fluid, and

$$\Delta_s = \delta_s + \partial\delta_s/\partial \ln A. \qquad (2.150)$$

This is associated with an extra restoring force in the following manner.

The local value of τ changes along the direction of propagation of the surface wave according to

$$\frac{\partial \tau}{\partial x_1} = \frac{\partial \tau}{\partial \ln A} \frac{\partial \ln A}{\partial x_1}.$$

The surface dilation $\delta \ln A$ is just $\delta A/A = \partial u_1/\partial x_1$. Thus

$$\partial \tau/\partial x_1 = \Delta_s \partial^2 u_1/\partial x_1^2. \tag{2.151}$$

The gradient of τ represents a force parallel to the surface, just as a gradient of pressure represents a volume force, but we must look carefully at the signs. Imagine an element of area $\delta x_1 \delta x_2$. The net surface force acting on this element of area is

$$[\tau(x_1 + \delta x_1) - \tau(x_1)]\delta x_2 \approx (\tau + \delta\tau - \tau)\delta x_2 = \delta\tau\delta x_2.$$

Thus the force per unit area is $\partial\tau/\partial x_1$, i.e. precisely (2.151), and the equilibrium of forces parallel to the surface requires that

$$\tau_{31}^{(2)} + \partial\tau/\partial x_1 = \tau_{31}^{(1)},$$

whence, by (2.151),

$$\Delta\tau_{31} = \tau_{31}^{(2)} - \tau_{31}^{(1)} = -\Delta_s \partial^2 u_1/\partial x_1^2 = \Delta_s \kappa^2 u_1(0). \tag{2.152}$$

The two matching conditions (2.148) and (2.152) display the surface forces which are the components of $\boldsymbol{m}_s \cdot \boldsymbol{u}(0)$, where $\boldsymbol{m}_s$ is the surface tensor introduced in § 2.8. This yields

$$\boldsymbol{m}_s = \mathscr{I} \begin{vmatrix} \Delta_s \kappa^2 & 0 & 0 \\ 0 & 0 & 0 \\ 0 & 0 & \tau\kappa^2 \end{vmatrix} \tag{2.153}$$

We are now ready to apply the formal analysis of § 2.8. We only need to change (2.114) which, by using (2.139) and (2.153), becomes

$$\boldsymbol{G}_s^{-1} = \frac{\rho \mathscr{I}}{(\kappa^2 - b_t b_l)} \times \begin{vmatrix} \omega^2 b_l + \dfrac{\Delta_s \kappa^2(\kappa^2 - b_t b_l)}{\rho} & 0 & \mathrm{i}\kappa[\omega^2 + 2t^2(b_t b_l - \kappa^2)] \\ 0 & t^2 b_t(\kappa^2 - b_t b_l) & 0 \\ -\mathrm{i}\kappa[\omega^2 + 2t^2(b_t b_l - \kappa^2)] & 0 & \omega^2 b_t + \dfrac{\tau\kappa^2(\kappa^2 - b_l b_t)}{\rho} \end{vmatrix} \tag{2.154}$$

By putting $\alpha_s = \tau/\rho$, $\theta_s = \Delta_s/\rho$, this yields

$$\det G_s^{-1} = \rho^3 t^2 b_t/(\kappa^2 - b_t b_l)$$
$$\times [4t^4\kappa^2 b_t b_l + \omega^2\kappa^2(\alpha_s b_l + \theta_s b_t) \tag{2.155}$$
$$+ \alpha_s\theta_s\kappa^4(\kappa^2 - b_t b_l) - (\omega^2 - 2t^2\kappa^2)^2],$$

from which we can again extract most of the physical results of interest.

For example, by defining

$$\begin{aligned} \phi_s &= \theta_s/\alpha_s; & \kappa_b &= t^2/\alpha_s; & \omega_b &= t\kappa_b; \\ \Omega_s &= \omega/\omega_b; & K_s &= \kappa/\kappa_b; & \omega &= Xt\kappa, \end{aligned} \tag{2.156}$$

the s.m.d.r. can be written in the form

$$4f_t f_l + (f_l + \phi_s f_t)X^2 K_s + \phi_s(1 - f_t f_l)K_s^2 = 4(1 - \tfrac{1}{2}X^2)^2, \tag{2.157}$$

where

$$f_t = (1 - X^2)^{1/2}; \qquad f_l = [1 - (t^2/l^2)X^2]^{1/2}. \tag{2.158}$$

This equation will be studied in the following way. Take a given value of κ, expressed in dimensionless form as K_s. Look for a real root X of (2.157). This yields $X(K_s)$. The frequency of the surface wave is then

$$\omega = \omega_b \Omega_s; \qquad \Omega_s = X(K_s)K_s. \tag{2.159}$$

Notice how the situation has changed. At very long waves $K_s \to 0$ and we recover the Rayleigh mode, for which $X = \xi$. But now the dispersion relation, for higher values of κ, becomes dispersive. In order to study this we need experimental information on τ and Δ_s, i.e. ϕ_s. Unfortunately such data are hardly available for the time being. One way of measuring τ is to use Laplace's formula in the form (2.147). If one can produce very small granules of material, then the excess pressure inside can compress the crystal lattice; this can be detected with X-ray diffraction and τ can thus be estimated. However, it seems that the very small sizes required can only be achieved with noble metals, for which small colloidal granules can be produced. Even for these materials the existing experimental information hardly permits anything more than educated guesswork. On this meagre basis one can surmise that such indications as there are point to values of ϕ_s less than about 0.1. If this is true then the main effects associated with surface stresses are essentially due to the normal-stress condition (2.148) and we can set $\phi_s = 0$ to a good approximation. However, if more experimental information eventually becomes available, then the more complete formulae (2.154)–(2.159) are ready for use. For the time being we shall study the case $\phi_s = 0$, which is sufficiently interesting. It is then easy to see that (2.158) has a real root such that

$$\xi \leqslant X(K_s) \leqslant 1$$

up to a maximum cutoff value,

$$K_{cc} = (1 - t^2/l^2)^{-1/2}; \qquad \kappa_{cc} = K_{cc}\kappa_b. \tag{2.160}$$

At this point we find $X = 1$, i.e. $\omega = t\kappa_{cc}$. The surface wave (fig. 23) exhibits a *capillary cutoff* where it reaches the transverse bulk threshold

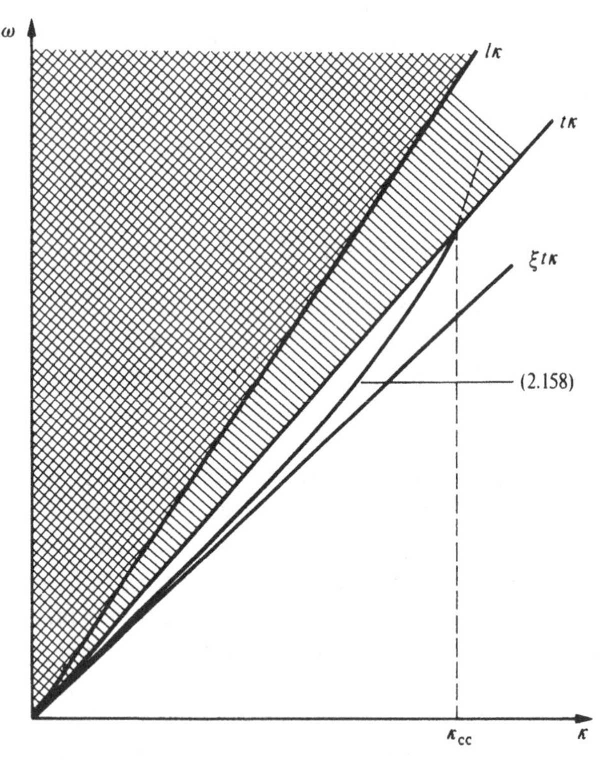

Fig. 23. The Rayleigh mode, $\omega = \xi t\kappa$, is modified by the effect of surface stress so that it becomes dispersive and merges with the continuum at the cutoff value κ_{cc} of κ.

and from then on it merges with the continuum. This has a clear physical interpretation: the effect of the surface stress is to introduce an additional restoring force which tends to flatten out the curvature associated with the surface wave. This hardens the vibrations, and the frequencies of the surface mode increase above the Rayleigh value until they go beyond the lowest bulk threshold.

The really interesting question is: can this result, obtained with an elastic continuum model, suggest any significant effects for real crystalline surfaces? Since the effects of capillarity involve introducing a cutoff, we must compare κ_{cc} with the Debye cutoff that we would introduce in trying to use the elastic continuum surface as a model of a solid surface. If we find, for example, that κ_{cc} is larger, then it looks as if the effect is negligible, but if κ_{cc} is smaller, then we may expect appreciable effects because the area of the – isotropic, average – two-dimensional Brillouin zone over which stationary modes exist depends on κ_{cc}^2. As a matter of fact, estimates of κ_{cc} for averaged models of solid surfaces tend to give results comparable with the Debye cutoff value, with some tendency towards the lower side. Thus the situation is marginal and somewhat suggestive, and this at least indicates that when doing lattice dynamics calculations of surface modes one ought to pay attention to changes in the interaction constants. Moreover, there are other considerations of physical interest. For example, we have insisted (§ 2.7) that in anisotropic elastic models the nature of the surface modes depends on the propagation direction. Concentrating only on stationary modes, the velocity of the surface wave may become degenerate with that of the lowest bulk threshold for some special direction (S in fig. 24) and very close to it for a range of neighbouring directions. The capillary effect may then introduce a *forbidden sector* of points in $\boldsymbol{\kappa}$-space for which no true stationary surface modes exist, as indicated qualitatively in fig. 24. It is clear that this will produce a redistribution of the spectral strength. Of course the quantitative importance of this depends on where the maximum (Debye-like) cutoff comes into the picture and this can only be found by quantitative evaluation.

Another physically interesting situation may arise in connection with chemisorption on, say, a metal surface. Consider again a simple isotropic model. The point is that, apart from possible effects associated with the different masses of the adsorbed species, a change in adsorption coverage entails a change in γ and τ. The effect of this is to move κ_{cc} up and down. Remember that, from (2.156) and (2.160),

$$\kappa_{cc} = \left(1 - \frac{t^2}{l^2}\right)^{-1/2} \frac{\rho t^2}{\tau},$$

and an increase of τ by a given numerical factor entails a reduction of the allowed domain in two-dimensional κ-space by the square of this factor. It is not at all unreasonable to expect that adsorption can have quite appreciable effects just via the induced changes in capillarity.

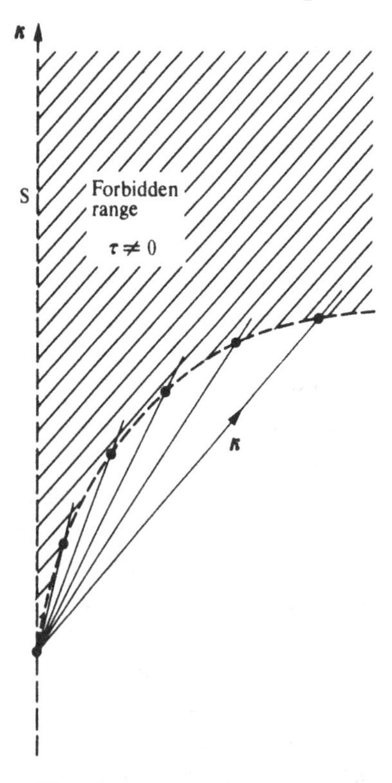

Fig. 24. Anisotropic elastic solid. Qualitative picture of the effects of surface stress on allowed and forbidden stationary surface waves for different propagation directions. Points: locus of κ_{cc}. The two-dimensional Debye-type cutoff is not shown in this picture.

But we have only been considering one aspect of the problem. When we say that surface modes have disappeared we are simply saying that the same total number of degrees of freedom produces a different spectrum. Clearly if some value $\kappa > \kappa_{cc}$ is still permitted by the Debye cutoff, it is not enough to say that there is no surface mode for this value of κ. We must also investigate the changes in the spectrum above the bulk threshold; and, indeed, why not also for $\kappa < \kappa_{cc}$? We have only concentrated on the possible disappearance of the surface mode, which is the more apparent part, but the real question we must investigate is how different the new spectrum is, anywhere, for any value of κ, because even for $\kappa < \kappa_{cc}$ we have no assurance that $N_{\kappa,t}(\omega)$, or $N_{\kappa,s}(\omega)$ will be the same for $\omega > t\kappa$. But this is just the problem we can easily study on the basis of (2.118). It suffices to use (2.155), which includes capillarity effects. For the reasons explained above we shall discuss the case $\theta_s = 0$; this will illustrate the point. Pure threshold or 'band edge' effects arise from the hard-wall term, which does not change. Thus we shall simply concentrate on the behaviour of arg det $\boldsymbol{G}_s^{-1}$, for given $\boldsymbol{\kappa}$, as a function of ω. This is the part of the phase function which changes when capillarity effects are included. To be precise, we shall study the matching part of the phase function,

$$\Phi_m = \arg\det \boldsymbol{G}_s^{-1},$$

as a function of ω for given κ. The behaviour of $\Phi_m(\omega)$ depends on whether $\kappa \gtrless \kappa_{cc}$ (fig. 24). Let us first look at $\kappa > \kappa_{cc}$ (fig. 25*b*). The surface mode has disappeared, which implies a change in the spectrum for $\omega \leqslant t\kappa$, but this is not the only effect. The former surface branch has now merged with the continuum, producing resonances which also cause drastic changes of the spectrum for all $\omega > t\kappa$. Now look at $\kappa < \kappa_{cc}$ (fig. 25*a*). Here the surface mode has only moved a little but, although not as drastic as those occurring when $\kappa > \kappa_{cc}$, significant changes also occur for all $\omega > t\kappa$.

What does this imply physically? Since $\Phi_m(\omega)$, for given κ, varies between the same values for ω varying between 0 and ∞, the total number of modes between 0 and ∞ is the same. Actually, on making a physical model – in this case the two-dimensional Debye model – we are interested in frequencies up to a maximum value ω_M, as we saw in § 2.7. In practice we expect that by using the new phase function to determine the new value of ω_M, this will not differ a great deal from (2.122). But a look at fig. 24, which is actually quantitative (it has been calculated for an averaged isotropic model of Ag), tells us that, although the same number of modes may be accommodated up to nearly the same maximum frequency, these modes are now distributed throughout the spectrum in a different way. These changes may seem fairly insignificant at a given κ on the low side, but they grow as κ grows and this affects an extension of $\boldsymbol{\kappa}$-space which, remember, increases quadratically. A calculation of the surface entropy, for example, may prove sensitive to the effect of such changes when integrated over the whole spectrum. It is easy to imagine that these effects might be quite considerable in specially favourable situations, like anisotropy and adsorption, as indicated above. This seems a suggestive open program for which this and the previous section give an adequate theoretical basis.

2.10. Elastic surface waves in fluids

Although we are primarily concerned with solid surfaces, the theory of surface waves in fluids can be given in strict parallel with the analysis in the previous sections for solids. It is interesting to expound this in a way

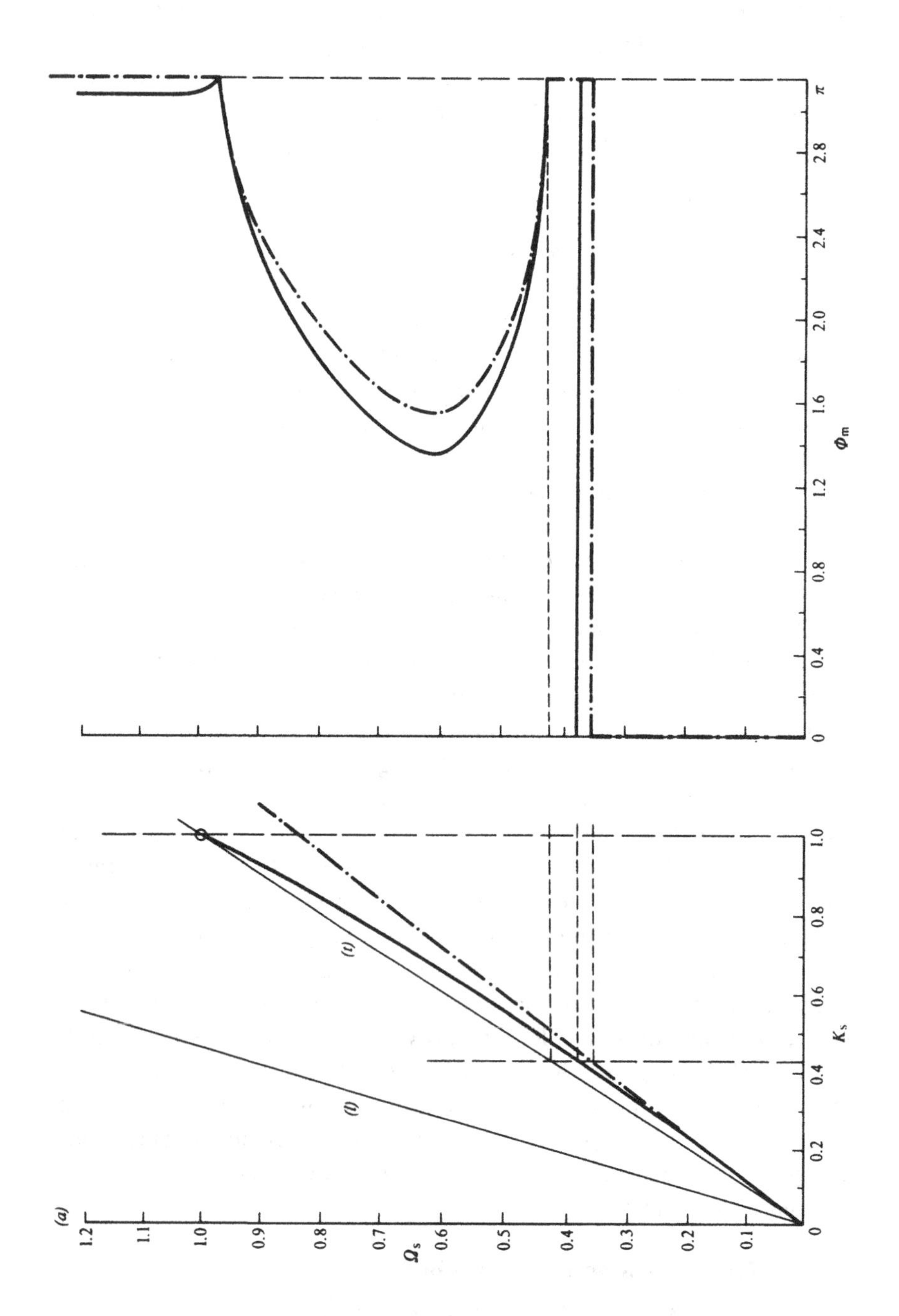
(a)
(l)
(t)
Ω_s
K_s
Φ_m
0
0.1
0.2
0.3
0.4
0.5
0.6
0.7
0.8
0.9
1.0
1.1
1.2
1.6
2.0
2.4
2.8
π

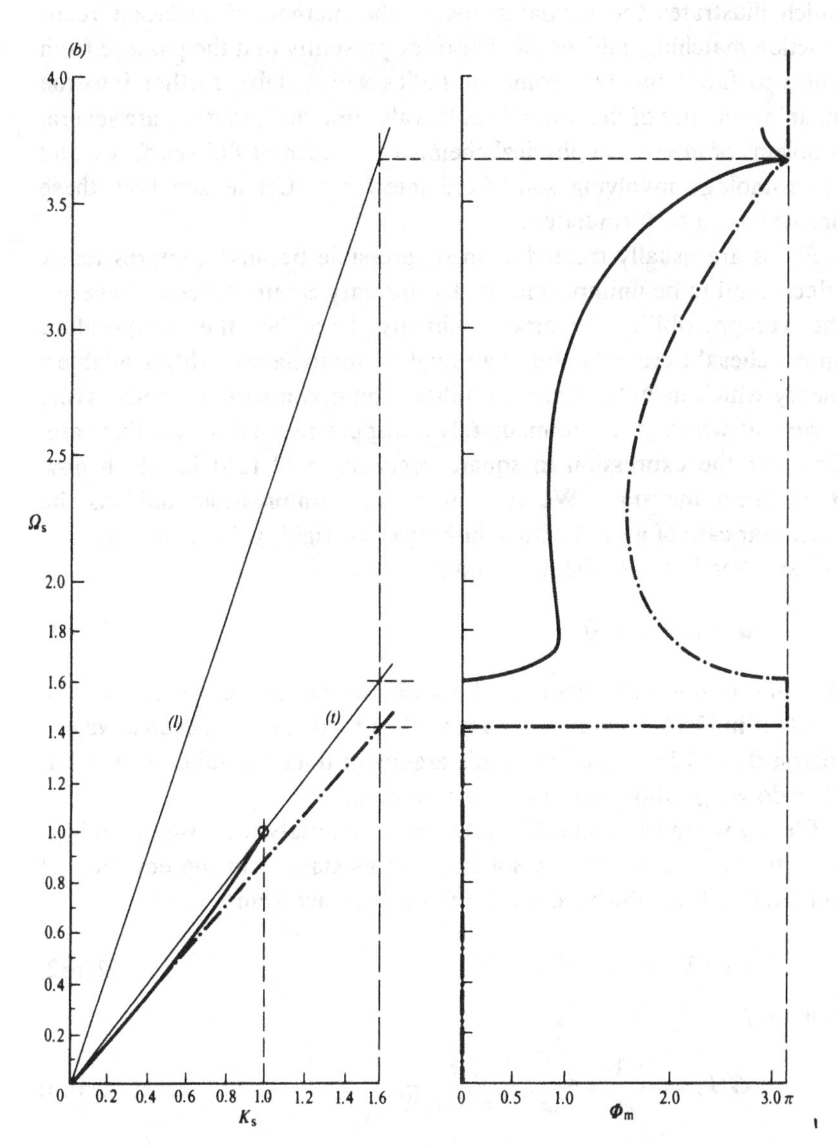

Fig. 25. Matching part of the phase function, Φ_m, as a function of ω for given κ in dimensionless plot, Ω_s against K_s. Solid line: $\tau \neq 0$. Dash-dotted line: $\tau = 0$. (*a*) $K_s = 0.426$, below capillary cutoff. (*b*) $K_s = 1.600$, above capillary cutoff. The results correspond to an averaged isotropic model of Ag, with $l = 3.60 \times 10^5$ cm s^{-1}, $t = 1.59 \times 10^5$ s^{-1}, $\tau = 1415$ dyn cm^{-1} (1 dyn $= 10^{-5}$ N), $\kappa_b = 1.876 \times 10^8$ cm^{-1}, $\omega_b = 2.983 \times 10^{13}$ s^{-1}.

which illustrates the formal scope of the method of surface Green-function matching, and we shall also see presently that the passage from solids to fluids involves some subtleties which delve further into the meaning and use of the formalism. It is also practical, as there are several problems of interest in physical chemistry concerning fluid surfaces, and in seismology involving solid–fluid interfaces. Let us see how these problems can be formulated.

Fluids are usually treated as incompressible because compressibility effects tend to be unimportant under ordinary circumstances. However, the compressibility becomes infinitely large as the temperature approaches the critical point so we might want to have readily available a theory which includes compressibility. There is a sort of hand-waving argument which yields immediately a dispersion relation for this case. Consider the expression in square brackets in (2.155) in which now $\theta_s = 0$ from the start. We can regard the compressible fluid as the particular case of a solid with vanishing shear rigidity, i.e. $\mu = 0$ or $t = 0$. We are then left with the dispersion relation

$$\alpha_s \kappa^2 b_l - \omega^2 = 0. \tag{2.161}$$

But this is not very rigorous, because the factor before the square brackets in (2.155) tends to zero with t^2 as $t \to 0$. Besides, even if we can reason that (2.162) is correct, this argument does not take us very far. How do we go about the rest of the problem?

Clearly we must formulate rather more precisely what we mean by a fluid as the case $\mu = 0$ of a solid. So let us start from the equation of motion (2.13) in which we set $\mu = 0$, i.e. we start from

$$-l^2 \nabla\nabla \cdot \boldsymbol{u} + \ddot{\boldsymbol{u}} = \boldsymbol{f}, \tag{2.162}$$

whence

$$\hat{\boldsymbol{G}}(\boldsymbol{k}, \omega) = \frac{1}{\omega^2}\boldsymbol{I} + \frac{l^2}{\omega^2}\frac{\boldsymbol{k}\boldsymbol{k}}{(l^2k^2 - \omega^2)}, \tag{2.163}$$

which indeed is (2.108) for $\mu = 0$. However, the move from the solid to the fluid is not quite so simple. For example, with $\boldsymbol{k} = (\kappa, 0, q)$ as usual, we can calculate the surface projections

$$\rho \boldsymbol{G} = \mathscr{I} \begin{vmatrix} \kappa^2/2\omega^2 b_l & 0 & 0 \\ 0 & 0 & 0 \\ 0 & 0 & -b_l/2\omega^2 \end{vmatrix}, \tag{2.164}$$

and

$$\rho' \boldsymbol{G}^{\pm} = \mathcal{I} \begin{vmatrix} 0 & 0 & -\mathrm{i}\,\kappa b_l/2\omega^2 \\ 0 & 0 & 0 \\ -\mathrm{i}\,\kappa b_l/2\omega^2 & 0 & 1/2l^2 \end{vmatrix}, \tag{2.165}$$

Not only are these results radically different from those obtained by taking $\mu \to 0$ in (2.110) and (2.111); but what is worse, the matrices being singular, have no inverses, which is a formidable difficulty indeed for a formalism in which we constantly invoke, say, $\boldsymbol{G}^{-1}$. The way out of the difficulty must come from physical considerations. Indeed, it suffices to look at the basic linear form of (2.19), and actually just at the first two terms which are the only ones contributing to the central property (2.20). With $\mu = 0$, only the case $i = 3$ gives a non-zero contribution. Thus, we can find the s.m.d.r. by just using a fictitious surface stimulus F_3, which will be eliminated from the expression for the excess surface pressure at $z = 0$:

$$\lambda\left(\frac{\partial u_1}{\partial x_1} + \frac{\partial u_3}{\partial x_3}\right) = \gamma \frac{\partial^2 u_3}{\partial x_1^2}. \tag{2.166}$$

We are combining Laplace's formula with the statement that our fluid is compressible and therefore the hydrostatic pressure is $-\lambda$ div $\boldsymbol{u}$. Furthermore, the displacements are given by the response of the medium with Green function (2.163) to a surface stimulus F_3. Evaluating all this at $z = 0$ we find

$$\rho l^2[\mathrm{i}\kappa G_{13} + {}'G_{33}^{(-)}] + \gamma\kappa^2 G_{33} = 0. \tag{2.167}$$

This involves only the non-vanishing matrix elements shown in (2.164) and (2.165), whence we have the secular equation

$$\omega^2 = (\gamma/\rho)\kappa^3(1 - \omega^2/l^2\kappa^2)^{1/2}, \tag{2.168}$$

which reduces to the standard formula for capillary waves in the incompressible limit $l \to \infty$. This is indeed the same as (2.161), which is now derived from a rigorous argument but, more than that, we can now look at the actual wave and learn something useful. The wave amplitudes are easily evaluated as

$$u_i(z) = \frac{1}{2\pi} \int \hat{G}_{i3}(q) F_3 \,\mathrm{e}^{\mathrm{i}qz} \,\mathrm{d}z \quad (i = 1, 3). \tag{2.169}$$

By (2.164) this yields

$$\left.\begin{aligned} u_1 &= u_0\kappa\, \mathrm{e}^{-b_l z}\, \mathrm{e}^{\mathrm{i}(\kappa x-\omega t)}, \\ u_3 &= \mathrm{i}\, u_0 b_l\, \mathrm{e}^{-b_l z}\, \mathrm{e}^{\mathrm{i}(\kappa x-\omega t)}. \end{aligned}\right\} \tag{2.170}$$

We shall soon see that (2.168) guarantees that b_l is real. Thus (2.170) represents a stationary surface wave with amplitudes localised near $z = 0$. Furthermore, *this wave is purely longitudinal*, as is easily seen by taking the curl of (2.170), which is zero. One is tempted to go back to (2.133) and to say that all that happens is that $t \to 0$ implies $b_t \to \infty$, and thus the transverse part vanishes. Actually this would not be in itself a bad physical argument. The Rayleigh wave has a mixed character because the matching conditions at the surface mix longitudinal and transverse parts in a way which is typical of surface problems, but in the fluid case we are switching off all possible transverse interactions from the model, to start with. However, this argument would not take us very far because there are no surface tension effects in the Rayleigh problem and there is no way in which we could obtain the dispersion relation (2.168) from (2.116). But now that we know from a direct argument that the surface mode for our model is purely longitudinal, we can use this to rewrite the problem in an equivalent way which avoids the formal difficulties associated with (2.162) and (2.164). Instead of starting from the equation of motion (2.162), we introduce again a velocity potential, as in § 2.8, such that $\boldsymbol{v} = \dot{\boldsymbol{u}} = \nabla \Phi$, which now satisfies the equation

$$-l^2\nabla^2\Phi + \ddot{\Phi} = \Psi_{\text{ext}}, \tag{2.171}$$

where Ψ_{ext} is an appropriate external potential generating the external driving force. The incompressible limit of this $(l \to \infty)$ yields $\nabla^2\Phi = 0$, which was used in § 2.8, while the grad of (2.171) is just (2.162). This has the practical advantage that our Green function is now a scalar, namely,

$$\hat{G}(\boldsymbol{k}, \omega) = 1/(l^2k^2 - \omega^2). \tag{2.172}$$

Notice that we are free to choose whichever definition of Green function we prefer to use, provided it generates a field with the correct properties. In this case (2.172) is the Green function which yields, as response to an external stimulus Ψ_{ext}, a velocity potential from which we can derive a longitudinal displacement field. Indeed, we can easily calculate the density of modes of such a system. We have

$$N(\omega^2) = -\frac{1}{\pi}\,\mathrm{Im}\int_0^\infty \frac{4\pi k^2\,\mathrm{d}k}{(2\pi)^3(lk - \omega + \mathrm{i}\,\eta)(lk + \omega - \mathrm{i}\,\eta)}.$$

The only factor in the denominator which can vanish in the limit $\eta \to 0$ is $(lk - \omega + \mathrm{i}\,\eta)$. Thus

$$N(\omega^2) = -\frac{1}{2\pi^3}\,\mathrm{Im}\int_0^\infty \left[p\frac{1}{lk-\omega} - \mathrm{i}\,\pi\delta(lk-\omega)\right]\frac{k^2\,\mathrm{d}k}{(lk+\omega)},$$

whence

$$N(\omega) = 2\omega N(\omega^2) = \omega^2/2\pi^2 l^3, \tag{2.173}$$

which is indeed the correct formula for the density of modes of a system with longitudinal waves of speed l. It is, of course, the same formula that one obtains starting from (2.162) and (2.163), as can be easily checked.

The point of this exercise was to stress that (2.172), although a great deal simpler than (2.163), is as legitimate a Green function to use here where we know the surface eigenmode is also purely longitudinal; and we get rid of the formal difficulties associated with (2.164). The rest is straightforward application of the formalism. Thus

$$G = 1/2l^2 b_l, \qquad {}'G^{(\pm)} = 1/2l^2.$$

The linear form A for the present problem is $l^2\,{}'G$. Then $\Delta A = 1$ again. If the medium is in $z > 0$, then

$$G_s^{-1} = -A^{(-)}G^{-1} + m_s = l^2 b_l + m_s.$$

Now the boundary condition reflecting the role of Φ is again written down in terms of γ as in § 2.8, and in particular (2.142) also applies to this case. But the object of which we must calculate the discontinuity this time is $l^{2\prime}G$. Thus, from (2.142), we have

$$m_s = -l^2\rho\omega^2/\gamma\kappa^2,$$

and

$$G_s^{-1} = l^2\left(b_l - \frac{\rho}{\gamma}\frac{\omega^2}{\kappa^2}\right).$$

Needless to say, this yields the s.m.d.r. (2.168) at once, but the important thing is that now we have the full canonical formulation worked out without any pathological terms, and we have the explicit formula for G_s, from which we could proceed in the standard way. We shall not carry the analysis any further. The above is sufficient to illustrate the main point, namely, that the move to a fluid as the limit of a solid with vanishing shear rigidity must be exercised with great care. We shall simply look at the form of the dispersion relation (2.168). This can be solved for ω^2, taking the square of both sides and discarding the physically unrealisable root thus introduced. This yields

$$2\omega^2 = -\frac{\alpha_s^2\kappa^4}{l^2} + \sqrt{\left\{\frac{\alpha_s^4\kappa^8}{l^4} + 4\alpha^2\kappa^6\right\}},$$

which is easy to study explicitly. It is convenient to introduce the definition $\kappa_a = l^2/\alpha_s$. The two limits of interest are then

$$\omega \approx \alpha_s^{1/2}\kappa^{3/2} \quad (\kappa \ll \kappa_a); \qquad \omega \approx l\kappa \quad (\kappa \gg \kappa_a),$$

and it turns out that $\omega < l\kappa$ for all κ (fig. 26). This is the main feature. It means that no mode, be it bulk or surface, can propagate with a speed larger than l. It is the analogue of the retardation effects in e.m. theory.

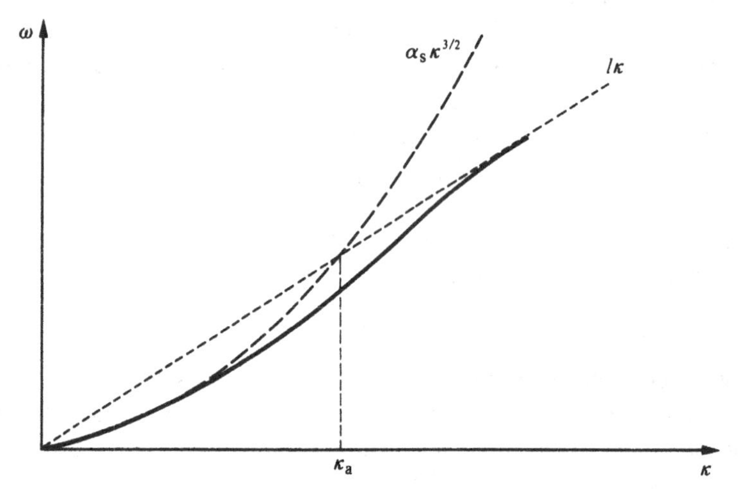

Fig. 26. Dispersion relation for the surface modes of a fluid. Dashed line: incompressible ordinary capillary waves. Solid line: compressible.

The crossover between $l\kappa$ and the capillary branch is just the value κ_a introduced above, which moves up to infinity as $l \to \infty$, the incompressible limit. Ordinarily κ_a is in practice much too large to have any significant effect, but the fact that the curve $l\kappa$ is the upper asymptote is an important theoretical result because l tends precisely to zero as T tends to the critical temperature. Of course, so does γ, but the question is how do the critical behaviours of these two parameters compare, and here again not much is known with certainty.

There is another aspect of fluids which makes a treatment in parallel with solids all the more interesting. This is the effect of viscosity. The general form of the stress tensor in this case is

$$\boldsymbol{\tau} = \lambda \boldsymbol{I} \nabla \cdot \boldsymbol{u} + w(\nabla \boldsymbol{v} + \tilde{\nabla}\tilde{\boldsymbol{v}} - \tfrac{2}{3}\boldsymbol{I}\nabla \cdot \boldsymbol{v}) + w'\boldsymbol{I}\nabla \cdot \boldsymbol{v}, \qquad (2.174)$$

where w and w' are the two viscosity coefficients. In the incompressible case we would simply retain w, but here we shall develop the formal theory, including both shear and bulk viscosity. Thus we start from the linearised equation of motion

$$\rho \ddot{\boldsymbol{u}} - \lambda \nabla\nabla \cdot \boldsymbol{u} - w \nabla^2 \boldsymbol{v} - (w + \tfrac{2}{3} w') \nabla\nabla \boldsymbol{v} = \boldsymbol{f}. \tag{2.175}$$

Since $\boldsymbol{v} = \dot{\boldsymbol{u}}$, it is clear that, with appropriate $-\mathrm{i}\omega$ factors wherever necessary, the problem is mathematically similar to the one which describes the solid. Indeed, defining $\zeta = w/\rho$, $\zeta' = w'/\rho$ and

$$c_{\mathrm{v}}^2 = l^2 - \mathrm{i}\omega(\zeta + \tfrac{1}{3}\zeta'),$$

we find (compare with (2.108))

$$\hat{\boldsymbol{G}}(\boldsymbol{k}, \omega) = \frac{1}{\rho(\omega^2 + \mathrm{i}\,\omega\zeta k^2)}\left[-\boldsymbol{I} + \frac{c_{\mathrm{v}}^2 \boldsymbol{k}\boldsymbol{k}}{(c_{\mathrm{v}}^2 k^2 - \mathrm{i}\,\omega\zeta k^2 - \omega^2)}\right], \tag{2.176}$$

and, for $\boldsymbol{k} = (\kappa, 0, q)$, the surface projection is

$$\boldsymbol{G} = \frac{\mathcal{I}}{\rho}\begin{vmatrix} \dfrac{(\kappa^2 - Q_l Q_t)}{2\omega^2 Q_l} & 0 & 0 \\ 0 & \dfrac{\mathrm{i}}{2\omega\zeta Q_t} & 0 \\ 0 & 0 & \dfrac{(\kappa^2 - Q_l Q_t)}{2\omega^2 Q_t} \end{vmatrix}, \tag{2.177}$$

where

$$Q_l = \left(\kappa^2 - \frac{\omega^2}{c_{\mathrm{v}}^2 - \mathrm{i}\,\omega\zeta}\right)^{1/2}; \qquad Q_t = \left(\kappa^2 - \mathrm{i}\frac{\omega}{\zeta}\right)^{1/2}. \tag{2.178}$$

Also

$$'G^{(\pm)} = \frac{\mathcal{I}}{\rho}\begin{vmatrix} \pm\dfrac{\mathrm{i}}{2\omega\zeta} & 0 & \dfrac{\mathrm{i}\,\kappa(Q_t - Q_l)}{2\omega^2} \\ 0 & \pm\dfrac{\mathrm{i}}{2\omega\zeta} & 0 \\ \dfrac{\mathrm{i}\,\kappa(Q_t - Q_l)}{2\omega^2} & 0 & \dfrac{1}{2(c_{\mathrm{v}}^2 - \mathrm{i}\,\omega\zeta)} \end{vmatrix}. \tag{2.179}$$

So far these results could almost be written down at once by appropriate transliteration of those of § 2.7. It is obvious, for example, how to write down the linear form $\hat{A}$ for this case. But these are only the starting elements of the formal analysis and now we must look at the physics of the problem, which is clearly different. First, it is obvious that viscosity effects will produce damping of any type of modes, bulk or surface. Secondly, we must investigate the boundary conditions at the interface. If we are interested in the free and clean surface of a pure liquid it suffices to include the normal stress condition (2.148), remembering (i) that for a fluid $\tau=\gamma$, and (ii) that we must use (2.174) for τ_{ij}. But the case of real interest in practice, whether for a free fluid surface or for the interface of separation between two non-miscible fluids, is that in which there is an adsorbed layer at the interface. This could be a very difficult problem indeed, and it might require a formal rheological analysis of the adsorbate. But, in practice, the situation that often corresponds to the cases of actual interest in physical chemistry can be roughly characterised as follows. The frequencies and diffusional time constants involved are such that diffusional interchange between the adsorbed layer and the adjacent fluid layers is usually negligible. The adsorbed layer then expands and contracts by itself, so to speak, with its own surface elasticity modulus δ_s. We can also usually assume that specific surface viscosity effects are negligible compared with surface elasticity. The situation is then the strict analogue of that of § 2.9, with the following differences: we start from (2.174)–(2.179) and we use the boundary conditions (2.148) and (2.152), with γ instead of τ, and δ_s instead of Δ_s. For a free surface this yields

$$G_s^{-1}=\rho\mathcal{S}/(\kappa^2-Q_lQ_t)$$

$$\times\begin{vmatrix} \omega^2Q_l+\dfrac{\delta_s\kappa^2(\kappa^2-Q_lQ_t)}{\rho} & 0 & \mathrm{i}\kappa[\omega^2-2\,\mathrm{i}\,\omega\zeta(Q_lQ_t-\kappa^2)] \\ 0 & -\mathrm{i}\omega\zeta Q_t(\kappa^2-Q_lQ_t) & 0 \\ -\mathrm{i}\kappa[\omega^2-2\,\mathrm{i}\,\omega\zeta(Q_lQ_t-\kappa^2)] & 0 & \omega^2Q_t+\dfrac{\gamma\kappa^2(\kappa^2-Q_lQ_t)}{\rho} \end{vmatrix}. \tag{2.180}$$

With this result we could study a viscous fluid with an adsorbed layer on its surface. The case of two different fluids with an interface is similarly studied. Actually there is a further difference which characterises the fluid case, namely, the presence of gravitational effects. These ought to be included in principle, for the sake of a reasonably complete formal theory, and also sometimes in practice. It is very easy to include the effect of gravitational forces in this analysis. Suppose these act in the x_3-direction and stick to the convention that medium 1 is in $x_3<0$. It suffices to add the change in gravitational force, $\rho_2 g\boldsymbol{u}^{(2)}-\rho_1 g\boldsymbol{u}^{(1)}$ at $x_3=0$, to the change in normal stress $\tau^{(2)}_{33}-\tau^{(2)}_{33}$ in the normal stress condition (2.148). Then $\gamma\kappa^2$ is replaced everywhere by $\gamma\kappa^2+g(\rho_1-\rho_2)$. Application of the formal theory is now straightforward. For the general case, i.e., two different (non-miscible) fluids with an adsorbed layer at the interface, including gravitational effects, this leads to the following s.m.d.r.:

$$\mathrm{i}\,\omega\kappa^2[w_1d_1(\kappa^2-Q_{l2}Q_{t2})-w_2d_2(\kappa^2-Q_{l1}Q_{t1})][a_1(\kappa^2-Q_{l2}Q_{t2})$$
$$-a_2(\kappa^2-Q_{l1}Q_{t1})]=\omega^2LT, \qquad (2.181)$$

where, for each medium $i=1,2$,

$$a_i=\rho_ic_{vi}^2(\kappa^2-Q_{li})^2+\mathrm{i}\,w_i\omega(\kappa^2+Q_{li}^2-2Q_{li}Q_{ti});$$
$$d_i=\kappa^2+Q_{ti}^2-2Q_{li}Q_{ti},$$

and L and T are given by

$$-\omega L=\delta_s\kappa^2(\kappa^2-Q_{l1}Q_{t1})(\kappa^2-Q_{l2}Q_{t2})$$
$$+[\rho_1Q_{l1}(\kappa^2-Q_{l2}Q_{t2})+\rho_2Q_{l2}(\kappa^2-Q_{l1}Q_{t1})]\omega^2, \qquad (2.182)$$

and

$$-\omega T=[\gamma\kappa^2+g(\rho_1-\rho_2)](\kappa^2-Q_{l1}Q_{t1})(\kappa^2-Q_{l2}Q_{t2})$$
$$+[\rho_1Q_{t1}(\kappa^2-Q_{l2}Q_{t2})+\rho_2Q_{t2}(\kappa^2-Q_{l1}Q_{t1})]\omega^2. \qquad (2.183)$$

It is interesting to look at some limiting cases. The simplest is the non-viscous one, which corresponds to $\delta_s\to0$, $w_i\to0$, $w_iQ_{ti}\to0$, $w_iQ_{ti}^2\to-\mathrm{i}\rho_i\omega$, $Q_{li}\to b_{li}$, $a_i\to\rho_i\omega^2$, and $d_i\to\infty$. Then the s.m.d.r. reduces to

$$\omega^2=\frac{\gamma\kappa^2+g(\rho_1-\rho_2)}{\rho_1b_{l2}+\rho_2b_{l1}}b_{l1}b_{l2}.$$

Neglecting compressibility ($b_{li}=\kappa$) we have the textbook formula known as the *Kelvin relation.*

If we neglect compressibility but retain all other effects (most often the case of practical interest) then $c_{vi}\to\infty$, $Q_{li}\to\kappa$, $\rho_ic_{vi}^2(\kappa^2-Q_{li})^2\to\rho_i\omega^2$, $a_i\to\mathrm{i}w_i\omega(\kappa-Q_{ti})^2$, and $d_i\to(\kappa-Q_{ti})^2$. Then (2.182) and (2.183) become

$$L_{\mathrm{inc}}=-\delta_s\kappa^2/\omega^2+\mathrm{i}\,w_1(\kappa+Q_{t1})+\mathrm{i}\,w_2(\kappa+Q_{t2}), \qquad (2.184)$$

and

$$T_{\mathrm{inc}}=-\frac{\gamma\kappa^2}{\omega}-\frac{g(\rho_1-\rho_2)}{\omega}+(\rho_1+\rho_2)\frac{\omega}{\kappa}+\mathrm{i}\,w_1(\kappa+Q_{t1})$$
$$+\mathrm{i}\,w_2(\kappa+Q_{t2}), \qquad (2.185)$$

and the s.m.d.r. is

$$[w_1(\kappa-Q_{t1})-w_2(\kappa-Q_{t2})]^2+L_{\mathrm{inc}}T_{\mathrm{inc}}=0. \qquad (2.186)$$

The actual wave amplitudes can be written down by proceeding exactly as in § 2.7. Using (2.131) and (2.176) we find, for the matrix elements $\hat{G}_{ij}(z;\kappa;\omega)$,

$$\left.\begin{aligned}
\hat{G}_{11} &= \frac{1}{2\omega^2}\left[\frac{\kappa^2}{Q_l}\,\mathrm{e}^{-Q_l z} - Q_t\,\mathrm{e}^{-Q_t z}\right];\\
\hat{G}_{13} = \hat{G}_{31} &= \frac{\mathrm{i}\kappa}{2\omega^2}[\mathrm{e}^{-Q_l z} - \mathrm{e}^{-Q_t z}];\\
\hat{G}_{33} &= \frac{1}{2\omega^2}\left[-Q_l\,\mathrm{e}^{-Q_l z} + \frac{\kappa^2}{Q_t}\,\mathrm{e}^{-Q_t z}\right].
\end{aligned}\right\} \qquad (2.187)$$

These formulae are identical to (2.132), with b_l, b_t replaced by Q_l, Q_t, remember that the form of the wave is always given by (2.130). Thus, the form of the spatial dependence of the surface wave comes from the column vectors $(\hat{G}_{11}, \hat{G}_{13})$ and $(\hat{G}_{31}, \hat{G}_{33})$ which, as in § 2.7, occur neatly as the sum of longitudinal and transverse parts, and this holds on either side of the interface. However, the precise combination of the matrix elements $\hat{G}_{ij}$ which gives the components u_i depends on the ratio F_3/F_1, i.e. on G_s, and this is where the situation now is different. If we follow this up we come across an interesting situation. Suppose the densities and viscosities of the two fluids are equal. (This is not too hypothetical. In cases of real physicochemical interest this equality turns out to be very neatly satisfied.) Under these conditions the square bracket in (2.186) vanishes and the s.m.d.r. factorises out into two equations, namely $L_{\text{inc}} = 0$, and $T_{\text{inc}} = 0$. Using in each instance the corresponding form of the dispersion relation obtained from (2.184) or (2.185), it turns out that with (2.184) the resulting combination of matrix elements (2.187) picks out only the terms associated with Q_l, while the combination of terms associated with Q_t is identically zero. This means that the *surface wave* in question is purely longitudinal. Similarly, with (2.185) we obtain a purely transverse surface wave. What happens is that the coupling of longitudinal and transverse parts at the surface is due to the existence of transverse interactions in the media, e.g. shear rigidity in solids, or viscosity in fluids. If these vanish, then we obtain purely longitudinal surface waves. This was the situation with the non-viscous fluid (2.170) regarded as the limit $\mu = 0$ of the solid. But, if the transverse interactions – and the inertial response – have equal coefficients on both sides, then longitudinal modes couple independently on both sides of the interface, giving longitudinal surface modes, and also transverse modes couple independently, giving transverse surface modes. This is the result we have just obtained.

This decoupling is more general. If both fluids are compressible, and the compressibilities are also equal, then the l.h.s. of (2.181) vanishes and we obtain one dispersion relation, namely $L = 0$, for longitudinal surface modes, and another one, namely $T = 0$, for transverse surface waves. Notice that L (2.182) depends only on the surface elasticity modulus, δ_s, associated with the boundary condition on the tangential stress parallel to the surface, while T (2.183) depends only on the surface tension, γ, associated with the normal stress boundary condition.

3

Electronic theory of simple metal surfaces

3.1. The density of states and related basic concepts for a bounded electron gas

Except in the discussion of inhomogeneous surface layers in semiconductors (§ 1.7) we have so far assumed that the homogeneous bulk material extends right up to the surface, although the general formalism of chapter 2 does not depend on any particular model of this kind. We have seen that even with this simplification the effects of the surface itself raise a number of non-trivial problems. We are now going to concentrate on the study of the inhomogeneity associated with the surface. Notice that even the discussion of § 1.7 was on a macroscopic scale. What we shall do now is rather different. The discussion in this chapter will be limited to the static properties of metallic surfaces, but we shall aim at a fairly detailed quantum mechanical picture of the state of affairs in a 'surface region' on a microscopic scale.

From a phenomenological point of view the main static physical properties of a metallic surface are the *surface energy*, the *work function*, and, to some extent, the surface charge. The latter manifests itself rather indirectly in different types of experiments and is ultimately the result of a detailed charge redistribution which must be selfconsistently related to the surface potential barrier. Briefly, we need a detailed quantum mechanical study of the ground state of a surface system and this involves, as we shall see, some basic concepts like the density of states – already discussed in chapter 2 – and the *phase shifts* due to the potential barrier, regarded as a perturbation of the bulk. We shall first discuss in detail the *jellium model* – i.e. an electron gas embedded in a continuous positive background charge – and then study the inclusion of the discrete crystal lattice.

3.1.1. Bounded one-dimensional electron gas

Let us first concentrate on the concepts of phase shifts and density of states, remembering the definitions of § 2.3. We shall build up the discussion by increasing the levels of complexity. Thus we start with the simplest conceivable model, namely (fig. 27) a one-dimensional gas of

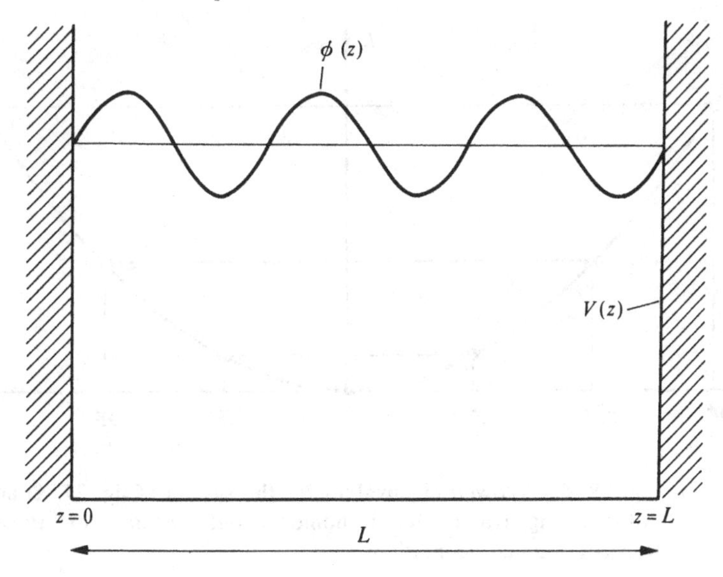

Fig. 27. Wavefunctions in a one-dimensional system bounded by infinite barriers.

independent electrons, bounded, within a length L, by two infinite potential barriers. The normalised wavefunctions are

$$\psi_m(z) = \surd(2/L)\sin(m\pi z/L), \tag{3.1}$$

with discrete energy levels

$$E_m = \tfrac{1}{2}(m\pi/L)^2. \tag{3.2}$$

Here m is a positive integer, not to be confused with the electron mass. Here and henceforth we shall use atomic units $e = \hbar = m = 1$. The energy levels given by (3.2) are shown in fig. 28 as solid circles on the curve $E = \frac{1}{2}q^2$, corresponding to $q = m\pi/L$.

Consider now a length L' of 'homogeneous bulk material', i.e. with *Born–von Karman* boundary conditions at $z = 0, L'$. This system admits both sine and cosine wavefunctions which can now be combined into complex exponentials, i.e.

$$\psi_m = (1/\surd L')\exp(\mathrm{i}2\pi mz/L'), \tag{3.3}$$

with eigenvalues

$$E_m = \tfrac{1}{2}(2\pi m/L')^2, \tag{3.4}$$

where now m (still an integer) can be positive, negative, or *even zero*. This simple fact is important. Let us take $L' = 2L$. The eigenvalues (3.4) are

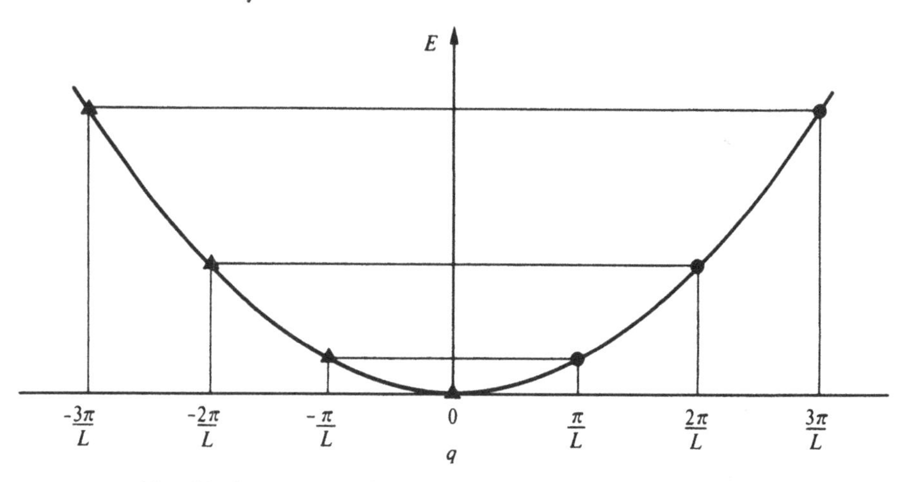

Fig. 28. ●, Allowed eigenvalues for the system of fig. 27. ● and ▲, Allowed eigenvalues for a homogeneous system with Born–von Karman boundary conditions.

then both the solid circles and the triangles of fig. 28 and can be regarded as given by the same formula (3.2) but allowing now for $m \leqslant 0$. Thus, in order to obtain as many states as in (3.4) we would need twice the states contained in the length L with infinite barriers at $Z = 0, L$ and we would have to add another state at $E = 0$. In other words, the relation between the values of the density of states of the two systems is

$$N_t^\infty(E) = \tfrac{1}{2}[N^{2L}(E) - \delta(E)]. \tag{3.5}$$

Now take the limit of very large L. The density of states $N^{2L}(E)$ is then

$$N^{2L}(E) = 2\frac{\mathrm{d}q}{\mathrm{d}E}\frac{L}{\pi}, \tag{3.6}$$

since the allowed states in q-space are separated by an infinitesimal interval π/L. The factor 2 is due to the fact that the same value of E corresponds to $\pm q$. We have not yet considered spin; thus

$$N_t^\infty(E) = \frac{L}{\pi}\frac{\mathrm{d}q}{\mathrm{d}E} - \frac{1}{2}\delta(E). \tag{3.7}$$

Notice that the first term of this formula is of order L, while the second term is of order unity. Thus (3.7) holds only as long as (3.6) is correct to order L^{-1}, and this is true if the separation between states in q-space is the first-order infinitesimal π/L to order L^{-3}. This is the situation we have here, but we shall encounter cases where this is not true because of energy level shifts of order L^{-2}. In surface problems it is important to keep track of such subtle details.

It is now clear that the first term of (3.7) is just the density of states of a length L of 'homogeneous bulk material', while the second term contains the effect due to the surfaces, i.e. to the *two* surfaces at $z = 0, L$. Thus, in the nomenclature of § 2.3, for a semi-infinite one-dimensional system with one surface, the surface density of states is

$$N_s^\infty(E) = -\tfrac{1}{4}\delta(E), \tag{3.8}$$

i.e. the effect of the surface is to remove $\frac{1}{4}$ state from the bottom of the free electron band.

We have only considered energies, and really the only (formal) reason why we have defined (3.8) as a surface quantity is that its order of magnitude is L^{-1} compared with the bulk term. Of course we know from the formal analysis of § 2.3 that the changes on the bulk density of states are associated with terms which are localised near the surface, but we have not yet established the connection with the present elementary analysis and it will be instructive to look at the problem in real space. We want to have an explicit picture of the spatial extension of the states associated with (3.8).

It is useful to introduce the concept of local density of states, defined in general as

$$N(E, \boldsymbol{r}) = \sum_i |\psi_i(\boldsymbol{r})|^2 \delta(E - E_i). \tag{3.9}$$

Here

$$N(E, z) = N_t^\infty(E)|\psi_E(z)|^2, \tag{3.10}$$

which by (3.1) and (3.7) is, if we pass to the continuum limit,

$$N(E, z) = \frac{1}{\pi}\frac{\mathrm{d}q}{\mathrm{d}E}[1 - \cos(2z\sqrt{(2E)})]. \tag{3.11}$$

Notice that the term in $\delta(E)$ does not contribute to (3.11) because the (1-cosine) term vanishes for $E = 0$. It is clear that the constant term in (3.11) is just what yields the homogeneous bulk term, while the z-dependent term represents the effect of the surface. Let us find the charge density which goes with (3.11). For a Fermi system, accounting for spin degeneracy, we have

$$n(z) = 2\int_{-\infty}^{E_F} N(E, z)\,\mathrm{d}E. \tag{3.12}$$

Here, from (3.11),

$$n(z) = \frac{2q_F}{\pi}\left[1 - \frac{\sin(2q_F z)}{2q_F z}\right], \tag{3.13}$$

where $E_F = \frac{1}{2}q_F^2$. If we put n_b for the bulk charge density and $n_b + n^s(z)$ for $n(z)$, then the surface term is

$$n_s(z) = -\frac{2q_F}{\pi}\frac{\sin(2q_F z)}{2q_F z}, \tag{3.14}$$

which is localised near the surface within a distance of order q_F^{-1}, and exhibits the typical *Friedel-type* oscillations (fig. 29). Notice that $n_s(z) \to 0$

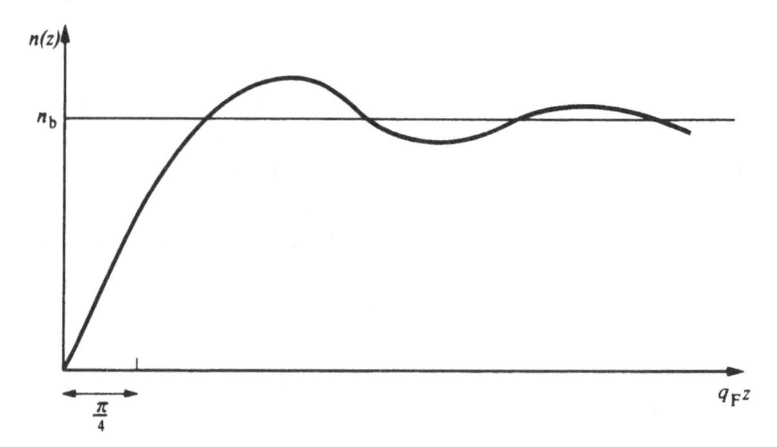

Fig. 29. Electronic density distribution near a hard wall – infinite barrier – at $z = 0$.

as $z \to 0$; this means there is some charge defect near the surface. By (3.14) the total value of this charge defect is

$$\int_0^\infty n_s(z)\,\mathrm{d}z = -\tfrac{1}{2}. \tag{3.15}$$

Indeed, if we work directly from the density of states we find that this is exactly the charge associated with (3.8), inserting the factor 2 for spin degeneracy.

It is instructive to look at the same problem in yet another way which involves another useful concept. We define the *integrated density of states* $N_d(E)$ as

$$N_d(E) = \int_0^d N(E, z)\,\mathrm{d}z. \tag{3.16}$$

Notice that this is also a position-dependent quantity. Using (3.11), we have

$$N_d(E) = \frac{d}{\pi\sqrt{(2E)}}\left[1 - \frac{\sin(2d\sqrt{(2E)})}{(2d\sqrt{(2E)})}\right]. \tag{3.17}$$

Again the first term is just the homogeneous bulk term integrated from 0 to d. This formula resembles (3.13) in appearance but is in fact more akin to (3.11). It is the energy dependence that is of interest here. First we notice that, for any value of d,

$$-\int_0^{\infty} \frac{d}{\pi\sqrt{(2E)}} \frac{\sin(2d\sqrt{(2E)})}{(2d\sqrt{(2E)})} \mathrm{d}E = -\tfrac{1}{4}. \tag{3.18}$$

We find again the defect of $\frac{1}{4}$ state that we have already associated with the effect of the surface. But now we can also see how this is distributed in energy. Fig. 30 shows $N_d(E)/d$ as a function of E. Of course for large E

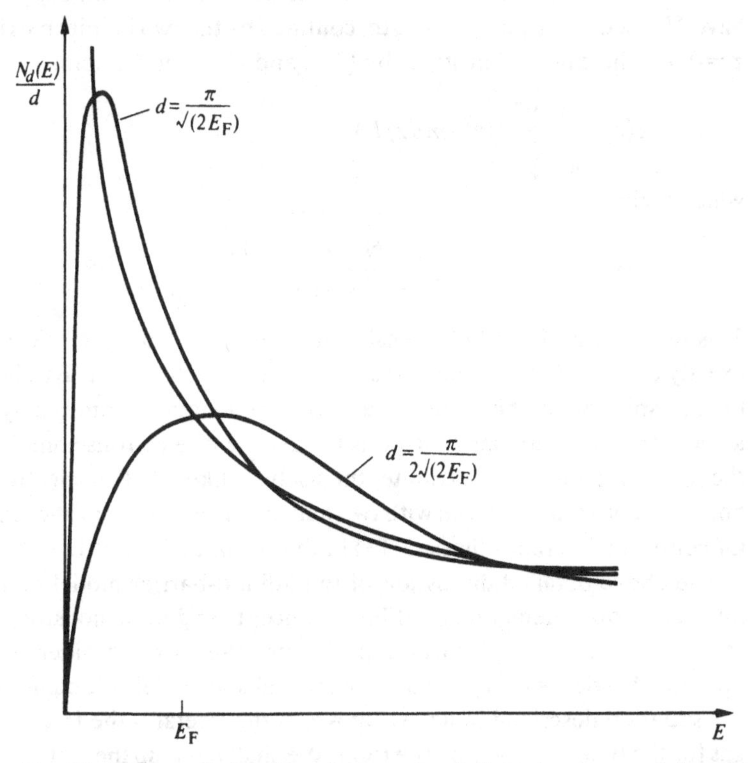

Fig. 30. Integrated density of states for infinite-barrier model and different values of d. If $d \to \infty$, then $N_d(E)/d \to 1/\pi\sqrt{(2E)}$, which is the curve with a divergence at $E=0$.

this yields again the uniform bulk density of states $1/\pi\sqrt{(2E)}$, identical to the first term of (3.11), while for $E \to 0$, $N_d(E)/d \to (2d^2/3\pi)\sqrt{(2E)}$. That is to say, the effect of the surface is only significant in the low energy range, i.e. near the bottom of the free electron band. For given d, this affects a range of energies such that $2d\sqrt{(2E)} \leqslant 1$. Thus, by taking d very large, the differences between $N_d(E)/d$ and $1/\pi\sqrt{(2E)}$ accumulate in a very narrow range of energies up to values of order d^{-2}. Therefore, in the limit $d \to \infty$ we have a defect of $\frac{1}{4}$ state at the bottom of the band. In practice, for $d \gg q_F^{-1}$, the differences between $N_d(E)/d$ and $1/\pi\sqrt{(2E)}$ are concentrated in a range of energies very much less than E_F, thus the defect of $\frac{1}{4}$ state is spatially located within distances of order q_F^{-1}.

We have thus converged on the result of (3.7) by looking at the problem from two different angles, namely, spatial dependence and energy dependence of the surface term. This demonstrates the role of the density of states and related basic concepts which will be useful in more general situations. As a trivial application let us keep L finite and suppose we have N electrons inside this length, confined by the two infinite barriers at $z = 0, L$. The charge density is, by (3.1) and allowing for spin,

$$n(z) = \frac{4}{L} \sum_{m=1}^{N/2} \sin^2 (m\pi z/L),$$

which yields

$$n(z) = \frac{(N+1)}{L}\left\{1 - \frac{\sin[(N+1)\pi z/L]}{(N+1)\sin(\pi z/L)}\right\}. \tag{3.19}$$

This means a uniform bulk density of $(N+1)/L$. The extra electron is exactly cancelled out by the defect of $\frac{1}{4}$ state at each barrier. By allowing for the spin factor this removes exactly one electron from the system, leaving the total number N. That is to say, the N electrons contained in the box of length L distribute themselves like $(N+1)$ electrons of homogeneous bulk material with two half electrons removed at each wall. Of course (3.19) reproduces (3.13) in the continuous limit $L \to \infty$.

The above detailed discussion of the infinite-barrier model served to introduce some elementary but basic concepts and to demonstrate their role in the analysis of surface problems. We now consider a finite potential barrier, which gives a far more realistic model. We contemplate the situation described in fig. 31, in which the 'metal' – the free electron gas for the time being – is on the right. We shall build up the system in two stages, as in § 2.3. First we introduce an infinite barrier or hard wall, which will serve as a reference. Then we lower the potential step to a finite value and displace it if necessary – this point will be very important – until we reach what we now regard as the actual barrier. In the first step of this process we introduce two infinite barriers separated by a distance L. We have seen that if L is sufficiently large the surface effects are confined to the different surfaces without mutual interference. In the next step we retain the second infinite barrier – the one at $z = L$ – exactly as it was,

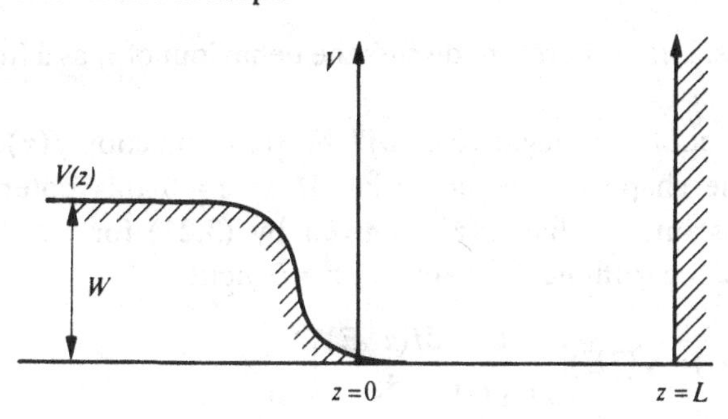

Fig. 31. A smooth surface potential barrier near $z = 0$ and a hard wall at $z = L$.

while we change the first one, by lowering and displacing it, to obtain the actual finite barrier. The new wavefunctions satisfy the Schrödinger equation

$$-\tfrac{1}{2}\,d^2\psi(z)/dz^2 + V(z)\psi(z) = E\psi(z), \tag{3.20}$$

with the potential of fig. 31.

For z sufficiently deep inside, where $V(z)$ is practically constant, the wavefunction behaves like (3.1) except for an energy dependent phase shift $\eta(E)$, i.e.

$$\psi(z) \sim \sin(qz + \eta). \tag{3.21}$$

If the barrier becomes infinitely high, then $\eta \to 0$ for all energies and (3.21) becomes (3.1). This wavefunction must still satisfy the condition $\psi(L) = 0$, whence the eigenvalue condition

$$qL + \eta = m\pi, \tag{3.22}$$

whereas with an infinite barrier at $z = 0$ we have

$$qL = m\pi. \tag{3.23}$$

Fig. 32 shows the eigenvalues of (3.22) displaced by $-\eta(q)/L$ with respect to those of (3.23). If we know the phase shift we know the new

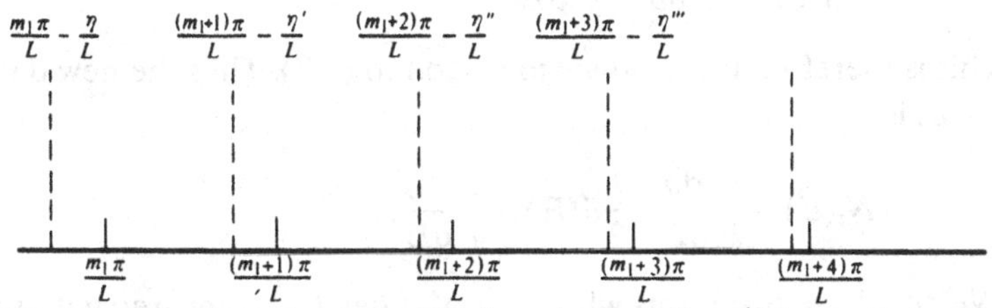

Fig. 32. The allowed values of q for a hard-wall model are displaced when the infinite barrier is lowered and becomes a finite barrier which produces a phase shift η.

density of states. Let us therefore discuss the behaviour of η as a function of $E=\frac{1}{2}q^2$.

The wavefunction for negative z will be some function $f(z)$ which depends on the shape of the potential. If at $z=0$ this potential is practically constant, so that $\psi(z)$ is given by (3.21) for $z\geq 0$, then matching of the logarithmic derivative at $z=0$ yields

$$\tan\eta(E)=\sqrt{(2E)}\left[\frac{1}{f(z,E)}\frac{\mathrm{d}f(z,E)}{\mathrm{d}z}\right]^{-1}_{z=0}. \tag{3.24}$$

Now, the energy E here is the kinetic energy of the electrons inside, and it is easily seen that the logarithmic derivative $(f'/f)_0$ is practically independent of E for E sufficiently small compared with the barrier height. Thus

$$\eta(E\to 0)\to \text{const.}\times E^{1/2}. \tag{3.25}$$

This observation will be useful again later on. For the moment it tells us that the change in the density of states due to the move from the first infinite barrier to the actual finite barrier is vanishingly small for $E\to 0$. Thus the analysis of the infinite-barrier model is equally applicable in this limit. In particular, *with a finite barrier we also have a defect of $\frac{1}{4}$ state at the bottom of the band.*

At all finite energies the spacing between eigenvalues in the q-axis, which was π/L for the infinite barrier, is now, to order L^{-2},

$$E(m+1)-E(m)\approx\frac{\pi}{L}-\frac{\mathrm{d}}{\mathrm{d}q}(\eta/L)\frac{\pi}{L}=\frac{\pi}{L}\left(1-\frac{1}{L}\frac{\mathrm{d}\eta}{\mathrm{d}q}\right). \tag{3.26}$$

With the spacing π/L in q-space we had a density of states $(L/\pi)\,\mathrm{d}q/\mathrm{d}E$ in the limit $L\to\infty$. The correction to the level separation appearing in (3.26) adds a correction to the density of states which, again in the limit $L\to\infty$, is

$$+\frac{L}{\pi}\frac{\mathrm{d}q}{\mathrm{d}E}\times\frac{1}{L}\frac{\mathrm{d}\eta}{\mathrm{d}q}=\frac{1}{\pi}\frac{\mathrm{d}\eta}{\mathrm{d}E}.$$

This is therefore the term we must add to (3.7). Thus the new density of states is

$$N_t(E)=\frac{L}{\pi}\frac{\mathrm{d}q}{\mathrm{d}E}-\frac{1}{2}\delta(E)+\frac{1}{\pi}\frac{\mathrm{d}\eta}{\mathrm{d}E}. \tag{3.27}$$

We now have two terms which are of order L^{-1} compared with the bulk term, and are due to the presence of the surface barrier which produces the phase shift $\eta(E)$. If we now let the second barrier recede to infinity, we must remember to take $\frac{1}{2}$ of the term in the δ-function, as we did on going from (3.7) to (3.8). Thus if we have a semi-infinite system bounded by a finite potential barrier, the surface density of states is

$$N_s(E) = -\frac{1}{4}\delta(E) + \frac{1}{\pi}\frac{d\eta}{dE}. \tag{3.28}$$

There is an arbitrary additive term in this formula, which depends on the origin chosen for z. Let us again take L very large but still finite. Suppose we displace the origin of z by an amount z_0, so that the length L is reduced to $L - z_0$. The bulk term then loses $(z_0/\pi)\, dq/dE$, but the wavefunction inside is now written as

$$\psi(z) \sim \sin\,[q(z+z_0)+\eta], \tag{3.29}$$

i.e. the phase shift is now $\eta + qz_0$. Therefore N_s increases by exactly the amount lost by the bulk term, so that the total density of states remains constant.

Let us now look again at the space-dependent quantities, namely, charge density and local density of states. The analysis with a finite potential barrier is now not quite so simple. First we need the correct normalisation factor for (3.21), for which we write down explicitly the Schrödinger equation (3.20) for two different energies, E and E':

$$-\frac{1}{2}\frac{\partial^2\psi(z,E)}{dz^2} + V(z)\psi(z,E) = E\psi(z,E); \tag{3.30a}$$

$$-\frac{1}{2}\frac{\partial^2\psi(z,E')}{dz^2} + V(z)\psi(z,E') = E'\psi(z,E'). \tag{3.30b}$$

Multiplying (3.30a) by $\psi(z, E')$, and (3.30b) by $\psi(z, E)$, and subtracting we have

$$\psi(z,E)\psi(z,E') = \frac{1}{E-E'}\frac{1}{2}\frac{\partial}{\partial z}\left\{\psi(z,E)\frac{\partial\psi(z,E')}{\partial z} - \psi(z,E')\frac{\partial\psi(z,E)}{\partial z}\right\}.$$

Taking the limit $E' \to E$ and integrating up to a positive value of $z = d$, we have

$$\int_{-\infty}^{d} \psi^2\, dz = \frac{1}{2}\left[\frac{\partial\psi}{\partial E}\frac{\partial\psi}{\partial z} - \psi\frac{\partial^2\psi}{\partial z\,\partial E}\right]_{z=d}. \tag{3.31}$$

Since the r.h.s. is evaluated for positive z we can write ψ as in (3.21) with a normalisation factor $C(E)$. Thus

$$\int_{-\infty}^{d} \psi^2(z,E)\, dz = \frac{C^2}{2}\left\{\left[d + \sqrt{(2E)}\frac{d\eta}{dE}\right] - \frac{1}{2\sqrt{(2E)}}\sin\,(2d\sqrt{(2E)} + 2\eta)\right\}. \tag{3.32}$$

The normalisation constant is now found by recalling that (from (3.22)) the sine in (3.32) vanishes for $z = L$. This yields

$$C = [2/L + \surd(2E)\,\mathrm{d}\eta/\mathrm{d}E]^{1/2}. \tag{3.33}$$

Thus the effect of the finite barrier is not only to produce an energy-dependent phase shift in the wavefunction but also to introduce an energy dependence in the normalisation constant.

We need this constant to evaluate (3.9) for this case, i.e. (3.10) with N_t replaced by the expression given in (3.27). Using (3.33), we obtain

$$N(E, z) = \frac{1}{\pi}\frac{\mathrm{d}q}{\mathrm{d}E}\{1 - \cos(2z\surd(2E) + 2\eta)\} \quad (z > 0). \tag{3.34}$$

The contribution of the δ-function term is $\delta(E)\sin(qz + \eta)$, which vanishes because of (3.25). Thus the charge density (3.12) for $z > 0$ is

$$n(z) = \frac{2}{\pi}\int_{-\infty}^{E_F}\frac{\mathrm{d}q}{\mathrm{d}E}\{1 - \cos(2z\surd(2E) + 2\eta)\}\,\mathrm{d}E. \tag{3.35}$$

In order to evaluate this formula explicitly we would need to know $\eta(E)$, which in general can only be achieved numerically. However, some key features of $n(z)$ can be studied analytically. Thus, for sufficiently large z such that $2z\surd(2E_F) \gg 1$ a power series expansion in $1/z\surd(2E_F)$ can be obtained by partial integration of (3.35). This yields

$$n(z) \approx \frac{2}{\pi}\surd(2E_F)\left\{1 - \frac{\sin(2z\surd(2E_F) + 2\eta_F)}{2z\surd(2E_F)}\right\} + O(1/2E_F z^2), \tag{3.36}$$

where $\eta_F = \eta(E_F)$. The shape of this density profile is very similar to that of (3.13). We have again the constant bulk term and the oscillating surface term which decays for large z, where to order L^{-1} it does not contribute to the mean charge density.

Let us look more closely at the localisation of the surface density of states by studying again the integrated density of states. By (3.16), (3.34), (3.33) and (3.32)

$$\begin{aligned} N_d(E) &= \int_{-\infty}^{d} N(E, z)\,\mathrm{d}z \\ &= \frac{d}{\pi}\frac{\mathrm{d}q}{\mathrm{d}E}\left\{1 - \frac{\sin(2d\surd(2E))}{2d\surd(2E)}\cos 2\eta\right\} \\ &\quad + \left\{\frac{1}{\pi}\frac{\mathrm{d}\eta}{\mathrm{d}E} - \frac{1}{\pi}\frac{\cos(2d\surd(2E))}{4E}\sin 2\eta\right\}. \end{aligned} \tag{3.37}$$

The first term on the r.h.s. is just like (3.17) except for the cos 2η factor.

Look at its behaviour for large d. The position-dependent term, i.e. the change upon the constant bulk term, is only significant if E is sufficiently small so that $E \lesssim d^{-2}$. But we have seen in (3.25) that $\eta \to 0$ for small E, so that for large d we can let $\cos 2\eta \to 1$, whereupon the first term on the r.h.s. of (3.37) behaves indeed like (3.17). Thus the discussion following (3.17) applies here and this yields again a defect of $\frac{1}{4}$ state localised in energy near the bottom of the band, $E=0$. Notice that, while the integrated density of states (3.17) vanishes for $E \to 0$, it gives a finite contribution (3.18) due to the presence of the factor d; otherwise (3.18) would yield $-1/4d$, which vanishes for $d \to \infty$. The same happens with the first term on the r.h.s. of (3.37). The second term requires a separate discussion. For this let us take a fixed value of d and look at the two opposite cases:

(a) Large energies: $d\sqrt{(2E)} \gg 1$. Then because of the rapid oscillations of the cosine factor the entire term can be replaced by $\pi^{-1}\, \mathrm{d}\eta/\mathrm{d}E$.

(b) Small energies. Here on the contrary the oscillatory term becomes comparable with $\pi^{-1}\, \mathrm{d}\eta/\mathrm{d}E$ and even dominates for energies such that $d\sqrt{(2E)} \lesssim 1$. For very small energies, in the limit $E \to 0$, by (3.25) we have $\mathrm{d}\eta/\mathrm{d}E = \eta/2E$. For the same reason we can expand $\sin 2\eta$ in powers of 2η and $\cos(2d\sqrt{(2E)})$ in powers of $2d\sqrt{(2E)}$. The important thing is that the leading term in the expansion of $\cos(2d\sqrt{(2E)}) \sin 2\eta$ exactly cancels the term $\eta/2E$, while the first non-vanishing term is of order $E^{1/2}$, so that the entire bracket in the second term of (3.37) vanishes for $E \to 0$. But now we do not have the factor d which multiplies the first bracket and we do not obtain a finite contribution as in (3.18). We have thus proved that of the two terms making up $N_d(E)$ in (3.37) the first one contributes to the surface part with a δ-function which subtracts $\frac{1}{4}$ state from the bottom of the band in the limit $d \to \infty$, while in the same limit the contribution of the second term is $\pi^{-1}\, \mathrm{d}\eta/\mathrm{d}E$. As with the infinite barrier, the limit $d \to \infty$ is in practice effectively reached for $d \gg q_F^{-1}$. This can be appreciated in fig. 33. Since the main interest is in the behaviour for small energies, $d^{-1}N_d(E)$ has been evaluated for a phase shift of the form $\eta = u_0\sqrt{E}$, u_0 being a characteristic scattering length. Notice (i) that the oscillations are faster – as a function of E – for d larger; (ii) the behaviour when $E \to 0$ is similar for all values of d; (iii) the range of energies close to $E=0$ which is not dominated by the strong oscillations extends only up to energies of order d^{-2}.

We have discussed in detail the charge density and the behaviour of the local and integrated densities of states, identifying the surface terms which ultimately make up the surface density of states. It is now important to discuss the relation that all this has with the principle of

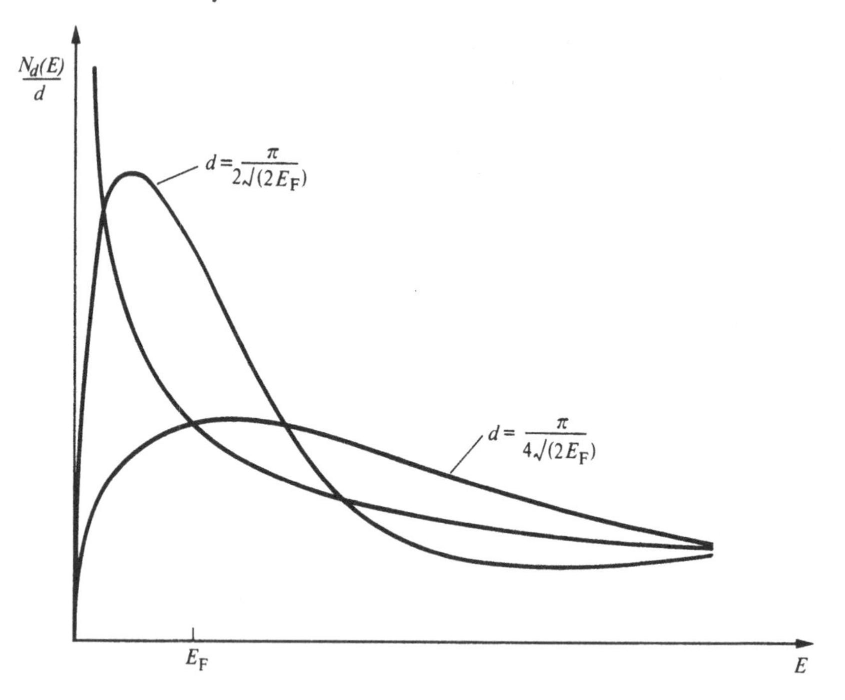

Fig. 33. Integrated density of states for different values of d corresponding to a finite barrier with phase shift $\eta = u_0\sqrt{\partial E}$. If $d \to \infty$, then $N_d(E)/d \to 1/\pi\sqrt{(2E)}$, as in fig. 30.

charge neutrality, because this yields an important condition to be satisfied by the phase shift. We shall still assume that the positive charge is simply a continuous jellium uniformly distributed, with constant density up to its termination. The implications of this assumption, and attempts at improving it, will be discussed at the end of this chapter. What we shall discuss now is the location of the jellium edge relative to the surface barrier. For this we need some physical criterion, and we shall use the requirement that in the bulk there must everywhere exist charge neutrality to order L^{-1}, so that the electrostatic potential associated with the total charge can only create electric fields localised near the surface. Notice that if this condition were not satisfied, a net uncompensated charge of order L^{-1} spreading through a length of order L would create a finite electric field everywhere inside the system; which could not adequately describe the physical situation in a metal, where electrostatic fields are not supposed to penetrate beyond a (short) screening distance. We shall thus use the physical criterion of bulk charge neutrality to order L^{-1}.

Now, in units of the electronic charge, the positive jellium charge density is

$$n_+ = -(2/\pi)\sqrt{(2E_F)}, \tag{3.38}$$

and we must find out where this charge extends to so that overall charge neutrality is achieved. For this we shall use the surface density of states obtained in (3.28). We recall that the term $(L/\pi)\,dq/dE$ in the total density of states is associated with a constant electronic charge density which extends from $z = 0$ to $z = L$. This is the density which cancels (3.38) locally to order L^{-1} in the bulk. If we take the origin $z = 0$ exactly at the jellium edge, then the total charge associated with (3.28) must vanish. Thus, with this origin for z,

$$\int_{-\infty}^{E_F} N_s(E)\,dE = -\frac{1}{4} + \frac{1}{\pi}\eta(E_F) = 0. \tag{3.39}$$

Therefore charge neutrality requires $\eta(E_F) = \frac{1}{4}\pi$ if the origin $z = 0$ is taken at the jellium edge. This can be used to fix the relative positions of the jellium edge and the surface barrier. The simplest case is that of the infinite barrier. Let l be the distance from the barrier to the jellium edge. The wavefunction inside is then $\psi(z) \sim \sin[q(z+l)]$, i.e., $\eta = l\sqrt{(2E)}$ here, and (3.39) yields

$$l = \pi/4\sqrt{(2E_F)}. \tag{3.40}$$

Indeed, if the jellium is removed this distance l from the surface barrier, the amount of jellium charge missing from this length is (by (3.38)) $ln_+ = -\frac{1}{2}$, which, by accounting for spin degeneracy, exactly cancels out the effect of the missing $\frac{1}{4}$ state associated with the δ-function term.

3.1.2. Bounded three-dimensional electron gas

By and large surface problems are inherently three-dimensional and serious mistakes can be made by relying too literally on one-dimensional models. However, these can be very useful in setting up a convenient theoretical framework to introduce some basic concepts and demonstrate general principles. Thus in § 3.1.1 we have discussed phase shifts, density of states, local and integrated density of states, charge density, charge neutrality, and their mutual relationships. We now move on to three-dimensional systems, using otherwise the same simplified model as before. In particular we start again with two infinite barriers, at $z = 0, L$, while a constant potential extends to $\pm\infty$ in the (x, y) plane. Thus, using

Born–von Karman boundary conditions in this plane we start from the wavefunctions

$$\psi_{\boldsymbol{\kappa},m}(\boldsymbol{r})=\sqrt{\frac{2}{L}}\frac{e^{i\boldsymbol{\kappa}\cdot\boldsymbol{\rho}}}{2\pi}\sin\left(\frac{m\pi z}{L}\right). \tag{3.41}$$

These are now normalised so that

$$\int_{-\infty}^{\infty} d^2\boldsymbol{\rho}\int_0^L dz\,\psi^*_{\boldsymbol{\kappa},m}(\boldsymbol{r})\psi_{\boldsymbol{\kappa}',m'}(\boldsymbol{r})=\delta_{mm'}\delta(\boldsymbol{\kappa}-\boldsymbol{\kappa}'), \tag{3.42}$$

while the energy eigenvalues are

$$E_{\boldsymbol{\kappa},m}=\tfrac{1}{2}[\kappa^2+(m\pi/L)^2]. \tag{3.43}$$

So far the only difference with the one-dimensional problem is the addition of $\frac{1}{2}\kappa^2$ to the eigenvalues of (3.2). Thus the discussion leading to (3.5) can be repeated literally but parametrised for a given κ, and we find

$$N^{\infty}_{\boldsymbol{\kappa},\mathrm{t}}(E)=\tfrac{1}{2}[N^{2L}_{\boldsymbol{\kappa}}(E)-\delta(E-\tfrac{1}{2}\kappa^2)]. \tag{3.44}$$

For very large L, E can be treated as a continuous variable, so we have, instead of (3.7),

$$N^{\infty}_{\boldsymbol{\kappa},\mathrm{t}}(E)=\frac{L}{\pi}\frac{1}{\sqrt{(2E-\kappa^2)}}-\frac{1}{2}\delta\left(E-\frac{1}{2}\kappa^2\right). \tag{3.45}$$

Thus, for a semi-infinite free-electron system bounded by one hard wall the surface density of states is

$$N^{\infty}_{\boldsymbol{\kappa},\mathrm{s}}(E)=-\tfrac{1}{4}\delta(E-\tfrac{1}{2}\kappa^2), \tag{3.46}$$

instead of (3.8). So far all the formulae are identical, except that they depend on $\boldsymbol{\kappa}$ through the energy difference $E-\frac{1}{2}\kappa^2$. It is as if we take a new energy origin at $\frac{1}{2}\kappa^2$ and we talk about the *$\boldsymbol{\kappa}$-sub-band*. In particular, the $\frac{1}{4}$ state is now missing from the bottom of the $\boldsymbol{\kappa}$-sub-band. It is clear that the discussion of § 3.1.1 concerning localisation in real space or in energy can be carried over for each given $\boldsymbol{\kappa}$. But now we must integrate over the two-dimensional $\boldsymbol{\kappa}$-space, and this introduces new forms of functional dependence on E. We shall see presently that this does not alter the general qualitative picture concerning localisation and characterisation of surface terms, but of course it is this that produces the correct topological features pertaining to the surface of a true three-dimensional system.

Let us start with the local density of states (3.9), i.e. putting $d^2\kappa = d(\pi\kappa^2)=2\pi\kappa\, d\kappa$, we have

$$N(E,z)=\int_0^{\sqrt{(2E)}}\frac{d^2\kappa}{(2\pi)^2}N^{\infty}_{\boldsymbol{\kappa},\mathrm{t}}(E)|\psi_{\boldsymbol{\kappa},E}(\boldsymbol{\rho},z)|^2. \tag{3.47}$$

The $\boldsymbol{\rho}$-dependence need not be indicated because of translational invariance in the (x, y) plane. Since

$$|\psi_{\kappa,E}(\boldsymbol{\rho}, z)|^2 \propto \sin^2[z(2E-\kappa^2)^{1/2}],$$

the δ-function term of (3.44) makes no contribution, and we have

$$N(E, z) = \frac{\sqrt{(2E)}}{2\pi^2}\left[1 - \frac{\sin(2z\sqrt{(2E)})}{2z\sqrt{(2E)}}\right], \tag{3.48}$$

again with a constant bulk term and an oscillating and evanescent surface term that depends on z. The corresponding charge density is

$$n(z) = 2\int_{-\infty}^{E_F} N(E, z)\,\mathrm{d}E$$

$$= \frac{k_F^3}{3\pi^2}\left[1 + 3\frac{\cos(2k_F z)}{(2k_F z)^2} - 3\frac{\sin(2k_F z)}{(2k_F z)^3}\right]. \tag{3.49}$$

Notice that the *Friedel-type oscillations* are qualitatively similar but functionally different from both the long-range oscillations appearing in the screening of a point impurity in three dimensions, and from the one-dimensional surface formula (3.13). For the integrated density of states we find

$$N_d(E) = \int_0^d N(E, z)\,\mathrm{d}z = d\frac{\sqrt{(2E)}}{2\pi^2}\left[1 - \frac{\mathrm{si}\,(2d\sqrt{(2E)})}{2d\sqrt{(2E)}}\right]. \tag{3.50}$$

Notice that now instead of $\sin(2d\sqrt{(2E)})$ as in (3.17) we have

$$\mathrm{si}\,x = \int_0^x \frac{\sin x'}{x'}\,\mathrm{d}x' \quad (x = 2d\sqrt{(2E)}).$$

It is instructive to make the following comparison. Consider the total density of states when we have *only one surface*. This is, for large L,

$$N_t^{\infty}(E) = \int_0^{\sqrt{(2E)}} \frac{L}{\pi}\frac{1}{\sqrt{(2E-\kappa^2)}}\frac{\mathrm{d}^2\kappa}{(2\pi)^2} - \frac{1}{4}\int_0^{\infty} \delta\!\left(E - \frac{1}{2}\kappa^2\right)\frac{\mathrm{d}^2\kappa}{(2\pi)^2}, \tag{3.51a}$$

i.e.

$$N_t^{\infty}(E) = L\sqrt{(2E)}/2\pi^2 - 1/8\pi. \tag{3.51b}$$

This ought to be equal to (3.50) for $d \to L$. Actually (fig. 34) there is a small difference. The functions are equal except for a narrow range of energies up to order L^{-2}. This is due to the fact that in our analysis we have worked precisely up to order L^{-1} only, which is sufficient for the limit $L \to \infty$. However, it is important to keep track of first-order infinitesimals and to ensure that the analysis is exact to this order because

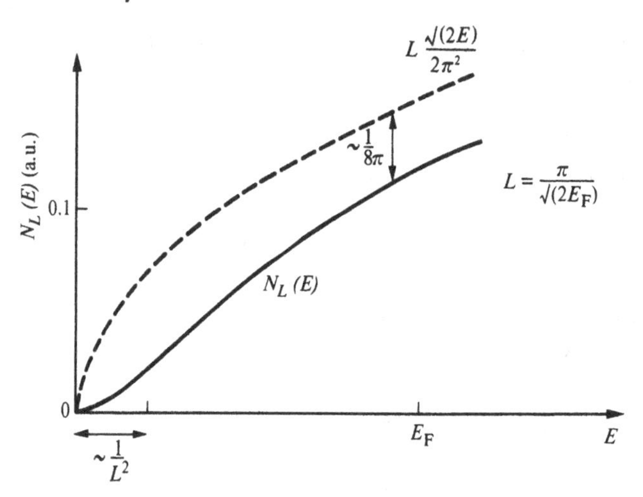

Fig. 34. Three-dimensional integrated density of states ((3.50), solid line) and total density of states ((3.51*b*), dashed line).

integrations over the entire system, from $z=0$ to $z=L$, could introduce errors of finite magnitude.

Now we take the second step. We lower the infinite barrier to some finite value. The wavefunction inside then takes the form

$$\psi_{\kappa,m}(\gamma)=\sqrt{\frac{2}{L}}\frac{e^{i\kappa\cdot\rho}}{2\pi}\sin\left[\frac{m\pi z}{L}+\eta_\kappa(E)\right]. \tag{3.52}$$

Clearly the phase shift is just $\eta_\kappa(E)=\eta(E-\frac{1}{2}\kappa^2)$. Thus we can directly obtain the equivalent of (3.27) for this case. Actually we shall give the formula for *one* surface, i.e. assume that the second barrier, which has remained infinitely high throughout the analysis, has been removed to infinity. Thus

$$N_{\kappa,t}(E)=\frac{L}{\pi}\frac{1}{\sqrt{(2E-\kappa^2)}}-\frac{1}{4}\delta\left(E-\frac{1}{2}\kappa^2\right)+\frac{1}{\pi}\frac{d\eta(E-\frac{1}{2}\kappa^2)}{dE}. \tag{3.53}$$

Formally we keep the factor L to stress the relative order of magnitude of the different terms. The surface contributions are of order L^{-1} relative to the bulk term. Alternatively, if we refer to the density of states *per unit volume*, then the bulk term is finite and the surface terms are affected by a factor A/V, as in the discussion of § 2.3. In other words, the surface density of states per unit area, for a surface with a finite potential barrier that causes a phase shift $\eta_\kappa(E)$, is (compare with (3.28))

$$N_{\kappa,s}(E)=-\frac{1}{4}\delta\left(E-\frac{1}{2}\kappa^2\right)+\frac{1}{\pi}\frac{d\eta(E-\frac{1}{2}\kappa^2)}{dE}. \tag{3.54}$$

We stress that this formula contains an arbitrary additive term which depends on the choice of origin $z=0$, and this is related to the discussion of charge neutrality and relative position of surface barrier and jellium edge. First we write the complete surface density of states

$$N_s(E)=\int N_{\kappa,s}(E)\frac{d^2\kappa}{(2\pi)^2}=\frac{1}{2\pi^2}\left[\eta(E)-\frac{\pi}{4}\right], \tag{3.55}$$

which contains, with respect to the surface term appearing in (3.51*b*), the extra term $\eta(E)/2\pi^2$ due to the finiteness of the barrier height. Then we notice that, for the three-dimensional system, the uniform jellium charge density is $n_+=-k_F^3/3\pi^2$. Now, if we take the origin $z=0$ at the jellium edge, the condition of charge neutrality (3.39) yields, by (3.55),

$$\frac{1}{E_F}\int_0^{E_F}\eta(E)\,dE=\frac{\pi}{4}. \tag{3.56}$$

This can again serve to fix the relative position of surface barrier and jellium edge. For the simple infinite-barrier case the same argument leading from (3.39) to (3.40) can be used starting from (3.56); this gives

$$l=3\pi/8k_F. \tag{3.57}$$

This is, in three dimensions, the distance that the jellium edge must be removed from an infinite surface barrier to ensure charge neutrality. Notice finally that (3.56) can be rewritten as

$$\frac{1}{k_F^2}\int_0^{k_F^2}\eta\left(E_F-\frac{1}{2}\kappa^2\right)d\kappa^2=\frac{\pi}{4}. \tag{3.58}$$

This has a direct meaning as the three-dimensional generalisation of (3.39). In one dimension, charge neutrality fixes the value of the phase shift at the Fermi level. In three dimensions it fixes the two-dimensional average of the phase shift for all wavefunctions with total energy equal to the Fermi energy.

Let us now study the local density of states and related concepts. We can avoid a great deal of repetition by relying heavily on the analysis of § 3.1.1. For the local density of states we must evaluate

$$N(E,z)=\frac{1}{4\pi^2}\int_0^{2E}\frac{1}{\sqrt{(2E-\kappa^2)}}\left\{1-\cos\left[2z\sqrt{(2E-\kappa^2)}+2\eta\left(E-\frac{1}{2}\kappa^2\right)\right]\right\}d\kappa^2, \tag{3.59}$$

in the region where the wavefunction takes the form (3.52). Again, a detailed evaluation would require the knowledge of the phase shift and this would mean numerical computation, but we can get a good idea of the asymptotic behaviour of (3.59) for large z. In this limit

$$N(E, z) \approx \frac{\sqrt{(2E)}}{2\pi^2}\left\{1 - \frac{\sin[2z\sqrt{(2E)} + 2\eta(E)]}{2z\sqrt{(2E)}}\right\}, \tag{3.60}$$

which differs from (3.48) only in the addition of the phase shift to the argument of the sine function. For the charge density we have

$$\begin{aligned} n(z) &= 2\int_0^{E_F} N(E, z)\, \mathrm{d}z \\ &= \frac{1}{2\pi^2}\int_0^{E_F} \mathrm{d}E \int_0^{\sqrt{(2E)}} \left\{1 - \cos\left[2z\sqrt{(2E-\kappa^2)}\right.\right. \\ &\quad \left.\left. + 2\eta\left(E - \frac{1}{2}\kappa^2\right)\right]\right\} \mathrm{d}\sqrt{(2E-\kappa^2)} \\ &= \frac{1}{\pi^2}\int_0^{E_F} \mathrm{d}E \int_0^{\sqrt{(2E)}} \sin^2[kz + \eta(k)]\, \mathrm{d}k \\ &= \frac{1}{\pi^2}\int_0^{k_F} (k_F^2 - k^2)\sin^2[kz + \eta(k)]\, \mathrm{d}k. \end{aligned} \tag{3.61a}$$

The leading term in the asymptotic behaviour for large z can now be evaluated, yielding

$$n(z) \approx \frac{k_F^3}{3\pi^2}\left[1 + 3\frac{\cos(2k_F z + 2\eta_F)}{(2k_F z)^2}\right], \tag{3.61b}$$

which again has an evanescent oscillating surface term differing from the leading term of (3.49) only in the addition of η_F. For the integrated density of states, (3.59) yields

$$\begin{aligned} N_d(E) = \frac{d}{2\pi^2}\int_0^{\sqrt{(2E)}} &\left\{1 - \frac{\sin(2d\sqrt{(2E-\kappa^2)})}{2d\sqrt{(2E-\kappa^2)}}\cos\eta + \frac{1}{d}\frac{d\eta}{d\sqrt{(2E-\kappa^2)}}\right. \\ &\left. - \frac{\cos(2d\sqrt{(2E-\kappa^2)})}{2d\sqrt{(2E-\kappa^2)}}\sin 2\eta\right\} \mathrm{d}\sqrt{(2E-\kappa^2)}, \end{aligned} \tag{3.62}$$

which of course becomes (3.50) for $\eta = 0$, corresponding to the infinite barrier. In particular, for an abrupt potential barrier of height W, with the origin $z = 0$ at the plane of the abrupt barrier, it is easy to find

$$\eta(E) = \tan^{-1}\sqrt{\frac{E}{W-E}} = \sin^{-1}\sqrt{\frac{E}{W}},$$

which enables us to evaluate (3.62). The result for $N_d(E)$ is shown in

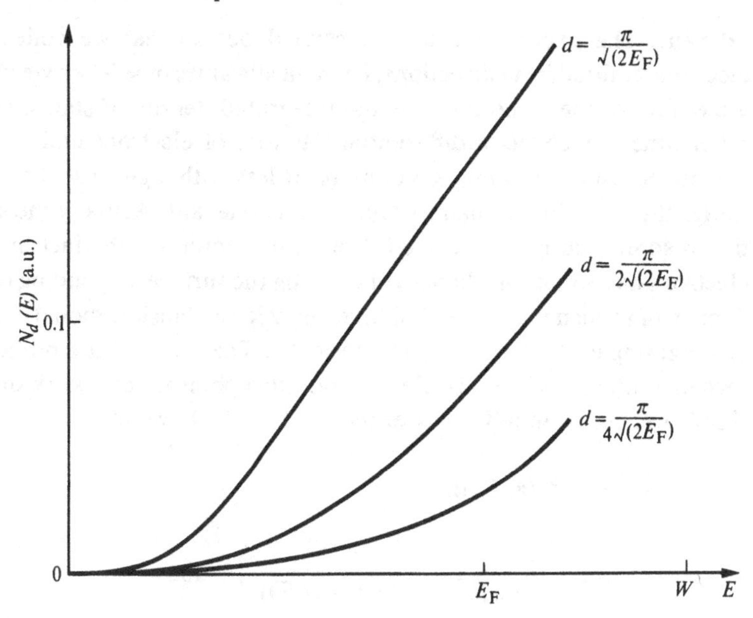

Fig. 35. Integrated density of states evaluated from (3.62) for different values of d by using an abrupt-potential model with $W = \frac{3}{2}E_F$.

fig. 35. Notice that the phase shift behaves like $u_0\sqrt{E}$ for $E \to 0$, whence $N_d(E)$ varies as $E^{3/2}$ for $E \to 0$. It is also instructive to compare figs. 34 and 35. The difference for large d is due to the effect of the phase shift in (3.62).

We have now discussed the three-dimensional case, stressing the similarities and the differences with the one-dimensional problem. There is one quantity which deserves a separate comment, and that is the integrated density of states. Considerable differences are apparent on comparing figs. 30 and 35. These differences can have interesting physical implications in relation to experimental situations which, at this stage, we can only discuss in a broad idealised manner. Let us imagine an experiment in which owing to some external excitation, electrons are emitted from a metal. Let us furthermore imagine a situation in which the intensity of emitted electrons is proportional to the initial density of states. Naturally the idea of the experiment is to probe into the density of states of a metal with a surface. But the emitted electrons have escaped from a surface layer which spreads as far as a certain escape length. Thus the theoretical concept involved in the analysis of this type of experiment is actually the integrated density of states, where the length d is some appropriate measure of an escape length. It is here that the difference between the one-dimensional and the three-dimensional formulae is relevant. The experiment can be carried out so that we collect the electrons emitted in all directions, i.e. with all values of $\boldsymbol{\kappa}$. Then we obtain a measure of the three-dimensional integrated density of states. But at other times we collect a differential intensity of electrons emitted in a narrow pencil of solid angle, i.e. more or less with a given $\boldsymbol{\kappa}$. Then it is rather the one-dimensional formula that is relevant. Actually the situation is somewhat more involved. We must account for the fact that the electrons have different chances of crossing the surface, depending on the depth from which they start. A simple but at least plausible model consists of weighting each depth with a factor $e^{-z/d}$. This somewhat modifies the formulae (3.37) and (3.32). To illustrate this point we can work out the simple case of an infinite-barrier model, for which we find

$$\int_0^\infty e^{-z/d} N(z,E)\,dz = \left\{ \begin{array}{ll} \dfrac{d}{\pi\sqrt{(2E)}}\left[1-\dfrac{d^{-1}}{d^{-2}+(2\sqrt{(2E)})^2}\right] & \text{1-dim.} \\[2ex] \dfrac{d\sqrt{(2E)}}{2\pi^2}\left[1-\dfrac{1}{d}\displaystyle\int_0^\infty e^{-z/d}\,\dfrac{\sin(2z\sqrt{(2E)})}{2z\sqrt{(2E)}}\,dz\right] & \text{3-dim.} \end{array} \right\} \quad (3.63)$$

Compare with (3.17) and (3.50). Of course the theoretical treatment outlined here is rather rudimentary; a serious theory would have to pay proper attention to the dynamics of complicated processes involved in the experiment, but the above discussion does bear out a significant basic feature and illustrates the role of dimensionality in this sort of problem.

3.1.3. Surface Green-function matching analysis

It is now instructive to relate the above analysis to the formulation in terms of surface Green-function matching developed in chapter 2. It is a matter of the way we look at a surface problem. Sometimes we stress that the presence of a surface changes the boundary conditions and this is the key feature whose consequences we study. The analysis then emphasises technical questions of matching, say, e.m. fields, as in chapter 1. In discussing the infinite-barrier model we have stressed the role of the boundary conditions and then, when studying finite barriers, although the analysis naturally involved matching equations, we have really emphasised the viewpoint in which we regard the finite barrier as a perturbation which produces a phase shift. In chapter 2 we introduced a formal technique for matching in terms of surface Green functions and we explicitly discussed its equivalence to wavefunction matching. But then the argument (§ 2.2) leading up to the construction of $\hat{G}_s$ is very reminiscent, even in its diagrammatic interpretation, of the standard sort of arguments used in perturbation or scattering theory. Indeed a formula like (2.39) is very similar in formal appearance to the result of summing up to infinite order a Born series for a perturbed resolvent. Here the surface is regarded as a perturbation of the bulk. It is now instructive to establish the connection with the phase-shift analysis, which we have used in §§ 3.1.1, 2 because it is more intuitive for the problems just discussed. What we want to do is to obtain a practical formula which extracts the phase shift from the surface Green function. It will suffice to discuss the one-dimensional case.

Consider first the infinite-barrier model. From the preceding analysis we know that if $N(E)$ is the density of states for the infinite bulk medium, then for the semi-infinite medium bounded by an infinite barrier we have

$$N_t^\infty(E)=\tfrac{1}{2}N(E)-\tfrac{1}{4}\delta(E). \tag{3.64}$$

On the other hand we can use the general result (2.66). We know from the general argument of § 2.3 that G_s makes no contribution to the density of states for the infinite-barrier model. Thus for the case of one semi-infinite medium, we are only interested in N_t^∞ of (2.55) and we have the alternative formula,

$$N_t^\infty(E)=\frac{1}{2}N(E)-\frac{1}{2\pi}\frac{\mathrm{d}}{\mathrm{d}E}\arg G(E). \tag{3.65}$$

We recall that the variable E on which the Green functions depend must be understood as the limit $E^+=E+\mathrm{i}0$, and G is the surface projection of $\hat{G}$ as discussed in chapter 2. For the free-electron Hamiltonian, in atomic units,

$$\hat{G}(E^+,q)=\left\langle q\left|\frac{1}{E^+-H}\right|q\right\rangle=\frac{1}{E-\frac{1}{2}q^2}, \tag{3.66a}$$

whence,

$$G(E^+)=-\mathrm{i}/2\surd E^+. \tag{3.66b}$$

Thus, for $E<0$, $G=-1/2\surd E$, while for $E>0$, $G=-\mathrm{i}/2\surd E$. Therefore

$$\frac{\mathrm{d}}{\mathrm{d}E}\arg G(E^+)=\frac{\pi}{2}\delta(E), \tag{3.67}$$

which when used in (3.65) again yields (3.64). Notice that it is the hard-wall part that removes $\frac{1}{4}$ state from the bottom of the band. We know again from the general argument of § 2.3 that, as in general scattering theory, with a finite barrier we have the hard-wall part and the scattering part. The former remains unchanged and thus yields, for the finite-barrier case, the loss of $\frac{1}{4}$ state at the bottom of the band, in agreement with the preceding analysis (§§ 3.1.1, 2). It is now the scattering part that reflects the details of the finite barrier and it is from this that we shall obtain the phase shift.

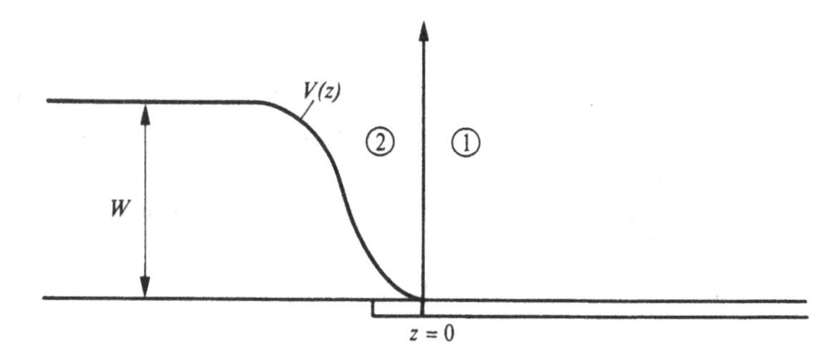

Fig. 36. The two media as defined in the text for studying a metal surface by the surface Green-function matching method.

Consider the situation in fig. 36, with medium 1 – the metal being studied – on the right. In the spirit of the discussion following (2.56) we write

$$N_t(E) = N_1^\infty(E) + N_2^\infty(E) - \frac{1}{\pi}\frac{d}{dE}\arg G_s^{-1}(E^+). \tag{3.68}$$

We are interested in the range of energies below the vacuum level, i.e. the constant level on the left in fig. 36. In theory the spectrum formally indicated in (3.68) could obtain discrete eigenvalues in the energy range of interest, corresponding to bound states of the potential well on the left of the plane $z = 0$, but in practice these can be eliminated by a suitable choice of the position of the plane $z = 0$. In any event we could always identify such spurious discrete eigenvalues as a recognisable artifact and adopt the convention that they are not included in our definition of $N_t(E)$, the density of states of the semi-infinite medium. Thus, instead of (3.68) we can write, *for the energy range of interest*,

$$N_t(E) = N_1^\infty(E) - \frac{1}{\pi}\frac{d}{dE}\arg G_s^{-1}(E^+). \tag{3.69}$$

Comparing with (3.27) we obtain the phase shift

$$\eta(E) = -\arg G_s^{-1}(E) + \text{const.} \tag{3.70}$$

As a simple illustration consider an abrupt potential step of height W and take $z=0$ at the position of this potential barrier. Then

$$G_1(E)=-\frac{\mathrm{i}}{2\sqrt{E^+}};\quad G_2(E)=-\frac{\mathrm{i}}{2\sqrt{(E^+-W)}}, \tag{3.71}$$

i.e. in the energy range of interest

$$G_1(E)=-\frac{\mathrm{i}}{2\sqrt{E}}\quad (E>0);\quad G_2(E)=-\frac{1}{2\sqrt{(W-E)}}\quad (W-E>0). \tag{3.72}$$

Then

$$G_s^{-1}=-(\sqrt{(W-E)}-\mathrm{i}\sqrt{E}), \tag{3.73}$$

whence

$$\arg G_s^{-1}=-\tan^{-1}\sqrt{[E/(W-E)]}+\pi, \tag{3.74}$$

i.e.

$$\arg G_s^{-1}=-\eta(E)+\pi, \tag{3.75}$$

in agreement with (3.70). The additive constant of (3.70) can be expressed in general terms by looking at the behaviour of $\eta(E)$ for $E\to 0$ (3.25). This gives

$$\eta(E)=-\arg G_s^{-1}(E)+\arg G_s^{-1}(0). \tag{3.76}$$

We have thus established the connection between the two methods and derived equation (3.76) in which the phase shift is obtained from the surface Green function. This analysis can be completely carried over to the three-dimensional case by parametrising to a fixed $\boldsymbol{\kappa}$, as was done in general terms in chapter 2 and explicitly in §§ 3.1.1, 2 by putting $E-\frac{1}{2}\kappa^2$ everywhere instead of E for the free-electron model. If we had a band structure associated with crystallinity in the (x, y) plane we would have instead a phase shift of the form $\eta(\boldsymbol{\kappa}, E)$. Otherwise the formal analysis goes through without change.

3.1.4. Surface energy and density of states in the one-electron approximation

We shall see in due course that the study of the surface energy and related questions involves in a non-trivial manner the many-body interactions in the electron gas. However, as a first step we shall study these problems within the one-electron approximation. This is not only instructive as an intermediate step in a pedagogic progression, but also produces some results which will be of practical use later on in the study of the interacting electron gas. We shall also now see the relevance of the concepts and results met in the previous sections.

Let us then start with an independent particle model, with one-electron wavefunctions satisfying the Schrödinger equations,

$$[-\tfrac{1}{2}\nabla^2 + V_{\text{eff}}]\psi_i = E_i\psi_i. \tag{3.77}$$

Here V_{eff} is some effective surface potential which for $z \to \infty$ tends to the constant potential met by the independent electrons still embedded in a positive jellium. The total energy of the electron assembly is

$$W_{\text{tot}} = 2 \sum_i^{\text{occ}} E_i, \tag{3.78}$$

i.e.

$$W_{\text{tot}} = 2\int_{-\infty}^{E_F} EN_t(E)\,\mathrm{d}E, \tag{3.79}$$

for a degenerate Fermi gas. Using (3.27) for N_t we obtain the total energy split into bulk and surface contributions. To keep everything in order let us maintain L finite, although as large as necessary. Looking naively at the problem we would say that the bulk term to be subtracted from the total energy is just the one due to $(L/\pi)\,\mathrm{d}q/\mathrm{d}E$. However, we must remember the subtleties of charge neutrality and choice of origin discussed in §§ 3.1.1, 2. The surface energy will be correctly obtained from (3.79) by using only $-\delta(E)/4 + \pi^{-1}\,\mathrm{d}\eta/\mathrm{d}E$ as the surface part of $N_t(E)$ provided the origin $z = 0$ is taken at the jellium edge and the Fermi level E_F remains unchanged to order L^{-1} when the surface is created. We saw in § 3.1.2 the physical criterion to guarantee this. On this basis we can write for the surface energy

$$W_s = 2\int_{-\infty}^{E_F} EN_s(E)\,\mathrm{d}E, \tag{3.80}$$

where $N_s(E)$ is given by (3.28) and the above conventions are understood. This implies also that (3.39) is satisfied. Thus (3.80) could be formally rewritten as

$$W_s = 2\int_{-\infty}^{E_F} (E - E_F)N_s(E)\,\mathrm{d}E. \tag{3.81}$$

This is not just formal rewriting of (3.80); it also constitutes the generalisation to a hypothetical case in which the creation of the surface introduces a small change in the bulk density, since an excess of electrons at the Fermi energy must be exactly cancelled by a defect of electronic charge associated with the surface density of states, i.e. precisely by

$$-2\int_{-\infty}^{E_F} N_s(E)\,dE.$$

A conceivable physical situation could be that of thin films.

We shall start our analysis from (3.81), which can be rewritten more explicitly as

$$\begin{aligned} W_s &= 2\int_{-\infty}^{E_F} (E-E_F)\left[-\frac{1}{4}\delta(E)+\frac{1}{\pi}\frac{d\eta}{dE}\right]dE \\ &= \frac{E_F}{2}+2\int_0^{E_F}\frac{d\eta}{dE}\frac{E-E_F}{\pi}\,dE. \end{aligned} \tag{3.82}$$

For the one-dimensional case we can use rule (3.39), whereupon (3.82) yields

$$W_s = \frac{2}{\pi}\int_0^{E_F} E\frac{d\eta}{dE}\,dE. \tag{3.83}$$

The analysis for three dimensions goes through in much the same manner, with the result that

$$W_s = \frac{1}{\pi^2}\int_0^{E_F}(E-E_F)\left[\eta(E)-\frac{\pi}{4}\right]dE, \tag{3.84}$$

which can again be simplified by using rule (3.56), yielding

$$W_s = \frac{1}{\pi^2}\int_0^{E_F} E\left[\eta(E)-\frac{\pi}{4}\right]dE. \tag{3.85}$$

Notice that (3.84) would hold even if there were a small change in bulk density when the surface was created.

3.2. The interacting electron gas in the jellium model

We shall now tackle the problems due to many-body interactions in the electron gas. When a surface is introduced, so that the electron gas becomes inhomogeneous in the surface region, the problem becomes quite difficult even if the positive ionic charge is spread out in a uniform jellium. We shall still carry on with this model and leave the discussion of the effects of crystallinity of the lattice for later. We shall base our discussion on the *Hohenberg–Kohn* density functional formulation. This is basically an extension of the *Fermi–Thomas* method in that one adopts a local point of view and seeks a description in terms of the local electron density. We shall use this approach to discuss different physical approximations.

3.2.1. Density functional formulation

The first step consists of proving that the physical properties of the system can be uniquely described in terms of the charge – or electronic – density. The argument applies generally to any system whose ground state density we shall denote by $n_0(\boldsymbol{r})$. Suppose some external one-electron potential $v_0(\boldsymbol{r})$ acts on the electrons of the assembly. It is intuitively obvious that the ground state density and $v_0(\boldsymbol{r})$ are in one to one correspondence, but it is not quite the same if the ground state wavefunction Ψ_0 is involved. We would expect, for example, that n_0 corresponds uniquely to a given v_0, so that Ψ_0 is uniquely determined, whereas a given charge density n_0 could obviously correspond to different wavefunctions Ψ_0. It suffices to assume that Ψ_0 depends on $3N$ variables if N is the number of electrons.

The formal proof is actually fairly simple. Suppose the ground state is non-degenerate. Let $|\Psi_1\rangle$ and $|\Psi_2\rangle$ be the N-electron ground-state vectors corresponding to two different external one-body potentials, $v_1(r)$ and $v_2(r)$. It is obvious that $|\Psi_1\rangle$ and $|\Psi_2\rangle$ cannot be equal if $v_1(\boldsymbol{r})$ and $v_2(\boldsymbol{r})$ differ by more than an additive constant. Indeed, suppose for example that $|\Psi_1\rangle$ were an eigenstate of H_0+v_1 and H_0+v_2. Then we would have

$$(H_0+v_1)|\Psi_1\rangle = E_1|\Psi_1\rangle; \quad (H_0+v_2)|\Psi_1\rangle = E_2|\Psi_1\rangle,$$

whence

$$(v_1-v_2)|\Psi_1\rangle = (E_1-E_2)|\Psi_2\rangle,$$

which is only possible if $v_1(\boldsymbol{r}) = v_2(\boldsymbol{r}) + \text{const}$. Let us then consider two external potentials differing by more than an additive constant, and let $n_1(\boldsymbol{r})$ and $n_2(\boldsymbol{r})$ be the corresponding densities, and $|\Psi_1\rangle$ and $|\Psi_2\rangle$ the corresponding eigenstates. Since only one state vector can correspond to the ground state in each instance we have

$$\langle\Psi_1|H_0+v_1|\Psi_1\rangle < \langle\Psi_2|H_0+v_1|\Psi_2\rangle, \tag{3.86a}$$

as well as

$$\langle\Psi_2|H_0+v_2|\Psi_2\rangle < \langle\Psi_1|H_0+v_2|\Psi_1\rangle. \tag{3.86b}$$

Here

$$H_0 = -\sum_i \frac{1}{2}\nabla_i^2 + \sum_{i\neq j}\frac{1}{2}\frac{1}{|\boldsymbol{r}_i-\boldsymbol{r}_j|}$$

and

$$v_{1,2}(\boldsymbol{r})=\sum_i v_{1,2}(\boldsymbol{r})\delta(\boldsymbol{r}-\boldsymbol{r}_i).$$

Adding (3.86*a*) and (3.86*b*), we obtain

$$\int [v_1(\boldsymbol{r})-v_2(\boldsymbol{r})]\,[n_1(\boldsymbol{r})-n_2(\boldsymbol{r})]\,\mathrm{d}\boldsymbol{r}<0, \tag{3.87}$$

where we have used

$$\langle\Psi_2|v_1|\Psi_2\rangle=\int v_1(\boldsymbol{r})n_2(\boldsymbol{r})\,\mathrm{d}\boldsymbol{r},\quad \text{etc.}$$

Suppose $n_1(\boldsymbol{r})$ and $n_2(\boldsymbol{r})$ were equal; then (3.87) could never be satisfied. It is therefore impossible to have two different potentials $v(\boldsymbol{r})$ corresponding to the same density $n(\boldsymbol{r})$. In other words, $v(\boldsymbol{r})$ *is a unique functional of* $n(\boldsymbol{r})$. Hence $n(\boldsymbol{r})$ uniquely determines the Hamiltonian $H_0+v(\boldsymbol{r})$ and therefore the ground state, and the ground state properties. Thus every physical property of interest can be expressed as a functional of $n(\boldsymbol{r})$. For example, the ground state energy is

$$E_0=\int n_0(\boldsymbol{r})v_0(\boldsymbol{r})\,\mathrm{d}\boldsymbol{r}+\left\langle\Psi_0\left|-\frac{1}{2}\sum_i\nabla_i^2\right|\Psi_0\right\rangle+\left\langle\Psi_0\left|\frac{1}{2}\sum_{i\neq j}\frac{1}{|\boldsymbol{r}_i-\boldsymbol{r}_j|}\right|\Psi_0\right\rangle, \tag{3.88}$$

where we explicitly separate out the kinetic energy terms from those due to electron–electron interactions. But we have seen that Ψ_0 is uniquely determined by $n_0(\boldsymbol{r})$. Thus E_0 can be written down as a functional of $n_0(\boldsymbol{r})$ which we shall express as

$$E_0[n_0(\boldsymbol{r})]=\int n_0(\boldsymbol{r})v_0(\boldsymbol{r})\,\mathrm{d}\boldsymbol{r}+\mathscr{F}[n_0(\boldsymbol{r})]. \tag{3.89}$$

This defines the functional

$$\mathscr{F}[n_0(\boldsymbol{r})]=\left\langle\Psi_0\left|-\frac{1}{2}\sum_i\nabla_i^2\right|\Psi_0\right\rangle+\left\langle\Psi_0\left|\frac{1}{2}\sum_{i\neq j}\frac{1}{|\boldsymbol{r}_i-\boldsymbol{r}_j|}\right|\Psi_0\right\rangle. \tag{3.90}$$

Now, according to the variational principle, E_0 is a minimum in the ground state, i.e. with the density $n_0(\boldsymbol{r})$ corresponding to a given potential $v_0(\boldsymbol{r})$. The conditions of minimum expressed in terms of variations of $n_0(\boldsymbol{r})$ read

$$\delta E[n(\boldsymbol{r})]_{n_0}=\int v_0(\boldsymbol{r})\delta n(\boldsymbol{r})\,\mathrm{d}\boldsymbol{r}+\delta\mathscr{F}[n(\boldsymbol{r})]_{n_0}=0. \tag{3.91}$$

In these variations we must observe the subsidiary condition which expresses particle number conservation, i.e.

$$\int n(\boldsymbol{r})\,\mathrm{d}\boldsymbol{r} = N, \tag{3.92}$$

or

$$\int \delta n(\boldsymbol{r})\,\mathrm{d}\boldsymbol{r} = 0. \tag{3.93}$$

Using the customary scheme of a Lagrangian multiplier we find

$$v_0(\boldsymbol{r}) + \{\delta\mathscr{F}[n(\boldsymbol{r})]/\delta n(\boldsymbol{r})\}_{n_0} = \mu, \tag{3.94}$$

where μ is the chemical potential, which is determined by (3.92), and $\delta\mathscr{F}/\delta n$ means the functional derivative. Alternatively, (3.94) can be derived from (3.91) and (3.92) by minimisation of $E-\mu N$.

In the functional $\mathscr{F}$ it is customary to separate out the long-range Coulomb energy given by

$$\frac{1}{2}\int \frac{n(\boldsymbol{r})n(\boldsymbol{r}')}{|\boldsymbol{r}-\boldsymbol{r}'|}\,\mathrm{d}\boldsymbol{r}\,\mathrm{d}\boldsymbol{r}'.$$

We then write

$$\mathscr{F}[n(\boldsymbol{r})] = \frac{1}{2}\int \frac{n(\boldsymbol{r})n(\boldsymbol{r}')}{|\boldsymbol{r}-\boldsymbol{r}'|}\,\mathrm{d}\boldsymbol{r}\,\mathrm{d}\boldsymbol{r}' + \mathscr{G}[n(\boldsymbol{r})]. \tag{3.95}$$

This defines the functional $\mathscr{G}$, which therefore contains only kinetic, exchange, and correlation terms, the Hartree-energy terms having been separated out. Using (3.95) in (3.94), we obtain

$$v_0(\boldsymbol{r}) + \int \frac{n(\boldsymbol{r}')}{|\boldsymbol{r}-\boldsymbol{r}'|}\,\mathrm{d}\boldsymbol{r}' + \left\{\frac{\delta\mathscr{G}[n(\boldsymbol{r})]}{\delta n(\boldsymbol{r})}\right\}_{n_0} = v_{\mathrm{H}}(\boldsymbol{r}) + \left\{\frac{\delta\mathscr{G}[n(\boldsymbol{r})]}{\delta n(\boldsymbol{r})}\right\}_{n_0} = \mu, \tag{3.96}$$

where the Hartree potential v_{H} has been defined to include the external potential and the mean Coulomb potential due to electron–electron interaction. This equation can be used as a basis for the study of the surface properties of the interacting electron gas. The hard part of the problem lies in the determination of the functional $\mathscr{G}$. The different approximations in the solution of the surface problem can be characterised as different approximations made in the choice of the functional $\mathscr{G}$. This will be done presently in a systematic way.

3.2.2. General results for the jellium model

Before discussing the successive stages of approximation it is interesting to look at some general results which can be proved for the jellium model.

We shall now do this, while introducing the main phenomenological concepts pertaining to the ground-state properties of a metal surface. First we study the relation between the *work function* Φ of a metal and its chemical potential μ. The work function is defined as the energy needed to remove an electron from the Fermi level inside the metal and put it at rest – i.e. with zero kinetic energy – at infinity. If the metal is on the right ($z>0$) and we denote by $\phi_e(-\infty)$ the electrostatic potential energy of an electron at $-\infty$, we can write

$$\Phi = -E_J(N)+E_J(N-1)+\phi_e(-\infty). \tag{3.97}$$

Here $E_J(N)$ means the ground-state energy of the system of N electrons with a given jellium J and $E_J(N-1)$ is that of $N-1$ electrons *with the same jellium*. Notice the occurrence of $\phi_e(-\infty)$ in (3.97).

The problem now is the evaluation of $E_J(N-1)-E_J(N)$, for which we rewrite the energy functional formally as a power series expansion in the density, expanding about the ground-state density $n_0(\boldsymbol{r})$ which satisfies (3.96) for our surface system. We then put

$$n(\boldsymbol{r}) = n_0(\boldsymbol{r})+\delta n(\boldsymbol{r}), \tag{3.98}$$

where $\delta n(\boldsymbol{r})$ is a general hypothetical change in density which need not be restricted by any condition like (3.93). Thus we write

$$E[n(\boldsymbol{r})] = E[n_0(\boldsymbol{r})]+\int\left\{\frac{\delta E[n(\boldsymbol{r})]}{\delta n(\boldsymbol{r})}\right\}_{n_0}\delta n(\boldsymbol{r})\,\mathbf{dr}$$
$$+\frac{1}{2}\int\left\{\frac{\delta^2 E[n(\boldsymbol{r})]}{\delta n(\boldsymbol{r})\,\delta n(\boldsymbol{r}')}\right\}_{n_0}\delta n(\boldsymbol{r})\,\delta n(\boldsymbol{r}')\,\mathbf{dr}\,\mathbf{dr}'+\cdots \tag{3.99}$$

But we know from (3.89), (3.95), and (3.96) that

$$\left\{\frac{\delta E[n(\boldsymbol{r})]}{\delta n(\boldsymbol{r})}\right\}_{n_0} = \mu,$$

which enables us to rewrite (3.99) as

$$E[n(\boldsymbol{r})] = E[n_0(\boldsymbol{r})]+\mu\int\delta n(\boldsymbol{r})\,\mathbf{dr}$$
$$+\frac{1}{2}\int\left\{\frac{\delta^2 E[n]}{\delta n(\boldsymbol{r})\,\delta n(\boldsymbol{r}')}\right\}_{n_0}\delta n(\boldsymbol{r})\,\delta n(\boldsymbol{r}')\,\mathbf{dr}\,\mathbf{dr}'+\cdots \tag{3.100}$$

It is clear that this includes energy changes associated with changes in the total number of particles. In order to use this to evaluate $E_J(N)-E_J(N-1)$ we must study the relative orders of magnitude of the different terms in the expansion. We are interested in the case

$$\int\delta n(\boldsymbol{r}) = -1.$$

It is then clear that the second term on the r.h.s. of (3.100) is of order N^{-1} relative to the first one. Furthermore, the factor in front of $\delta n(\boldsymbol{r})\,\delta n(\boldsymbol{r}')$ in the third term is only significant for distances $|\boldsymbol{r}-\boldsymbol{r}'|$ small compared with the size of the system.† This introduces another factor N^{-1} with respect to the order of magnitude of the second term. Thus the successive terms in the expansion of (3.100) decrease like N^{-1}. We can then neglect the third while consistently keeping all our approximations to order N^{-1} or, equivalently, to order L^{-1}. Furthermore, on the l.h.s. of (3.100) we have precisely $E_J(N-1)$, since the jellium has not been changed. Thus

$$-E_J(N)+E_J(N-1)=-\mu, \tag{3.101}$$

and, from (3.97),

$$\Phi=-\mu+\phi_e(-\infty). \tag{3.102}$$

Notice that (3.96) can be used deep inside the metal, where surface effects are negligible. This yields

$$\phi_e(+\infty)+\{\delta\mathscr{G}[n(\boldsymbol{r})]/\delta n(\boldsymbol{r})\}_b=\mu, \tag{3.103}$$

where the subscript b indicates evaluation at n_b, the uniform bulk electron density. Remember that the definition of chemical potential contains an arbitrary additive constant which depends on the choice of an energy reference level, but the functional $\mathscr{G}$ in (3.103) does not depend on this choice. We can define

$$\bar{\mu}=\{\delta\mathscr{G}[n(\boldsymbol{r})]/\delta n(\boldsymbol{r})\}_b \tag{3.104}$$

as the *internal chemical potential*, which is the value of μ when we choose $\phi_e(+\infty)=0$. Then (3.103) and (3.102) yield in general

$$\Phi=-\bar{\mu}+\phi_e(-\infty)-\phi_e(+\infty). \tag{3.105}$$

The difference $\phi_e(-\infty)-\phi_e(+\infty)$ represents the electrostatic potential jump D_e across the surface barrier. This is the voltage across a surface dipole due to charge redistribution. Some electronic charge spills out leaving a charge deficiency – i.e. a positive space charge – behind, thus forming a dipolar layer which increases the height of the potential well in which the electrons are trapped inside the metal. We rewrite (3.105) as

$$\Phi=D_e-\bar{\mu}. \tag{3.106}$$

Here Φ is expressed as the sum of two separate contributions: D_e is a surface property, while $\bar{\mu}$ depends only on the properties of the bulk material.

† The long-range Coulomb energy contribution is negligible if $\delta n(\boldsymbol{r})$ spreads over the entire surface.

Let us now look for an alternative equation for Φ in order to obtain a useful relation involving different surface quantities. For this we compare the energies of two different *and electrically neutral* systems of N and $N-1$ electrons respectively. This implies that the jellium is now different. We shall call J the jellium corresponding to N electrons and J′ that corresponding to $N-1$ electrons. Both systems have identical surfaces but they differ (fig. 37) in a small change in the linear dimension L, corresponding exactly to one electronic charge. We can then write

$$E_J(N) - E_{J'}(N-1) = \tilde{\varepsilon}(n_b), \tag{3.107}$$

where $\tilde{\varepsilon}$ is the mean energy per electron in the homogeneous bulk, *excluding the electrostatic energy*, which in the jellium model, is equal to the Hartree potential. This is exactly cancelled out by the electrostatic energy of the excess positive charge so that only $\tilde{\varepsilon}$ occurs in the energy

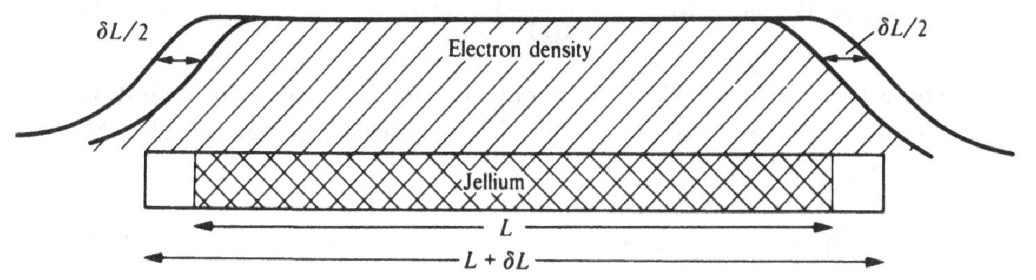

Fig. 37. Electronic charge and positive background charge for two metals with the same bulk charge density but different total charges, namely N and $N-1$.

difference of (3.107). However, this does not yet enable us to evaluate the energy difference of (3.101). What we need to know now is $E_J(N-1) - E_{J'}(N-1)$, corresponding to two systems with the same number of electrons but with a different jellium. This can be achieved by displacing positive charge from infinity to the jellium edge, thereby increasing its length (fig. 37). In order to evaluate (3.107) we use (3.100) again. This time $\delta n(\boldsymbol{r})$ is the difference in electronic charge density due to the change in length of the jellium. Furthermore, $E[n]$ in (3.100) must give the total energy of the ground state of the new system – $E_J(N-1)$ in this case – while the r.h.s. involves only electronic-energy terms. Thus, for correct use of (3.100) here we must add to the r.h.s. the change of energy involved in displacing positive charge from infinity to the jellium edge. To first order this is $-\phi_e(0)$. On the other hand the first-order term appearing on the r.h.s. of (3.100) is now zero because this time

$$\int \delta n(\boldsymbol{r})\, \mathbf{d}\boldsymbol{r} = 0, \tag{3.108}$$

while the next term is of second order. Finally we obtain

$$E_J(N-1)-E_{J'}(N-1)=-\phi_e(0), \tag{3.109}$$

which together with (3.107) and (3.97) yields

$$\Phi=\phi_e(-\infty)-\phi_e(0)-\tilde{\varepsilon}(n_b). \tag{3.110}$$

We have obtained a different expression for Φ, this time involving the potential difference between the vacuum and the jellium edge, as well as $\tilde{\varepsilon}(n_b)$, which is a bulk property. We thus have two different expressions for Φ. Of these, (3.106) is more physical and can be regarded as a formula for Φ, while (3.110) is rather a condition to be satisfied by the surface potential. From (3.110) and (3.105) this is

$$\phi_e(0)-\phi_e(+\infty)=\left\{\frac{\delta\mathscr{G}[n(\boldsymbol{r})]}{\delta n(\boldsymbol{r})}\right\}_b-\tilde{\varepsilon}(n_b), \tag{3.111}$$

where the r.h.s. is exclusively a bulk property. Since $\tilde{\varepsilon}$ by definition excludes the mean Coulomb potential, and so does $\mathscr{G}$, we can write for the homogeneous bulk

$$\mathscr{G}[n]=\int n_b\tilde{\varepsilon}(n_b)\,\mathbf{dr}, \tag{3.112}$$

whence

$$\left\{\frac{\delta\mathscr{G}[n]}{\delta n}\right\}_b=\frac{\mathrm{d}}{\mathrm{d}n_b}(n_b\tilde{\varepsilon}_b), \tag{3.113}$$

which used in (3.111) yields

$$\phi_e(0)-\phi_e(+\infty)=n_b\,\mathrm{d}\tilde{\varepsilon}(n_b)/\mathrm{d}n_b. \tag{3.114}$$

Another important physical property is the *surface energy*. This is by definition the energy needed to create the surface while the number of electrons remains constant. In order to calculate this quantity we must evaluate the total energy of the system with a surface and all the selfconsistent charge redistribution that goes with it and then subtract the energy of the same number of electrons – with corresponding jellium – forming homogeneous bulk material. In order to obtain an expression for the total energy of the electrons in the surface system we can use (3.89) and (3.95) where the external potential acting on the electrons is now that due to the jellium charge $n_J(\boldsymbol{r})=-n_b\theta(z)$. In this way we evaluate the total energy of the electrons with all interactions included; both with the jellium and among themselves. Thus

$$E[n(\boldsymbol{r})]=\int \frac{n_{\mathrm{J}}(\boldsymbol{r}')n(\boldsymbol{r})}{|\boldsymbol{r}-\boldsymbol{r}'|}\,\mathbf{d}\boldsymbol{r}\,\mathbf{d}\boldsymbol{r}' + \frac{1}{2}\int \frac{n(\boldsymbol{r})n(\boldsymbol{r}')}{|\boldsymbol{r}-\boldsymbol{r}'|}\,\mathbf{d}\boldsymbol{r}\,\mathbf{d}\boldsymbol{r}' + \mathscr{G}[n(\boldsymbol{r})]. \tag{3.115}$$

To this we must still add the selfenergy of the positive background charge:

$$\frac{1}{2}\int \frac{n_{\mathrm{J}}(\boldsymbol{r})n_{\mathrm{J}}(\boldsymbol{r}')}{|\boldsymbol{r}-\boldsymbol{r}'|}\,\mathbf{d}\boldsymbol{r}\,\mathbf{d}\boldsymbol{r}'. \tag{3.116}$$

Adding (3.115) and (3.116) we obtain for the total energy of the entire surface system – i.e. the metal with its surface in the jellium model –

$$E_{\mathrm{t}}=\frac{1}{2}\int \frac{[n(\boldsymbol{r}')+n_{\mathrm{J}}(\boldsymbol{r}')][n(\boldsymbol{r})+n_{\mathrm{J}}(\boldsymbol{r})]}{|\boldsymbol{r}-\boldsymbol{r}'|}\,\mathbf{d}\boldsymbol{r}\,\mathbf{d}\boldsymbol{r}' + \mathscr{G}[n(\boldsymbol{r})]$$
$$=\frac{1}{2}\int V_{\mathrm{e}}(\boldsymbol{r})[n(\boldsymbol{r})+n_{\mathrm{J}}(\boldsymbol{r})]\,\mathbf{d}\boldsymbol{r} + \mathscr{G}[n(\boldsymbol{r})], \tag{3.117}$$

where we use the definition

$$V_{\mathrm{e}}(\boldsymbol{r})=\int \frac{[n(\boldsymbol{r}')+n_{\mathrm{J}}(\boldsymbol{r}')]}{|\boldsymbol{r}-\boldsymbol{r}'|}\,\mathbf{d}\boldsymbol{r}'.$$

Notice that $n(\boldsymbol{r})+n_{\mathrm{J}}(\boldsymbol{r})$ is zero in the homogeneous bulk region, where local charge neutrality exists everywhere. This is precisely the space charge which only differs from zero in the inhomogeneous surface layer. Thus for the bulk material whose energy we must subtract from (3.117) the first term on the r.h.s. is zero. On the other hand the formula for $\mathscr{G}$ in the homogeneous bulk system was given in (3.112), where the integration extends to $z>0$. Thus the formula for the surface energy is

$$E_{\mathrm{s}}=\frac{1}{2}\int V_{\mathrm{e}}(\boldsymbol{r})[n(\boldsymbol{r})+n_{\mathrm{J}}(\boldsymbol{r})]\,\mathbf{d}\boldsymbol{r} + \mathscr{G}[n(\boldsymbol{r})] - \int_{z>0} n_{\mathrm{b}}\tilde{\varepsilon}(n_{\mathrm{b}})\,\mathbf{d}\boldsymbol{r}. \tag{3.118}$$

The first term is the electrostatic surface energy. The rest contains the contributions of kinetic, exchange, and correlation energy. Notice that the electrostatic-energy term does not depend on any particular choice of origin because the condition

$$\int [n(\boldsymbol{r})+n_{\mathrm{J}}(\boldsymbol{r})]\,\mathbf{d}\boldsymbol{r}=0$$

is always satisfied no matter where we put $z=0$. It is clear that the sum of the three terms in (3.118) has a factor A/V with respect to the bulk energy.

It is interesting to relate E_s to the electrostatic surface potential. Such a relation, together with (3.114), will enable us to establish some general properties of E_s in the jellium model before attempting any explicit evaluation. The idea is to look for the change in E_s due to a change in n_b which we can think of as caused by a small deformation of the jellium (fig. 38). The same total charge is distributed over an infinitesimally longer

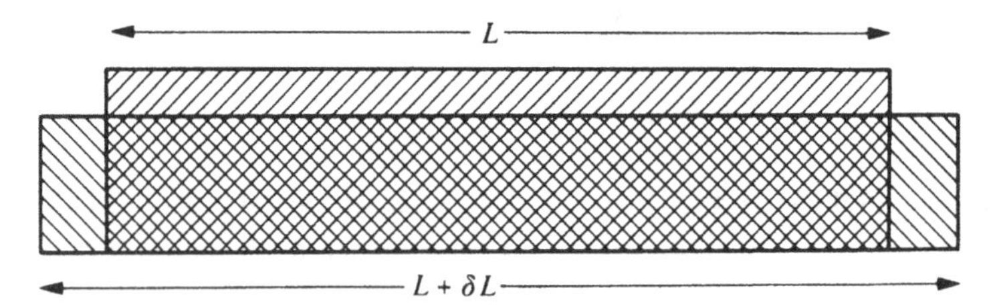

Fig. 38. Positive background charge for two different metallic systems with the same total charge. One can be thought of as the result of a small deformation of the other.

length. If we let E_t and E_s be the total and surface energies *per unit area*, then the initial value of E_t is given by

$$E_t = Ln_b\tilde{\varepsilon}(n_b) + 2E_s(n_b), \tag{3.119}$$

accounting for the two surfaces. If L becomes $L + \delta L$, so that n_b becomes $n_b + \delta n_b$, then E_t becomes

$$E_t + \delta E_t = (L + \delta L)(n_b + \delta n_b)\tilde{\varepsilon}(n_b + \delta n_b) + 2E_s(n_b + \delta n_b), \tag{3.120}$$

so that to first order

$$\delta E_t = L\frac{\mathrm{d}[n_b\tilde{\varepsilon}(n_b)]}{\mathrm{d}n_b}\delta n_b + \delta L n_b\tilde{\varepsilon}(n_b) + 2\frac{\mathrm{d}E_s}{\mathrm{d}n_b}\delta n_b. \tag{3.121}$$

But the total charge remains constant, i.e.

$$L\,\delta n_b + n_b\,\delta L = 0.$$

Thus (3.121) becomes

$$\delta E_t = \delta L\left\{-n_b\frac{\mathrm{d}[n_b\tilde{\varepsilon}(n_b)]}{\mathrm{d}n_b} + n_b\tilde{\varepsilon}(n_b) - 2\frac{n_b}{L}\frac{\mathrm{d}E_s}{\mathrm{d}n_b}\right\}, \tag{3.122}$$

which expresses the clear separation between bulk and surface contributions, the latter being of order L^{-1}. We can use (3.122) to evaluate $\mathrm{d}E_s/\mathrm{d}n_b$ if we know how to find δE_t independently. For this we can imagine a process in which we bring some positive charge from infinity and add to the jellium to produce the new value of its density. Simultaneously we ought to readjust the electronic charge distribution, but to first order this does not change the value of the energy, and we only need to count the electrostatic work required to displace the positive charge. This is the essence of the Hellmann–Feynman theorem. Thus

$$\delta E_t = -\int_0^L \phi_e(z)\,\delta n(z)\,dz + \phi_e(0)\int_0^L \delta n(z)\,dz.$$

Let us put $\phi_e(+\infty)$ for the bulk potential. This is only temporaril·· an inaccurate notation, as we shall soon take the limit $L\to\infty$. We can then rewrite δE_t, identically as

$$\delta E_t = \{\phi_e(0)-\phi_e(+\infty)\}L\,\delta n_b - \delta n_b \int_0^L [\phi_e(z)-\phi_e(+\infty)]\,dz. \qquad (3.123)$$

While the form of (3.122) is useful in displaying the factor L^{-1} for the surface term, in order to compare with (3.123), it is convenient to put $\delta L = -(L/n_b)\,\delta n_b$. Identifying volume terms in these two equations we recover (3.114), which is not surprising; but now we can go a step farther. Identifying surface terms we find, taking the limit $L\to\infty$,

$$\int_0^\infty [\phi_e(z)-\phi_e(+\infty)]\,dz = -\frac{dE_s}{dn_b}. \qquad (3.124)$$

Notice that we have removed the factor 2 of (3.122), where we still had two surfaces, and in (3.124) we give an equation for one surface. This is the formula we were looking for. It is useful in providing the basis for studying the general trend in the values that E_s takes for different metals, corresponding to different values of n_b in the jellium model. For this we shall now make some estimates based on explicit approximations.

Notice that (3.114) fixes the value of the integrand of (3.124) for $z=0$. On the other hand, since this integrand vanishes for $z\to+\infty$, we can make a reasonable approximation in (3.124) by introducing an approximate measure of the thickness of the inhomogeneous surface layer. Let us assume that this length is proportional to k_{FT}^{-1}. Since k_{FT} is proportional to $n_b^{1/6}$, we can write

$$dE_s/dn_b \approx -a n_b^{-1/6} n_b\, d\tilde{\varepsilon}(n_b)/dn_b, \qquad (3.125)$$

where a is some suitable factor, with appropriate dimensions, relating k_{FT}^{-1} to $n_b^{-1/6}$. With this we can make an approximate estimate of the surface energy in terms of bulk properties only. First we must estimate $\tilde{\varepsilon}$. As a reasonable approximation we shall use the low-density *Wigner approximation* for the correlation energy. Then

$$\tilde{\varepsilon}(n_b) \approx \frac{3}{10}(3\pi^2 n_b)^{2/3} - \frac{3}{4}\left(\frac{3n_b}{\pi}\right)^{1/3} - \frac{0.44}{7.8+(3/4\pi n_b)^{1/3}}. \qquad (3.126)$$

The three terms appearing here are the kinetic, exchange, and correlation energies respectively. In this case the correlation energy is rather small, except for very low densities. Since we are only interested now in a rough estimate we shall ignore correlation for the time being. With this simplification

$$n_{\mathrm{b}} \frac{\mathrm{d}\tilde{\varepsilon}}{\mathrm{d}n_{\mathrm{b}}} \approx \frac{1}{5}(3\pi^2 n_{\mathrm{b}})^{2/3} - \frac{1}{4}\left(\frac{3n_{\mathrm{b}}}{\pi}\right)^{1/3},$$

so that

$$\frac{\mathrm{d}E_{\mathrm{s}}}{\mathrm{d}n_{\mathrm{b}}} \approx -a\left[\frac{1}{5}(3\pi^2)n_{\mathrm{b}}^{1/2} - \frac{1}{4}\left(\frac{3}{\pi}\right)^{1/3} n_b^{1/6}\right],$$

whence

$$E_{\mathrm{s}} \approx a\left[\frac{3}{14}\left(\frac{3}{\pi}\right)^{1/3} n_{\mathrm{b}}^{7/6} - \frac{2}{15}(3\pi^2)^{2/3} n_{\mathrm{b}}^{3/2}\right]. \tag{3.127}$$

According to this E_{s} can become negative for large densities. In this estimate the critical value of n_{b} at which (3.127) becomes negative is

$$n_{\mathrm{crit}} = \left[\frac{45}{28}\left(\frac{1}{3\pi^5}\right)^{1/3}\right]^3 = 0.00452 \text{ a.u.},$$

corresponding to $r_{\mathrm{s}} = 3.75$, with $\frac{4}{3}\pi r_{\mathrm{s}}^3 = n_{\mathrm{b}}^{-1}$. This figure must not be taken too literally, but the fact that including the correlation energy would raise the critical density a little is not significant. The present argument stresses an important – and disastrous – property of the jellium model, namely, that E_{s} becomes negative for large densities. This is clearly a pathological feature of the model, because if it were true, high-density metals would be liable to spontaneous break-up, thus forming new surface area and reaching lower energies. The degree of seriousness of this shortcoming depends on the approximation used for $\tilde{\varepsilon}$. (For example, if kinetic energy only was included, then E_{s} would be negative at all densities.) However, despite this shortcoming, a detailed study of the jellium model is in itself very instructive and to a large extent useful, both because some results (depending on the property under consideration and the approximations used) can be of practical value, and because it is important to understand well the part of surface physics which can be explained only as a consequence of the properties of the inhomogeneous electron gas, before introducing crystallinity.

Let us pause to see where we stand. We have introduced the density functional formulation within the jellium model. We have derived several results which are generally valid for this model irrespective of approximations, and we have just made a rough estimate of E_{s} in order to demonstrate a general feature. Now we shall explicitly implement this formal analysis by working out the results in different approximations, systematically and to increasing levels of sophistication.

3.2.3. The Fermi–Thomas approximation

In this approximation (F.T. henceforth) the functional $\mathcal{G}[n(\boldsymbol{r})]$ is evaluated in a local approximation which includes only the kinetic energy. Then

$$g[n(\boldsymbol{r})]=\int n(\boldsymbol{r})\tilde{\varepsilon}(n)\,\mathrm{d}\boldsymbol{r}, \tag{3.128}$$

so that $\tilde{\varepsilon}$ now includes only the mean kinetic energy per particle. This approximation assumes that the density is roughly uniform, which is obviously a poor assumption for the surface layer under ordinary circumstances, although it is not such a bad one for very high densities. Now, in this approximation

$$\tilde{\varepsilon}(n)=t(n)=\frac{3}{10}(3\pi^2 n)^{2/3}. \tag{3.129}$$

Using this in (3.96) and rewriting $v_{\mathrm{H}}(\boldsymbol{r})$ as $V_{\mathrm{e}}(\boldsymbol{r})$ – as corresponds to the jellium model – we have

$$V_{\mathrm{e}}(\boldsymbol{r})+\tfrac{1}{2}[3\pi^2 n(\boldsymbol{r})]^{2/3}=\mu. \tag{3.130}$$

Including all charges, positive and negative, the electrostatic potential satisfies the Poisson equation

$$\nabla^2 V_{\mathrm{e}}(\boldsymbol{r})=-4\pi[n(\boldsymbol{r})-n_{\mathrm{b}}\theta(z)]. \tag{3.131}$$

This, together with appropriate boundary conditions which we shall specify presently, determines the electronic density and potential. In the jellium model we only have z-dependence, so that we start from

$$V_{\mathrm{e}}(z)+\tfrac{1}{2}[3\pi^2 n(z)]^{2/3}=\mu, \tag{3.132a}$$

and

$$\mathrm{d}^2 V_{\mathrm{e}}(z)/\mathrm{d}z^2=-4\pi[n(z)-n_{\mathrm{b}}\theta(z)], \tag{3.132b}$$

which give

$$-\tfrac{1}{2}(3\pi^2)^{2/3}\,\mathrm{d}^2[n(z)]^{2/3}/\mathrm{d}z^2=-4\pi[n(z)-n_{\mathrm{b}}\theta(z)]. \tag{3.133}$$

This can be rewritten as

$$\frac{1}{2}(3\pi^2)^{2/3}\frac{\mathrm{d}[n(z)]^{2/3}}{\mathrm{d}z}\,\mathrm{d}\left\{\frac{\mathrm{d}[n(z)]^{2/3}}{\mathrm{d}z}\right\}$$
$$=4\pi[n(z)-n_{\mathrm{b}}\theta(z)]\,\mathrm{d}[n(z)]^{2/3}, \tag{3.134}$$

which can now be formally integrated from $+\infty$ to some $z>0$. The boundary conditions are $n(z)\to n_b$ and $dn(z)/dz\to 0$ for $z\to+\infty$. This yields

$$\frac{1}{4}(3\pi^2)^{2/3}\left\{\frac{d[n(z)]^{2/3}}{dz}\right\}^2=4\pi\left\{\frac{2}{5}n^{5/3}-n_b n^{2/3}+\frac{3}{4}n_b^{5/3}\right\} \quad (z>0), \tag{3.135}$$

and taking z to the surface $z=0$:

$$\frac{1}{4}(3\pi^2)^{2/3}\left\{\frac{d[n(z)]^{2/3}}{dz}\right\}_s^2=4\pi\left[\frac{2}{5}n_s^{5/3}-n_b n_s^{2/3}+\frac{3}{5}n_b^{5/3}\right]. \tag{3.136}$$

This equation can now be used to integrate (3.134) between $z=0$ and any $z<0$:

$$\frac{1}{4}(3\pi^2)^{2/3}\left\{\frac{d[n(z)]^{2/3}}{dz}\right\}^2-4\pi\left\{\frac{2}{5}n_s^{5/3}-n_b n_s^{2/3}+\frac{3}{5}n_b^{5/3}\right\}$$
$$=4\pi\{\tfrac{2}{5}[n(z)]^{5/3}-\tfrac{2}{5}n_s^{5/2}\} \quad (z<0). \tag{3.137}$$

On the other hand $n(z)\to 0$ and $dn(z)/dz\to 0$ for $z\to-\infty$, whence, from (3.137)

$$n_s^{2/3}=\tfrac{3}{5}n_b^{2/3}. \tag{3.138}$$

Of course the electronic charge density at $z=0$ is smaller than the bulk value. Using this in (3.137) we have

$$\frac{5}{8}(3\pi^2)^{2/3}\left\{\frac{d[n(z)]^{2/3}}{dz}\right\}^2=4\pi[n(z)]^{5/3} \quad (z<0). \tag{3.139}$$

Incidentally, it is easy to see that (3.138) is just a particular case of (3.114). It suffices to use (3.132a) to evaluate $\phi_e(0)-\phi_e(+\infty)$ and to replace $\tilde{\varepsilon}(n_b)$ by $t(n_b)$.

Both (3.135) and (3.139) can be solved by using (3.138). For this it is convenient to introduce a coordinate $\bar{z}$ defined by

$$\bar{z}=\frac{3}{2}\left[\frac{32\pi}{5(3\pi^2)^{2/3}}\right]^{1/2}z, \tag{3.140}$$

with which we find

$$n(\bar{z})=[-\tfrac{1}{6}\bar{z}+n_s^{-1/6}]^{-6} \quad (\bar{z}<0), \tag{3.141}$$

and

$$\int_{n_s}^{n}\frac{dn}{n^{1/3}\sqrt{\{n^{5/3}-\frac{5}{2}n_b n^{2/3}+\frac{3}{2}n_b^{5/3}\}}}=\bar{z} \quad (\bar{z}>0). \tag{3.142}$$

This shows that $n(z)\to z^{-6}$ for z very large and negative. The asymptotic behaviour for positive z can be obtained from (3.142) if we note that in the limit $n \to n_b$ the radicand behaves like

$$n^{5/3} - \tfrac{5}{2}n_b n^{2/3} + \tfrac{3}{2}n_b^{3/5} \to \tfrac{5}{2}n_b^{-1/3}(n - n_b)^2.$$

Using this in (3.142), the asymptotic behaviour of n for large positive z is

$$n(z) \to n_b - C \exp[-\bar{z} \to (\tfrac{5}{2})^{1/2} n_b^{1/6}], \tag{3.143}$$

i.e. the electronic charge density relaxes exponentially to the bulk value, the inhomogeneous term decaying with a decay constant proportional to $n_b^{1/6}$.

Equations (3.141) and (3.142) can be written in dimensionless form by defining $n^* = n/n_b$ and $z^* = \frac{1}{6}\bar{z}n_b^{1/6}$. Then $n_s^* = n_s/n_b = (\frac{3}{5})^{3/2}$ and

$$n^*(z^*) = [-z^* + (\tfrac{5}{3})^{1/4}]^{-6} \quad (z^* < 0), \tag{3.144}$$

while

$$\int_{n_s^*}^{n^*} \frac{dn^*}{n^{*1/3}\sqrt{(n^{*5/3} - \frac{5}{2}n^{*2/3} + \frac{3}{2})}} = 6z^* \quad (z^* > 0). \tag{3.145}$$

The behaviour of $n^*(z^*)$ is shown in fig. 39. The thickness of the surface layer is proportional to $n_b^{-1/6}$, i.e. to the reciprocal of k_{FT}, which is the characteristic screening length in a F.T. approximation.

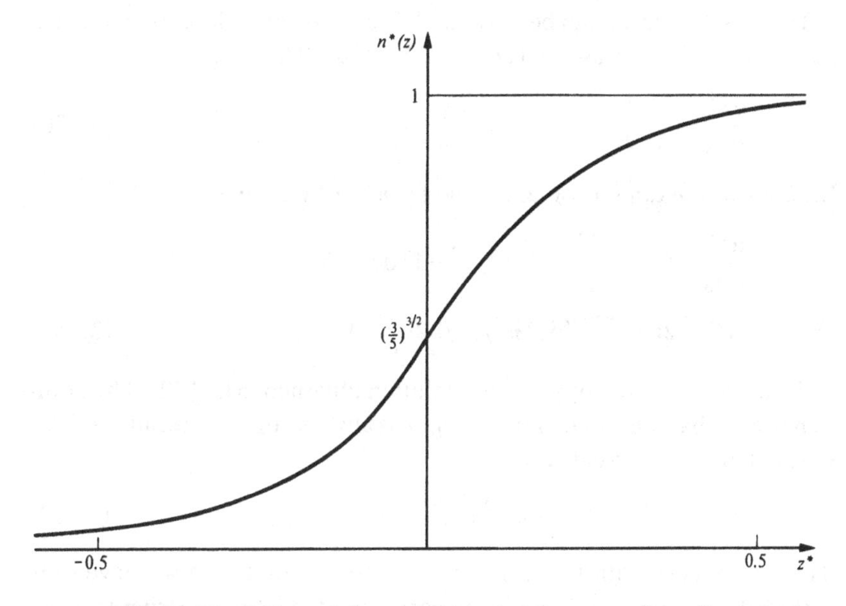

Fig. 39. The F.T. result for a metal surface in terms of the dimensionless variables n^* and z^* defined in the text.

Having discussed the charge density profile, let us study the work function in this approximation. The simplest way to do this is to use (3.132a) by taking asymptotic limits. For $z \to -\infty$, $n(z) \to 0$, hence

$$V_e(-\infty) = \mu, \tag{3.146}$$

while for $z \to +\infty$, $n \to n_b$, and

$$V_e(+\infty) + \tfrac{1}{2}(3\pi^2 n_b)^{2/3} = \mu. \tag{3.147}$$

Therefore

$$V_e(-\infty) - V_e(+\infty) - \tfrac{1}{2}(3\pi^2 n_b)^{2/3} = 0. \tag{3.148}$$

But the last term, the kinetic energy, is exactly $\tilde{\varepsilon}(n_b)$ in this approximation. Hence, by (3.110),

$$\Phi = 0 \quad \text{(F.T.)}. \tag{3.149}$$

In the F.T. approximation the work function is always zero, irrespective of the bulk density. This is obviously a severe limitation of the F.T. model. For very high densities, when the F.T. approximation tends to be valid, this result would indicate that the work function is very small compared with the mean kinetic energy per electron. In fact, experimental evidence shows that Φ for all metals does not vary much, always staying within the order of 3–4 eV.

The surface energy can be evaluated directly from its definition, but it is easier in this case to use (3.124) and (3.132a). This yields

$$\frac{\mathrm{d}E_s}{\mathrm{d}n_b} = \int_0^\infty \frac{1}{2}(3\pi^2)^{2/3}\{n(z)^{2/3} - n_b^{2/3}\}\,\mathrm{d}z. \tag{3.150}$$

In terms of the dimensionless variables n^*, z^* we have

$$\frac{\mathrm{d}E_s}{\mathrm{d}n_b} = Cn_b^{1/2}\int_0^\infty [n^*(z^*)^{2/3} - 1]\,\mathrm{d}z^*$$

$$(C = 2(3\pi^2)^{2/3}[5(3\pi^2)^{2/3}/32\pi]^{1/2}). \tag{3.151}$$

This can be evaluated by using the solution obtained in (3.142). The point to notice is that the value of this integral is always negative because $n^* < 1$ everywhere. The actual result is

$$E_s = -0.654\, n^{3/2} \text{ a.u.} \tag{3.152}$$

The discrepancy with (3.127) is due to the fact that here we have only included the kinetic energy. This corresponds to the negative term of (3.127), whose dependence on n_b is the same as in (3.152). The main point is that the F.T. estimate is always negative, and this is the worst shortcoming of this approximation.

3.2.4. The Fermi–Thomas–Dirac–Gombas approximation

Strictly speaking the next step would be the *Fermi–Thomas–Dirac* (F.T.D.) approximation, but this can easily be obtained as a by-product of the F.T.D.G. approximation which we shall now study in detail. The functional $\mathscr{G}$ is still treated locally, as in (3.128), but the idea now is to evaluate the mean energy per particle including kinetic, exchange, and correlation energy in the *Wigner approximation*. Thus we start from

$$\tilde{\varepsilon}(n)=\frac{3}{10}(3\pi^2 n)^{2/3}-\frac{3}{4}\left(\frac{3n}{\pi}\right)^{1/3}-\frac{0.44}{7.8+(3/4\pi n)^{1/3}}, \tag{3.153}$$

which differs from (3.126) in that n here can be $n(\boldsymbol{r})$. The minimum condition for the energy functional is then

$$V_e(\boldsymbol{r})+\mathrm{d}[n\tilde{\varepsilon}(n)]/\mathrm{d}n=\mu, \tag{3.154}$$

and V_e and n are also related by the Poisson equation (3.131). In the jellium model these reduce to

$$V_e(z)+\mathrm{d}[n\tilde{\varepsilon}(n)]\mathrm{d}n=\mu, \tag{3.155}$$

and to (3.132*b*). These equations can be solved by proceeding as before (with the F.T. model), i.e. (3.155) and (3.132*b*) are combined into a single equation for $n(z)$ which is then multiplied by $\mathrm{d}\{\mathrm{d}[n\tilde{\varepsilon}(n)]/\mathrm{d}n\}$ and integrated from $+\infty$ to an arbitrary $z\geqslant 0$. This yields the first integral

$$\frac{1}{2}\left\{\frac{\mathrm{d}}{\mathrm{d}z}\frac{\mathrm{d}[n\tilde{\varepsilon}(n)]}{n}\right\}^2 =4\pi\left\{(n-n_b)\frac{\mathrm{d}[n\tilde{\varepsilon}(n)]}{\mathrm{d}n}-[n\tilde{\varepsilon}(n)-n_b\tilde{\varepsilon}(n_b)]\right\}\quad(z\geqslant 0). \tag{3.156}$$

In particular, for $z=0$,

$$\frac{1}{2}\left\{\frac{\mathrm{d}}{\mathrm{d}z}\frac{\mathrm{d}[n\tilde{\varepsilon}(n)]}{\mathrm{d}n}\right\}_s^2 =4\pi\left\{(n_s-n_b)\frac{\mathrm{d}[n_s\tilde{\varepsilon}(n_s)]}{\mathrm{d}n_s}-[n_s\tilde{\varepsilon}(n_s)-n_b\tilde{\varepsilon}(n_b)]\right\}. \tag{3.157}$$

This can now be used as a boundary condition when integrating the equation for n from $z=0$ to an arbitrary $z<0$, which yields the first integral

$$\frac{1}{2}\left\{\frac{\mathrm{d}}{\mathrm{d}z}\frac{\mathrm{d}[n\tilde{\varepsilon}(n)]}{\mathrm{d}n}\right\}^2$$

$$=4\pi\left\{n^2\frac{\mathrm{d}\tilde{\varepsilon}(n)}{\mathrm{d}n}-n_\mathrm{b}\left[\frac{\mathrm{d}[n_\mathrm{s}\tilde{\varepsilon}(n_\mathrm{s})]}{\mathrm{d}n_\mathrm{s}}-\tilde{\varepsilon}(n_\mathrm{b})\right]\right\}\quad (z<0). \quad (3.158)$$

It would seem that this could now be integrated by using the fact that $n\to 0$, $\mathrm{d}n/\mathrm{d}z\to 0$ for $z\to -\infty$. This would yield

$$\left\{\frac{\mathrm{d}[n\tilde{\varepsilon}(n)]}{\mathrm{d}n}\right\}_\mathrm{s}=\tilde{\varepsilon}(n_\mathrm{b}), \quad (3.159)$$

whereupon (3.158) would reduce to

$$\frac{1}{2}\left\{\frac{\mathrm{d}}{\mathrm{d}z}\frac{\mathrm{d}[n\tilde{\varepsilon}(n)]}{\mathrm{d}n}\right\}^2=4\pi n^2\frac{\mathrm{d}\tilde{\varepsilon}(n)}{\mathrm{d}n}\quad (z<0). \quad (3.160)$$

However, it turns out that this equation has no real roots for $n(z)$ smaller than a certain critical value. This can be seen from fig. 40, where $\tilde{\varepsilon}(n)$ has

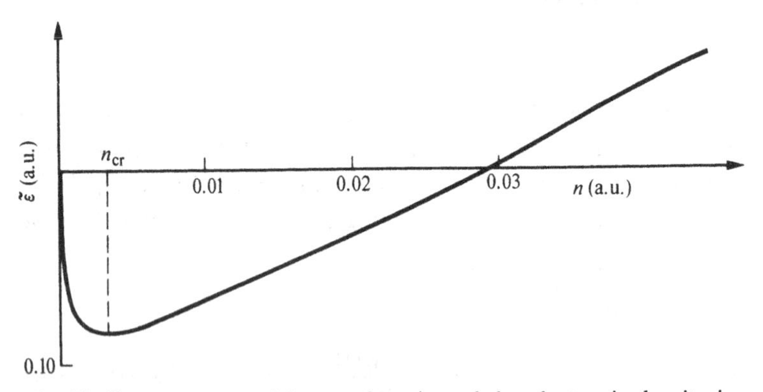

Fig. 40. Energy per particle as a function of the electronic density in atomic units in the F.T.D.G. approximation.

a minimum for $n=n_\mathrm{cr}$ and therefore a negative slope for $n<n_\mathrm{cr}$, while the l.h.s. of (3.160) is positive definite. That is to say, in the F.T.D.G. model there is no solution in which $n(z)$ varies continuously from n_s down to zero for $z\to-\infty$. But a formal solution can be constructed provided it is discontinuous (fig. 41). Let us look for such a solution. Its key feature is that it decreases continuously until it reaches the value n_cr and at that point it drops abruptly to zero. A necessary assumption of course is $n_\mathrm{b}>n_\mathrm{cr}$, which we shall imply in the following argument.

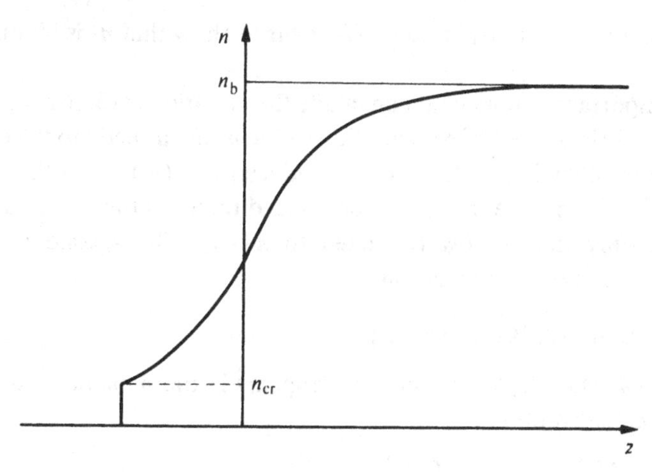

Fig. 41. Schematic plot of the solution for $n(z)$ in the F.T.D.G. approximation. Notice the discontinuity for $n = n_{cr}$.

Now, differentiating (3.155), we obtain

$$\frac{d}{dz}\frac{d[n\tilde{\varepsilon}(n)]}{dn} = -\frac{d}{dz}V_e(z),$$

and using this in (3.160), we find

$$\frac{1}{2}\left[\frac{dV_e(z)}{dz}\right]^2 = 4\pi n^2 \frac{d\tilde{\varepsilon}(n)}{dn}. \tag{3.161}$$

A formal objection could be raised against this procedure, since (3.159) and (3.160) have been derived by assuming a continuous variation of $n(z)$. However, this is easily dispensed with, for (3.159) can be directly obtained from (3.114) by using (3.155) and without requiring continuity of $n(z)$. According to (3.161) the abrupt change in density must take place precisely at the point where $n(z) = n_{cr}$, since there we must have $dV_e/dz = 0$, otherwise we would have a residual electric field (given by the value of $-dV_e/dz$) which would remain constant all the way to $z \to -\infty$. This requires $d\tilde{\varepsilon}(n)/dn = 0$ at that point. Thus, such a solution is formally consistent, but it remains to be proved that it is compatible with the minimum condition for the energy functional. To see this we rewrite

$$E[n] = \frac{1}{2}\int_{z(n_a)}^{\infty} d\boldsymbol{r} \int_{z'(n_a)}^{\infty} d\boldsymbol{r}' \frac{[n(\boldsymbol{r}) - n_b\theta(z)][n(\boldsymbol{r}') - n_b\theta(z')]}{|\boldsymbol{r} - \boldsymbol{r}'|}$$

$$+ \int_{z(n_a)}^{\infty} n(\boldsymbol{r})\tilde{\varepsilon}[n(\boldsymbol{r})]\, d\boldsymbol{r}, \tag{3.162}$$

where $z(n_a)$ denotes the value of z when n reaches some value n_a

dropping there abruptly to zero. We want to show that n_a is identical to n_{cr}.

It is important to notice that on taking the variation of (3.162) there are now two distinct types of contributions. One is due directly to the change $\delta n(z)$; the other is due to a change induced in $z(n_a)$ when the charge density is varied. The first variation leads directly to the same equation (3.155), only this is now restricted to $n \geqslant n_a$. The second variation produces the extra contribution

$$n_a\,\delta z(n_a)[V_e(z_a)+\tilde{\varepsilon}(n_a)]. \tag{3.163}$$

But the variation $n_a\,\delta z(n_a)$ now also appears in the subsidiary condition (charge conservation),

$$\delta\int_0^{z(n_a)} n(z)\,\mathbf{dr} = \int_0^{z(n_a)} \delta n(z)\,\mathbf{dr} + n_a\,\delta z(n_a) = 0. \tag{3.164}$$

Thus the contribution of (3.163) and (3.164) yields the following condition for n_a:

$$V_e(z_a)+\tilde{\varepsilon}(n_a)=\mu. \tag{3.165}$$

But we have seen that (3.155) holds for all $n \geqslant n_a$, so that, using this in (3.165) we find

$$\mathrm{d}\tilde{\varepsilon}(n_a)/\mathrm{d}n_a = 0, \tag{3.166}$$

which is exactly the condition to be satisfied by n_{cr}. We have thus proved that $n_a = n_{cr}$. As regards the explicit evaluation of n_{cr} it depends of course on the approximation used for $\tilde{\varepsilon}$. In the F.T.D. model only kinetic exchange energy is included. This yields $n_{cr} = 0.00213$ a.u., which corresponds to $r_s = 4.86$. In the F.T.D.G. model, which includes correlation energy in the Wigner approximation, we find $n_{cr} = 0.0027$ a.u., corresponding to $r_s = 4.46$.

Let us summarise the main features of the F.T.D. and F.T.D.G. approximations:

(i) They do not admit a continuous solution for $n(z)$.

(ii) However, a formal solution minimising $E[n]$ can be constructed provided it has an abrupt cutoff when it reaches a critical value satisfying (3.166). The actual value of n_{cr} turns out to be slightly less than the value of n_b for Na. Since the condition $n_b > n_{cr}$ is necessary for the existence of such a solution, this limits the range of metals for which the above solution could be used at all.

(iii) The density $n(z)$ satisfies (3.156) in the range $z > 0$, and (3.160) in the range $z < 0$ for which $n(z) \geqslant n_{cr}$.

Under these conditions the solution of (3.160) is

$$\int_{n_s}^{n} \frac{d^2[n\tilde{\varepsilon}(n)]/dn^2}{\sqrt{[8\pi n^2\, d\tilde{\varepsilon}(n)/dn]}}\, dn = z < 0 \quad (n_{cr} < n(z) < n_b). \tag{3.167}$$

The value of n_s satisfies (3.157). Likewise, integrating (3.156),

$$\int_{n_s}^{n} (8\pi)^{-1/2} \frac{d^2[n\tilde{\varepsilon}(n)]/dn^2}{\sqrt{\{(n-n_b)\, d[n\tilde{\varepsilon}(n)]/dn - [n\tilde{\varepsilon}(n) - n_b\tilde{\varepsilon}(n_b)]\}}}\, dn = z > 0. \tag{3.168}$$

These two equations determine $n(z)$.

It is instructive to look at two extreme situations. One is when $n_b = n_{cr}$, in which case it is easy to see that $n_s = n_b = n_{cr}$, i.e. the cutoff is at the surface. Indeed we can find from (3.167) the length l to which the electronic charge extends, on replacing n by n_{cr}. This yields

$$\int_{n_s}^{n_{cr}} \frac{d^2[n\tilde{\varepsilon}(n)]/dn^2}{\sqrt{[8\pi n^2\, d\tilde{\varepsilon}(n)/dn]}}\, dn = -l. \tag{3.169}$$

Obviously the solution is $n_{cr} = n_s$ and $l = 0$. Thus the electronic charge distribution is constant, equal to n_b, until its abrupt cutoff at the jellium edge. The other extreme is when $n_b \to \infty$. Then the dominant contribution to $\tilde{\varepsilon}$ is the kinetic energy and we recover the F.T. case except for the small abrupt jump to zero exhibited by $n(z)$ at $z = -l$. For the sake of illustration, the F.T. and F.T.D.G. are compared in fig. 42 for the particular case $r_s = 2$. It is also interesting to look at the behaviour of the

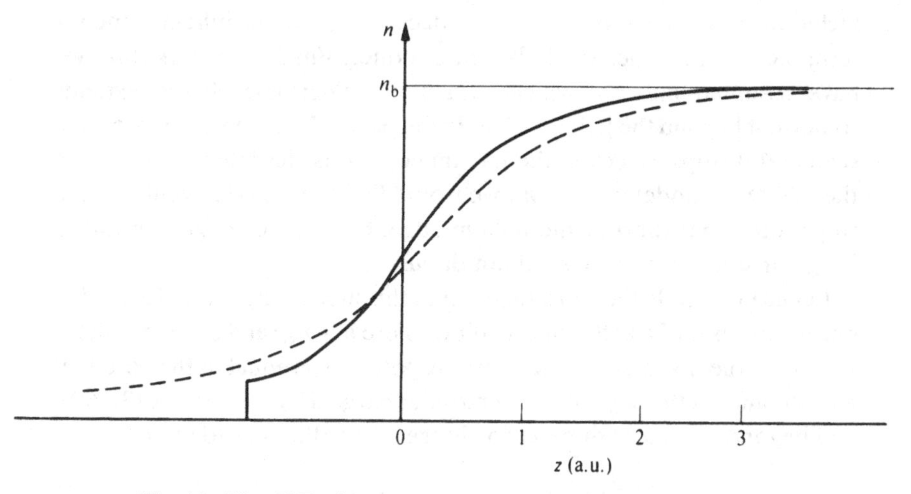

Fig. 42. The F.T. (dotted line) and F.T.D.G. (solid line) solutions compared for $r_s = 2$.

solution of (3.167) and (3.168) inside the metal, where $z \to +\infty$ and $n \to n_b$. Then we can expand

$$n\tilde{\varepsilon}(n) = n_b\tilde{\varepsilon}(n_b) + (n - n_b)\left\{\frac{\mathrm{d}[n\tilde{\varepsilon}(n)]}{n}\right\}_b + \frac{1}{2}(n - n_b)^2\left\{\frac{\mathrm{d}^2[n\tilde{\varepsilon}(n)]}{\mathrm{d}n^2}\right\}_b + \ldots,$$

and using this in the radicand of (3.168), we obtain

$$z \approx \int_{n_s}^{n} \frac{\{\mathrm{d}^2[\tilde{\varepsilon}(b)]/\mathrm{d}n^2\}_b}{\sqrt{\{4\pi(n-n_b)^2\{\mathrm{d}^2[n\tilde{\varepsilon}(n)]/\mathrm{d}n^2\}_b\}}}\,\mathrm{d}n$$

$$= \sqrt{\left\{\frac{\{\mathrm{d}^2[n\tilde{\varepsilon}(n)]/\mathrm{d}n^2\}_b}{4\pi}\right\}} \int_{n_s}^{n} \frac{\mathrm{d}n}{-n+n_b} \quad (z \to +\infty).$$

This shows that

$$n \to n_b - C\exp\left[-z\sqrt{\left\{\frac{4\pi}{\{\mathrm{d}^2[n\tilde{\varepsilon}(n)]/\mathrm{d}n^2\}_b}\right\}}\right] \quad (z \to +\infty). \qquad (3.170)$$

For example, in the F.T.D. approximation

$$\left\{\frac{\mathrm{d}^2[n\tilde{\varepsilon}(n)]}{\mathrm{d}n^2}\right\}_b = \frac{1}{3}(3\pi^2)^{2/3}n_b^{-1/3} - \frac{1}{3}\left(\frac{3}{\pi}\right)^{1/3} n_b^{-2/3}.$$

For very large n_b the first term dominates and using this in (3.170) we recover (3.143), the F.T. solution. Otherwise we see that the effect of including exchange is to decrease the decay length of the inhomogeneous term, i.e. $n(z)$ reaches the bulk value sooner. Finally we stress that we have assumed $n_b > n_{cr}$, which means the electronic charge extends somewhat beyond the jellium edge. In the particular case $n_b = n_{cr}$ we have seen that it stops exactly at the jellium edge. It is clear that if we studied the F.T.D.G. model for $n_b < n_{cr}$ we would find that the electronic charge stops somewhat short of the jellium edge, but it actually gives a rather irregular solution and we shall not discuss it here.

Let us now study the work function in this model, using (3.155) for the particular cases of z at the cutoff point, where $n = n_{cr}$; and $z \to +\infty$, where $n \to n_b$. In the first case the electrostatic potential is equal to the potential at $-\infty$, since at the edge of the electronic charge $\mathrm{d}V_e/\mathrm{d}z = 0$ (see (3.160)) and beyond this point there is no charge. This allows us to write

$$V_e(-\infty) + \left\{\frac{\mathrm{d}[n\tilde{\varepsilon}(n)]}{\mathrm{d}n}\right\}_{cr} = \mu.$$

From the second case we obtain

$$V_e(+\infty)+\left\{\frac{d[n\tilde{\varepsilon}(n)]}{dn}\right\}_b=\mu.$$

Hence the work function (3.105) is

$$\Phi = V_e(\infty)-V_e(-\infty)-\left\{\frac{d[n\tilde{\varepsilon}(n)]}{dn}\right\}_b=-\tilde{\varepsilon}(n_{cr}). \tag{3.171}$$

Thus in the F.T.D.G. approximation we do obtain a finite work function – an improvement upon the F.T. model – but it does not depend on the metallic bulk density. The value of $\tilde{\varepsilon}(n_{cr})$ is (fig. 40) the minimum of $\tilde{\varepsilon}(n)$, equal to -1.29 eV in the F.T.D. approximation and to -2.26 eV for the F.T.D.G. model. In fact a work function of 2.26 eV is not too bad an approximation to the experimental values for many metals, but of course this model is still too crude to account for differences between real metals. Since it also pays the price of a rather artificial electronic charge distribution with an abrupt cutoff, altogether the F.T.D.G. approximation is not really very satisfactory for the study of the inhomogeneous electron gas in the surface region.

For completeness we shall now discuss the surface energy in this model, before moving on to better approximations. We recall that for $n_b=n_{cr}$ the electronic charge density is uniform right up to the jellium edge. This would obviously give

$$E_s(n_{cr})=0. \tag{3.172}$$

Using this and (3.124) we can find $E_s(n_b)$ for all $n_b>n_{cr}$. Thus

$$\frac{dE_s}{dn_b}=-\int_0^{\infty}[\phi_e(z)-\phi_e(-\infty)]\,dz=\int_0^{\infty}\left(\frac{d[n\tilde{\varepsilon}(n)]}{dn}-\left\{\frac{d[n\tilde{\varepsilon}(n)]}{dn}\right\}_b\right)dz, \tag{3.173a}$$

so that

$$E_s(n_b)=\int_{n_{cr}}^{n_b}\frac{dE_s}{dn}\,dn. \tag{3.173b}$$

In order to extract E_s from these equations we would have to use the solution given by (3.167) and (3.168), and this would require numerical computation. However it is possible to establish some general features of E_s without doing this.

First we notice that $n(z)\leq n_b$ for $z>0$. The sign of the integrand of (3.173a) can be ascertained by looking at the behaviour of $d[n\tilde{\varepsilon}(n)]/dn$, shown in fig. 43 for the F.T.D. case. Notice that the minimum is not at n_{cr}.

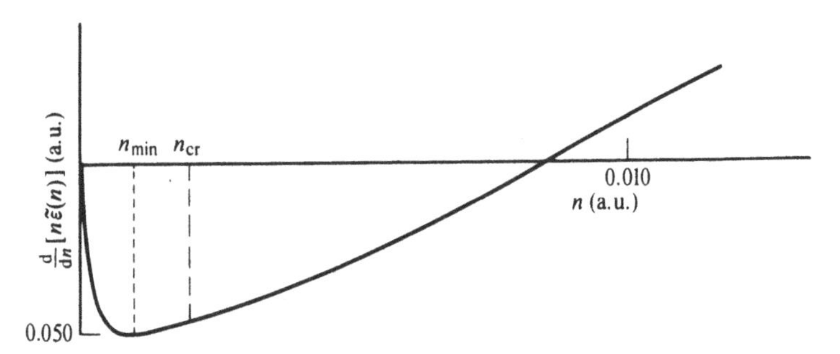

Fig. 43. The function $d[n\tilde{\varepsilon}(n)]/dn$ as a function of n in the F.T.D. approximation. The value n_{cr} is that at which $d\tilde{\varepsilon}(n)/dn$ vanishes in the same approximation.

For values of $n_b > n_{cr}$ we are in the situation in which $n_b \geqslant n \geqslant n_{cr}$ and the integrand of (3.173a) is less than 0. Thus $E_s(n_b)$ decreases as n_b increases above n_{cr}. The behaviour of $E_s(n_b)$ for this model is shown in fig. 44. It is seen that it tends to agree a little better than the F.T. model with the general conclusions of § 3.2.2, but this is at the price of introducing an artificial cutoff in the density profile.

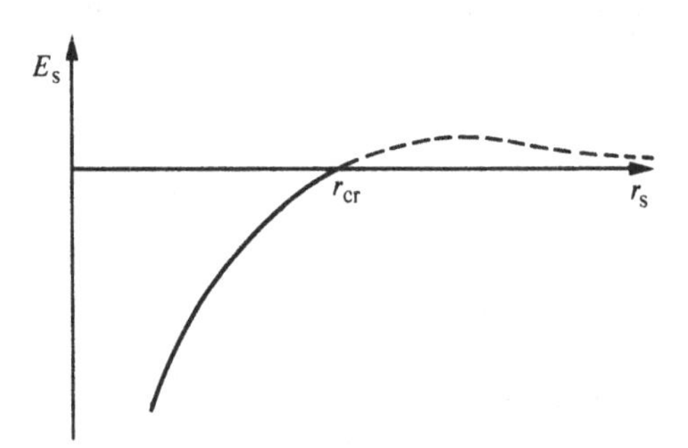

Fig. 44. Qualitative behaviour of E_s as a function of r_s in the F.T.D.G. approximation. The surface energy vanishes for the value (r_{cr}) of r_s corresponding to n_{cr}.

In conclusion, the F.T.D.G. approximation on the whole makes some improvement upon the simpler F.T. model, but it is still basically unsatisfactory for the surface problem.

3.2.5. Beyond the F.T.D.G. approximation

So far we have studied different local approximations for the functional $\mathscr{G}[n]$. Now we shall go a step farther and study the first stages beyond the strict lowest-order local approximations for $\mathscr{G}[n]$. We write the Hamiltonian of the interacting electron assembly in terms of the one-particle creation and annihilation operators $\psi^\dagger(\boldsymbol{r})$ and $\psi(\boldsymbol{r})$ as

$$H = -\frac{1}{2}\int \psi^\dagger(\boldsymbol{r})\nabla^2\psi(\boldsymbol{r})\,\mathbf{d}\boldsymbol{r} + \int \psi^\dagger(\boldsymbol{r})V(\boldsymbol{r})\psi(\boldsymbol{r})\,\mathbf{d}\boldsymbol{r} + \frac{1}{2}\int \psi^\dagger(\boldsymbol{r})\psi^\dagger(\boldsymbol{r}')\frac{1}{|\boldsymbol{r}-\boldsymbol{r}'|}\psi(\boldsymbol{r}')\psi(\boldsymbol{r})\,\mathbf{d}\boldsymbol{r}\,\mathbf{d}\boldsymbol{r}'. \tag{3.174}$$

Suppose we know the ground state $|\Psi_0\rangle$ corresponding to a given charge density $n(\boldsymbol{r})$. Then (remember that $\psi^\dagger$ and ψ are now operators, not wavefunctions)

$$\begin{aligned} E[n] = \langle\Psi_0|H|\Psi_0\rangle &= -\frac{1}{2}\int \langle\Psi_0|\psi^\dagger(\boldsymbol{r})\nabla^2\psi(\boldsymbol{r})|\Psi_0\rangle\,\mathbf{d}\boldsymbol{r} \\ &+ \int V(\boldsymbol{r})\langle\Psi_0|\psi^\dagger(\boldsymbol{r})\psi(\boldsymbol{r})|\Psi_0\rangle\,\mathbf{d}\boldsymbol{r} \\ &+ \frac{1}{2}\int \mathbf{d}\boldsymbol{r}\left\{\int \mathbf{d}\boldsymbol{r}'\frac{1}{|\boldsymbol{r}-\boldsymbol{r}'|}\langle\Psi_0|\psi^\dagger(\boldsymbol{r})\psi^\dagger(\boldsymbol{r}')\psi(\boldsymbol{r}')\psi(\boldsymbol{r})|\Psi_0\rangle\right\} \\ &= \int V(\boldsymbol{r})n(\boldsymbol{r})\,\mathbf{d}\boldsymbol{r} + \frac{1}{2}\int \frac{n(\boldsymbol{r})n(\boldsymbol{r}')}{|\boldsymbol{r}-\boldsymbol{r}'|}\,\mathbf{d}\boldsymbol{r}\,\mathbf{d}\boldsymbol{r}' + \int \mathfrak{g}[n(\boldsymbol{r}),\boldsymbol{r}]\,\mathbf{d}\boldsymbol{r}. \end{aligned} \tag{3.175}$$

Here $\mathfrak{g}[n(\boldsymbol{r}),\boldsymbol{r}]$ is now, for each point $\boldsymbol{r}$, a functional density defined by

$$\mathfrak{g}[n,\boldsymbol{r}] = -\frac{1}{2}\langle\Psi_0|\psi^\dagger(\boldsymbol{r})\nabla^2\psi(\boldsymbol{r})|\Psi_0\rangle + \frac{1}{2}\int \frac{\mathbf{d}\boldsymbol{r}'}{|\boldsymbol{r}-\boldsymbol{r}'|}\left\{\langle\Psi_0|\psi^\dagger(\boldsymbol{r})\psi^\dagger(\boldsymbol{r}')\psi(\boldsymbol{r}')\psi(\boldsymbol{r})|\Psi_0\rangle - n(\boldsymbol{r})n(\boldsymbol{r}')\right\}. \tag{3.176}$$

For a homogeneous system

$$\mathfrak{g}[n,\boldsymbol{r}] = n_b\tilde{\varepsilon}(n_b). \tag{3.177a}$$

The question now is how to treat $\mathfrak{g}$ in an inhomogeneous system. If the spatial variations are very slow – in practice, very small within lengths of order k_F^{-1} – then one tries a local approximation so that, in terms of an inhomogeneous density $n(\boldsymbol{r})$, $\mathfrak{g}$ is simply

$$\mathfrak{g}[n,\boldsymbol{r}] = n\tilde{\varepsilon}(n). \tag{3.177b}$$

This is what is done in the F.T., F.T.D. and F.T.D.G. approximations. We can regard (3.177*b*) as a first extension of (3.177*a*), but it is still a local approximation. To go a step farther, it is plausible to treat $\mathfrak{g}$ as a power series expansion involving spatial derivatives of $n(\mathbf{r})$, thus involving simultaneously the values of the density at a given point, and at neighbouring points within short distances (of order k_F^{-1} in practice). That is,

$$g[n, \boldsymbol{r}] = F(n, \nabla n, \nabla\nabla n, \ldots), \tag{3.178}$$

where F is some functional form to be determined. This expansion must be of the form

$$\begin{aligned} F(n, \nabla n, \ldots) &= F_0(n) + \boldsymbol{F}_1(n) \cdot \nabla n + F_2(n)\nabla n \cdot \nabla n \\ &\quad + \boldsymbol{F}_3(n) : \nabla\nabla n + \{\boldsymbol{F}_4(n) \cdot \nabla n\}\{\boldsymbol{F}_5(n) \cdot \nabla n\} + \cdots \end{aligned} \tag{3.179}$$

Here $\boldsymbol{F}_3$ is a second rank tensor, while $\boldsymbol{F}_1$, $\boldsymbol{F}_4$ and $\boldsymbol{F}_5$ are vectors. The zero-order term, $F_0(n)$, corresponds to the local approximation (3.177*b*). Thus we can write

$$\begin{aligned} F(n, \nabla n, \ldots) &= n\tilde{\varepsilon}(n) + \boldsymbol{F}_1(n) \cdot \nabla n + F_2(n)\nabla n \cdot \nabla n \\ &\quad + \boldsymbol{F}_3(n) : \nabla\nabla n + \{\boldsymbol{F}_4(n) \cdot \nabla n\}\{\boldsymbol{F}_5(n) \cdot \nabla n\} + \cdots \end{aligned} \tag{3.180}$$

This can be considerably simplified by symmetry considerations. For example, $\boldsymbol{F}_1$ is a vector which measures the first-order change in energy due to a density gradient. But this change cannot depend on the direction along which we take an infinitesimal displacement for the isotropic jellium model. Thus $\boldsymbol{F}_1$ must vanish. For the same reason the tensor $\boldsymbol{F}_3$ must be of the form $F_3\boldsymbol{I}$, where $\boldsymbol{I}$ is the unit tensor, while $\boldsymbol{F}_4$ and $\boldsymbol{F}_5$ must vanish. Thus we are left with

$$g[n, \boldsymbol{r}] = n\tilde{\varepsilon}(n) + F_2(n)(\nabla n)^2 + F_3(n)\nabla^2 n + \ldots, \tag{3.181}$$

where ∇^2 means the scalar $\nabla \cdot \nabla$. The question of the convergence of this series is a rather technical one, which is discussed in the specialised literature and will not be discussed here. We shall assume that this series converges and that (3.181) is therefore meaningful. Thus we shall start our analysis from the functional

$$\mathscr{G}[n] = \int n\tilde{\varepsilon}(n)\, \mathbf{d}\boldsymbol{r} + \int F_2(n)(\nabla n)^2\, \mathbf{d}\boldsymbol{r} + \int F_3(n)\nabla^2 n\, \mathbf{d}\boldsymbol{r} + \cdots \tag{3.182}$$

The last term can be rewritten by partial integration as follows

$$\begin{aligned} \int F_3(n)\nabla^2 n\, \mathbf{d}\boldsymbol{r} &= \int F_3(n)\nabla \cdot \nabla n\, \mathbf{d}\boldsymbol{r} = -\int \nabla F_3(n) \cdot \nabla n\, \mathbf{d}\boldsymbol{r} \\ &= -\int \frac{\mathrm{d}F_3}{\mathrm{d}n}(\nabla n)^2\, \mathbf{d}\boldsymbol{r}, \end{aligned}$$

which shows that this term is exactly of the form of the second term on the r.h.s. of (3.182). Therefore it can be absorbed by suitable redefinition of

$F_2(n)$, and (3.182) is then reduced to

$$G[n]=\int n\tilde{\varepsilon}(n)\,\mathbf{d}\boldsymbol{r}+\int F_2(n)(\boldsymbol{\nabla} n)^2\,\mathbf{d}\boldsymbol{r}+\cdots \tag{3.183}$$

We stress that this is actually an expansion up to third order, the first non-vanishing terms involving things like $(\boldsymbol{\nabla} n)^4$, $(\boldsymbol{\nabla} n)^2\boldsymbol{\nabla}^2 n$, $(\boldsymbol{\nabla}^2 n)^2$ etc., i.e. fourth-order terms, as can be seen from symmetry arguments. The problem now is to determine the function $F_2(n)$ and this can be done by studying the response of a homogeneous system to a long-wave external disturbance, in two independent ways: (i) by using a response-function analysis, and (ii) by formally using (3.183) and then comparing.

From (3.183) the equations describing the response of the electron system to an external potential V_{ext} are

$$V_{\text{H}}(\boldsymbol{r})+\frac{\text{d}[n\tilde{\varepsilon}(n)]}{\text{d}n}+\frac{\text{d}F_2}{\text{d}n}(\boldsymbol{\nabla} n)^2-2\,\text{div}\,[F_2(n)\boldsymbol{\nabla} n]=\mu, \tag{3.184}$$

and

$$\boldsymbol{\nabla}^2 V_{\text{ind}}=-4\pi\,\delta n, \tag{3.185}$$

where δn is the induced density change, V_{ind} is the induced potential, $n=n_0+\delta n$ and $V_{\text{H}}=V_{\text{ext}}+V_{\text{ind}}$. Linearising (3.184) and remembering that $d[n\tilde{\varepsilon}(n)]/\text{d}n=\mu$ for $n=n_0$, we have

$$V_{\text{H}}+\left\{\frac{\text{d}^2[n\tilde{\varepsilon}(n)]}{\text{d}n^2}\right\}_{n_0}\delta n-2F_2(n_0)\nabla^2\,\delta n=0. \tag{3.186}$$

The linear response of the system is described by (3.185) and (3.186). On Fourier transforming, we have

$$\frac{V_{\text{ext}}(k)}{V_{\text{H}}(k)}=\frac{1+(k^2/4\pi)\{\text{d}^2[n\tilde{\varepsilon}(n)]/\text{d}n^2+2F_2(n)k^2\}_{n_0}}{(k^2/4\pi)\{\text{d}^2[n\tilde{\varepsilon}(n)]/\text{d}n^2+2F_2(n)k^2\}_{n_0}}. \tag{3.187}$$

But the ratio on the l.h.s. of this equality is just the dielectric function $\varepsilon(k)$. Hence, in the long-wave limit,

$$\begin{aligned}\varepsilon(k)&=1+\frac{4\pi}{k^2\{\text{d}^2[n\tilde{\varepsilon}(n)]/\text{d}n^2+2F_2(n)k^2\}_{n_0}}\\&\xrightarrow[k\to 0]{}1+\frac{4\pi}{k^2\{\text{d}^2[n\tilde{\varepsilon}(n)]/\text{d}n^2\}_{n_0}}\left(1-\frac{2F_2(n_0)}{\{\text{d}^2[n\tilde{\varepsilon}(n)]/\text{d}n^2\}_{n_0}}k^2\right).\end{aligned} \tag{3.188}$$

This is so far just a formal expression for $\varepsilon(k)$ because it contains $F_2(n_0)$, which is precisely what we want to find.

Another way to proceed is to use a standard linear response function in which we evaluate $\varepsilon(k)$ independently, and this can be done by including some approximation for exchange and correlation. Although several reasonable dielectric functions are available for this, for the time being we shall restrict ourselves to what is called the gradient expansion term for the kinetic energy. In this case the gradient expansion term is obtained by evaluating the kinetic energy of an independent electron gas of the same density as the real system under consideration. That is to say, we pretend that the Hamiltonian to be considered is

$$H' = -\frac{1}{2}\sum_i \nabla_i^2 + \sum_i V'(\boldsymbol{r}_i), \tag{3.189}$$

which does not include electron–electron interactions. The density functional formulation can also be used for this hypothetical system so that the same density $n(\boldsymbol{r})$ as that of the real system now determines $V'(\boldsymbol{r})$ and a ground state $|\Psi_0'\rangle$, which is different from the real ground state $|\Psi_0\rangle$ because the real Hamiltonian (3.174) does include electron–electron interactions. The kinetic energy of (3.189) is then

$$\langle \Psi_0'| -\frac{1}{2}\sum_i \nabla_i^2 |\Psi_0'\rangle. \tag{3.190}$$

Now, given the density $n(\boldsymbol{r})$ corresponding to a real Hamiltonian H which includes electron–electron interactions, we *formally define* (3.190) as the kinetic energy *associated with H*. Of course this is only a definition. We know (3.190) is not the real kinetic energy of H because the real ground state $|\Psi_0\rangle$ is different from $|\Psi_0'\rangle$. However, the difference between (3.190) and the real kinetic energy,

$$\langle \Psi_0| -\frac{1}{2}\sum_i \nabla_i^2 |\Psi_0\rangle,$$

is compensated for by reabsorption in the definition of the correlation energy, which then includes, strictly speaking, a part of the kinetic energy. All this is simply a device to evaluate F_2 in this approximation. What we do is to first rewrite (3.184)–(3.188) as if there were no electron–electron interactions. Actually only the exchange and correlation energies are omitted, since the mean Coulomb energy in this analysis is treated as an external potential. Thus in (3.188) we use for $\tilde{\varepsilon}(n)$ simply the kinetic energy (3.129) and then we compare this result with the random phase approximation (r.p.a.) for the longitudinal dielectric function, which also does not include exchange and correlation, and in which the mean Coulomb interaction is also an external potential acting on the electrons. This has the form

$$\varepsilon_{\text{rpa}}(k) = 1 + (k_{\text{FT}}^2/k^2) f(k/2k_{\text{F}}), \tag{3.191}$$

where

$$f(x)=\frac{1}{2}+\frac{1-x^2}{4x}\ln\left|\frac{1+x}{1-x}\right|.$$

Comparing with (3.188) we obtain the form of $F_2(n)=1/72n$. With this we have the gradient expansion for the kinetic energy,

$$G[n]=\int n\tilde{\varepsilon}(n)\,\mathrm{d}\boldsymbol{r}+\frac{1}{72}\int\frac{(\nabla n)^2}{n}\,\mathrm{d}\boldsymbol{r}+\cdots \tag{3.192}$$

Actually, by continuing the expansion of (3.191) in powers of k we can obtain further terms in the expansion (3.192). This has an important implication: the gradient expansion method for $G[n]$ can never reproduce the features associated with the singular behaviour of $\varepsilon_{\mathrm{rpa}}(k)$ for $k=2k_{\mathrm{F}}$. In particular, we could never in this way obtain the long-range *Friedel oscillations.* This also holds for the F.T., F.T.D. and F.T.D.G. models, which differ in the approximations used for $\tilde{\varepsilon}(n)$, but are all equivalent to the lowest-order approximation in a gradient expansion. Furthermore, we can be sure that the same will happen no matter how many terms we go up to in a gradient expansion. Nevertheless, although we know we shall not find the long-range oscillations in $n(z)$, it is interesting to study this approximation in detail and to see how much this improves on the previous model.

What we are looking at is the effect of the term in $(\nabla n)^2$ on $n(z)$ within the jellium model. The minimum condition for the energy functional is now

$$V_{\mathrm{e}}(z)+\frac{\mathrm{d}[n\tilde{\varepsilon}(n)]}{\mathrm{d}n}-\frac{1}{72n^2}\left(\frac{\mathrm{d}n}{\mathrm{d}z}\right)^2-\frac{1}{36}\frac{\mathrm{d}}{\mathrm{d}z}\left(n\frac{\mathrm{d}n}{\mathrm{d}z}\right)=\mu, \tag{3.193}$$

which we shall solve together with the Poisson equation (3.132*b*) with the usual boundary conditions. A complete solution of these equations is rather complicated, but some general features are fairly easy to see. For example, taking the limit $z\to+\infty$ in (3.193), so that $n\to n_{\mathrm{b}}$ and $\mathrm{d}n/\mathrm{d}z\to 0$, we have

$$V_{\mathrm{e}}(+\infty)+\left\{\frac{\mathrm{d}[n\tilde{\varepsilon}(n)]}{\mathrm{d}n}\right\}_{\mathrm{b}}=\mu, \tag{3.194}$$

while for $z\to-\infty$, with $n\to 0$ and $\mathrm{d}n/\mathrm{d}z\to 0$:

$$V_{\mathrm{e}}(-\infty)-\lim_{z\to-\infty}\left[\frac{1}{72n^2}\left(\frac{\mathrm{d}n}{\mathrm{d}z}\right)^2+\frac{1}{36}\frac{\mathrm{d}}{\mathrm{d}z}\left(\frac{1}{n}\frac{\mathrm{d}n}{\mathrm{d}z}\right)\right]=\mu. \tag{3.195}$$

So far we have not chosen the energy reference level. Let us now choose

$V_e(-\infty)=0$. Then

$$\mu = -\lim_{z\to-\infty}\left[\frac{1}{72n^2}\left(\frac{dn}{dz}\right)^2+\frac{1}{36n}\frac{d}{dz}\left(\frac{1}{n}\frac{dn}{dz}\right)\right]. \tag{3.196}$$

By making the change of variable $\ln n = \nu$ this becomes

$$\frac{1}{72}\left(\frac{d\nu}{dz}\right)^2+\frac{1}{36}\frac{d^2\nu}{dz^2}=-\mu. \tag{3.197}$$

It is understood that this only holds in the asymptotic limit of very large $|z|$, where we want to find the behaviour of $n(z)$. This equation (3.197) can be readily integrated, leading to $\nu = z\sqrt{(-72\mu)}+\text{const.}$, whence

$$n(z)\underset{z\to-\infty}{\to} C \exp z\sqrt{(-72\mu)}. \tag{3.198}$$

We found, as expected, an exponential decay, but notice that the decay constant is proportional to $-\mu$, which requires a chemical potential $\mu<0$. This need not cause any perplexity. Remember that by definition μ (not the internal potential $\bar{\mu}$) contains an arbitrary additive constant which depends on the zero of energy. In fact it is obvious from (3.102) that $\mu<0$ if we take $\phi_e(-\infty)=0$, which is just what we are doing here, since in the jellium model ϕ_e and V_e are identical.

For the opposite limit, $z\to+\infty$, we can write $V_e(z)=V_e(+\infty)+\Delta V$, $n(z)=n_b+\Delta n$ and, from (3.193), (3.194) and (3.195)

$$\Delta V(z)+\left\{\frac{d^2[n\tilde{\varepsilon}(n)]}{dn^2}\right\}_b\Delta n+\frac{1}{72n_b^2}\left(\frac{d\,\Delta n}{dz}\right)^2+\frac{1}{36n_b}\frac{d^2\,\Delta n}{dz^2}=0, \tag{3.199}$$

and

$$\frac{d^2\,\Delta V(z)}{dz^2}=-4\pi\,\Delta n(z). \tag{3.200}$$

It is now easy to see that the asymptotic behaviour is

$$n(z)\xrightarrow[z\to+\infty]{} n_b - C\,e^{-\alpha z}, \tag{3.201a}$$

with

$$\alpha^2=18n_b\left(\sqrt{\left[\left\{\frac{d^2[n\tilde{\varepsilon}(n)]}{dn^2}\right\}_b^2+\frac{4\pi}{9n_b}\right]}-\left\{\frac{d^2[n\tilde{\varepsilon}(n)]}{dn^2}\right\}_b\right). \tag{3.201b}$$

The difference between this and the F.T.D.G. estimate (3.170) is due to the contribution of the last two terms on the l.h.s. of (3.199). Of course for sufficiently large n_b this difference is negligible, but this requires

$$\left\{\frac{d^2[n\tilde{\varepsilon}(n)]}{dn^2}\right\}_b \gg \sqrt{\left(\frac{4\pi}{9n_b}\right)},$$

which in practice entails values of n_b considerably larger than real metallic densities.

To sum up, the main features of the solution of (3.193) and (3.194) are:

(i) It yields a charge density which shows no long-range oscillations. The inhomogeneous part of this charge density decays exponentially on both sides away from the surface region according to (3.198) and (3.201).

(ii) This is consistent with a positive work function $\Phi = -\mu$.

A more detailed solution would be rather complicated and will not be attempted here. We shall instead follow a different line, which can yield at least an acceptable approximation. The idea is to return to the energy functional and use the fact that $n(z)$ must minimise it, i.e. to use some trial form and to adjust the variational parameters by minimisation. We start from the following explicit form which includes the first non-vanishing term in the gradient expansion of the kinetic energy:

$$E[n]=\frac{A}{2}\int V_e(z)[n(z)-n_b\theta(z)]\,dz+A\int n\tilde{\varepsilon}(n)\,dz+\frac{A}{72}\int\frac{1}{n}\left(\frac{dn}{dz}\right)^2 dz. \tag{3.202}$$

This shows the factor A – surface area – and includes the self energy associated with the uniform jellium. The surface energy per unit area (§ 3.2.2) is then

$$\frac{E_s}{A}=\frac{1}{2}\int V_e(z)[n(z)-n_b\theta(z)]\,dz + \int[n\tilde{\varepsilon}(n)-n_b\tilde{\varepsilon}(n_b)\theta(z)]\,dz+\frac{1}{72}\int\frac{1}{n}\left(\frac{dn}{dz}\right)^2 dz. \tag{3.203}$$

(Notice that we have subtracted the energy of the homogeneous system extending up to $z=0$.) This is the equation we shall minimise after postulating the following trial form for $n(z)$:

$$n(z)=\begin{cases} n_b(1-a\,e^{\alpha z}) & (z<c),\\ bn_b\,e^{-\beta z} & (z>c).\end{cases}$$

The parameters (a, b, c) can be determined from the conditions of continuity of $n(z)$ and of $dn(z)/dz$, and from the charge neutrality principle. The actual variational parameters are the decay constants (α, β). The analysis is considerably simpler if we put $\alpha=\beta$, and is still good enough for the present purpose, which is to describe an exponentially damped inhomogeneity located about the surface region, in line with the general features to be expected here. We then start from

$$n(z)=\begin{cases} n_b(1-\frac{1}{2}e^{-\alpha z}) & (z>0),\\ \frac{1}{2}n_b\,e^{\alpha z} & (z<0).\end{cases} \tag{3.204}$$

This satisfies the conditions of continuity and charge neutrality and we have only one variational parameter. We can now evaluate the different contributions to the energy:

(i) *Hartree energy*. The Coulomb potential created by (3.204) and by the jellium charge is

$$V(z)=\begin{cases} -\dfrac{4\pi n_b}{\alpha^2}+\dfrac{2\pi n_b}{\alpha^2}\mathrm{e}^{-\alpha z} & (z>0),\\ -\dfrac{2\pi n_b}{\alpha^2}\mathrm{e}^{\alpha z} & (z<0). \end{cases} \tag{3.205}$$

The energy associated with the total charge is

$$E_{s,H}=\pi n_b^2/2\alpha^3. \tag{3.206}$$

(ii) *Kinetic energy*. This is

$$\begin{aligned} E_{s,kin} &= \int_{-\infty}^{\infty}\frac{3}{10}(3\pi^2)^{2/3}\{n(z)^{5/3}-n_b^{5/3}\theta(z)\}\,dz \\ &= \int_0^{\infty}\frac{3}{10}(3\pi^2)^{2/3}n_b^{5/3}\left[\left(1-\frac{1}{2}\mathrm{e}^{-\alpha z}\right)^{5/3}-1\right]dz \\ &\quad +\int_{-\infty}^{0}\frac{3}{10}(3\pi^2)^{2/3}\frac{n_b^{5/3}}{2^{5/3}}\mathrm{e}^{5\alpha z/3}\,dz. \end{aligned}$$

We define the integral

$$I_1=\int_0^{\infty}\left[\left(1-\frac{1}{2}\mathrm{e}^{-x}\right)^{5/3}-1\right]dx,$$

and hence obtain

$$E_{s,kin}=\frac{3}{10}(3\pi^2)^{2/3}\frac{n_b^{5/3}}{\alpha}\left(I_1+\frac{3}{5}\times 2^{5/3}\right). \tag{3.207}$$

(iii) *Exchange energy*. This is

$$\begin{aligned} E_{s,x} &= -\frac{1}{4}\left(\frac{3}{\pi}\right)^{1/3}\int_{-\infty}^{\infty}[\mathrm{n}^{4/3}-\mathrm{n}_b^{4/3}\theta(z)]\,dz \\ &= -\frac{3}{4}\left(\frac{3}{\pi}\right)^{1/3}n_b^{4/3}\int_0^{\infty}\left[\left(1-\frac{1}{2}\mathrm{e}^{-\alpha z}\right)^{4/3}-1\right]dz \\ &\quad -\frac{3}{4}\left(\frac{3}{\pi}\right)^{1/3}n_b^{4/3}\int_{-\infty}^{0}2^{-4/3}\,\mathrm{e}^{4\alpha z/3}\,dz. \end{aligned}$$

Here we define the integral

$$I_2=\int_0^\infty \left[\left(1-\frac{1}{2}\mathrm{e}^{-x}\right)^{4/3}-1\right]\mathrm{d}x$$

and write $E_{\mathrm{s,x}}$ as

$$E_{\mathrm{s,x}}=-\frac{3}{4}\left(\frac{3}{\pi}\right)^{1/3}\frac{n_{\mathrm{b}}^{4/3}}{\alpha}\left(I_2+\frac{3}{4}\times 2^{4/3}\right). \tag{3.208}$$

(iv) *Correlation energy.* This is

$$E_{\mathrm{s,c}}=\int_{-\infty}^{\infty}\left[\frac{0.44n^{1/3}}{7.8n^{1/3}+(3/4\pi)^{1/3}}-\frac{0.44n_{\mathrm{b}}^{1/3}}{7.8n_{\mathrm{b}}^{1/3}+(3/4\pi)^{1/3}}\theta(z)\right]\mathrm{d}z$$

$$=-\int_0^{\infty}\left[\frac{0.44n_{\mathrm{b}}^{1/3}(1-\frac{1}{2}\mathrm{e}^{-\alpha z})^{1/3}}{7.8(1-\frac{1}{2}\mathrm{e}^{-\alpha z})^{1/3}+(3/4\pi)^{1/3}}-\frac{0.44n_{\mathrm{b}}^{1/3}}{7.8n_{\mathrm{b}}^{1/3}+(3/4\pi)^{1/3}}\right]\mathrm{d}z$$

$$+\int_{-\infty}^{0}\left[\frac{0.44n_{\mathrm{b}}^{1/3}\frac{1}{2}\mathrm{e}^{\alpha z}}{7.8(\frac{1}{2}\mathrm{e}^{\alpha z})^{1/3}+(3/4\pi)^{1/3}}\right]\mathrm{d}z.$$

It turns out that this varies as $1/\alpha$. Thus we write

$$E_{\mathrm{s,c}}=E_{\mathrm{c}}^{\mathrm{s}}(n_{\mathrm{b}})/\alpha. \tag{3.209}$$

So far we have only used the F.T.D.G. with the trial form (3.204) for $n(z)$. Now we add the next term.

(v) *Gradient expansion term for the kinetic energy.* This is

$$E=\frac{1}{72}\int_{-\infty}^{\infty}\frac{1}{n}\left(\frac{\mathrm{d}n}{\mathrm{d}z}\right)^2\mathrm{d}z$$

$$=\frac{1}{72}\int_0^{\infty}\frac{n_{\mathrm{b}}\alpha^2\mathrm{e}^{-2\alpha z}}{4(1-\frac{1}{2}\mathrm{e}^{-\alpha z})}\mathrm{d}z+\frac{1}{72}\int_{-\infty}^{0}\frac{n_{\mathrm{b}}\alpha^2}{2}\mathrm{e}^{\alpha z}\,\mathrm{d}z.$$

Now we define the integral

$$I_3=\int_0^{\infty}\frac{\mathrm{e}^{-2x}}{(1-\frac{1}{2}\mathrm{e}^{-x})}\mathrm{d}z,$$

and write

$$E_{\nabla}=\frac{n_{\mathrm{b}}\alpha}{144}\left(\frac{1}{2}I_3+1\right). \tag{3.210}$$

Adding (3.206)–(3.210) we find the surface energy as a function of the variational parameter α:

$$\frac{E_s}{A}=\frac{\pi n_b^2}{2\alpha^3}+\frac{3}{10}(3\pi^2)^{2/3}\frac{n_b^{5/3}}{\alpha}\left(I_1+\frac{3}{5}\times 2^{5/3}\right)$$
$$-\frac{3}{4}\left(\frac{3}{\pi}\right)^{1/3}\frac{n_b^{4/3}}{\alpha}\left(I_2+\frac{3}{4}\times 2^{4/3}\right)+\frac{E_c^s(n_b)}{\alpha}+\frac{n_b\alpha}{144}\left(\frac{1}{2}I_3+1\right). \tag{3.211}$$

Evaluating the definite integrals, we obtain

$$\frac{E_s}{A}=1.571\frac{n_b^2}{\alpha^3}-1.642\frac{n_b^{5/3}}{\alpha}+0.250\frac{n_b^{4/3}}{\alpha}+\frac{E_c^s(n_b)}{\alpha}$$
$$+9.63\times 10^{-3}n_b\alpha. \tag{3.212}$$

This shows the explicit dependence on α and on n_b, except for the correlation term, which is rather complicated and is here denoted by $E_c^s(n_b)$. However, it usually amounts to a small fraction of the exchange energy and we need not worry too much about it for the purposes of the present approximation. The value of α which minimises (3.212) is then given by

$$\alpha^2=\frac{-1.642n_b^{5/3}-0.250n_b^{4/3}-E_c^s(n_b)+\sqrt{\{[1.642n_b^{5/3}-0.250n_b^{4/3}-E_c^s(n_b)]^2+0.182n_b^3\}}}{1.93\times 10^{-2}n_b}. \tag{3.213}$$

Now we can see the relative importance of the different contributions. For very high densities the dominant terms in (3.212) are the Hartree and kinetic contributions. Then

$$\alpha^2\approx 2.87n_b^{1/3}\quad (n_b\text{ large}). \tag{3.214a}$$

In the opposite extreme – low densities – the dominant contributions come from the exchange, correlation and gradient terms. If we neglect $E_c^s(n)$ we can easily obtain an approximate estimate:

$$\alpha^2\approx 25n_b^{1/3}\quad (n_b\text{ small}). \tag{3.214b}$$

Thus when n_b decreases, α decreases as $n_b^{1/3}$, but on the other hand the factor in front increases from about 3 to about 25. These two variations tend to balance each other and for actual metallic densities, in the range $1.5\times 10^{-3}<n_b<35\times 10^{-3}$ a.u., the value of α stays fairly constant between 1.33 and 1.22 a.u. The main conclusion is that for ordinary metallic densitities the different contributions from the various terms of (3.212) tend to be comparable. This suggests that for a correct treatment of the surface problem one ought to take account of all these contributions, including the kinetic energy associated with the gradient expansion. Using (3.213) we can estimate the surface energy (3.212) and this provides a good test of the present approximation since the result can be compared with the conclusions of the independent analysis carried out in § 3.2.2. The results of this evaluation show that the surface energy is positive for small and intermediate densities – e.g. for Na, with $r_s \approx 4$, we obtain $E_s/A \approx 116\ \mathrm{erg/cm^2}$ ($1\ \mathrm{erg} = 10^{-7}\ \mathrm{J}$); for Li, with $r_s \approx 3.28$, $E_s/A \approx 136\ \mathrm{erg/cm^2}$ – whereas it becomes negative for high densities when $r_s \approx 2.6$. This is quite in line with the general conclusions of § 3.2.2.

Thus, although many simplifications have been employed, this model seems basically reasonable and it is interesting to see what it predicts for the other surface properties. Let us discuss in particular the work function Φ, which we can calculate from (3.105) after obtaining the surface dipole term which in this case (3.205) is $4\pi n_b/\alpha^2$. Table 2 shows for various

TABLE 2. *Parameters and results for the calculation of the work function in the kinetic energy approximation for the gradient expansion*

	r_s	α† (a.u.)	$\frac{1}{2}k_F^2$(eV)	$\bar{\mu}_{xc}$(eV)	$\bar{\mu}$(eV)	D_e† (eV)	Φ (eV)
Al	2.07	1.24	11.69	−9.32	2.37	6.00	3.63
Mg	2.65	1.22	7.13	−7.51	−0.38	2.95	3.33
Na	3.99	1.27	3.15	−5.29	−2.15	0.79	2.94
Cs	5.63	1.33	1.58	−3.97	−2.39	0.26	2.65

† Source: J. R. Smith, *Phys. Rev.* **181**, 522 (1969).

metallic densities the values of the surface dipole and of the internal chemical potential $\bar{\mu}$ as the sum of two components according to

$$\bar{\mu} = \left\{\frac{\mathrm{d}[n\tilde{\varepsilon}(n)]}{\mathrm{d}n}\right\}_b = \frac{1}{2}k_F^2 + \bar{\mu}_{xc}. \tag{3.215}$$

Notice that D_e depends very strongly on n_b, being rather large for Al($r_s = 2.07$) and very small for Cs($r_s = 5.63$). The internal chemical potential, $\bar{\mu}$, is also strongly dependent on n_b and even changes sign for $r_s \approx 3$, so that it is the effect of D_e that maintains $\Phi > 0$ for all densities. In fact Φ does not change a great deal from one metal to another, although the values resulting from this approximation represent a definite improvement upon the constant F.T.D.G. value of 2.26 eV.

In conclusion we can say:

(i) The kinetic-energy approximation for the gradient expansion gives, even to the first non-vanishing order in this expansion, an estimate of the surface energy which is in good agreement with the general results of § 3.2.2.

(ii) It also gives a positive work function with some dependence on the metallic density. As a matter of fact these results are not bad compared with experimental data. This is a reflection of the ability of this model to produce reasonable results for the surface dipole D_e at high densities only. At low densities D_e is very small and Φ is dominated by $\bar{\mu}$, which is simply a bulk property.

(iii) The main shortcoming of this approximation is that it cannot provide a method – even if carried to higher order – of producing long-range oscillations of $n(z)$. The fact that even with a comparatively poor description of the charge density profile we can obtain reasonable results for the main surface properties indicates that the density functional approach is adequate to deal with surface problems.

3.2.6. Exact treatment of the kinetic energy functional

We have just seen that there would be no point in continuing the gradient expansion to any finite order, no matter how high. The analytical features associated with the singular behaviour of the dielectric function for $k = 2k_F$ could only be reproduced by exact summation of the series up to infinite order. The *Kohn–Sham method* provides a way of treating the functional $T[n]$ exactly. We recall that, given a real Hamiltonian with electron–electron interactions, we define the kinetic energy functional $T[n(z)]$ as (3.190), which is not the actual kinetic energy of the interacting electrons. The energy functional for the hypothetical electron gas without mutual interactions and subject to an external potential $V(\boldsymbol{r})$ is

$$E_0[n] = T[n] + \int n(\boldsymbol{r}) V(\boldsymbol{r})\, \mathbf{dr}, \tag{3.216}$$

whence we have minimum condition

$$\delta T[n]/\delta n + V(\boldsymbol{r}) = \mu. \tag{3.217}$$

Now, since here the electrons are only subject to the external potential, this procedure is entirely equivalent to solving the set of one-electron equations

$$-\tfrac{1}{2}\nabla^2\psi_i + V(\boldsymbol{r})\psi_i(\boldsymbol{r}) = E_i\psi_i(\boldsymbol{r}), \tag{3.218a}$$

and filling up the lower energy levels until we have the total number of electrons. Then the charge density is

$$n(\boldsymbol{r}) = 2\sum_i^{\text{occ}} |\psi_i(r)|^2, \tag{3.218b}$$

and the kinetic-energy functional is

$$T[n]=2\sum_i^{\text{occ}} E_i-\int V(\boldsymbol{r})n(\boldsymbol{r})\,\mathbf{d}\boldsymbol{r} \tag{3.219}$$

This equivalence can be used to reduce the problem of the interacting electron gas to a system of independent one-electron equations.

For the real interacting electron system we have

$$E[n]=\int n(\boldsymbol{r})V(\boldsymbol{r})\,\mathbf{d}\boldsymbol{r}+\frac{1}{2}\int\frac{n(\boldsymbol{r})n(\boldsymbol{r}')}{|\boldsymbol{r}-\boldsymbol{r}'|}\,\mathbf{d}\boldsymbol{r}\,\mathbf{d}\boldsymbol{r}'$$
$$+T[n]+\int n\tilde{\varepsilon}_{\text{xc}}(n)\,\mathbf{d}\boldsymbol{r}. \tag{3.220}$$

Here we make an explicit approximation for exchange and correlation, which are described in terms of a local energy density. It is the exact treatment of the functional $T[n]$ that we want; this is sufficient to produce the long-range oscillations, and we may expect that neglecting the gradient expansion term in the exchange and correlation energy is not too much of an approximation, although this will be discussed presently.

The minimisation of (3.220) leads to

$$V_{\text{H}}(\boldsymbol{r})+\frac{\mathrm{d}[n\tilde{\varepsilon}_{\text{xc}}(n)]}{\mathrm{d}n}+\frac{\delta T[n]}{\delta n}=\mu, \tag{3.221}$$

which again will be solved together with (3.132*b*). But solving (3.221) is equivalent to solving the system of one-electron equations

$$-\tfrac{1}{2}\nabla^2\psi_i+V_{\text{H}}(\boldsymbol{r})\psi_i+V_{\text{xc}}(\boldsymbol{r})\psi_i=E_i\psi_i, \tag{3.222}$$

where

$$V_{\text{xc}}(\boldsymbol{r})=\mathrm{d}[n\tilde{\varepsilon}_{\text{xc}}(n)]/\mathrm{d}n, \tag{3.223}$$

whence we can obtain $n(\boldsymbol{r})$ by evaluating (3.218*b*). Notice that the total potential acting on the electrons in (3.222), $V_{\text{H}}(\boldsymbol{r})+V_{\text{xc}}(\boldsymbol{r})$, depends on the actual charge density $n(\boldsymbol{r})$, so that we are confronted with a standard selfconsistent calculation in which exchange and correlation are treated in a local approximation and (3.222) must be solved selfconsistently together with (3.132*b*), (3.223) and (3.218*b*). Before going any further it is interesting to make some remarks about this approximation. The Kohn–Sham model for the exchange part is

$$\tilde{\varepsilon}_{\text{x}}(n)=-\frac{3}{4}\left(\frac{3n}{\pi}\right)^{1/3}, \tag{3.224}$$

which yields

$$V_x(\boldsymbol{r}) = \frac{\mathrm{d}[n\tilde{\varepsilon}_x(n)]}{\mathrm{d}n} = -\left(\frac{3}{\pi}\right)^{1/3} n^{1/3}. \tag{3.225}$$

This is two-thirds the customary *Slater local exchange* potential. This difference has a definite physical meaning. The Slater potential represents an average of the potentials experienced by all the electrons in the assembly, while (3.225) gives in the same approximation the exchange potential experienced by the electrons at the Fermi level. The Slater potential could be regarded, in a certain sense, as a potential which includes some correlation energy. This suggests replacing $V_{xc}(\boldsymbol{r})$ by an expression of the form

$$-\alpha(n)\left(\frac{3}{\pi}\right)^{1/3} n^{1/3}, \tag{3.226}$$

where we expect $\alpha(n)$ to be somewhat larger than unity for ordinary metallic densities.

In the application to the surface problem, within the jellium model, an effective one-electron potential of the form

$$V_{\text{eff}}(z) = V_{\text{H}}(z) + V_{\text{xc}}(z) \tag{3.227}$$

is usually employed, in which $V_{\text{H}}(z)$ is the electrostatic potential created by both the electronic and the jellium charges and satisfies the Poisson equation (3.132*b*); and V_{xc} is given by (3.223), where the Wigner approximation (3.126) is used for ε_{xc}. The wavefunctions obeying (3.222) are then of the form

$$\psi_i(\boldsymbol{r}) \sim \mathrm{e}^{\mathrm{i}\boldsymbol{k}\cdot\boldsymbol{\rho}}\psi_q(z), \tag{3.228}$$

where $\psi_q(z)$ satisfies the equation

$$-\tfrac{1}{2}\,\mathrm{d}^2\psi_q(z)/\mathrm{d}z^2 + V_{\text{eff}}(z)\psi_q(z) = (E - \tfrac{1}{2}\kappa^2)\psi_q(z). \tag{3.229}$$

The solution of this type of equation was discussed in § 3.1. For large $z > 0$ we have

$$\psi_q(z) \sim \sin(qz + \eta), \quad E = \tfrac{1}{2}(\kappa^2 + q^2). \tag{3.230}$$

The phase shift is of course a function of q only, i.e. of the part of the energy not associated with κ, $E_\perp = E - \frac{1}{2}\kappa^2 = \frac{1}{2}q^2$. The general properties of the phase shifts and their relation to the density of states were discussed in § 3.1. In particular they yield the following formula for the charge density (3.61*a*):

$$n(z) = \frac{1}{\pi^2}\int_0^{k_{\text{F}}} \left(k_{\text{F}}^2 - q^2\right)|\psi_q(z)|^2\,\mathrm{d}q. \tag{3.231}$$

Thus the theoretical framework is ready for us to solve selfconsistently the surface problem by treating the kinetic energy exactly while using a local approximation for exchange and correlation.

Table 3 reproduces some results from actual numerical calculations for four different metals, so that we can compare these with the results of treating the kinetic energy by a gradient-expansion approach (table 2).

TABLE 3. *Surface dipole, work function and surface energy obtained in the exact treatment of the kinetic energy*

	D_e (eV)	Φ (eV)	E_s/A(erg cm^{-2})
Al	6.24	3.87	−730
Mg	3.28	3.66	110
Na	0.91	3.06	160
Cs	0.42	2.49	70

Source: N. D. Lang & W. Kohn, *Phys. Rev.* B **1**, 4555 (1970).

Notice that the values of D_e are in very good agreement for Al but in marked disagreement for Cs. Since the value of the surface dipole reflects a measure of the effective thickness of the surface layer we can surmise from here that the gradient-expansion method gives a fairly good picture of the surface charge for high densities but tends to be seriously in error for low densities, where it gives a narrower surface layer. The reason for this lies in the absence of long-range oscillations and the effect that these have on the surface charge. We know from the analysis of § 3.1 that for a potential with phase shifts $\eta(q)$ the charge density for large $z > 0$ behaves like

$$n(z) \sim n_b\left\{1+\frac{3\cos[2k_Fz+2\eta(k_F)]}{(2k_Fz)^2}\right\}. \tag{3.232}$$

Fig. 45 shows the density profile for $r_s = 2$ and $r_s = 5$. For low densities the amplitude of the oscillation becomes quite appreciable, reaching 10% of the bulk density when $r_s = 5$, while for high densities the constant value n_b is reached sooner and the oscillations have very small amplitudes. This can be qualitatively understood if we remember that the effective thickness of the surface layer is of order $k_{FT}^{-1} \propto n_b^{-1/6}$. The length k_F^{-1}, associated with the range of the oscillations, varies as $n_b^{-1/3}$ and therefore predominates at low densities. This explains why the results obtained in § 3.2.5 are better for high densities ($r_s \approx 2$) and worse for low densities.

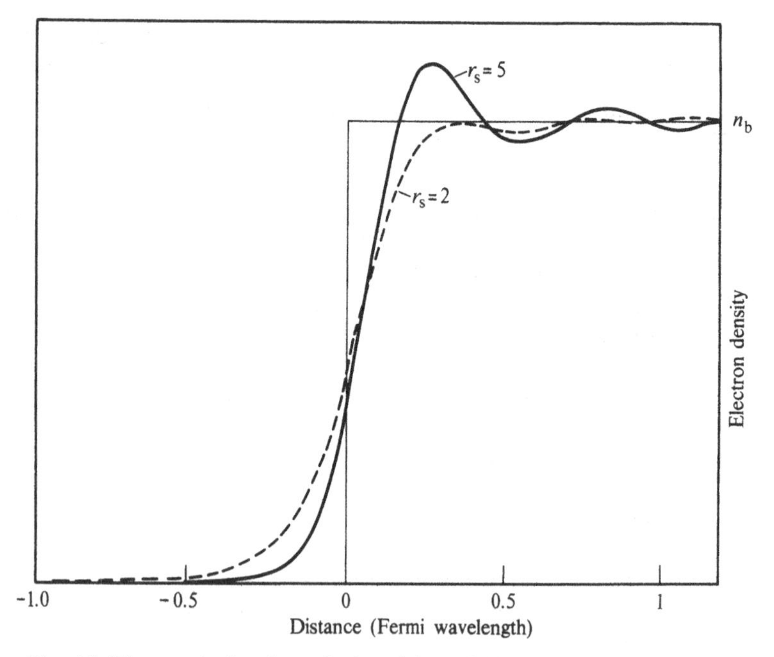

Fig. 45. Electronic density calculated for a jellium model by using the Kohn–Sham method. Results for $r_s = 2$ and $r_s = 5$. One Fermi wavelength is $2\pi/k_F$. At $r_s = 2$ this is 6.55 a.u. (3.46 Å), at $r_s = 5$ it is 16.37 a.u. (8.66 Å). N. D. Lang, *Solid St. Commun.* **7**, 1047 (1969).

A similar situation is found in the study of the surface energy, starting from the functional

$$E[n] = \frac{1}{2}\int V_e(z)[n(z) - n_b\theta(z)]\,\mathbf{dr} + T[n] + \int n\tilde{\varepsilon}_{xc}(n)\,\mathbf{dr}, \tag{3.233}$$

which includes the jellium selfenergy. The surface energy per unit area is then

$$\frac{E_s}{A} = \frac{1}{2}\int_{-\infty}^{\infty} V_e(z)[n(z) - n_b\theta(z)]\,dz + \frac{1}{A}T[n] - \int_0^{\infty} n_b t(n_b)\,dz + \int_{-\infty}^{\infty}[n\tilde{\varepsilon}_{xc}(n) - n_b\tilde{\varepsilon}_{xc}(n_b)\theta(z)]\,dz. \tag{3.234}$$

Here $t(n_b)$ is the kinetic energy functional per particle. If we know $n(z)$, (3.234) can be readily evaluated except for $T[n]$. But this can be obtained by using (3.219) where now the potential V is the effective potential, $V_e(z) + V_{xc}(z)$, i.e.

$$T[n] = 2\sum_i^{\text{occ}} E_i - \int V_{\text{eff}}(\mathbf{r})n(\mathbf{r})\,\mathbf{dr}. \tag{3.235}$$

At this point we can make use of the results derived in § 3.1. We saw there that the first term on the r.h.s. of (3.235) has two contributions, namely, the surface part equal to

$$\frac{1}{\pi^2}\int_0^{E_F}(E-E_F)\left[\eta(E)-\frac{\pi}{4}\right]dE$$

per unit area, and the bulk part, which would be the contribution of homogeneous bulk material extending from 0 to ∞. In our case, where the E_is are the eigenvalues of (3.222), the energy per unit area of the homogeneous system is

$$\int_0^\infty n_b t(n_b)\,dz+\int_0^\infty V_{eff}(\infty)n_b\,dz$$
$$\equiv\int_0^\infty n_b t(n_b)\,dz+\int V_{eff}(\infty)n(z)\,dz.$$

In the last integral $n(z)$ can identically replace n_b because $V_{eff}(\infty)$ is a constant and the total charge is invariant. Using all this in (3.235) we have

$$\frac{1}{A}T[n]=\int_0^\infty n_b t(n_b)\,dz+\frac{1}{\pi^2}\int_0^{E_F}(E-E_F)\left[\eta(E)-\frac{\pi}{4}\right]dE$$
$$-\int[V_{eff}(z)-V_{eff}(\infty)]n(z)\,dz. \tag{3.236}$$

Thus the surface energy can be evaluated if we know $n(z)$ and the one-electron wavefunctions of (3.206). Some representative results are included in table 3, where we see that E_s is positive for low densities and becomes negative for $r_s\approx 2.4$. This is again in agreement with the general considerations of § 3.2.2, and is in line with the results obtained by means of the gradient-expansion method, although there are appreciable quantitative discrepancies for low densities. For example for Na one finds 160 erg cm^{-2} (now) as against 116 erg cm^{-2} (gradient expansion) and, for Li, 210 erg cm^{-2} (now) as against 136 erg cm^{-2} (gradient expansion). These discrepancies are also associated with the effect of the long-range oscillations.

As regards the practical value of this approximation, the values obtained for the work function compare fairly well with experimental evidence, while the results for surface energy are not all as good. This is connected with the tendency to produce negative values, which we have seen is a general feature of the jellium model. To sum up, the position with this approximation is the following:

(i) It gives a better description of the charge density profile than the previous approximations, as it reproduces the long-range oscillations. This improves the situation considerably at low densities.

(ii) The amplitude of the oscillations is smaller, in practice negligible, at high densities. This explains why the gradient-expansion approximation is already fairly acceptable in this range.

3.2.7. Gradient expansion for exchange and correlation energy

We have studied a sequence of approximations leading up to the exact treatment of the kinetic energy but, in all of these, exchange and correlation have been treated in local approximations. It is now interesting to see how much the picture can change if we also include some non-local effects in the exchange and correlation energies. We shall study the corrections obtained when using the gradient-expansion method up to the first non-vanishing term. The same symmetry considerations hold, so that the exchange and correlation function we start from is of the form

$$E_{xc}[n]=\int n\tilde{\varepsilon}_{xc}(n)\,\mathrm{d}\boldsymbol{r}+\int F_{xc}(n)(\nabla n)^2\,\mathrm{d}\boldsymbol{r}+\cdots \tag{3.237}$$

The coefficient $F_{xc}(n)$ can be found by the same method as that used in § 3.2.5 for the expansion of the kinetic energy. In (3.188) we obtained a formal expression for the dielectric function by comparing it with (3.187). Now we do the same but use in (3.188) the following equality:

$$\begin{aligned}\left\{\frac{\mathrm{d}^2[n\tilde{\varepsilon}(n)]}{\mathrm{d}n^2}+2F_2(n)k^2\right\}_{\mathrm{b}}\\ =\left\{\frac{\mathrm{d}^2[nt(n)]}{\mathrm{d}n}\right\}_{\mathrm{b}}+\frac{1}{36n_{\mathrm{b}}}k^2+\left\{\frac{\mathrm{d}^2[n\tilde{\varepsilon}_{xc}(n)]}{\mathrm{d}n^2}\right\}_{\mathrm{b}}+2F_{xc}(n_{\mathrm{b}})k^2,\end{aligned} \tag{3.238}$$

where we have put $\tilde{\varepsilon}(n)=t(n)+\tilde{\varepsilon}_{xc}(n)$ and $F_2(n)=F_{\mathrm{kin}}(n)+F_{xc}(n)=1/72n+F_{xc}(n)$, using (3.192). This yields

$$\begin{aligned}\varepsilon(k)&=1+\frac{4\pi/k^2}{\{\mathrm{d}^2[nt(n)]/\mathrm{d}n^2\}_{\mathrm{b}}+k^2/36n_{\mathrm{b}}+\{\mathrm{d}^2[n\tilde{\varepsilon}_{xc}(n)]/\mathrm{d}n^2\}_{\mathrm{b}}+2F_{xc}(n_{\mathrm{b}})k^2}\\ &=1+\frac{\Pi_0}{1+(k^2/4\pi)\Pi_0\{\mathrm{d}^2[n\tilde{\varepsilon}_{xc}(n)]/\mathrm{d}n^2+2F_{xc}(n)k^2\}_{\mathrm{b}}},\end{aligned} \tag{3.239a}$$

where we have defined

$$\Pi_0=\frac{4\pi/k^2}{\{\mathrm{d}^2[nt(n)]/\mathrm{d}n^2\}_{\mathrm{b}}+k^2/36n_{\mathrm{b}}}. \tag{3.239b}$$

This can now be used to find $F_{xc}(n_{\mathrm{b}})$ by comparing with some independent estimate of the dielectric function which includes exchange and correlation. For example, we can use the *Singwi–Tosi–Land–Sjölander* formula

$$\varepsilon(k) = 1 + Q_0(k)/[1 - \mathscr{P}(k)Q_0(k)], \tag{3.240}$$

where $Q_0 = -4\pi\chi_{\text{rpa}}(k)/k^2$ and $\mathscr{P}(k)$ is a correction factor which includes exchange and correlation effects. In comparing (3.240) with (3.239) we must stress that (2.238) has been obtained with a gradient expansion approximation and we must therefore compare (3.239) with the long-wave limit of (3.240). Notice also that if we take $\mathscr{P}(k) = 0$ – i.e. no exchange and correlation effects in (3.240) – then this comparison leads to

$$1 + \Pi_0(k) = 1 + Q_0(k), \tag{3.241}$$

whence in the long-wave limit we recover $F_{\text{kin}} = 1/72n$, as in § 3.2.2. We can now include $\mathscr{P}(k)$ and in the long-wave limit – which in order to take account of all the terms involved requires an expansion up to order k^6 – we find

$$-\frac{k^2}{4\pi}\left\{\frac{\mathrm{d}^2[n\tilde{\varepsilon}_{\text{xc}}(n)]}{\mathrm{d}n^2} + 2F_{\text{xc}}(n)k^2\right\} = \mathscr{P}(k), \tag{3.242}$$

where it is understood that the r.h.s. is also expanded in powers of k. This relates the first terms in the expansion of $\mathscr{P}(k)$ in powers of k to those in the expansion of $E_{\text{xc}}[n]$. Thus the final outcome depends on what one uses for $\mathscr{P}(k)$. For example the *Vashishta–Singwi* approximation, for $k < 2k_{\text{F}}$, is of the form

$$\mathscr{P}(k) = A\{1 - \exp[-B(k/k_{\text{F}})^2]\}, \tag{3.243}$$

where A and B are slowly varying functions of the density. Using this in (3.242) and expanding in powers of k we find

$$F_{\text{xc}}(n_{\text{b}}) = \pi AB^2/k_{\text{F}}^4 \approx 0.020 r_{\text{s}}^4. \tag{3.244}$$

The numerical coefficient has been estimated by using for A and B the average of the values they take between $r_{\text{s}} = 1$ and $r_{\text{s}} = 6$.

Having obtained $F_{\text{xc}}(n)$ we can now proceed to study the different physical properties of interest. We start from the total functional

$$E[n] = \int n(\boldsymbol{r})V(\boldsymbol{r})\mathbf{d}\boldsymbol{r} + \frac{1}{2}\int \frac{n(\boldsymbol{r})n(\boldsymbol{r}')}{|\boldsymbol{r}-\boldsymbol{r}'|}\mathbf{d}\boldsymbol{r}\,\mathbf{d}\boldsymbol{r}'$$
$$+ T[n] + \int n\tilde{\varepsilon}_{\text{xc}}(n)\,\mathbf{d}\boldsymbol{r} + \int F_{\text{xc}}(n)(\nabla n)^2\,\mathbf{d}\boldsymbol{r}. \tag{3.245}$$

Notice that by removing the last term we are left with (3.220) of § 3.2.6.

The minimum condition is then

$$\delta T[n]/\delta n + V_{\mathrm{H}}(\boldsymbol{r}) + V_{\mathrm{xc}}(\boldsymbol{r}) = \mu, \tag{3.246a}$$

where we now define

$$V_{\mathrm{xc}}(\boldsymbol{r}) = \frac{\mathrm{d}n\tilde{\varepsilon}_{\mathrm{xc}}(n)}{\mathrm{d}n} + \frac{\mathrm{d}[F_{\mathrm{xc}}(n)]}{\mathrm{d}n}(\nabla n)^2 - 2\nabla \cdot [F_{\mathrm{xc}}(n)\nabla n]. \tag{3.246b}$$

Notice again that (3.246*a*) is equal to (3.221) if the exchange and correlation potential is redefined according to (3.246*b*). Thus the kinetic energy can be calculated again from a system of one-electron equations, and the density can be obtained from a selfconsistent calculation involving (3.222) with V_{xc} defined by (3.246*b*). It is not necessary to go over the details of a discussion which follows the same lines as before, but it is interesting to look at some representative results obtained in this approximation. Table 4 gives the change – upon the approximation of § 3.2.7 – in E_{s} for four different metals; compare with table 3. The corrections for

TABLE 4. *Changes in surface energy due to the introduction of the gradient expansion for exchange and correlation up to second order*

	Al	Zn	Mg	Na
δE_{s} (erg cm^{-2})	365	260	155	45

Source: J. Ferrante & J. R. Smith, *Solid St. Commun.* **23**, 527 (1977). These represent a recalculation of the values given by J. H. Rose Jr, H. B. Shore, D. J. Geldart & M. Rasolt, *Solid St. Commun.* **19**, 619 (1976).

Φ turn out to be rather small, typically of order 0.01 to 0.02 eV. Thus the gradient expansion term for exchange and correlation changes the surface dipole very little, so that we have a very similar description of the surface charge distribution. However, the surface energy does vary appreciably, as one would expect. At the very least, even if there is no change in $n(\boldsymbol{r})$, we must have the new contribution due to F_{xc}. It is interesting to notice that the changes in E_{s} are all positive, although they are not sufficiently great to make E_{s} over-all positive for Al.

To sum up, it appears that a local approximation for exchange and correlation is fairly reasonable in producing an acceptable picture of the surface charge distribution. The main point is that the exact treatment of the kinetic energy is sufficient to produce the long-range oscillations. The inclusion of the gradient does not change the surface charge distribution significantly. It does modify the estimate of the surface energy rather more, but in accordance with the general features of all jellium models, this cannot produce a positive value of E_s for all densities. In order to take a step in this direction it is necessary to include the effects of the lattice ions and this we shall now do.

3.3. Effects of crystallinity

Let us study the effect of introducing a discrete crystal lattice instead of a positive jellium. The main thing we want to consider is the surface energy, whose behaviour as a function of bulk electronic density constitutes the worst feature of the jellium model, but we shall also discuss how the picture of the surface region is modified. We shall consider the problem in two stages.

3.3.1. Linear corrections to surface energy and work function

We could imagine the jellium model as a simplification obtained by starting from the positive ionic charges and spreading out each one of these into the same amount of positive charge uniformly distributed in a prism of appropriate height, with base the two-dimensional unit cell in a plane parallel to the surface. This is shown in fig. 46 for the (100) face of a

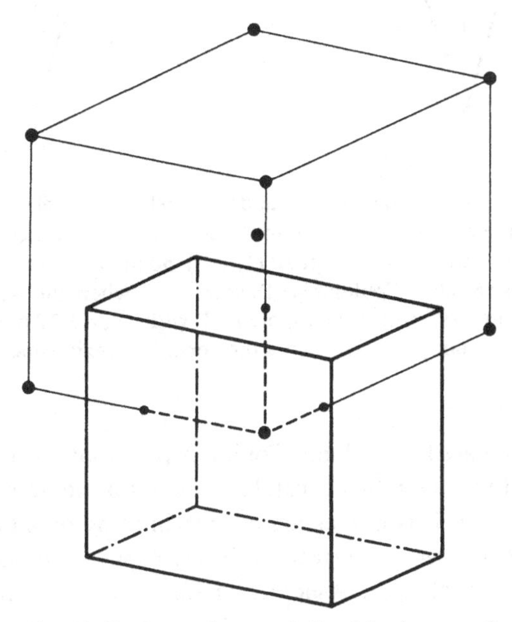

Fig. 46. Surface prism, as defined in the text, for the (1, 0, 0) face of a b.c.c. structure. The square face is formed by the surface atoms in the unit cell of the crystal. The height is equal to one half the length of the side of the unit cell in the bulk crystal.

base-centred cubic (b.c.c.) structure. Thus the crystalline case can be obtained from the jellium model by adding a potential $\delta V(r)$ equal to the difference between the potentials due to the two charge distributions. If we average $\delta V(r)$ over the (x, y) plane then we obtain $\delta V(z)$. This could be, for example, the potential shown in fig. 47. Notice that in the region between the atomic planes this potential is parabolic, as corresponds to the electrostatic potential of a uniform one-dimensional charge density.

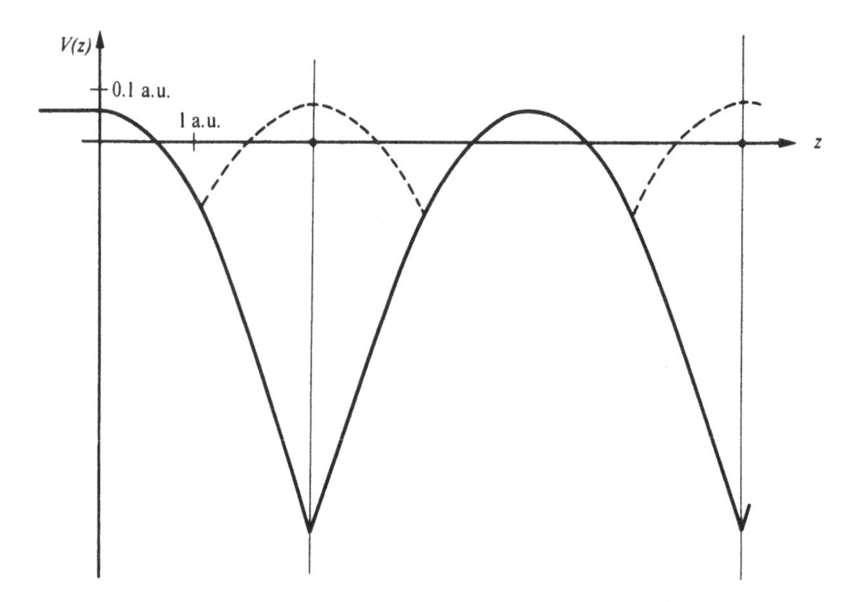

Fig. 47. Perturbing potential, to be added to the jellium model, shown as a function of the coordinate normal to the surface after averaging over the plane parallel to the surface for Al (1, 1, 1). Solid line: case in which the crystal potential is Coulombic. Dashed line: when the above is replaced by an Ashcroft pseudopotential with $R_s = 1.12$ a.u. The nucleus of this pseudopotential defines the perturbed region about each ionic plane.

Of course δV, calculated from ionic Coulomb potentials, would be hopelessly strong but we know that it can be replaced by some weaker effective pseudopotential. It is interesting to look in some detail at the implications that this has for the surface problem. Suppose we use, for example, a pseudopotential of the Ashcroft type, i.e. with the centre of each ion:

$$V(r) = \begin{cases} 0, & r < R_c; \\ -z/r, & r > R_c. \end{cases} \tag{3.247}$$

If we remove the strong attractive part, the resulting – weaker – pseudopotential can be treated perturbatively in the standard manner. In fig. 47 we see (dotted lines) the pseudopotential averaged again in the (x, y) plane, as a function of z, for the $(1, 1, 1)$ face of the f.c.c. structure. Now, this figure shows *only* the electrostatic potential – or pseudopotential – to be added to that of the one-dimensional uniform charge. Beyond the jellium edge δV tends to a constant 'vacuum level'. But remember that *this does not include the total selfconsistent one-electron potential* previously calculated in the jellium model, which includes, for example, electron–electron interactions in the inhomogeneous electron gas. It is only δV that we are concerned with here, and as regards its z-dependence, an important feature is the magnitude of the energy difference between its vacuum level and its average value inside the crystal. The main effect, on changing from the Coulomb potential to the pseudopotential, comes from the large change in δV in the core regions, $r < R_c$, of the ions. The cancellation of the strong attractive part raises the average value of δV inside the crystal so that its distance from its own vacuum level is greatly reduced. Thus not only is δV weaker in the bulk but it is also less strongly inhomogeneous in the surface region. It is this fact that enables us to treat it as a small perturbation when we redo the surface calculation. We shall therefore assume for the time being that the corrections due to δV can be evaluated to first order.

In order to study the corrections to the surface energy it is convenient to imagine the real surface as created in two stages: (i) We first cut the bulk material in half while keeping the electronic charge density frozen in its bulk distribution (fig. 48*a*), and then separate the two halves a

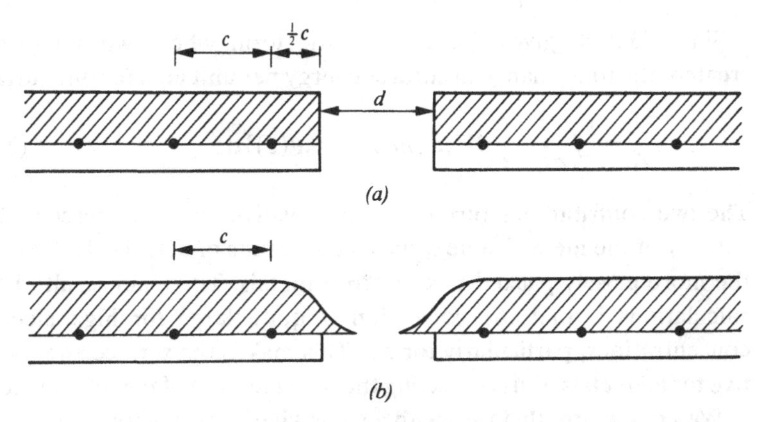

Fig. 48. Diagram showing the two stages in the creation of the surface. (*a*) The material is cleaved and the charge distribution remains frozen. (*b*) The charge distribution is allowed to relax.

distance d apart and let $d \to \infty$. (ii) In the second stage we allow the electronic charge density to relax to its selfconsistent surface distribution. If we let this process take place in the presence of a total potential which includes δV, it will suffice to first order to assume that the relaxed charge density is simply the solution obtained in the jellium model.

The contributions of the crystalline correction to the surface energy in these two stages are then obtained as follows: (i) We must evaluate the energy needed to move the two half crystals apart. If we take a plane of separation which does not cut through the core regions of the ions, then it will be enough to evaluate the change in electrostatic interaction energy. Roughly, we may say with a lattice constant c the interaction between two ions with charge Z is of order Z^2/c and this energy per unit area is of order $(Z^2/c)c^{-2}$. Thus we expect the first contribution to the change in surface energy to be of the form

$$E_1/A = fZ^2/c^3, \tag{3.248}$$

where f is a dimensionless number which depends on crystal structure and surface orientation. (ii) If $E[n_0]$ is the energy of the system with $n_0(z)$ equal to the selfconsistent charge distribution of the jellium model, then to first order this is changed into

$$E[n, \delta V] = E[n_0] + \int \delta V(\boldsymbol{r}) n_0(z)\, \mathbf{dr}. \tag{3.249}$$

Thus the second contribution to the change in the surface energy per unit area is, to first order,

$$\frac{E_2}{A} = \int \delta V(\boldsymbol{r})[n_0(z) - n_b\theta(z)]\, dz. \tag{3.250}$$

Since (3.248) gives the electrostatic term when two surfaces are created, the total change in surface energy per unit area for one surface is

$$\frac{\delta E}{A} = \frac{1}{2} f \frac{Z^2}{a^3} + \int \delta V(\boldsymbol{r})[n_0(z) - n_b\theta(z)]\, dz. \tag{3.251}$$

The two contributions turn out to be positive and to increase with the valency of the metal. Table 5 shows the results of actual calculations for the more closely packed faces of four metals. The main result of these calculations is that the correction is quite large for high electronic concentrations, particularly for Al. This makes the surface energy positive for all metals, thus correcting the worst feature of the jellium models.

We are apparently faced with a somewhat contradictory situation. By including the effect of a weak pseudopotential to first order we obtain a very large change in E_s. Is the perturbative treatment not suspect under

TABLE 5. *The two corrections to the surface energy* (erg cm^{-2}) *due to the introduction of crystallinity to first order in perturbation theory*

	$\frac{1}{2}\frac{E_1}{A}$†	$\frac{E_2}{A}$†	$\frac{\delta E}{A}$	$\frac{E_s}{A}$‡
Al (111)	408	1050	1458	728
Mg (001)	130	300	430	540
Na (110)	33	35	68	228
Cs (110)	12	20	32	102

† Source: N. D. Lang & W. Kohn, *Phys. Rev.* B **1**, 4555 (1970).
‡ E_s/A here is the total energy obtained by adding these two corrections to the value of E_s/A given in table 3.

these circumstances? The point to realise is that it is the charge distribution that we must look at in order to decide whether or not a perturbative treatment is acceptable. A good test in fact consists in studying the extra surface dipole induced by the perturbation δV. This leads us to consider the work function.

Now, to lowest order it suffices to consider the effect of the Fourier components of the ionic potential with wavevector perpendicular to the surface, which is tantamount to averaging δV over the (x, y) plane. We are thus studying perturbations of the form shown in fig. 47. At this stage it is important to look at the potential used in relation to the crystal-face orientation. Consider, for example, the Coulomb potential. When the atomic planes are more closely packed, the distance between planes is larger and the variations of the crystalline potential stronger, hence the difference between the vacuum level (of δV) and its average value inside the crystal is larger. The situation can be very different when the pseudopotential is introduced (fig. 47). We then have large corrections to the Coulomb potential near the crystal planes, and these modify the average value in the bulk, so much so that even the sign of the above difference may change. Depending on the size of the pseudopotential core this may result in qualitative changes to the conclusions that would be reached by literally using a Coulomb potential.

The idea behind the quantitative estimate of the change in work function is to use perturbation theory in the definition (3.97), where Φ is expressed by comparing the energy of two different systems, one with N electrons and the other with $N-1$ electrons inside the metal and one electron outside, a macroscopic distance away. For the first system, in the presence of the perturbation δV and to first order in δV we have

$$E[N, \delta V] = E_J[N] + \int \delta V(\mathbf{r}) n_0(z)\, dz. \tag{3.252}$$

Consider now the second system in the unperturbed situation: one electron far away from the surface and $N-1$ electrons inside a jellium metal. It is clear that there is a deficiency of electronic charge in the surface region. Thus the charge density is of the form $n_0(z)+\delta n(z)$, where

$$\int \delta n(z)\, dz = -1.$$

Since the missing electron is a macroscopic distance away, the charge redistribution $\delta n(z)$ could be thought of as induced by a uniform external electric field normal to the surface. We then write for the energy of the second system, to first order in δV,

$$E[N-1, \delta V] = E_J[N-1] + \Phi_e(-\infty) + \delta V(-\infty) + \int \delta V[n_0(z) + \delta n(z)]\, dz. \tag{3.253}$$

Notice that the perturbation δV due to the ions could be finite at $-\infty$, and the total electrostatic energy of the electron removed to $-\infty$ is in general $\Phi_e(-\infty)+\delta V(-\infty)$. Using (3.252), (3.253) and (3.97) we find

$$\delta\Phi = \delta V(-\infty) + \int \delta V\, \delta n(z)\, dz. \tag{3.254}$$

Having decided on the pseudopotential δV to be used, we have only to find the charge redistribution. As indicated above this can be found by working out $\delta n(z)$ as induced by a uniform external electric field perpendicular to the surface and normalising it so that one electronic charge is missing. This can be evaluated in linear response theory by the method of § 3.2.6; notice that an electron a distance d away induces a surface charge of order d^{-2}. Knowing $\delta n(z)$ one can then use (3.254) to estimate $\delta\Phi$. Some results obtained in this way are shown in table 6, which also shows the value used for the core radius R_c of the pseudopotential (3.247).

It is interesting now to use the results of table 6 to estimate the effect of δV on the electronic charge distribution in the surface region, as reflected in the extra contribution to the surface dipole induced by δV. In order to estimate this we notice that $\delta\Phi$ has two contributions, namely, (i) the change due directly to the introduction of the lattice without allowing the electronic charge distribution to change and (ii) the change due to the redistribution of the electronic charge. The first contribution is due to the

TABLE 6. *Crystalline correction (first-order perturbation theory) to the work function* (eV) *for different surface orientations*

	R_c†	$\delta\Phi$ (111)	$\delta\Phi$ (110)	$\delta\Phi$ (100)
Al (f.c.c.)	1.12	0.19	−0.21	0.32
Na (b.c.c.)	1.67	−0.39	0.03	−0.29
Cs (b.c.c.)	2.93	−0.67	−0.23	−0.61

† Source: N. D. Lang & W. Kohn, *Phys. Rev.* B **3**, 1215 (1971). R_c (a.u.) is the core radius used for the pseudopotential.

potential difference, introduced by δV, between vacuum and average bulk values. In the situation described in fig. 47, for example, this tends to increase Φ. Table 7 shows the potential difference, $\delta\bar{V}$, calculated for the same cases as in table 6. Knowing $\delta\bar{V}$ we can evaluate the change in surface dipole δD_e through

$$\delta\Phi = \delta\bar{V} + \delta D_e. \tag{3.255}$$

Table 7 also shows the values of δD_e obtained in this manner by subtracting $\delta\bar{V}$ (table 7) from $\delta\Phi$ (table 6).

TABLE 7. *The contribution $\delta\bar{V}$ (see text) to the change in work function for different surface orientations and the change δD_e in surface dipole due to the crystalline perturbation to first order, evaluated by using (3.255) and table 6*

	$\delta\bar{V}$ (eV)			δD_e (eV)		
	111	110	100	111	110	100
Al	1.7	−3.0	−0.2	−1.5	2.8	0.5
Na	−1.5	0.0	−0.9	1.1	0.0	0.6
Cs	−1.8	−0.7	−1.3	1.1	0.5	0.7

We can now compare δD_e with the value of the surface dipole obtained in jellium models and with the Fermi energy. These values were given in tables 2 and 3; notice that table 3 represents a higher approximation for D_e. It is interesting to notice the following features:

(i) The perturbation is rather small for Al(100) and Na(110).

(ii) However, it is rather large for Cs(111) and Cs(100), where D_e is not only larger than the surface dipole obtained in the jellium model, but is also comparable with the Fermi energy.

(iii) The importance of the perturbation increases gradually from Al(111) and Al(110), through Na(100), and on to Na(111) and Cs(110).

Notice that the effect of the perturbation tends to be more important for metals of low electronic density, in contradiction to what one would have concluded by looking at the results for the surface energy. The most important feature is that the perturbation can become rather large in some cases and this may cast serious doubts on the generalised use of first-order perturbation theory. It is thus necessary to investigate the problem beyond the linear approximation.

3.3.2. Nonlinear corrections

An adequate framework for the study of the effect of δV is in principle provided by the functional

$$E[n, \delta V] = E_J[n] + \int \delta V(\mathbf{r}) n(\mathbf{r})\, \mathbf{dr}, \tag{3.256}$$

where $E_J[n]$ is the energy functional of the jellium model. The question is, which approximation do we use for $E_J[n]$? This problem was extensively discussed in § 3.2 and we saw that the exact treatment of the kinetic energy is more important than higher-order corrections to exchange and correlation. Thus we can take as a reasonable starting basis for $E_J[n]$ the functional of (3.220), where the kinetic energy is treated exactly and exchange and correlation are treated locally. The minimisation of (3.256) would require the selfconsistent solution of N one-electron equations like (3.222) including $\delta V(\mathbf{r})$, a formidable problem indeed. In order to avoid this, we could try using a plausible device in which $n(\mathbf{r})$ is sought within a family of functions with a parameter which is then varied to minimise (3.256). Notice that this is very similar in idea to the last part of § 3.2.5.

We seek the family of functions $n(\mathbf{r})$ which minimise the functional

$$E'[n, V_0] = E_J[n] + \int \theta(z) V_0 n(\mathbf{r})\, \mathbf{dr}. \tag{3.257}$$

Notice that this is not exactly the true energy functional $E[n, \delta V]$ of (3.256), but this functional retains the essential features, namely: (i) $E_J[n]$ is also taken from (3.222). Thus the kinetic energy is treated exactly and this picks out the extra contribution to the long-range oscillations induced by the surface perturbation. (ii) Although the pseudopotential is a weak perturbation in the bulk, it may represent a considerable change at the surface. In particular it is essential to retain an extra contribution to the height of the potential barrier. This is incorporated in (3.257) through the term $\theta(z) V_0$, in which the height of the abrupt step is the adjustable parameter. Now, solving the minimum condition for (3.257) yields a charge distribution which depends on the parameter V_0, has appropriate Friedel oscillations near the surface region, and tends to the constant value n_b for $z \to \infty$. Actually these functions depend only on z, so that they represent an average in the (x, y) plane. The idea then is to use the family of functions $n(z, V_0)$ thus obtained to evaluate the functional of (3.256), and to vary V_0, seeking the minimum of $E[n, \delta V]$, which we should then write rather as $E[n(z, V_0), \delta V]$; thus we have

$$\mathrm{d}E[n(z, V_0), \delta V]/\mathrm{d}V_0 = 0. \tag{3.258}$$

In the weak δV limit this procedure reproduces (3.249) of the perturbative treatment, since in this limit V_0 should be fairly small and hence $n(z)$ quite close to $n_0(z)$, the solution when $V = 0$. But the point is that this approach can be used beyond the linear approximation, and the new charge density $n(z)$ can be used to estimate the surface energy and the surface dipole δD_e induced by the lattice pseudopotential. Table 8 shows some results obtained in this manner, or rather with a slight variation,

TABLE 8. *Nonlinear corrections, D_e, to the surface dipole due to crystallinity and surface energy values, E_s, resulting from the same nonlinear calculation*

	δD_e† (eV)			
	111	110	100	E_s/A‡ (erg cm^{-2})
Al	−1.4	2.9	0.2	643 (111)
Na	1.1	0.0	0.7	226 (110)
Cs	1.4	0.7	—	85 (110)

† Source: J. P. Perdew & R. Monnier, *Phys. Rev. Lett.* **37**, 1286 (1976) and current work (private communication).
‡ In order to establish an adequate comparison with the results of table 5, the values of E_s/A shown here include all the crystalline corrections, but not the contribution coming from the gradient expansion of exchange and correlation energies to second order (table 4).

which consists in introducing a different adjustable parameter, namely z_0 – thus allowing for an adjustable effective location of the dipole – writing $-\delta\bar{V}\theta(z - z_0)$ instead of $V_0\theta(z)$ everywhere and minimising with respect to z_0. These results can be compared with those of tables 5 and 7 for the linear approximation.

This comparison deserves some comments. In general the values obtained for δD_e do not differ too much from those of the linear approximation. Even some of the larger differences need not be definitely significant, in view of the model used in the variational calculation. Notice that in choosing $n(z)$ to minimise (3.257) one takes account only of the average effect that the pseudopotential has on the height of the barrier, thereby neglecting the effect of the periodic oscillations in the bulk. For the particular case of Al(100) this introduces errors in the estimate of δD_e of the order of a few tenths of an electron volt, which is comparable to the differences between the values of δD_e given in tables 7 and 8. However, in most other cases these errors turn out to be unimportant.

More significant differences appear to affect the surface energy on comparing tables 5 and 8. Notice that E_s is precisely the quantity which is minimised in the variational procedure, and inclusion of the effects of the bulk oscillations could not reduce the differences between these two tables. This difference seems to be quite substantial for Cs.

However, over-all there are sufficiently many cases in which the changes introduced by the variational calculation do not seem to be so great. Thus, in spite of the comments at the end of § 3.3.1 there are indications that the linear approximation is basically reasonable, although it might be somewhat inaccurate in a case like Cs, for example.

3.3.3. The surface potential of a crystalline metal

In §§ 3.3.1, 2 we have consistently averaged in the (x, y) plane, thus getting rid of any crystallinity in the directions parallel to the surface, and we have argued that a fair approximation to the effect of crystallinity on the surface properties can be obtained in this way. While this may be a reasonable procedure in order to estimate, say, changes in surface energy or work function, it would be nice to have an idea, if only for the sake of completeness, of what the surface of a crystalline metal looks like. For example, rather than limiting ourselves to knowing the potential – averaged in the (x, y) plane – as a function of z, we would like to see what this function of z looks like when the surface is approached along various parallel lines, ending up at different points of the surface. A theory for this could be more than just a matter of curiosity. In principle it could be used as a basis for the study of various surface properties, including, say the chemisorption of simple species. Here we shall only outline a possible approach to the problem and indicate the sort of picture of the surface potential that this produces.

Let $V_u(\boldsymbol{r})$ be the unscreened bare potential of the ionic lattice and $V_s(\boldsymbol{r})$ the screened potential. What we want to study is the relation between V_s and V_u *in the presence of the surface*. It proves convenient to start from a relation which is exact and arises as follows: consider a crystal with a surface in the (x, y) plane and imagine that the ions, at positions $R = (\boldsymbol{\rho}, z)$, undergo infinitesimal displacements perpendicular to the surface, of the form $\delta z \exp(\mathrm{i}\boldsymbol{\kappa} \cdot \boldsymbol{\rho})$, in which $\kappa \to 0$. This produces an infinitesimal change δV_u in the unscreened potential of the ions, and the screening of this change must be equal to the infinitesimal change in the screened potential, δV_s. Thus, if we have a lattice with reciprocal lattice vectors $\boldsymbol{g}$ in the (x, y) plane and we Fourier transform (x, y) while explicitly retaining the z-dependence, then in the limit $\delta z \to 0$ we can write

$$\frac{\mathrm{d}}{\mathrm{d}z} V_\mathrm{s}(\boldsymbol{g}, z) = \lim_{\kappa \to 0} \sum_{\boldsymbol{g}'} \int \mathrm{d}z' \varepsilon^{-1}(\boldsymbol{g}+\boldsymbol{\kappa}, \boldsymbol{g}'+\boldsymbol{\kappa}; z, z') \frac{\mathrm{d}V_\mathrm{u}(\boldsymbol{g}'+\boldsymbol{\kappa}, z')}{\mathrm{d}z'}. \quad (3.259)$$

This equation is formally exact and involves the reciprocal longitudinal dielectric function which, in general, can be non-diagonal in momentum space. In principle ε depends on the selfconsistent screened potential itself, but to try and use (3.259) as a selfconsistent equation to be solved for V_s would be rather clumsy. Identifying ε for a surface system is precisely the difficulty. In a way the various elaborate attempts at solving the surface problem by resorting to different approximations constitute an effort to circumvent this difficulty. We do not know how to express the dielectric function of a surface system explicitly as a function of V_s. However, this scheme can be used to obtain at least a rough idea of what $V_\mathrm{s}(\boldsymbol{g}, z)$ looks like if we start from some approximation for ε, and this is what we are going to do.

First we assume ε is diagonal in $(\boldsymbol{g}, \boldsymbol{g}')$, as is often done in the bulk. Then (3.259) yields separate equations for each g-component, decoupled from each other. Thus

$$\frac{\mathrm{d}}{\mathrm{d}z} V_\mathrm{s}(\boldsymbol{g}, z) = \lim_{\kappa \to 0} \int \mathrm{d}z' \, \varepsilon^{-1}(\boldsymbol{g}+\boldsymbol{\kappa}; z, z') \frac{\mathrm{d}V_\mathrm{u}(\boldsymbol{g}+\boldsymbol{\kappa}, z')}{\mathrm{d}z'}. \quad (3.260)$$

Then we think of $V_\mathrm{u}(\boldsymbol{g}+\boldsymbol{\kappa}, z')$ as the sum of two terms, namely, the contribution of a uniform jellium, V_J, and the crystalline correction $\delta V_\mathrm{u}(\boldsymbol{g}+\boldsymbol{\kappa}, z)$. Now, $\mathrm{d}V_\mathrm{J}/\mathrm{d}z'$ is like the potential created by a surface charge at the jellium edge – i.e. $(\mathrm{d}V_\mathrm{J}/\mathrm{d}z')\,\delta z'$ is the infinitesimal potential created by the infinitesimal charge proportional to $\delta z'$ – whence

$$\frac{\mathrm{d}V_\mathrm{J}}{\mathrm{d}z'} = \frac{2\pi n_\mathrm{b}}{\kappa} \mathrm{e}^{-\kappa|z'|} \mathrm{e}^{\mathrm{i}\boldsymbol{\kappa}\cdot\boldsymbol{\rho}}. \quad (3.261)$$

Using a condensed but obvious notation we can then write (3.260) in the form

$$\frac{\mathrm{d}V_\mathrm{s}(z)}{\mathrm{d}z} = \varepsilon^{-1} \frac{\mathrm{d}V_\mathrm{J}}{\mathrm{d}z} + \varepsilon^{-1} \frac{\mathrm{d}(\delta V_\mathrm{u})}{\mathrm{d}z} \quad (g = 0), \quad (3.262a)$$

and

$$\frac{\mathrm{d}V_s(z)}{\mathrm{d}z} = \varepsilon^{-1}\frac{\mathrm{d}(\delta V_u)}{\mathrm{d}z} \quad (g \neq 0). \tag{3.262b}$$

Incidentally, if we go back to the jellium model, then these equations are reduced to

$$\frac{\mathrm{d}V_s(z)}{\mathrm{d}z} = \varepsilon^{-1}\frac{\mathrm{d}V_J}{\mathrm{d}z}. \tag{3.263}$$

This could be used to study the response of the bounded electron gas to a surface charge, in that if we used for $V_s(z)$ any of the selfconsistent solutions studied in this chapter, then we could try and obtain a property of the dielectric function of the surface system by using (3.263). However, let us return to the crystalline case, with which we are concerned here. For this we start from (3.262) and make the following approximation: we assume that the exchange and correlation effects in the inhomogeneous electron gas are sufficiently well described by a good solution of the jellium model and we treat the crystalline perturbation in the Hartree approximation.

Let us start with the case $\mathbf{g} = 0$. Fig. 49 shows δV_u and $\mathrm{d}(\delta V_u)/\mathrm{d}z$ for a Coulomb ionic potential. Concerning the latter (fig. 49*b*), notice that its derivative has a discontinuity at the origin. This is compensated by a

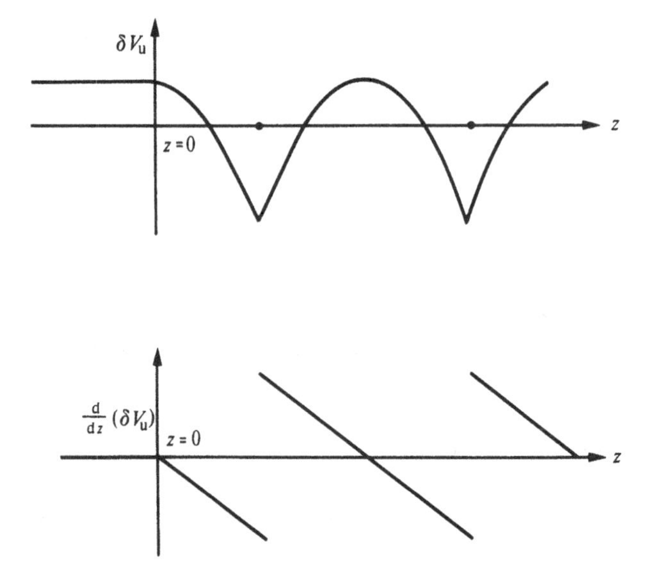

Fig. 49. The perturbing crystalline potential δV_u and its normal (to the surface) derivative after averaging in the plane parallel to the surface.

similar discontinuity in the derivative of (3.261). Thus the r.h.s. of (3.262a) behaves regularly, irrespective of the model used for the screening. The same happens if a pseudopotential is used. Then, the lowest-order approximation would consist in neglecting the response of the electron gas to δV_u and writing simply

$$\frac{dV_s(z)}{dz} \approx \varepsilon^{-1}\frac{dV_J}{dz} + \frac{d(\delta V_u)}{dz}. \tag{3.264}$$

In this case the total crystal potential is taken to be the sum of the selfconsistent potential, as calculated in a jellium model, and the unscreened perturbation due to the ionic pseudopotentials. In order to go beyond this approximation we can try to reduce (3.262a) to a problem of screening of surface charges. For this we recall that in the bulk the ionic pseudopotential is screened with the bulk dielectric function $\varepsilon(\boldsymbol{k})$, so that for the z-dependence, $V_b(z)$, of the bulk screened potential we have the Fourier transform

$$V_b(q) = V_u(q)/\varepsilon(q), \tag{3.265}$$

where only the dependence on $q = k_z$ is explicitly shown. Now, imagine the bulk potential V_b extending right up to $z = 0$. With this we are taking account of the screening of the ionic pseudopotentials, but now we must modify this in order to describe the inhomogeneous surface region. We could think of the inhomogeneous part, to be superimposed on $V_b(z)$, as the response of the system to some suitably chosen hypothetical surface stimulus, in the spirit of chapters 1 and 2. That is to say, in order to go beyond (3.264) we rewrite (3.262a) in the form

$$\frac{dV_s}{dz} = \frac{d}{dz}[V_b(z)\theta(z)] + \text{screened surface stimuli}. \tag{3.266}$$

The problem now is how to decide on suitable surface stimuli to use in (3.266), and we have discussed several examples of this sort of problem in chapter 1. What we do is impose correct electrostatic matching conditions, so that we have regular behaviour of the electrostatic field across the surface (now matching plane) $z = 0$.

At this stage the discontinuities referred to above become relevant. Notice that the first term on the r.h.s. of (3.266) has a discontinuous derivative at $z = 0$, like fig. 49b. Thus the stimulus to be used in the second term must include a surface stimulus whose behaviour is like that of dV_J/dz (3.261). The corresponding contribution to the r.h.s. of (3.266) is

$$\int \varepsilon^{-1}(z, z')\frac{2\pi n(0)}{\kappa}\,e^{-\kappa|z'|}\,dz'. \tag{3.267}$$

The difference with (3.263) and (3.261) is that here we have $n(0)$, the value of the actual electronic density at $z = 0$, instead of the uniform bulk value n_b. Furthermore, notice that the integration of the first term on the r.h.s. of (3.266) would produce a discontinuity in V_s. This requires including in the second term the response to a dipolar surface stimulus, which when suitably adjusted can cancel the above discontinuity. By combining this with (3.267) we can make up a sum of terms which together ensure the correct behaviour for V_s. Again in concise notation we have

$$\frac{dV_s}{dz} = \varepsilon^{-1}\frac{dV_u}{dz} \approx \frac{d}{dz}[V_b(z)\theta(z)] + \varepsilon^{-1}\nu_\sigma + \frac{d}{dz}(\varepsilon^{-1}\mathscr{D}), \tag{3.268}$$

where $\nu_\sigma = [2\pi n(0)/\kappa]\exp(-\kappa|z|)$ and $\mathscr{D} = \mathscr{V}\theta(z)$. Here $\mathscr{V}$ represents the potential difference between the value of $V_b(z)$ for $z = 0$ and its average bulk value. It is still very difficult to screen the surface stimuli ν_σ and $\mathscr{D}$ with the correct response function, but here it is not too unreasonable to approximate this by the screening provided by a jellium model, i.e. to put

$$\varepsilon^{-1}\nu_\sigma \approx \varepsilon_J^{-1}\nu_\sigma, \tag{3.269a}$$

and

$$\varepsilon^{-1}\mathscr{D} \approx \varepsilon_J^{-1}\mathscr{D}. \tag{3.269b}$$

Thus the problem of finding $V_s(\mathbf{g} = 0, z)$ is reduced to: (i) screening a surface charge, (ii) screening a surface dipole, and (iii) adjusting parameters by imposing continuity in the electrostatic field across the matching plane $z = 0$, as in the classical analysis of chapter 1. The screening of a surface charge in the jellium model can be solved by using (3.263), and the screening of a surface dipole by using an approximation equivalent to the model discussed in § 3.2.5.

Next we concentrate on $V_s(\mathbf{g}, z)$ for $\mathbf{g} \neq 0$. Basically the approach is the same as for $\mathbf{g} = 0$. However, it is interesting to note that in this case we can use a simpler but equivalent initial equation because the electrostatic potential created by a charge that varies as $\exp(i\mathbf{g} \cdot \boldsymbol{\rho})$ decays like $\exp(-g|z|)$ and is therefore strongly localised in the surface region. Let us go back a step to before (3.259). the same argument used to derive that equation can be applied when the ions are displaced in a direction parallel to the surface. This leads to

$$\mathbf{g}V(\mathbf{g}, z) = \lim_{\kappa \to 0} \sum_{\mathbf{g}'} \mathbf{g}' \int dz' \, \varepsilon^{-1}(\mathbf{g} + \boldsymbol{\kappa}, \mathbf{g}' + \boldsymbol{\kappa}; z, z') V_u(\mathbf{g} + \boldsymbol{\kappa}, z'). \tag{3.270}$$

If we assume, as before, that the dielectric function is diagonal in $(\mathbf{g}, \mathbf{g}')$

then we have the decoupled scalar equations,

$$V(\mathbf{g}, z) = \lim_{\kappa \to 0} \int \mathrm{d}z' \, \varepsilon^{-1}(\mathbf{g}+\boldsymbol{\kappa}; z, z') V_{\mathrm{u}}(\mathbf{g}, z'), \tag{3.271}$$

which only hold for $\mathbf{g} \neq 0$. Now we can approach (3.271) by using essentially the same ideas as we have used for $\mathbf{g} = 0$. This leads also to the introduction of adjustable surface stimuli. We extend the half metal by symmetrising about the plane $z = 0$ – this ensures the correct semiclassical limit, as discussed in chapter 1 – and we screen the surface stimuli and the bare-ion pseudopotential with the bulk dielectric function. Writing $V_{\mathrm{u}}^{\sigma}(\mathbf{g}, z)$ for the symmetrised bare pseudopotential we find

$$V_{\mathrm{s}}(\mathbf{g}, z) = \theta(z) \int \frac{\mathrm{d}q}{2\pi} \mathrm{e}^{\mathrm{i}qz} \frac{V_{\mathrm{u}}^{\sigma}(\mathbf{g}, q)}{\varepsilon(q)} + \int \mathrm{d}z' \, \varepsilon^{-1}(\mathbf{g} \cdot z, z') \mathscr{D}(z'). \tag{3.272}$$

Here we have also introduced a surface dipole, needed to ensure correct electrostatic matching. Making again the approximation $\varepsilon^{-1}\mathscr{D} \approx \varepsilon_{\mathrm{J}}^{-1}\mathscr{D}$, we have reduced – in (3.268), (3.269) and (3.272) – our problem to the screening of surface stimuli varying as $\exp(\mathrm{i}\mathbf{g} \cdot \boldsymbol{\rho})$.

It is interesting to look at the results obtained in this way. Consider, for example, Na(100) with an Ashcroft pseudopotential. If we assume that the replacement of ε^{-1} by $\varepsilon_{\mathrm{J}}^{-1}$ is reasonable, then the screening of the surface dipoles can be studied again by using an approximation equivalent to the model discussed in § 3.2.5. Fig. 50 shows the results obtained in this way for Na(100). First (fig. 50*a*) we have the potential V_{s}, averaged in the (x, y) plane, as a function of z, and then (fig. 50*b*) $V_{\mathrm{s}}(z)$ for two lines ending up at different (x, y) points on the surface.

Although the method just outlined enables us to get at least some idea of the crystalline surface potential, it cannot be compared in accuracy and degree of elaboration with the methods discussed in the preceding sections to estimate the changes in surface energy and work function due to the introduction of the crystal lattice. It is nevertheless instructive to see what this method would say about the changes in work function. In fact this depends only on the solution of the problem for $g = 0$. Remember that in (3.268) we have two essentially different contributions to the r.h.s. One of them, $\varepsilon^{-1}v_{\sigma}$, involves $n(0)$ instead of n_{b}. This reflects the fact that the charge density – everywhere and in particular at $z = 0$ – changes, owing to the screening of the ionic pseudopotential by the response of the electron gas. This process is schematically illustrated in fig. 51*a*. The other contribution – to the r.h.s. of (3.268) and hence to the work function – comes from the screened bulk potential. If this potential prevailed right up to $z = 0$, where $V_{\mathrm{b}}(0) = \mathscr{V}$, by itself this would give a contribution to the work function equal to $\mathscr{V}$. But the screening of this

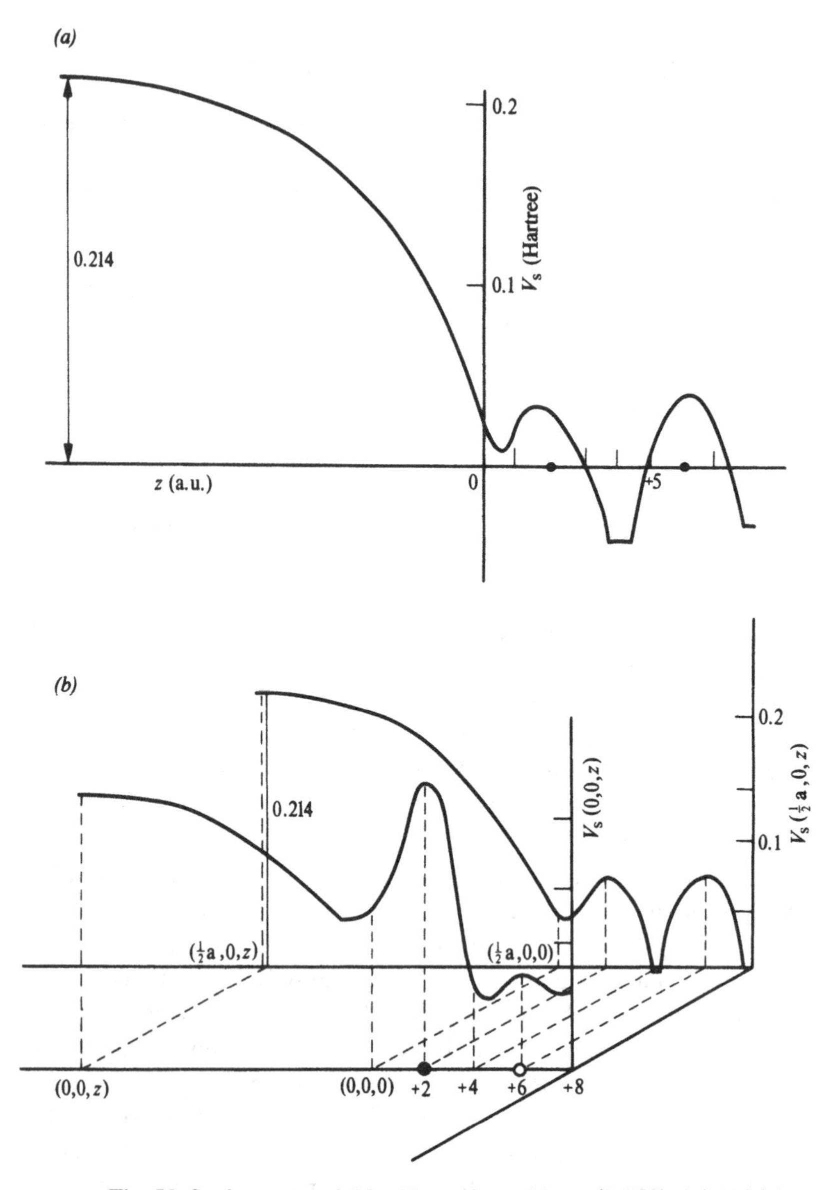

Fig. 50. Surface potential for Na evaluated from (3.272). (*a*) $V_s(z)$ for a (1, 0, 0) surface after averaging in the surface plane. ●, Atoms. (*b*) $V_s(z)$ along two different lines ending up at different points of the same surface. ●, Last atom. ○, Projection of second atomic layer. After C. Tejedor & F. Flores, *J. Phys.* F: *Metal Phys.* **6**, 1647 (1976).

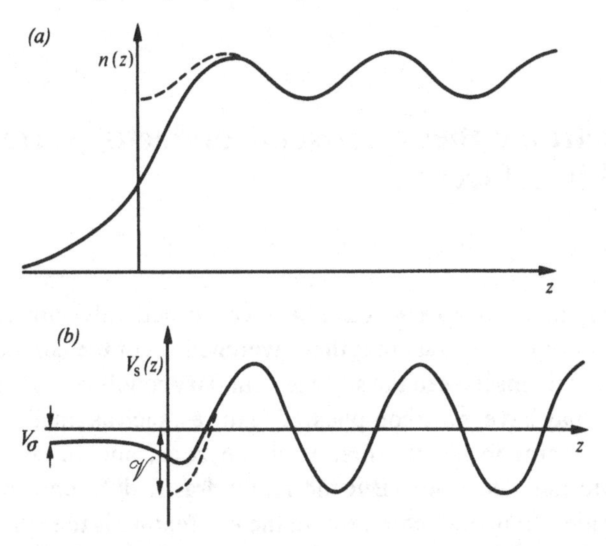

Fig. 51. Diagrams showing qualitatively the origin of the two effects which cause a change in the work function. (a) The induced volume charge modifies the surface charge distribution and hence the surface dipole. (b) The screened bulk potential, continued up to the surface, would produce a change $\mathscr{V}$ in the work function, but $\mathscr{V}$ is changed into V_σ because of the screening due to the surface charge.

term, represented by the addition of the screened surface dipole, reduces this contribution to a value V_σ (fig. 51b). In short, what happens if one tries to use this method to estimate corrections to the work function is that the effects illustrated in fig. 5 a and b tend to cancel each other out and the result, expressed as a correction $\delta\Phi$ to the work function, is so small that it cannot be taken as a very reliable quantitative estimate. This is essentially why this method is not very good for estimating $\delta\Phi$. But it does give, as we have seen, at least a plausible picture of the crystalline surface potential, which would otherwise be very complicated to obtain and would always involve heavy numerical computation. Moreover, the usefulness of the ideas involved in the present analysis will be seen in chapter 4 when we discuss metal–semiconductor interfaces.

Finally it is worth remarking that although we have estimated V_s in a linear approximation based on the replacement of ε^{-1} by ε_J^{-1}, we could in principle use the same approach and go a step farther by trying a better approximation for ε^{-1}. For example, a rather economical effort could consist in using a jellium model for this, but fudging a suitable value of the charge density so that one could account in an approximate way for the strong changes in electronic charge in the surface region due to the perturbation of the crystalline potential.

4

Electronic theory of semiconductor surfaces and interfaces

In the study of metal surfaces we have been concerned only with charge redistribution owing to the fact that the wavefunctions of the conduction band must adapt themselves to the surface boundary condition. Of course certain metals may have more complicated features, such as small gaps in some small domains inside the first Brillouin zone, and these would require a more detailed study. But the really drastic difference occurs when we consider semiconductors, where the key feature is the existence of the so-called *fundamental gap*, separating the valence and conduction bands. What makes this gap particularly important is that it contains the Fermi level and perhaps also new energy levels, corresponding to *surface states*, which would be forbidden in the bulk. This makes the situation interestingly different because now we must study not only the new spectrum, but also the occupation of the new states. It is clear that this will be linked with the problem of selfconsistency. These problems can be studied within the *nearly-free electron* (n.f.e.) model by using a pseudopotential approach. The analysis then becomes a natural continuation of the theory of chapter 3, with the additional basic feature of the existence of the fundamental gap. However there is an important alternative approach based on tight-binding methods, and we shall also discuss this, although rather briefly, just to emphasise basic principles. Finally, an interesting situation arises when a semiconductor and a metal match at an interface. Then we may have energy levels which are forbidden in the bulk on one side and allowed on the other side, and we shall also discuss this case, which combines the features of metals and semiconductors.

4.1. One-dimensional narrow-gap n.f.e. model

We start with the very elementary two-band model shown in fig. 52. The gap is created by a potential

$$V(z) = V\,e^{igz} + \text{c.c.} \tag{4.1}$$

We assume $V < 0$ (bonding gap) and also $(\frac{1}{2})(\frac{1}{2}g)^2 \ll |V|$ (narrow gap). We keep using atomic units ($\hbar = m = e = 1$), and we write $g = 2\pi/c$, where c is the lattice constant of the linear chain. For the present purpose we assume

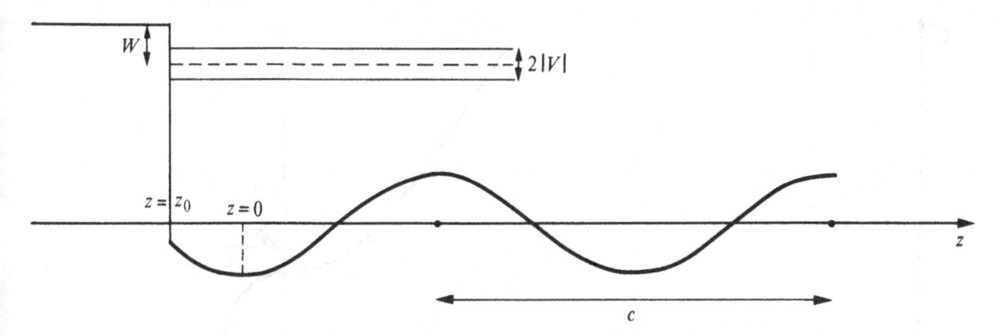

Fig. 52. Abrupt-potential-barrier model for a one-dimensional narrow-gap semiconductor.

an *abrupt potential model* (a.p.m.), in which the potential of the bulk extends to a point where it makes a discontinuous rise to the vacuum level. The matching 'plane' in this case is not a mere geometrical device: the choice of its location determines distinct physical features of the model thus defined, such as whether the potential just adjacent to the surface is attractive or repulsive. Moreover, this potential profile is clearly not selfconsistent. In spite of these shortcomings, this simple model can serve as a prototype for the study of the basic question of matching. Improvements will come later. Thus we concentrate on these two essential facts: (i) We have a system with a gap in the bulk spectrum. (ii) This is terminated at a surface. We want to study the consequences of introducing a surface, for which we first review briefly the description of the bulk.

If we measure the momentum q with respect to the edge of the Brillouin zone, $\frac{1}{2}g$, then the bulk wavefunctions have the form

$$\psi(z) = \alpha \exp[\mathrm{i}(q+\tfrac{1}{2}g)z] + \beta \exp[\mathrm{i}(q-\tfrac{1}{2}g)z]. \tag{4.2}$$

The coefficients α, β satisfy the equations

$$\begin{bmatrix} -E+\frac{1}{2}(q+\frac{1}{2}g)^2 & V \\ V & -E+\frac{1}{2}(q-\frac{1}{2}g)^2 \end{bmatrix} \begin{bmatrix} \alpha \\ \beta \end{bmatrix} = 0. \tag{4.3}$$

The secular determinant yields the eigenvalues

$$E = \tfrac{1}{2}(\tfrac{1}{2}g \pm q)^2 \pm [(V^2 + \tfrac{1}{4}q^2g^2)^{1/2} - \tfrac{1}{2}qg], \tag{4.4}$$

corresponding to the typical two-band dispersion relation with a gap (fig. 53). Far away from this gap, when $qg \gg |V|$, (4.4) reproduces the free-electron parabola, but deviations from it become appreciable near the gap, when $qg \approx |V|$. In this range it is more convenient to rewrite (4.4) as

$$E = \tfrac{1}{2}(\tfrac{1}{2}g)^2 + \tfrac{1}{2}q^2 \pm \sqrt{(V^2 + \tfrac{1}{4}q^2g^2)}. \tag{4.5a}$$

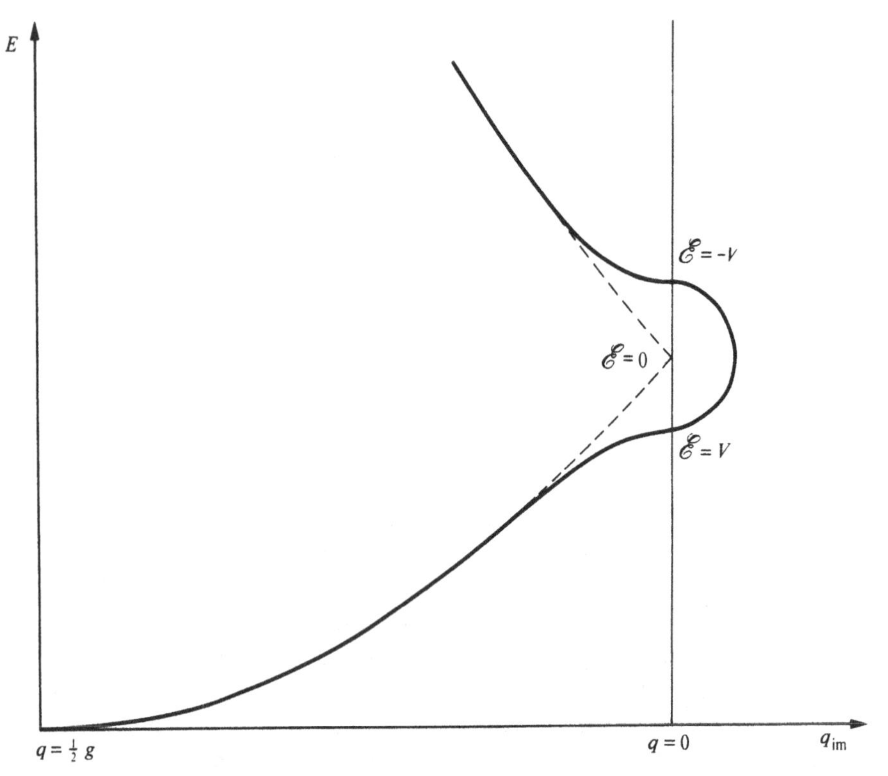

Fig. 53. Dispersion relation, E against q, for a narrow-gap two-band semiconductor. Dashed line: free electron parabola. The valence and conduction bands are connected through a real energy loop corresponding to imaginary values of q.

Furthermore, we write $\mathscr{E}$ for the energy measured from the midgap, $\frac{1}{2}(\frac{1}{2}g)^2$. Since in this range $q^2 \ll V^2/g^2$, the term in q^2 can be neglected and finally

$$\mathscr{E} = \pm\sqrt{(V^2+\tfrac{1}{4}q^2g^2)}. \tag{4.5b}$$

This is equivalent to replacing (4.3) by

$$\begin{bmatrix} -\mathscr{E}+\frac{1}{2}qg & V \\ V & -\mathscr{E}-\frac{1}{2}qg \end{bmatrix}\begin{bmatrix} \alpha \\ \beta \end{bmatrix}=0. \tag{4.6}$$

We shall concentrate on the energy range in and near the gap and consistently use the approximation (4.6).

Suppose we express (4.5b) as

$$q = \pm(2/g)\sqrt{(\mathscr{E}^2-V^2)}. \tag{4.7}$$

For $|\mathscr{E}|>|V|$ we have the bulk states with energies near the gap and q real.

This ensures regular behaviour for a proper wavefunction. But we could formally study the case in which $|\mathscr{E}|<|V|$. Real energies inside the gap thus correspond to imaginary values of q, i.e. to wavefunctions which either increase to infinity or decay exponentially, depending on the sign of z. In this way we can span with continuity the spectrum of real energies. The two bands are continuously connected by a real energy loop (fig. 53) with complex momentum.

4.1.1. Surface states

In a bulk system the states of the loop are obviously forbidden, but in a surface system nothing prohibits in principle an evanescent wavefunction, i.e. a function which decays into the bulk. If inside means $z>0$, then we must take the positive sign in (4.7) in order to ensure that the wavefunction decays inside the medium. Feeding this back into (4.6) yields the ratio α/β for such a state whose – unnormalised – wavefunction is

$$\psi(z)\propto\exp\left[-z\frac{2(v^2-\mathscr{E}^2)^{1/2}}{g}\right]\times[\exp(-\tfrac{1}{2}igz)+a\exp(\tfrac{1}{2}igz)], \tag{4.8}$$

where

$$a=[\mathscr{E}+i\surd(V^2-\mathscr{E}^2)]/V=a(\mathscr{E}).$$

If we assume the surface barrier is at $z=z_0$, then (4.8) can only describe the correct wavefunction for $z>z_0$. The problem is to match this to a different wavefunction with appropriate behaviour for $z<z_0$. This means that $\psi(z)$ must decay exponentially for $z\to-\infty$. But at the same time it must be an eigenfunction of the free-electron Hamiltonian, corresponding to the constant potential – vacuum level – outside. Moreover, since the energy level of the state that we are trying to construct is negative with respect to the vacuum level, it is clear that the wavefunction outside can only be (with all energies referred to the midgap)

$$\psi(z)\propto\exp\{z\surd[2(W-\mathscr{E})]\}. \tag{4.9}$$

A *surface state* (s.s.) exists if it is possible to find a value of $\mathscr{E}$ inside the gap for which (4.8) and (4.9) match at $z=z_0$, which of course in this model depends on z_0. Matching ψ and its derivative yields the matching equation

$$a(\mathscr{E})=\frac{1+ig/2\surd[2(W-\mathscr{E})]}{1-ig/2\surd[2(W-\mathscr{E})]}e^{-igz_0}. \tag{4.10a}$$

In the narrow-gap model we have $|\mathscr{E}|\ll W$ throughout the energy range of interest and we can write the matching equation as

$$\frac{1-\mathrm{i}\sqrt{[(|V|-\mathscr{E})/(\mathscr{E}+|V|)]}}{1+\mathrm{i}\sqrt{[(|V|-\mathscr{E})/(\mathscr{E}+|V|)]}}=\frac{1-\mathrm{i}g/2\sqrt{(2W)}}{1+\mathrm{i}g/2\sqrt{(2W)}}\,\mathrm{e}^{\mathrm{i}gz_0}. \tag{4.10b}$$

The l.h.s. is obtained simply by an identical rewriting of $-1/a$.

Since both sides of (4.10*b*) are complex numbers of unit modulus, the eigenvalue equation for the s.s. is given by the equality between their arguments. Notice that for a given value of z_0 the argument of the r.h.s. is a constant, independent of energy, while the argument of the l.h.s. varies between 0 and $-\pi$ as $\mathscr{E}$ spans the gap. This shows how the existence of a s.s. – and, in the event, the value of its energy level – depends on the assumed position of the abrupt surface barrier. Fig. 54 shows the s.s. eigenvalue as a function of the argument $[-gz_0+2\tan^{-1}(g/2\sqrt{(2W)})]$.

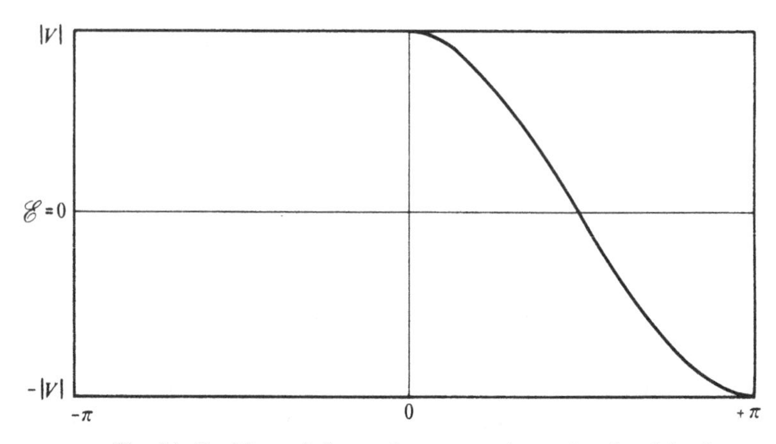

Fig. 54. Position of the surface state eigenvalue level in the gap as a function of the argument $-gz_0+2\tan^{-1}(g/2\sqrt{(2W)})$. See text.

Notice that for given W a s.s. eigenvalue exists only for one half of the range of values of gz_0. The question now is where to put z_0. In a simple way one would tend to choose $z_0=0$, corresponding to the midpoint between atoms. Let us provisionally make this choice (which will be discussed later). In this case there is always a s.s.: the eigenvalue depending on $gW^{-1/2}$. In order to obtain an explicit picture of the s.s. and its associated density distribution, let us take the particular case $g=2(2W)^{1/2}$. Then we have

$$|\psi(z)|^2 \propto [1+\cos(gz-\tfrac{1}{2}\pi)]\exp(-4|V|z/g)| \quad (z>0), \tag{4.11a}$$

and

$$|\psi(z)|^2 \propto \exp(2z\sqrt{(2W)}) \quad (z<0). \tag{4.11b}$$

If this state were occupied this would give a charge density that varied qualitatively as the density $\rho(z)$ shown in fig. 55. The ratio between the two consecutive maxima, roughly located at $z=\frac{1}{4}c$ and $z=\frac{5}{4}c$, is $\exp(-8\pi V/g^2)$. For values representative of covalent semiconductors this is roughly $\exp(-\frac{1}{5}\pi)\approx 0.53$. This indicates the degree of localisation of the state near the surface. A good fraction of the charge density associated with this state is localised within the first oscillation.

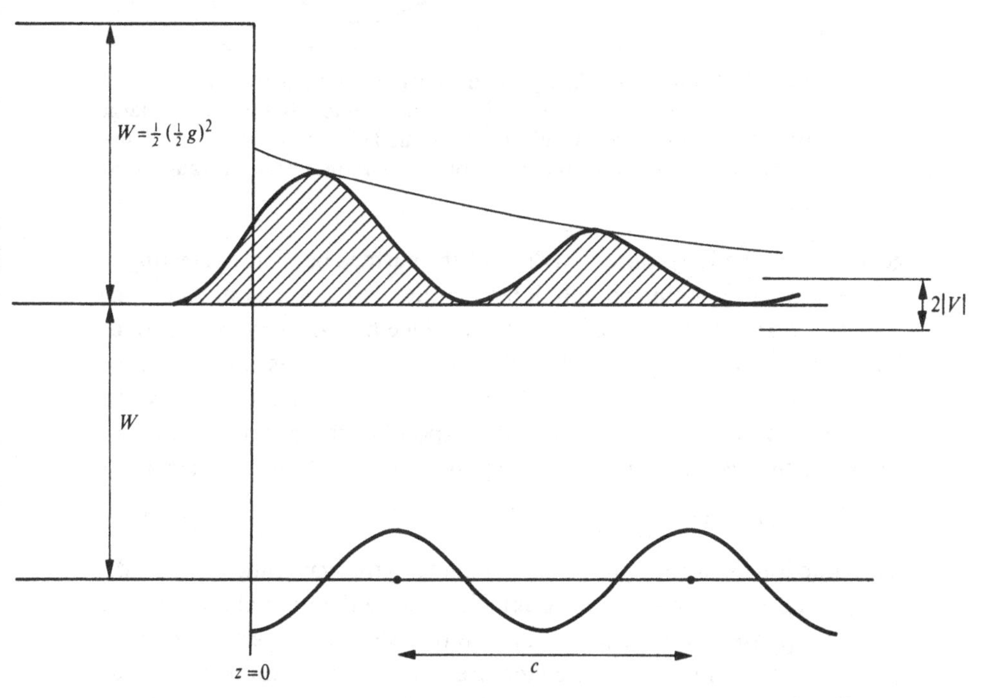

Fig. 55. Charge density (hatching) associated to a surface state wavefunction corresponding to an energy eigenvalue in the middle of the gap ($z_0=0$, $W=\frac{1}{2}(\frac{1}{2}g)^2$).

Now, the a.p.m. is far too simple. Even before doing an elaborate selfconsistent calculation we know the surface barrier will be smooth. So let us assume that we have a given non-abrupt barrier (fig. 56), and study the picture of the s.s. in this case. Now the choice of matching plane is a mere mathematical artifact. Shifting its position does not change the potential barrier. The point to notice is that the potential drop on going from the vacuum level to the mean level inside the crystal is usually much larger than the amplitude of the oscillations of the bulk potential – in practice a weak pseudopotential. This allows us to use a practical trick. We imagine the region just on the inner side of the interface as essentially

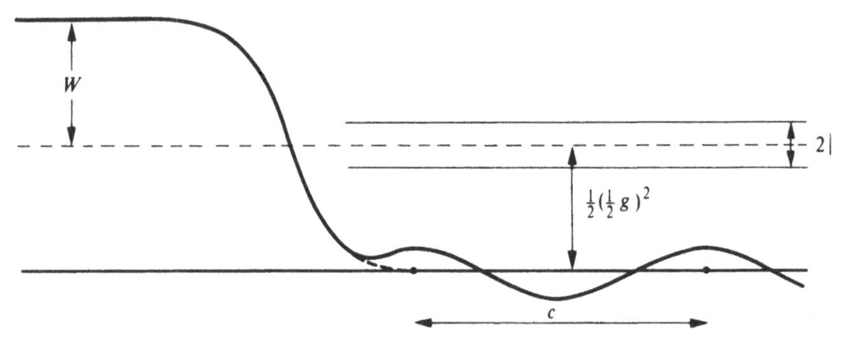

Fig. 56. A smooth-surface potential barrier for a narrow-gap semiconductor. The main feature in the surface region is the overall large potential drop which would be continued into the dashed line. The periodic oscillations of the bulk potential are important in causing the gap.

governed by the large potential drop of the surface barrier, neglecting in this region the crystalline oscillations. In other words, we assume that the wavefunction just emerging from the interface has the form corresponding to the smoothed potential of fig. 56, which ignores the crystalline oscillations. This is the potential discussed at length in chapter 3. If we call $\eta_0(E)$ the phase shift which would correspond to this potential, then we can write the wavefunction in the region just beyond the interface as

$$\psi(z) \propto \sin\left[(-q+\tfrac{1}{2}g)z+\eta_0\right]. \tag{4.12}$$

This function does not require any matching to the wavefunction outside. All the necessary information about the smoothed potential is embodied in its own phase shift η_0. But it does require matching to the wavefunction (4.8), which has the two required features, namely, the exponential decay and the Bragg reflection structure due to the periodic potential of the bulk. Matching (4.12) and (4.8) we find

$$\left(\frac{|V|-\mathscr{E}}{\mathscr{E}-|V|}\right)^{1/2} = \tan \eta_0(\mathscr{E}=0). \tag{4.13}$$

On the r.h.s. we have taken the value of η_0 particular to the midgap because the variation of η_0 for $\mathscr{E}$ inside the gap is negligible in the narrow-gap model. Notice that a given continuous surface barrier can be simulated by an effective abrupt potential wall with a suitable choice of its position, i.e. (4.13) is equivalent to (4.10) if we choose z_0 according to

$$2\eta_0(\mathscr{E}=0) = -gz_0 + 2\tan^{-1}(g/2\surd(2W)). \tag{4.14}$$

The r.h.s. is just the variable of fig. 54. For example the particular choice drawn in fig. 55 corresponds to $\eta_0(\mathscr{E}=0)=\frac{1}{4}\pi$.

4.1.2. Density of states in the continuum

The existence of s.s. with evanescent wavefunctions localised near the surface is the most outstanding feature of the semiconductor case, but we must also study the distortions of the continuum of the bulk bands. As in a metal, these come about because the wavefunctions of the bulk states must also adapt themselves to the surface boundary condition. This modifies the density of states and the spatial distribution of the bulk wavefunction near the surface, and all this now depends on the existence and position of the s.s. We now concentrate on the study of the bulk bands and, more particularly, on the energy range near the fundamental gap, where we can use (4.5*b*) and (4.6). It is only this energy range that displays semiconductor features, since away from the gap the system behaves just like a metal, for which we have the analysis of § 3.1.

By using (4.5*b*) and (4.6) the wavefunctions for the energy range of interest are

$$\psi_{\pm}(z) \propto \exp[\mathrm{i}(q-\tfrac{1}{2}g)z] + \frac{\mathscr{E} \pm \sqrt{(\mathscr{E}^2 - V^2)}}{V} \exp[\mathrm{i}(q+\tfrac{1}{2}g)z], \quad (4.15)$$

where q is given by (4.7) with $\pm$ sign for $\psi_{\pm}$ respectively. If we notice that

$$\frac{\mathscr{E} + \sqrt{(\mathscr{E}^2 - V^2)}}{V} = \frac{V}{\mathscr{E} - \sqrt{(\mathscr{E}^2 - V^2)}},$$

the two wavefunctions (4.15) corresponding to a given energy can be combined into the form

$$\psi(z) \propto \sin[(q-\tfrac{1}{2}g)z + \eta] + \frac{\mathscr{E} + \sqrt{(\mathscr{E}^2 - V^2)}}{V} \sin[(q+\tfrac{1}{2}g)z + \eta], \quad (4.16)$$

where q by definition is given by (4.7) with positive sign. For example, the values $\eta = \frac{1}{2}\pi$ and $\eta = 0$ yield the two independent wavefunctions corresponding to the symmetric and antisymmetric combinations of ψ_+ and ψ_-. Thus (4.16) represents the general form of the bulk wavefunction in the energy range of interest. It resembles (4.12) but with the essential difference that it contains a Bragg reflection pattern. We are now ready to follow a discussion similar to the one in § 3.1, for which it is convenient to first consider a finite linear chain with infinite barriers and then to introduce the actual finite barrier.

Now, for the infinite barrier case (fig. 57) the wavefunction satisfying the boundary condition of zero amplitude at $z = 0$ is

$$\psi(z) \propto \sin[(q-\tfrac{1}{2}g)z] + \frac{\mathscr{E} + \sqrt{(\mathscr{E}^2 - V^2)}}{V} \sin[(q+\tfrac{1}{2}g)z], \quad (4.17)$$

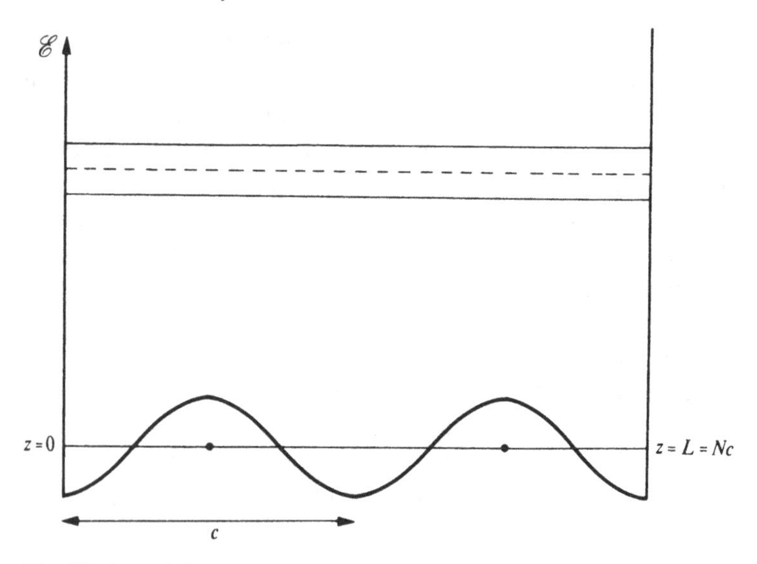

Fig. 57. A model narrow-gap one-dimensional semiconductor bounded by infinite barriers at $z=0$ and $z=L=Nc$.

which corresponds to $\eta=0$ in (4.16). The same boundary condition, $\psi=0$, must be satisfied at the other end. If there are N atoms in the chain with regular spacing c – i.e. $L=Nc$ is the length of the chain – then q must satisfy the condition (3.23). Notice that $m=0$ is not an allowed solution for states in the valence band. This is most easily seen by going back to (4.15) and putting $q=0$. If we take $\mathscr{E}$ in the valence band, approaching the gap, we find that for $\mathscr{E}=V$ the two wavefunctions of (4.15) coalesce into $\cos(\frac{1}{2}gz)$, which fails to satisfy the boundary condition at $z=0$. Thus the allowed values of q are given by $m=1,2,\ldots(N-1)$ in (3.23), and are shown as solid circles in fig. 58. The situation so far is as in the metallic case. As in § 3.1, we now imagine a length L' of homogeneous bulk material with Born–von Karman boundary conditions. Then the wavefunctions (4.16) are the solutions of the system with $\eta=0,\frac{1}{2}\pi$, for q satisfying

$$qL'=2m\pi. \tag{4.18}$$

The case $m=0$ is now allowed and corresponds to the band edge. The new eigenstates are indicated in fig. 58. Notice that there is only one eigenstate with $E=E_V$. By the same argument which led to (3.5) we find, for the density of states *in the valence band,*

$$N_V^{\infty}(E)=\tfrac{1}{2}[N_V^{2L}(E)-\delta(E)-\delta(E-E_V)]. \tag{4.19}$$

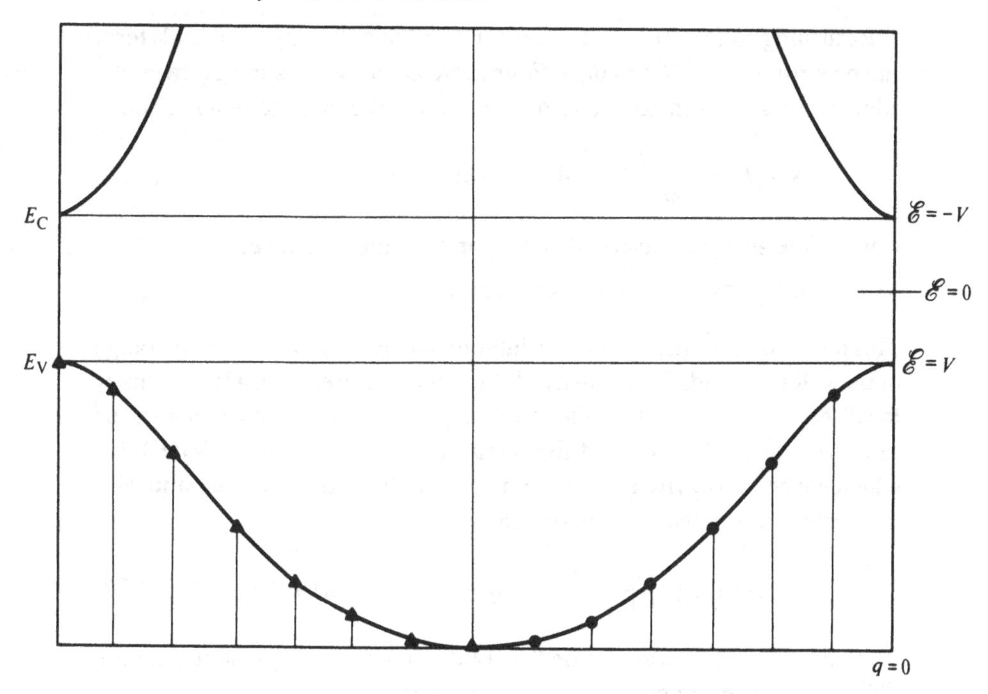

Fig. 58. Dispersion relation for a narrow-gap two band semiconductor. ●, Allowed eigenvalues for the situation of figure 57 (infinite barriers). ● and ▲, Allowed eigenvalues if the hard wall boundaries are replaced by Born–von Karman periodic boundary conditions at $z=0, L$.

The difference is that we now have two δ-function terms related to the bottom and the top of the valence band.

A similar argument can be applied to the conduction band, with an important difference. The stationary wave corresponding to the band edge E_C is then $\sin(\frac{1}{2}gz)$, which does vanish for $z=0$. Thus $m=0$ is an allowed value for the infinite-barrier system and we have (compare with (4.19) and with (3.5))

$$N_c^{\infty}(E)=\tfrac{1}{2}[N_c^{2L}(E)+\delta(E-E_c)]. \tag{4.20}$$

The sign difference between (4.20) and (4.19) for the band-edge terms can be understood by looking at (4.10) and fig. 54. If we take there $z_0=0$ and $W\to\infty$ (infinite barrier) then the s.s. moves to $\mathscr{E}=-V$, i.e. the bottom of the conduction band. Thus the introduction of the infinite barrier causes the disappearance of a state from the valence band and the appearance of a state at E_c. We shall now return to the study of the valence band and only indicate at the end the main results for the conduction band.

Returning to (4.19), we see that for L sufficiently large the bulk term can be written as $-(2L/\pi)\,\mathrm{d}q/\mathrm{d}E$, since we are now measuring q from the edge of the Brillouin zone and $\mathrm{d}q/\mathrm{d}E<0$ for the valence band. Then

$$N_{\mathrm{V}}^{\infty}(E)=-\frac{L}{\pi}\frac{\mathrm{d}q}{\mathrm{d}E}-\frac{1}{2}\,\delta(E)-\tfrac{1}{2}\delta(E-E_{\mathrm{V}}), \qquad (4.21)$$

whence the surface density of states for one single surface:

$$N_{\mathrm{s,V}}^{\infty}(E)=-\tfrac{1}{4}\delta(E)-\tfrac{1}{4}\delta(E-E_{\mathrm{V}}). \qquad (4.22)$$

This holds for one surface with an infinite barrier and for the energy range of the valence band. This density of states is associated with a local density localised near the surface. The analysis can be carried out much in the same way as § 3.1. We need the integrated density of states $N_d(E)$, for which we need to write the wavefunction (4.17) of the valence band with the correct normalisation factor. This is

$$\psi(z)=\sqrt{\left[\frac{2}{L(1+a)^2}\right]}\,\{\sin[(q-\tfrac{1}{2}g)z]+a\sin[(q+\tfrac{1}{2}g)z]\}. \qquad (4.23)$$

The function $a(\mathscr{E})$ was defined in (4.8), but for our present purpose $(|\mathscr{E}|>|V|)$ it is more convenient to rewrite it as

$$a=[\mathscr{E}+\surd(\mathscr{E}^2-V^2)]/V.$$

With this wavefunction, the density of states (4.21) yields (remember the general definition (3.10)) the local density of states

$$N(E,z)=-\frac{1}{\pi(1+a)^2}\frac{\mathrm{d}q}{\mathrm{d}E}\{(1+a^2)-\cos[(2q-g)z]$$
$$-a^2\cos[(2q+g)z]+2a\cos(gz)-2a\cos(2qz)\}. \qquad (4.24)$$

The terms in $(1+a^2)$ and in $2a\cos(gz)$ give the local density of states in the bulk. The cosine term describes the small periodic oscillation due to the crystalline pseudopotential. The contribution of the term in $\delta(E-E_{\mathrm{V}})$ vanishes, as well as that of the term in $\delta(E)$, because its prefactor vanishes as in the argument after (3.11). The corresponding integrated density of states is then (3.16)

$$N_d(E)=-\frac{d}{\pi}\frac{\mathrm{d}q}{\mathrm{d}E}\left\{1-\frac{1}{1+a^2}\frac{\sin[(2q-g)d]}{2q-g}\right.$$
$$\left.-\frac{a^2}{1+a^2}\frac{\sin[(2q+g)d]}{2q+g}+\frac{2a}{1+a^2}\frac{\sin(gd)}{gd}-\frac{2a}{1+a^2}\frac{\sin(2qd)}{2qd}\right\}. \qquad (4.25a)$$

In the energy range near the gap we can take $q \ll g$ and this simplifies to

$$N_d(E) = -\frac{d}{\pi}\frac{\mathrm{d}q}{\mathrm{d}E}\left[1 - \frac{2a}{1+a^2}\frac{\sin(2qd)}{2qd}\right]. \tag{4.25b}$$

This would not be true for very small $d \ll c = 2\pi/g$, but it becomes a good approximation when d approaches c. In passing from (4.25*a*) to (4.25*b*) we have also neglected the contribution of the term in $\sin(gd)$ which comes from the $\cos(gz)$ term in (4.24). As indicated above, this would only describe the small oscillations due to the weak pseudopotential. This is a distinct feature of the bulk, far away from the surface region, but it is comparatively unimportant for the study of the surface effects, which are mainly dominated by the large potential drop in the interface. The above approximation will be implicit in the rest of our discussion. With this we can now see again that a defect of $\frac{1}{4}$ state is localised near the surface. Indeed, for large d the oscillatory term in (4.25*b*) only contributes significantly for a small range of values of q, of order $1/2d$. On the other hand, in this range we can take $a \approx 1$, and the difference between $N_d(E)$ and the bulk value, $-(d/\pi)(\mathrm{d}q/\mathrm{d}E)$, is

$$\frac{d}{\pi}\int_{-\infty}^{E_V}\frac{\mathrm{d}q}{\mathrm{d}E}\frac{2a}{1+a^2}\frac{\sin(2qd)}{2qd}\,\mathrm{d}E \approx \frac{d}{\pi}\int_{-\infty}^{E_V}\frac{\mathrm{d}q}{\mathrm{d}E}\frac{\sin(2qd)}{2qd}\,\mathrm{d}E = -\frac{1}{4}. \tag{4.26}$$

Taking the lower limit as $-\infty$ and setting $a \approx 1$ is justified provided $\sin(2qd)/2qd$ contributes only in a range of values of q such that $q \ll 2|V|/g$ (remember (4.7)). But this requires $d \gg q/2|V|$, which means that the defect of $\frac{1}{4}$ state is localised within a distance of order $g/4|V|$ of the surface barrier. This rough estimate is instructive because it yields the same sort of localisation as we found in (4.11*a*) for a simple example of surface state.

Having studied the infinite barrier let us now discuss a finite barrier of arbitrary shape. As in § 3.1, imagine that the first infinite barrier becomes something like the potential profile of fig. 56, while retaining the second infinite barrier at $z = L$, a finite distance away so that we can use it as a device for counting states, but sufficiently distant that the two surfaces are not coupled. Now, remember we wrote down in (4.12) the wavefunction just emerging from the interface, where the oscillations of the pseudopotential are negligible compared with the large potential drop across the interface. In (4.16), on the other hand, we expressed the general form of the wavefunction well inside the system. This differs from (4.12) in two respects, namely (i) it exhibits a Bragg diffraction pattern, as the crystalline oscillations are now important, and (ii) it is governed by a different phase shift η which we have still to find. What we do now is to match

(4.12) and (4.16). This yields a relation between η and η_0 which, for $q \ll g$, is

$$\tan \eta(E) = (a-1)/(a+1) \tan \eta_0(E). \tag{4.27}$$

Note the essence of what we are doing: we express the boundary condition for the first barrier in two stages, so that we first embody the main effects of the large potential drop in η_0. But we also have the other boundary condition at $z = L$:

$$qL + \eta = m\pi. \tag{4.28}$$

If we know η_0, then (4.27) and (4.28) yield the allowed eigenvalues.

It is interesting to discuss the behaviour of $\eta(E)$ at the band edge E_V, when $a = 1$. In this case (4.27) shows that if $\eta_0(E_V) \neq m'\pi + \frac{1}{2}\pi$, then $\eta(E_V)$ is a multiple of π. For example, with an infinite barrier at $z = 0$ we have $\eta_0(E_V) = 0$ and $\eta(E_V) = 0$. Suppose we lower and displace the barrier. As long as $|\eta_0(E_V)| < \frac{1}{2}\pi$, $\eta(E_V)$ is constantly zero, and jumps to $\pm\pi$ when $|\eta_0(E_V)|$ goes through the value $\frac{1}{2}\pi$. It is interesting to note that if $\eta_0(E_V)$ is a half integral multiple of π, then by (4.13) the s.s. is just at E_V. The significance of this will be discussed later.

In order to study the density of states with a finite-barrier system we compare the allowed values of q for finite and infinite barriers, as in § 3.1 (fig. 32), noticing that if $\eta(E_V) = 0$ (which implies $|\eta_0(E_V)| < \frac{1}{2}\pi$) then the eigenvalue spectrum at the band edge is the same in both cases. This yields for the total density of states in the valence band

$$N_V(E) = -\frac{L}{\pi}\frac{\mathrm{d}q}{\mathrm{d}E} - \frac{1}{2}\delta(E) - \frac{1}{2}\delta(E - E_V) - \frac{1}{\pi}\frac{\mathrm{d}\eta}{\mathrm{d}E}, \tag{4.29}$$

whence, by the usual argument, the surface density of states for one single surface with a finite barrier is, for energies in the valence band,

$$N_{s,V}(E) = -\frac{1}{4}\delta(E) - \frac{1}{4}\delta(E - E_V) - \frac{1}{\pi}\frac{\mathrm{d}\eta}{\mathrm{d}E}. \tag{4.30}$$

The difference between this and (4.22) is due to the phase shift η produced by the finite barrier. Notice that the sign of this term is negative for the same reasons that $\mathrm{d}q/\mathrm{d}E$ is negative in the valence band. Now let $\eta_0(E) \to (\frac{1}{2}\pi) - 0$, so that the s.s. level tends to the valence band edge. We want to study the energy range near the band edge under these circumstances, for which we use (4.27). When $\eta_0(E_V)$ is very close to $\frac{1}{2}\pi$, $\eta(E)$ is practically equal to $-\frac{1}{2}\pi$ for all energies E for which $1-a$ differs appreciably from zero ($a \to 0$ for E negative and large). It is only in an extremely narrow energy range, close to E_V, when $1 - a \approx 0$, that $\eta(E)$

varies very rapidly between the value $-\frac{1}{2}\pi$ and the value 0 that it must have in the limit $E = E_V$. By (4.30), this has the effect of accumulating a loss of $\frac{1}{2}$ state at the valence band edge. But at the same time one full state – the s.s. – *appears* at the band edge. The net result is that, when $\eta_0(E_V) \to \frac{1}{2}\pi$, (4.30) yields a total value of $+\frac{1}{4}\delta(E - E_V)$ for the density of states at the valence band edge.

The above discussion suffices to characterise the situation for surface barriers in which $\eta_0(E)$ takes arbitrary values, since we can always displace the infinite barrier, initially at $z = 0$, by any multiple of c. Doing this will involve changing $\eta_0(E_V)$ by multiples of π, so that we can always reduce its value to lie in the range $|\eta_0(E_V)| \leq \frac{1}{2}\pi$. The displacement is also related to an arbitrariness in the definition of the surface density of states, as in § 3.1. Such a displacement induces a change in $\eta(E)$ and hence in its contribution to the density of states, but this is exactly compensated by an opposite change in the contribution of the bulk term.

We can now use (4.29) to derive a quantisation rule for the number of states in the valence band, for which it suffices to integrate:

$$\int_{-\infty}^{E_V} N_V(E)\, \mathrm{d}E = \frac{L}{\pi}\frac{g}{2} - 1 - \frac{1}{\pi}[\eta(E_V) - \eta(0)]$$

$$= \frac{L}{c} - 1 - \frac{1}{\pi}[\eta(E_V) - \eta(0)]. \tag{4.31}$$

But at $E = 0$, where the wavefunctions have metallic character – away from the semiconductor gap – we know from § 3.1 that $\eta(0) = 0$, whereas we have just seen that $\eta(E_V)$ takes only discrete values which are integral multiples of π. Thus (4.31) shows that the number of states – and therefore, spin included, the number of electronic charges – in the valence band is always an integer. The physical picture is as follows: with infinite barriers (fig. 57) the number of states in the valence band is $N - 1$, as was seen in the discussion after (4.20). If the surface barrier changes gradually, so that $\eta_0(E_V) < \frac{1}{2}\pi$, the number of states in the band remains constant, but the s.s. originally at E_c moves down the gap, approaching E_V. When $\eta_0(E_V) = \frac{1}{2}\pi$, the s.s. reaches E_V, so that one new state enters the valence band and the process is repeated as $\eta_0(E_V)$ reaches successive half-integral multiples of π.

Now that we have a new formula (4.30) for $N_{s,V}$ we must discuss its spatial localisation. This means that the effect of (4.30), which is of order $1/L$ compared with the volume term, must not change the bulk density of states to order $1/L$ for large distances inside the system. As with the infinite-barrier case, we study this by discussing the integrated density of states, following the path of the phase-shift term. We are interested in the wavefunction $\psi(z, E)$ which becomes (4.16) well away from the interface. Applying (3.32) to this case we find (assuming $q \ll g$)

$$\int_{-\infty}^{d} \psi^2(z, E)\,\mathrm{d}z = \frac{1}{2}C^2(E)\left[\left(d\frac{\mathrm{d}q}{\mathrm{d}E}+\frac{\mathrm{d}}{\mathrm{d}E}\right)\frac{g}{2}(a^2-1) - \frac{g}{2}\frac{\mathrm{d}a}{\mathrm{d}E}\sin(2qd+2\eta)\right]. \tag{4.32}$$

The normalisation constant $C(E)$ can be found putting $d = L$ in (4.32) and using (4.28). This yields

$$C^2(E) = 4/g(a^2-1)[L(\mathrm{d}q/\mathrm{d}E)+\mathrm{d}\eta/\mathrm{d}E].$$

Knowing C and using (4.29), (4.32) and (4.33) we have the integrated density of states in the valence band,

$$N_d(E) = \int_{-\infty}^{d} N_V(E)\psi^2(z, E)\,\mathrm{d}z$$

$$= -\left[\frac{d}{\pi}\frac{\mathrm{d}q}{\mathrm{d}E}+\frac{1}{\pi}\frac{\mathrm{d}\eta}{\mathrm{d}E}-\frac{1}{(a^2-1)}\frac{\mathrm{d}a}{\mathrm{d}E}\sin(2qd+2\eta)\right]. \tag{4.33}$$

The terms in $\delta(E-E_V)$ and $\delta(E)$, again, do not contribute because their prefactors vanish. Using the equality

$$\frac{\mathrm{d}a}{\mathrm{d}q} = \frac{g}{V}\frac{a^2}{1+a^2} \tag{4.34}$$

and the relation (4.27) between η and η_0, we can rewrite (4.33) in the more convenient form

$$N_d(E) = -\frac{d}{\pi}\frac{\mathrm{d}q}{\mathrm{d}E}\left\{1 - \frac{2}{(1+a^2)[1+a^2+2a\cos(2\eta_0)]}\frac{\sin(2qd)}{2qd}\right.$$
$$\times[(a^3+a)\cos(2\eta_0)+2a^2]+2/(1+a^2)[1+a^2+2a\cos(2\eta_0)]$$
$$\left.\times\frac{1-\cos(2qd)}{2qd}(a^3-a)\sin(2\eta_0)\right\}. \tag{4.35}$$

In going from (4.33) to (4.35) we have taken η_0 constant, neglecting its small energy dependence in the range of interest, near the gap. Notice also that (4.35) follows from (4.32), where it was assumed that $q \ll g$. As in the infinite-barrier case this means that (4.35) becomes valid when $d \geqslant c = 2\pi/g$. The constant term is of course the bulk term, if we ignore the integrated effect of the small oscillations due to the pseudopotential. The following term can be simplified for large $d \gg g/|V|$ noting that the factor $\sin(2qd)/(2qd)$ is then only significant for $q \approx 0$, when $a \approx 1$. Thus the second term on the r.h.s. of (4.35) can be written as

$$\frac{d}{\pi}\frac{\mathrm{d}q}{\mathrm{d}E}\frac{\sin(2qd)}{2qd}.$$

The integrated density of states represents an accumulated loss of $\frac{1}{4}$ state near $q=0$. We are once more in the same situation as with the infinite barrier (4.26) and, as before, we find that this loss of $\frac{1}{4}$ state, which is the term $-\frac{1}{4}\delta(E-E_V)$ in (4.30), in the limit $d\to\infty$, is spatially localised near the surface within distances of order $g/4|V|$. Notice how, for any value of d, if we look at very small values of q the first two terms in (4.35) add up to

$$-\frac{d}{\pi}\frac{\mathrm{d}q}{\mathrm{d}E}\left[1-\frac{\sin(2qd)}{2qd}\right],$$

so that for $q\to 0$ the factor after $\mathrm{d}q/\mathrm{d}E$ eliminates the singularity of $\mathrm{d}q/\mathrm{d}E$, which varies as $(V-\mathscr{E})^{-1/2}$, yielding a regular behaviour varying as $(V-\mathscr{E})^{1/2}$. Finally, in the last term of (4.35) the contribution of the term in $\cos(2qd)$ can be neglected for large $d \gg g/4|V|$, owing to its fast oscillations. We are left with the contribution

$$-\frac{d}{\pi}\frac{\mathrm{d}q}{\mathrm{d}E}\frac{2(a^3-a)}{(1+a^2)[1+a^2+2a\cos(2\eta_0)]\,2qd}\sin(2\eta_0).$$

This is easily identified as an alternative way of writing $-(1/\pi)\,\mathrm{d}\eta/\mathrm{d}E$, if we differentiate (4.27), assume η_0 is constant and use (4.34). Thus, we need only integrate up to distances beyond the order of $g/4|V|$ in order to recover the phase-shift term in the density of states, which means that this term is also localised within distances of that order. Notice, furthermore, that the factor $[1-\cos(2qd)]$ cancels the singularity which would otherwise exist in $\mathrm{d}q/\mathrm{d}E$.

The integrated density of states is shown in fig. 59, by taking $d=35$ Å (3.5 nm) and choosing $|V|=2$ eV and $(\frac{1}{2})(\frac{1}{2}g)^2=10$ eV as typical values for covalent semiconductors. In all cases it starts from $\mathscr{E}=V$ (valence band edge) varying as $(V-\mathscr{E})^{1/2}$, so that the singularity of $\mathrm{d}q/\mathrm{d}E$ (which varies as $(V-\mathscr{E})^{-1/2}$) is reflected only in the maximum which appears at an energy that differs from $\mathscr{E}=V$ by something of the order $g^2/4|V|d^2$. This is the energy corresponding to a value of q of order $1/d$, as can be seen from (4.5*b*). Notice how $N_d(E)$ is smaller when a s.s. exists in the gap than when the s.s. has just entered the band. It is clear that the strong peaks shown in fig. 59 for η_0 just below $\frac{1}{2}\pi$ are related to the new spectral strength contributed by the entrance of the s.s. in the band. The effect of the position of the s.s. level on the density of states in the continuum can also be seen by analysing the surface density of states for the valence

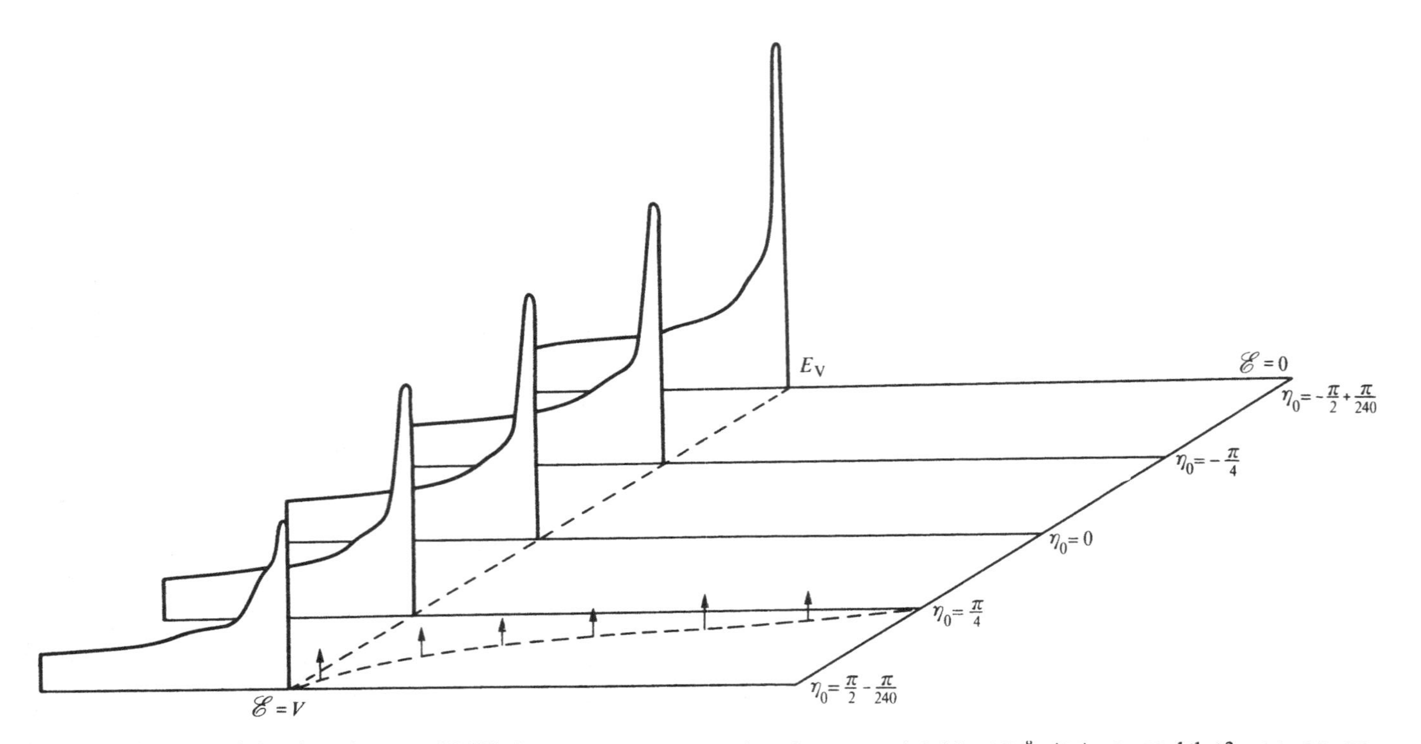

Fig. 59. Integrated density of states, $N_d(E)$ for a narrow-gap semiconductor model ($d = 35$ Å, $|V| = 2$ eV, $\frac{1}{2}(\frac{1}{2}g)^2 = 10$ eV). The energy range shown in the figure goes from the valence band, through the valence band edge $E_v(\mathscr{E} = V)$, to the midgap ($\mathscr{E} = 0$). The different curves are for different values of the phase shift η_0. The arrows show the position of the s.s. for the corresponding values of η_0.

band, $N_{s,V}$ in (4.30). If we concentrate only on states near the band edge and neglect the energy dependence of η_0 in this range, from the above discussion we can rewrite (4.30) as

$$N_{s,V}(E) = -\frac{1}{4}\delta(E-E_V) - \frac{(a^3-a)\sin(2\eta_0)}{\pi q(1+a^2)[1+a^2+2a\cos(2\eta_0)]}\frac{dq}{dE}.$$

We stress that this describes only the energy range near the band edge. This is shown in fig. 60 for different values of η_0. The $(V-\mathscr{E})^{-1/2}$ singularities are now apparent, as corresponds to the limit $d\to\infty$. The main point to notice is the strong dependence on η_0. For example, when $\eta_0\to\frac{1}{2}\pi$ a loss of $\frac{1}{2}$ state tends to accumulate at the band edge, which added to the δ-function term means a loss of $\frac{3}{4}$ state at E_V. But then, as we saw before, the s.s. is just at E_V, and the net value of $N_{s,V}$ is $+\frac{1}{4}\delta(E-E_V)$ at the band edge. The situation is quite different for $\eta_0\to-\frac{1}{2}\pi$, when the s.s. moves into the gap, away from E_V. As η_0 varies between $\pm\frac{1}{2}\pi$ it gives a contribution of $-\eta_0/\pi$ to the total density of states in the band. This is related to the quantisation of the number of states in the band, in line with the general argument following (4.30).

Having discussed the valence band in detail let us briefly indicate the main results for the conduction band. The analysis can be carried out in much the same manner, but just an indication of the results will suffice to give a picture of the situation. For the distortion of the continuum in the conduction band we find the following surface density of states term:

$$N_{s,C}(E) = -\frac{1}{4}\delta(E-E_C) - \frac{1}{\pi}\frac{d\eta}{dE}. \tag{4.36}$$

The sign of the δ-function term is not in contradiction with (4.20). In fact (4.36) yields the same density of states as (4.20) when $\eta_0\to 0$. The situation can be analysed by using (4.27) again. (The difference from the situation with the valence band is that now $a=-1$ at the band edge.) By a similar argument we find a net contribution of $+\frac{1}{4}$ state at E_C.

The integrated density of states is again given by (4.35), but a now has different values, corresponding to energies in the conduction band. If we take two opposite values of E we find that $a(\text{cond.})=-1/a(\text{val.})$. It is then easy to see that $N_d(\text{cond.}, E)$ is the same as $N_d(\text{val.}, -E)$ if we also replace η_0 by $\frac{1}{2}\pi-\eta_0$. Thus

$$N_d(\text{cond.}, E, \eta_0) = N_d(\text{val.}, -E, \tfrac{1}{2}\pi-\eta_0).$$

With this we can use for the conduction band all the results previously obtained for the valence band, *mutatis mutandis*. In particular, fig. 59 and 60 apply also to the conduction band after changing the sign on the energy axis and replacing η_0 by $\frac{1}{2}\pi-\eta_0$.

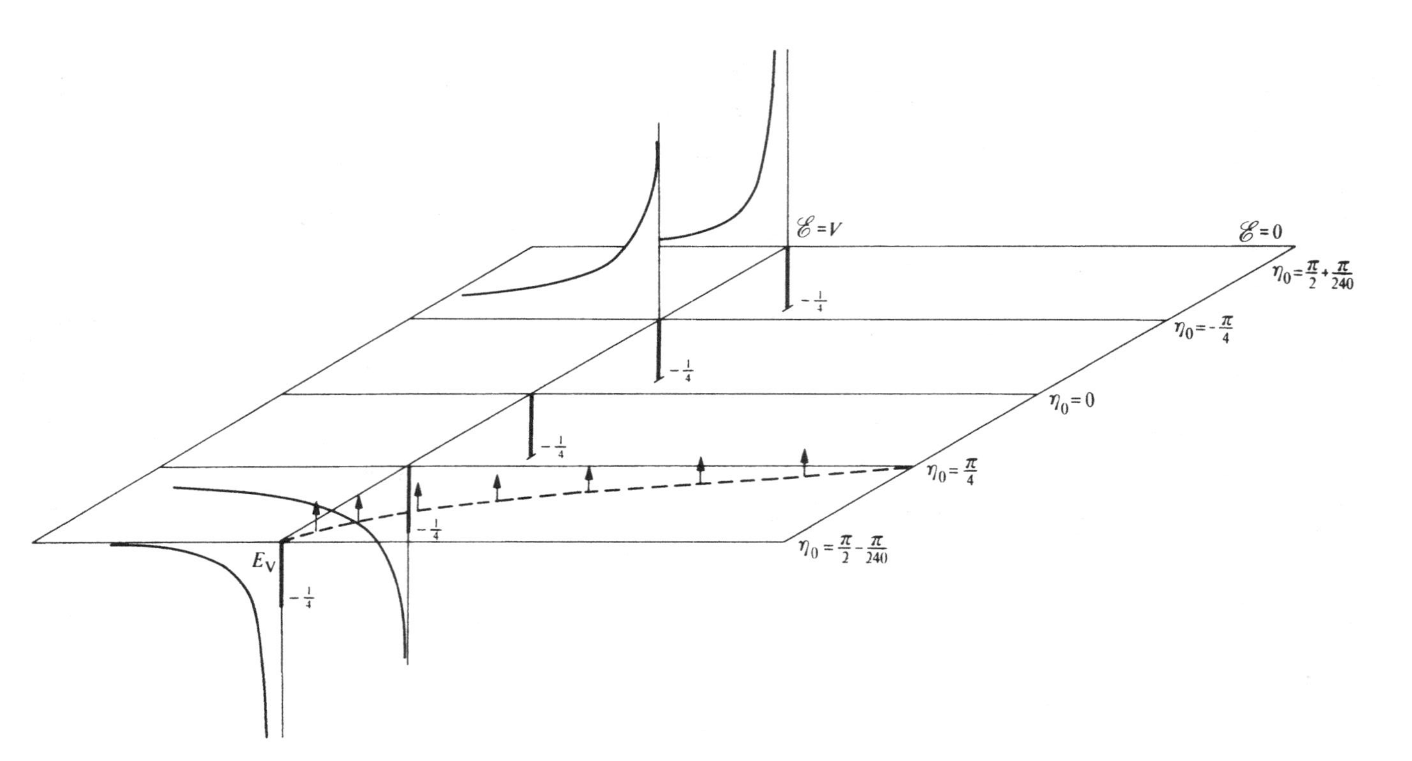

Fig. 60. Surface density of states near the valence band edge, $N_{s,v}(E)$, for the same semiconductor model as that of fig. 59, as a function of η_0. The arrows indicate the corresponding position of the s.s. eigenvalue in the gap.

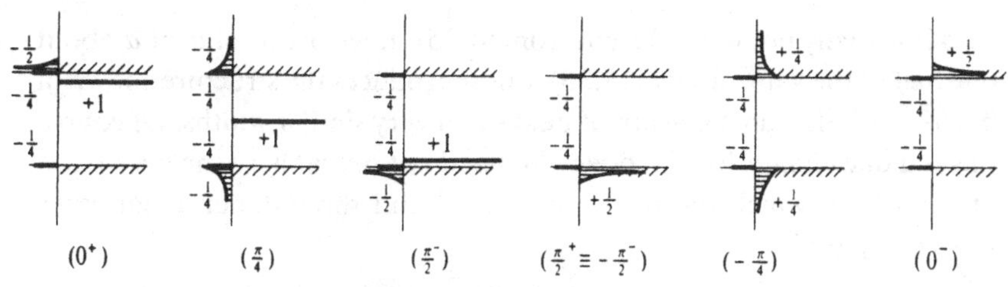

Fig. 61. Surface density of states in and around the gap of a narrow-gap semiconductor for different values of η_0. The bold lines are δ-function contributions. The δ-function contributions of strength +1 give the position of the s.s. eigenvalue, when there is one. The figure in brackets gives the value of η_0 in each case. After F. García-Moliner & F. Flores, *J. Phys. C: Solid St. Phys.* **9**, 1609 (1976).

A general picture of the situation is described in fig. 61, which shows the density of states near the two band edges as a function of the position of the s.s. or, rather, as a function of η_0. The important point is that all the modifications of the density of states in and about the gap add up to zero. Here we come across the first important result, which amounts to a conservation rule: if, on creating the semiconductor gap, we find that a s.s. appears, then the distortions of the continuum that the surface also creates near the edges of the adjacent bands subtract one full state overall so that the net change in the total number of states associated with the gap is zero. This is in fact a *Levinson theorem* for the energy range associated with a narrow gap. It depends on the symmetry between valence and conduction bands, and this in turn depends on the narrow-gap approximation.

As in § 3.1, the integrated density of states is the central concept from which we can build up a picture of the state of affairs at the surface, connecting the density of states with the spatial localisation of the s.s. and the distortions of the continuum. Besides this, the concept of integrated density of states also has an experimental relevance, in connection with directional photoemission experiments, as was discussed at the end of § 3.1.2. In fact we saw that rather than N_d itself we should introduce a weighting factor to account for the variation of escape probability. Thus, in a simple sense we ought to use something like

$$\int \mathrm{e}^{-z/d} N(z,E)\,\mathrm{d}z = -\frac{d}{\pi}\frac{\mathrm{d}q}{\mathrm{d}E}$$
$$\times\left\{1 - \frac{2[(a^3+a)\cos(2\eta_0)+2a^2]}{(1+a^2)[1+a^2+2a\cos(2\eta_0)][1+(2qd)^2]}\right.$$
$$\left. + \frac{4qd(a^3-a)\sin(2\eta_0)}{(1+a^2)[1+a^2+2a\cos(2\eta_0)][1+(2qd)^2]}\right\}.$$

This is actually not too different from (4.35). If we use a value of d about one half of the value used in (4.35), this reproduces the structure shown in fig. 59, with the same pattern of peaks and very similar widths. Of course this is a one-dimensional model. We need not repeat here the discussion of § 3.2.1, in which the one-dimensional and three-dimensional cases were compared.

4.1.3. Selfconsistency and surface potential

We have seen that the characteristic feature which distinguishes a semiconductor from a metal is the new structure in the density of states associated with the semiconductor gap. In the gap itself we can have surface states and then we have further distortions of the continuum about the gap, near the band edges. As is always the case with surface effects, all this spectral strength is localised in the space near the surface within distances of order $g/4|V|$. These are considerably longer distances than those in which the surface density of states is localised in the situation with a metal – which of course we also have in the rest of the bulk bands of the semiconductor, away from the gap. However, it is interesting to notice that there is a tendency towards a mutual cancellation of the effects associated with the gap. We have seen that this global cancellation exists in terms of density of states. It is then natural to enquire whether this cancellation may also take place in more detailed terms. We shall discuss this and then the questions of selfconsistency and charge neutrality, which are all related.

We first study the sum of all terms contributing to the local density of states, for which we use the integrated density of states given in (4.35). We recall that this formula can be used to study either the conduction or the valence band. In fact we are now interested in studying the distortion of the range of the continuum near the gap due to the surface, i.e. only in the surface part of this integrated density of states, which is

$$N_{s,d}(E) = \frac{d}{\pi} \frac{dq}{dE} \left\{ \frac{2}{(1+a^2)[1+a^2+2a\cos(2\eta_0)]} \frac{\sin(2qd)}{2qd} \right.$$
$$\times [(a^3+a)\cos(2\eta_0)+2a^2]$$
$$-\frac{2}{(1+a^2)[1+a^2+2a\cos(2\eta_0)]}$$
$$\left. \times \frac{1-\cos(2qd)}{2qd}(a^3-a)\sin(2\eta_0) \right\}. \tag{4.37}$$

This has been obtained by using the approximation (4.5b) for the dispersion relation, which is justified since the effects studied here are concentrated near the band edges. What we want to evaluate is

$$\bar{N}_{s,d}=\int_{-\infty}^{E_V} N_{s,d}(E)\,dE+\int_{E_c}^{\infty} N_{s,d}(E)\,dE. \tag{4.38a}$$

This integrates up to a given – but variable – distance d all the terms resulting from the distortion of both bulk bands. Using (4.5b) in the definition (4.8) of a, inserting (4.37) in the first integral of (4.38) and changing the integration variable from E to a, we find

$$\int_{-\infty}^{E_V} N_{s,d}(E)dE=\frac{d}{\pi}\int_0^{\infty}\frac{da}{a^2}\frac{V}{g}\left\{\frac{\sin(2qd)[(a^3+a)\cos(2\eta_0)+2a^2]}{[1+a^2+2a\cos(2\eta_0)]2qd}+\frac{[1-\cos(2qd)](a^3-a)\sin(2\eta_0)}{[1+a^2+2a\cos(2\eta_0)]2qd}\right\}. \tag{4.38b}$$

The second integral can be handled likewise and, adding, we have

$$\bar{N}_{s,d}=\frac{d}{\pi}\int_{-\infty}^{\infty}\frac{da}{a^2}\frac{V}{g[1+a^2+2a\cos(2\eta_0)]}\times\left\{\frac{\sin(2qd)[(a^3+a)\cos(2\eta_0)+2a^2]}{2qd}-\frac{1-\cos(2qd)}{2qd}(a^3-a)\sin(2\eta_0)\right\}. \tag{4.38c}$$

This integral can be evaluated by complex contour integration yielding

$$\bar{N}_{s,d}=\begin{cases}-1+\exp[-4|V|d\sin(2\eta_0)/g] & (0<\eta_0<\tfrac{1}{2}\pi);\\ 0 & (\tfrac{1}{2}\pi<\eta_0<\pi)\end{cases} \tag{4.39}$$

In the first case ($0<\eta_0<\frac{1}{2}\pi$) there is a surface state in the gap, whereas no s.s. appears when $\frac{1}{2}\pi<\eta_0<\pi$, as we saw in §§ 4.1.1, 4.1.2. Consider then the first case. The contribution of the s.s. to N_d can be obtained by using (3.32) and the asymptotic form of the s.s. wavefunction, which introduces the factor

$$\exp[-2|V\sin(2\eta_0)|/g]\sin(\tfrac{1}{2}gz+\eta_0).$$

On evaluating this contribution we find that it is exactly the negative of the value of (4.39) for $0<\eta_0<\frac{1}{2}\pi$. This yields the next important result:

$$\bar{N}_{s,d}(\text{contin.})+N_{s,d}(\text{surf. state})=0. \tag{4.40}$$

Thus the integrated effects cancel out on integrating up to a variable distance d. Hence *the cancellation is also local.* Of course this cancellation refers to the envelopes of the terms studied here. Consider, for example, the term associated with the s.s. In this case the effect of the term in $\sin^2(\eta_0+\frac{1}{2}gz)$ is reduced to the contribution of a factor $\frac{1}{2}$ because these oscillations are very fast in comparison with the slower decay of the exponential term. As a matter of fact, it can be seen that the local cancellation is still achieved, even when (4.40) takes account of the rapidly oscillating terms, but we shall not go into details. Notice also that (4.40) has only been explicitly proved for distances such that the wavefunction has already effectively reached its asymptotic form. However, in the narrow-gap approximation all wavefunctions of the energy range near the gap are practically equal in and around the interface. Thus it follows from the global cancellation beyond the interface that to the same approximation there is also local cancellation in the interface region. Of course the whole derivation of (4.40) is based on approximations, which ultimately hinge on one, namely, the assumption of a narrow-gap model. But here is a simple, yet plausible physical argument which leads to a rather suggestive result. The effects of the gap – the distinctive semiconductor feature – on the density of states cancel out globally and also locally. If all these states were occupied this would lead to an interesting situation: with the charge redistribution cancelling out, the semiconductor surface – i.e. its charge and potential barrier profiles – would for practical purposes be like a metal surface of the same electronic density. The trouble is that the conduction band is empty and we still do not know the occupation of the s.s. We shall therefore study this problem, together with the general question of selfconsistency.

We start with a linear chain with two electronic charges per atom (fig. 56) forming a covalent crystal. This means that two electrons are shared in a covalent bond between each pair of neighbouring ions, which corresponds to the situation when the crystal potential has a minimum – i.e. is attractive – in the region between atoms. We then create the surface and enquire about the occupation of the s.s. This can be studied by using the principle of charge neutrality and the quantisation rule derived in § 4.1.2 for an arbitrary surface barrier. For example, for the infinite-barrier model we have $N-1$ electronic states in the valence band, not counting spin. It is then clear that the s.s. – here appearing at the conduction band edge – must have occupation $\frac{1}{2}$ in order to achieve charge neutrality. The rule for s.s. occupation Q_s in a covalent system is then

$$Q_s(\text{covalent}) = \tfrac{1}{2}. \tag{4.41}$$

Of course this has been obtained for an infinite-barrier model, but we have seen that the number of electronic charges is quantised for all surface barriers. Thus the problem of whether or not Q_s changes on going from an infinite to a finite barrier depends on whether $|\eta_0|$ goes through $\frac{1}{2}\pi$. Let us look into this question.

Consider a finite barrier like that of fig. 56. We can think of the semiconductor system as constructed in two steps. First we take a metal with constant volume potential and with the same electronic density as the semiconductor under study. This hypothetical metal has its own selfconsistent solution, with a surface barrier going without crystalline oscillations from the vacuum level down to the constant bulk level. We can apply the analysis of chapter 3 to this system. In particular, taking the origin at the jellium edge, we can write for the asymptotic form of the wavefunction corresponding to the Fermi level,

$$\psi \propto \sin\left[q_F z + \eta_0(E_F)\right], \tag{4.42}$$

and we know $\eta_0(E_F) = \frac{1}{4}\pi$. Then we add the small pseudopotential oscillations. This opens the semiconductor gap in the bulk spectrum, and a s.s. appears in this gap. In principle we should now also study the distortion of the continuum and the consequent charge redistribution, which could change selfconsistently the potential barrier and hence the position of the s.s. level itself. But we have just seen that the changes in the spectrum due to the appearance of a s.s. and to the distortion of the valence and conduction bands cancel each other locally. Moreover, in the narrow-gap model, if $\eta_0(E_F) = \frac{1}{4}\pi$, we have seen that $N_d(\text{cond.}, +\mathscr{E}) = N_d(\text{val.}, -\mathscr{E})$, thus there is symmetry between the conduction and valence band edges, i.e. the distortions of the continuum near the two edges are equal. This means that the total spectral strength lost from the continuum is equally shared between the two bands. This is of crucial importance in our argument. Suppose we start from the surface potential of the hypothetical equivalent metal with its own selfconsistent potential barrier and charge distribution in the region just beyond the interface. We open the semiconductor gap and study the position and occupation of the s.s. *without allowing for any surface charge redistribution.* Owing to the fact that $\eta_0(E_F) = \frac{1}{4}\pi$ we find that (i) the s.s., by (4.13), is just at the midgap, and (ii) the occupation of this s.s. is also $\frac{1}{2}$ because $|\eta_0| < \frac{1}{2}\pi$ (remember the remark after (4.41)).

At this stage we should start the selfconsistent cycle by allowing for surface charge redistribution and so on. But the point is that there is no need for it. If the s.s. is half occupied and the spectral strength disappearing from the – fully occupied – valence band is $\frac{1}{2}$ state then these two charge distributions cancel out locally and there is no charge redistribution. The opening of the semiconductor gap introduces a s.s. but the half electronic charge in this state cancels out the surface charge missing from the valence band and the 'metallic' surface barrier is already selfconsistent for a narrow-gap semiconductor. Thus, except for the changes in the spectrum, in terms of surface potential and surface charge distribution, the narrow-gap covalent surface is like the surface of an equivalent metal with the same electronic density. In particular, the two key features are that: (i) the selfconsistent position of the s.s. is at the midgap, and (ii) the occupation rule for the s.s. is (4.41).

4.2. N.f.e. three-dimensional models

The study of one-dimensional models has served to introduce basic concepts and to raise key issues, but the actual results can only have at best an indicative value. We now analyse three-dimensional systems, while keeping to the study of ideal surfaces. We shall discuss in particular the (1, 1, 1) and (1, 1, 0) faces of covalent and ionic semiconductors with diamond or zincblende structure. The analysis will be carried out in terms of wavefunction matching, which is entirely equivalent to surface Green-function matching, as was seen in §§ 2.4, 2.5. In fact some of the results presently to be discussed have been obtained by using surface Green functions in the actual calculation, but we shall stick consistently to the wavefunction language.

4.2.1. Matching in three dimensions: general principles

Before explicitly carrying out a three-dimensional analysis it is convenient to give a quick summary of the main facts underlying the one-dimensional analysis and to stress some technical points which will soon be needed. First, we have only studied two-band models, while real crystals have a spectrum of bands with dispersion relations $E_n(q)$ in the first Brillouin zone (B.z.) and corresponding wavefunctions ψ_{nq} satisfying the Bloch property. The corresponding complex band structure is given by the analytic continuation of $E_n(q)$ to complex q values. Then the Bloch property shows that the corresponding wavefunctions grow or decay exponentially.

Consider now the band structure of a one-dimensional empty lattice. That is, take the free-electron spectrum, including the analytic continuation, and fold it inside the first B.z. For $E>0$ we have (fig. 62) the folded free-electron parabola. At $E=0$ we have the minimum of this parabola and also the maximum of a band of real energies corresponding to imaginary q. This is an instance of a general rule. The real and complex parts of the bands always have contacts at extrema of the (real) energy.

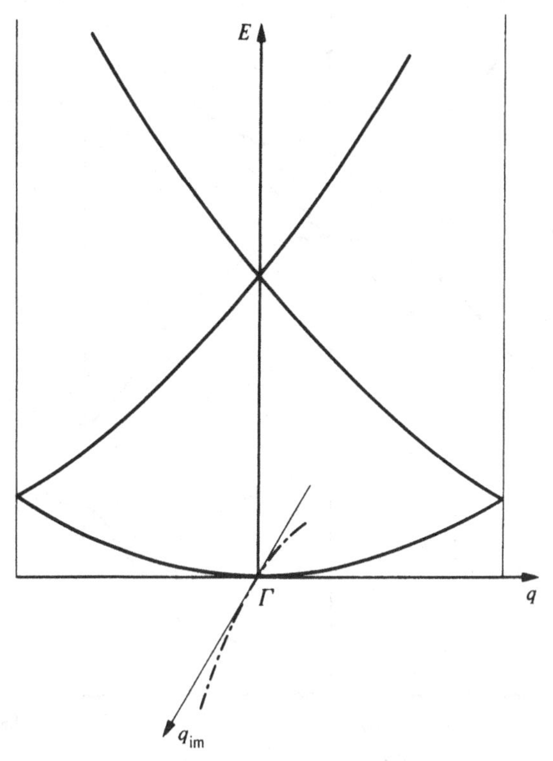

Fig. 62. Free electron parabola mapped (folded) inside the first Brillouin zone of a one-dimensional system. The dash-dotted line shows the negative real energy band corresponding to complex values of q.

Near such points we can expand, for a band labelled n,

$$E_n(q) = E_n(q_0) + (\partial^2 E_n/\partial q^2)_0(q-q_0)^2 + \ldots \tag{4.43}$$

$E_n(q_0)$ is a minimum/maximum for $(q-q_0)$ real/imaginary, and $E_n(q)$ has a saddle point at $q = q_0$ in the complex q plane. An example is the Γ-point in fig. 62. For each value of E we have two wavefunctions which are either oscillatory or exponentially growing or decaying. Because of this fact it is impossible to break up the energy spectrum into disconnected pieces. We can cover with continuity through the complex band spectrum the entire energy range from $-\infty$ to $+\infty$.

Now we switch on a perturbing crystalline potential. In the real band structure the positive energy range is broken into pieces through gaps appearing at the centre or edges of the B.z. These band edges correspond to extrema of the energy and it is at these points that the continuity in energy is restored through real energy loops of complex momentum (fig. 63). The situation is qualitatively as in fig. 53. Thus the general picture is

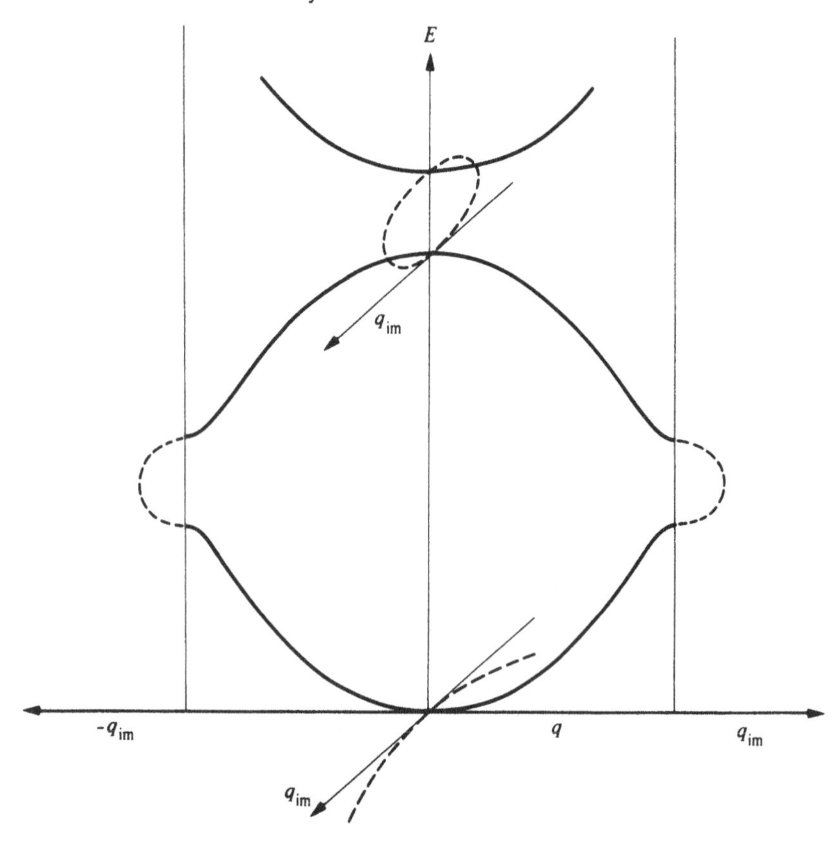

Fig. 63. Complex band structure of a hypothetical one-dimensional crystal showing the real energy loops (dashed lines) now appearing in gaps. These loops connect continuously the positive energy bands.

as follows: for any given value of E there are always two eigenfunctions of the crystal Hamiltonian. The real energy spectrum is spanned with continuity from $-\infty$ to $+\infty$ by real bands, by complex momentum loops connecting with the band edges at their extrema and by two real energy lines running to infinity (fig. 63).

Furthermore, the analytical continuation of a Hermitian Hamiltonian to complex q has the property $H^{+}(q) = H(q^*)$. This implies

$$E_n^*(q^*) = E_n(q). \tag{4.44a}$$

Another relation, namely,

$$E_n(q) = E_n^*(-q^*) \tag{4.44b}$$

follows directly from the Bloch theorem. Combining (4.44a) and (4.44b)

yields

$$E_n(q) = E_n(-q). \tag{4.45}$$

This quick survey summarises the main features of the complex band structure in one dimension.

Consider now the case of the wavevector in the ΓL line of a three-dimensional face-centred cubic (f.c.c.) lattice, which can be used to study the diamond and zincblende structures. The choice of a particular direction implies focusing attention on the normal to the surface which will eventually be studied. The band structure of the empty lattice for this case is shown in fig. 64. The situation seems more complicated than in the one-dimensional case, but we shall see that the three-dimensional problem can actually be described as a superposition of one-dimensional ones. The crucial thing to notice is that the new parabolae of the three-dimensional case – and their foldings inside the first B.z. – are related to

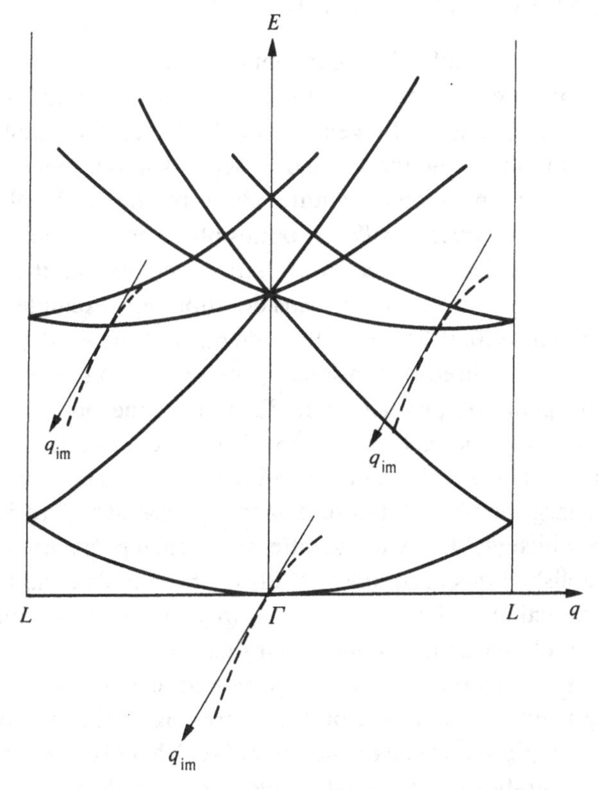

Fig. 64. Band structure for an empty f.c.c. lattice for q spanning the line ΓL. The figure also shows (dashed lines) the real energy parabolae of complex momentum.

reciprocal lattice vectors with components perpendicular to the direction under consideration For example, Γ is the wavevector (0, 0, 0) and L is $(\frac{1}{2}, \frac{1}{2}, \frac{1}{2})$. Consider the reciprocal lattice vectors

$$(1, 1, \bar{1}), (1, \bar{1}, 1), (\bar{1}, 1, 1), (\bar{1}, \bar{1}, 1), (\bar{1}, 1, \bar{1}), (1, \bar{1}, \bar{1}). \qquad (4.46)$$

Then the bands shown in fig. 64 in the reduced-zone scheme arise as follows: the non-degenerate parabola is $\frac{1}{2}\boldsymbol{k}^2$. The three parabolae $\frac{1}{2}|\boldsymbol{k}+\boldsymbol{G}|^2$, where $\boldsymbol{G}$ takes the values of each of the first three vectors of (4.46), together make up one of the two three-fold degenerate parabolae shown in the figure, while the other one is obtained by setting $\boldsymbol{G}$ equal to each one of the last three vectors of (4.46). Now look at the minima of the degenerate parabolae. In the reduced–zone scheme these minima appear at points between Γ and L, but in the extended zone scheme they unfold into the six points

$$\left(\frac{2}{3}, \frac{2}{3}, \frac{\bar{4}}{3}\right), \left(\frac{2}{3}, \frac{\bar{4}}{3}, \frac{2}{3}\right), \ldots, \left(\frac{4}{3}, \frac{\bar{2}}{3}, \frac{\bar{2}}{3}\right).$$

These points form a regular hexagon obtained by projecting the six vectors of (4.46) onto a plane perpendicular to the direction ΓL. The centre of the hexagon is Γ. In general we find that the number of independent parabolae – and therefore of independent one-dimensional problems whose superposition constitutes the three-dimensional problem – is equal to the number of *different* points obtained in the extended-zone scheme by projecting reciprocal lattice vectors on the plane perpendicular to the direction under consideration. For example, in the above case the vectors $(0, 0, \bar{2})$, $(0, \bar{2}, 0)$ ... do not yield new independent 'components' of the three-dimensional problem because their projections on the plane perpendicular to ΓL fall on the points already obtained by projecting the vectors of (4.46). This is a useful rule because a good approximation to a realistic bulk band structure may be achieved in practice with a large number of different bands associated with reciprocal lattice vectors which yield only a few different normal projections.

Having established this, a complete study of the complex band structure for the direction of interest in the empty lattice is obtained by studying separately each independent parabola as a one-dimensional problem. Let us now set out an explicit criterion for counting real energy lines. Starting from the minimum of each parabola we have two lines running to infinity (fig. 64). Then the number of such lines is equal to twice the number of parabolae, i.e. equal to twice the number of different projections $\bar{\boldsymbol{G}}$ in the sense just discussed. Thinking of the surface eventually to be studied we shall call these projections the *parallel components*, meaning parallel to the surface. In fig. 64 we have 2×7 lines running to infinity, corresponding to $\bar{\boldsymbol{G}}=0$ and to the parallel components of the six vectors of (4.46). Notice that, when it comes to studying wavefunctions, if $\boldsymbol{n}$ is the normal to the surface then the wavevector for a generic point of the two-dimensional band structure is formally three-dimensional, being of the form $\boldsymbol{\kappa}+q\boldsymbol{n}$, but only $\boldsymbol{\kappa}$ is real; q is complex. Furthermore, in the complex band structure the number of eigenfunctions for a given real energy value is obtained by the same rule. In the case of fig. 64 this is also 2×7, i.e. twice the number of different parallel components $\bar{\boldsymbol{G}}$ included in the band structure. Each one obeys its own second-order differential equation.

The empty lattice serves as a starting frame within which we now switch on the crystalline potential. The details of the new band structure are complicated and depend on the potential, but the key features can be easily recognised. The gaps are again spanned by real energy loops connecting the extrema of the real momentum band edges and, as in the one-dimensional case, the introduction of the crystal potential preserves the number of eigenfunctions for each energy value and the relationship to the number of independent parallel components $\bar{\boldsymbol{G}}$ included in the band structure. In practice one writes the Bloch wavefunctions as a plane-wave expansion,

$$\psi_{\boldsymbol{k}}(\boldsymbol{r})=\sum_{\boldsymbol{G}} a(\boldsymbol{G})\exp\left[\mathrm{i}(\boldsymbol{k}+\boldsymbol{G})\cdot\boldsymbol{r}\right], \tag{4.47}$$

involving a finite number of terms. By separating out the different parallel components, this can be formally rewritten as

$$\psi_{\boldsymbol{k}}(\boldsymbol{r})=\sum_{i} \phi_i(z)\exp(\mathrm{i}\bar{\boldsymbol{G}}_i\cdot\boldsymbol{\rho})\,\mathrm{e}^{\mathrm{i}\boldsymbol{\kappa}\cdot\boldsymbol{\rho}}. \tag{4.48}$$

Thus the number of different ϕ_is is equal to the number of different parallel components, i.e. one half the number of eigenfunctions corresponding to a given value of E. Consider, for example, the band structure of a covalent crystal in the ΓL direction including fifteen plane waves corresponding to $\boldsymbol{G}=0$ and to $\boldsymbol{G}$ equal to any vector of the sets $\{1, 1, 1\}$, $\{2, 0, 0\}$. We have seen that in this case we have seven different parallel components, and therefore fourteen eigenfunctions – with $\boldsymbol{k}$ real or complex – for any given energy.

The point of the above discussion is the following. A matching calculation must ensure proper matching at *all* points of the matching plane in real space. For this the number of independent surface-projected wavefunctions needed to ensure proper matching – consistent, of course, with the starting approximation used for the bulk band structure – must be unambiguous. The present discussion pays particular attention to this point of principle, while using simple models for the band structure. Suppose that we have worked out the complete band structure in a given approximation – analytical continuation to complex momentum included – and we look at an energy range in which there are no solutions with real momentum. All the eigenfunctions are then exponential and only half of them have the appropriate decay into the bulk needed to form a physically acceptable solution, as follows from a trivial extension of (4.44) and (4.45). Thus the number of wavefunctions of the form (4.48), with $\phi_i(z)$ now decaying into the bulk, is reduced to exactly the number of different parallel components $\bar{\boldsymbol{G}}_i$, henceforth denoted as p.

As a specific example consider an abrupt-potential model (a.p.m.) with the surface barrier at $z = z_0$ and the bulk in $z > z_0$. This model is purely hypothetical, and, as it happens, unrealistic, but it is a well-defined and tractable model which serves to demonstrate the principles and techniques involved in a correct matching calculation. The question of its possible relevance, if any, will be discussed later. Let ψ_{kj} be the different wavefunctions of type (4.48) with appropriate exponential decay for $z > z_0$. The surface state wavefunction inside the crystal is of the form

$$\psi^{ss}(\boldsymbol{r}) = \sum_j c_j^{X} \psi_{kj}(\boldsymbol{r}) \quad (z > z_0). \tag{4.49}$$

For the time being we have the p inknown coefficients c_j^{X}. But we must also look at the wavefunction outside. For a given parallel component $\bar{\boldsymbol{G}}_i$, the solution of the free-electron Hamiltonian with proper exponential decay for $z \to -\infty$ is

$$\psi_i^{V} \propto \exp[i(\boldsymbol{\kappa} + \bar{\boldsymbol{G}}_i) \cdot \boldsymbol{\rho}] \exp[z(2W - 2E - |\boldsymbol{\kappa} + \bar{\boldsymbol{G}}_i|^2)^{1/2}], \tag{4.50}$$

where W is the vacuum level. The s.s. wavefunction in the vacuum is then of the form

$$\psi^{ss}(\boldsymbol{r}) = \sum_i c_i^{V} \psi_i^{V}(\boldsymbol{r}) \quad (z < z_0). \tag{4.51}$$

This introduces p further unknowns, namely, the coefficients c_i^{V}. We thus have $2p$ unknowns which must be found by matching (4.49) and (4.51). Indeed, matching the wavefunctions themselves yields p equations, obtained by equating the p independent components, and another p equations result from matching the normal derivatives. This yields $2p$ equations for $2p$ unknowns, which guarantees correct matching everywhere on the matching plane.

4.2.2. Particular examples of matching

In order to see how these principles work in practice it is convenient to carry out some explicit exercises. We shall take up the (1, 1, 1) and (1, 1, 0) surfaces of covalent (diamond structure) and ionic (zincblende structure) semiconductors. The zincblende structure is shown in fig. 65. It

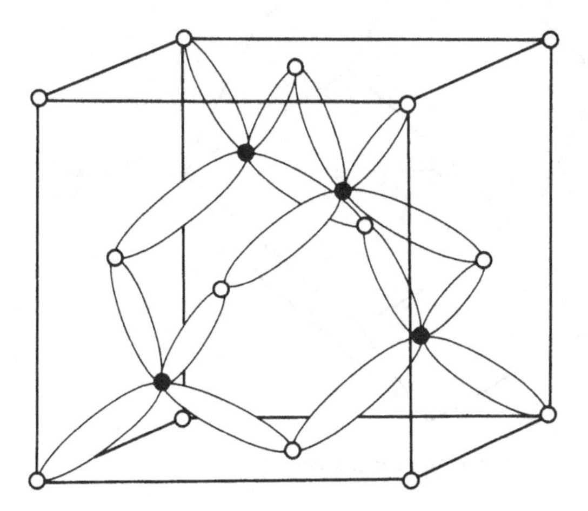

Fig. 65. Zincblende lattice.

consists of two interpenetrating f.c.c. lattices with different atoms. If the two atoms are identical then this becomes the diamond structure. In both cases there are two atoms and hence eight valence electrons per unit cell. The B.z. is that of the f.c.c. lattice. Another polyhedron of interest is the Jones zone (J.z.) (fig. 66). This is defined as the domain formed by the twelve planes normal to the set of vectors {2, 2, 0} and its own volume is just the extension of momentum space corresponding to eight electrons per unit cell. In these crystals the pseudopotential component $V(2, 2, 0)$ is significant, and all the states with extended momentum $\boldsymbol{k}$ inside the J.z. form the occupied valence band. Thus the fundamental gap appears in the extended momentum space at the J.z. boundary. Our attention will focus on surface states in this gap.

An accurate description of the bulk band structure may require many plane waves, but a not unreasonable low-order approximation is obtained in practice by including only those plane waves with momentum $\boldsymbol{k}+\boldsymbol{G}$ inside or near the J.z. edge. This involves a small number of terms in the plane-wave expansion while giving a reasonable qualitative picture of the bulk band structure. In this approximation the plane-wave states near the edge of the J.z. have the dominant weight in the buildup of the states near the gap. We shall make use of this in practice.

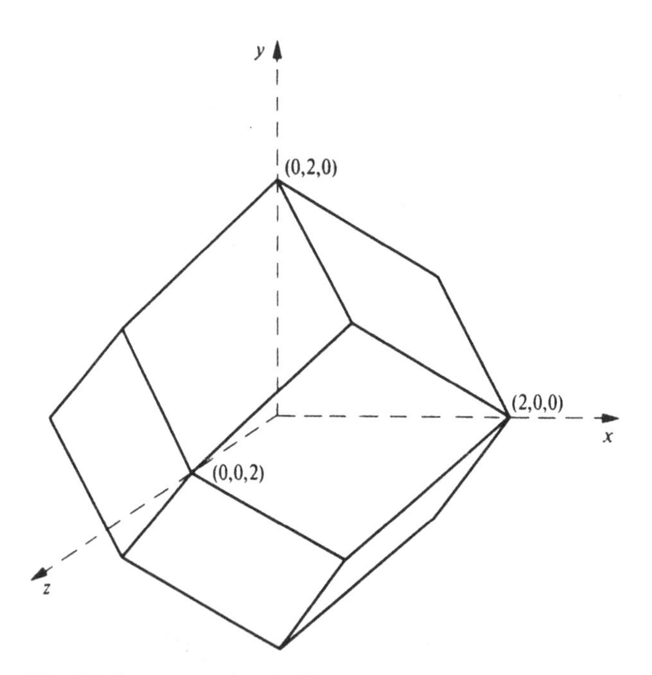

Fig. 66. Jones zone for the f.c.c. structure.

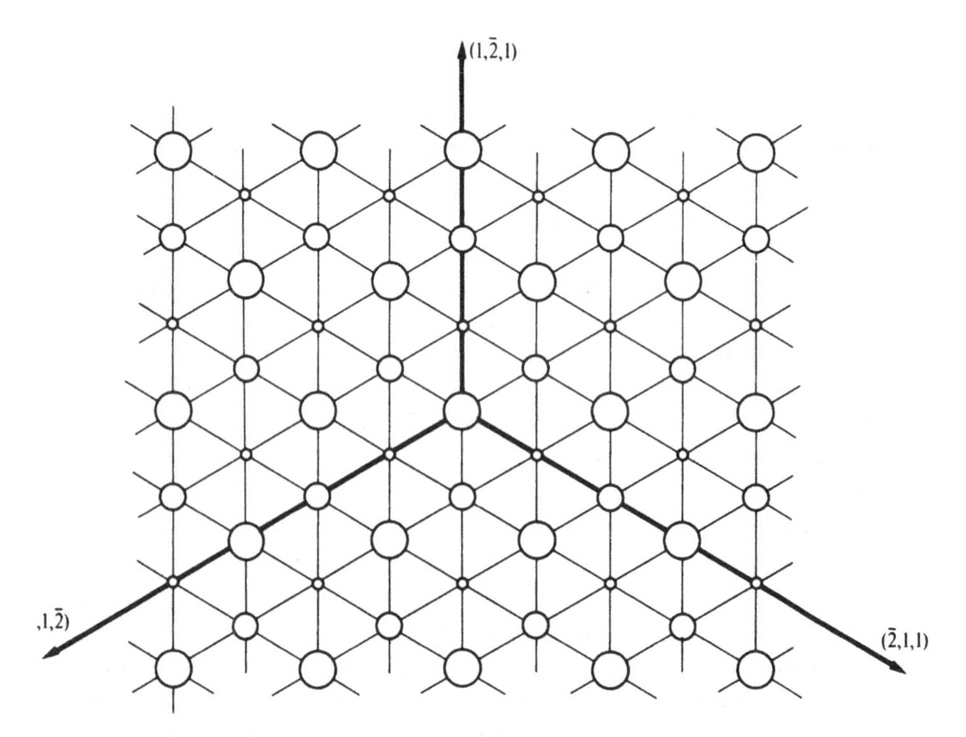

Fig. 67. Diamond lattice projected on the (1, 1, 1) surface. Atoms in successive atomic planes are indicated by circles of decreasing size.

Consider the (1, 1, 1) surface of a diamond structure (fig. 67). The two-dimensional B.z. is shown in fig. 68 (inner hexagon) together with the projection of the J.z. on this plane. Notice how the two-dimensional B.z. is formed: we first take the three-dimensional reciprocal lattice vectors $(1, 1, \bar{1})$, $(1, \bar{1}, 1)$, $(\bar{1}, 1, 1)$, then take their projections on the surface – which gives the directions $(1, \bar{2}, 1)$, $(\bar{2}, 1, 1)$, $(1, 1, \bar{2})$ – and then define the polygonal figure formed by the lines contained in the plane and normal to the above projections. Notice that the sets {2, 2, 0} and $\{\bar{2}, 0, 0\}$ have the same projections on the (1, 1, 1) plane and the shorter vectors {1, 0, 0} are not included in the reciprocal of the f.c.c. lattice. We now discuss in detail how to do the matching calculation for the point $\bar{J}$, the corner of the two-dimensional B.z. At this stage we introduce a different three-dimensional zone which proves more useful for a description of the band structure adapted to the study of the surface of interest. This zone is a prism with base the two-dimensional B.z. of fig. 68 and whose height is given by the reciprocal lattice vector (1, 1, 1). By taking account of the factor $2\pi/c$ it is easily seen that the volume of this zone, henceforth called the *surface prism*, is equal to the volume of the three-dimensional B.z.,

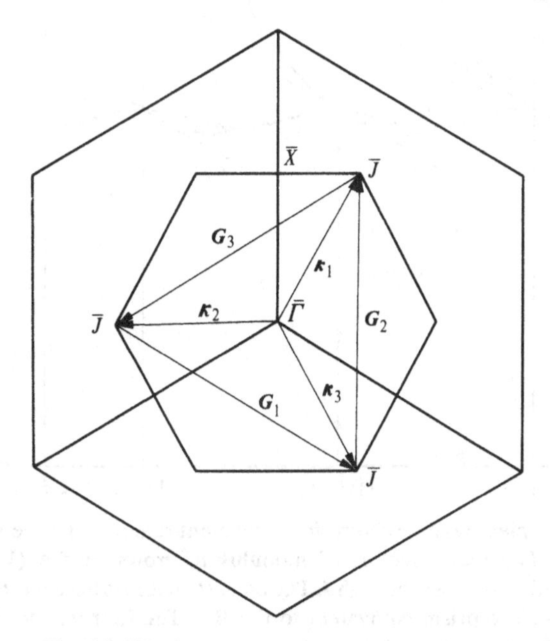

Fig. 68. The inner hexagon is the two-dimensional B.z. of the (1, 1, 1) face of the diamond lattice. The outer hexagon is the projection on this face of the corresponding J.z. The figure shows three equivalent $\bar{J}$ points with their corresponding values $\boldsymbol{\kappa}_i$, joined by the reciprocal lattice vectors $\boldsymbol{G}_i$.

and no two points of this domain are joined by a reciprocal lattice vector. The surface prism is thus a legitimate alternative to the B.z., equally valid as a description of the three-dimensional band structure, but more conveniently adapted to the surface under consideration. We situate the centre of inversion of this prism at the origin. Then, when we take the fixed value of $\boldsymbol{\kappa}$ corresponding to $\bar{J}$ and let the perpendicular component of the momentum grow in the (1, 1, 1) direction we span the segment which forms an edge of the surface prism with end points at $(7/6, \bar{1}/6, 1/2)$ and $(1/6, \bar{7}/6, \bar{1}/2)$, and middle point at $(2/3, \bar{2}/3, 0)$ in momentum space. Let us start by looking at the (bulk) band structure in the neighbourhood of this point (fig. 69). The first thing we must define is the number of plane waves to be included in our approximation for the

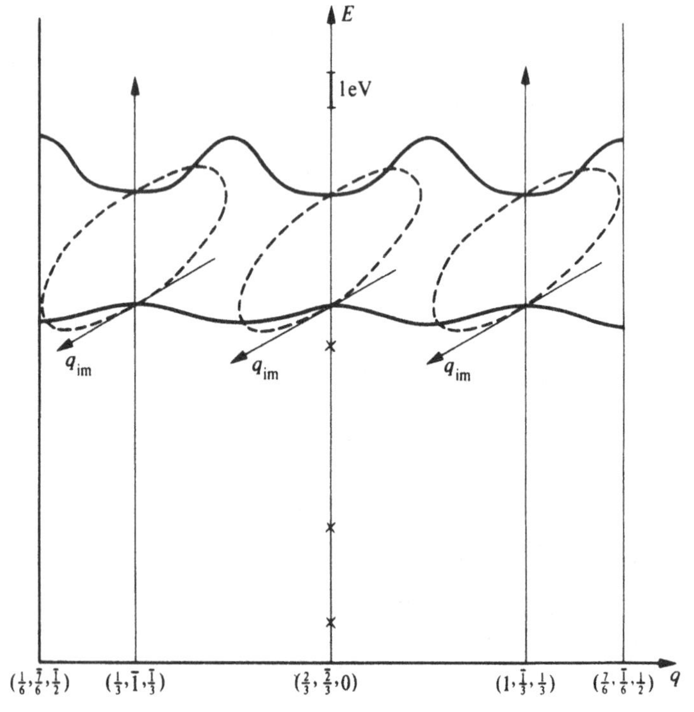

Fig. 69. Complex band structure for a covalent diamond lattice semiconductor. The wave vector, of modulus q, grows in the (1, 1, 1) direction and its tip goes through $\bar{J}$. The band structure is here plotted by using the surface prism equivalent to the B.z. The four valence band eigenvalues at the centre of the surface prism are the three crosses and the relative maximum immediately above: the valence band edge in this case. The two band edges on both sides of the fundamental gap are here shown explicitly, but this band structure is the result of an estimate based on the five-band approximation.

band structure. Starting from the plane wave of wavevector $(2/3, \bar{2}/3, 0)$ and adding reciprocal lattice vectors we generate the sequence of plane-wave states of wavevectors

$$\left(\frac{2}{3},\frac{\bar{2}}{3},0\right),\left(\frac{\bar{1}}{3},\frac{1}{3},1\right),\left(\frac{\bar{1}}{3},\frac{1}{3},\bar{1}\right),\left(\frac{2}{3},\frac{4}{3},0\right),\left(\frac{\bar{4}}{3},\frac{\bar{2}}{3},0\right),\left(\frac{\bar{4}}{3},\frac{4}{3},0\right),\ldots$$

The fourth and fifth vectors of this sequence correspond to points on the J.z. boundary in extended momentum space. Thus, consequently with the criterion described above – the J.z. approximation – we retain the first five plane waves of this sequence in the expansion of the Bloch wave functions, but we rearrange them in order of decreasing kinetic energy according to the following labelling

$$\begin{array}{ccccc} (1) & (2) & (3) & (4) & (5) \\ \left(\frac{2}{3},\frac{4}{3},0\right), & \left(\frac{\bar{4}}{3},\frac{\bar{2}}{3},0\right), & \left(\frac{\bar{1}}{3},\frac{1}{3},1\right), & \left(\frac{\bar{1}}{3},\frac{1}{3},\bar{1}\right), & \left(\frac{2}{3},\frac{\bar{2}}{3},0\right) \end{array} \tag{4.52}$$

Then (1) and (2) have energy

$$\tfrac{1}{2}k_0^2 = \tfrac{1}{2}[(\tfrac{2}{3})^2 + (\tfrac{4}{3})^2](2\pi/c)^2 \quad \text{etc.}$$

By taking (4.52) as a basis the Schrödinger equation takes the matrix form

$$\begin{bmatrix} -E+\frac{1}{2}k_0^2 & V_8 & V_3 & -V_3 & 0 \\ V_8 & -E+\frac{1}{2}k_0^2 & -V_3 & V_3 & 0 \\ V_3 & -V_3 & -E+\frac{1}{2}k_1^2 & 0 & V_3 \\ -V_3 & V_2 & 0 & -E+\frac{1}{2}k_1^2 & V_3 \\ 0 & 0 & V_3 & V_3 & -E+\frac{1}{2}k_2^2 \end{bmatrix} \begin{bmatrix} a_1 \\ a_2 \\ a_3 \\ a_4 \\ a_5 \end{bmatrix} = 0 \tag{4.53}$$

The a_is are the coefficients of the plane waves (4.52) in the expansion (4.47), V_8 is $V(2, 2, 0)$ and V_3 is $V(1, \bar{1}, \bar{1})$. We recall that the crystal pseudopotential component $V(\boldsymbol{G})$ is given by the product $S(\boldsymbol{G})v(\boldsymbol{G})$ of the structure factor $S(\boldsymbol{G})$ and the form factor $v(\boldsymbol{G})$. In a covalent crystal $v(\boldsymbol{G})$ is the Fourier transform of the only atomic potential, normalised to the volume of the unit cell. The structure factor for the diamond structure, with the origin at the middle point of a bond between two neighbouring atoms, is

$$S(\boldsymbol{G}) = 2\cos[\tfrac{1}{4}(l+m+n)\pi]; \quad \boldsymbol{G} = (l, m, n)2\pi/c. \tag{4.54}$$

In the notation of (4.52) and (4.53) the basis vectors are [1, 0, 0, 0, 0], [0, 1, 0, 0, 0], ... with amplitudes a_1, a_2, ... It is convenient here to change to the basis

$$\left[\frac{1}{\sqrt{2}}, \frac{1}{\sqrt{2}}, 0, 0, 0\right], \left[0, 0, \frac{1}{\sqrt{2}}, \frac{1}{\sqrt{2}}, 0\right], \left[0, 0, 0, 0, 1\right],$$
$$\left[\frac{1}{\sqrt{2}}, \frac{\bar{1}}{\sqrt{2}}, 0, 0, 0\right], \left[0, 0, \frac{\bar{1}}{\sqrt{2}}, \frac{1}{\sqrt{2}}, 0\right], \tag{4.55a}$$

with amplitudes $b_1, \ldots, b_5$. Then (4.53) factorises into the equations

$$E = \tfrac{1}{2}k_0^2 + V_8 \quad (b_2 = b_3 = b_4 = b_5 = 0), \tag{4.55b}$$

$$\begin{bmatrix} -E + \frac{1}{2}k_1^2 & \sqrt{2}\,V_3 \\ \sqrt{2}\,V_3 & -E + \frac{1}{2}k_2^2 \end{bmatrix} \begin{bmatrix} b_2 \\ b_3 \end{bmatrix} = 0 \quad (b_1 = b_4 = b_5 = 0), \tag{4.55c}$$

$$\begin{bmatrix} -E + \frac{1}{2}k_0^2 - V_8 & -2V_3 \\ -2V_3 & -E + \frac{1}{2}k_1^2 \end{bmatrix} \begin{bmatrix} b_4 \\ b_5 \end{bmatrix} = 0, \quad (b_1 = b_2 = b_3 = 0). \tag{4.55d}$$

These yield the eigenvalues of the bulk band structure at the point (2/3, $\bar{2}$/3, 0) shown in fig. 69 for $V_8 = -0.54$ eV, $V_3 = -2.02$ eV and $c = 5.43$ Å (0.543 nm). In order to study the band structure for $\boldsymbol{k}$ growing in the (1, 1, 1) direction and passing through this point – i.e. spanning the edge of the surface prism described above – we add the wavevector $q(1, 1, 1)$ to the wavevectors of (4.52). This yields the set,

$$\left(\frac{2}{3}+q, \frac{4}{3}+q, q\right), \left(\frac{\bar{4}}{3}+q, \frac{\bar{2}}{3}+q, q\right), \left(\frac{\bar{1}}{3}+q, \frac{1}{3}+q, 1+q\right),$$
$$\left(\frac{\bar{1}}{3}+q, \frac{1}{3}+q, \bar{1}+q\right), \left(\frac{2}{3}+q, \frac{\bar{2}}{3}+q, q\right),$$

and combining the corresponding plane-wave states as in (4.55a) we obtain a basis in which the Schrödinger equation has the matrix representation

$$\begin{bmatrix} -E + \frac{1}{2}k_0^2 + \frac{3}{2}q^2 + V_8 & 0 & 0 & 2\sqrt{2}\,q\omega & 0 \\ 0 & -E + \frac{1}{2}k_1^2 + \frac{3}{2}q^2 & \sqrt{2}\,V_3 & 0 & \sqrt{2}\,q\omega \\ 0 & \sqrt{2}\,V_3 & -E + \frac{1}{2}k_2^2 + \frac{3}{2}q^2 & 0 & 0 \\ 2\sqrt{2}\,q\omega & 0 & 0 & -E + \frac{1}{2}k_0^2 + \frac{3}{2}q^2 - V_8 & -2V_3 \\ 0 & \sqrt{2}\,q\omega & 0 & -2V_3 & -E + \frac{1}{2}k_1^2 + \frac{3}{2}q^2 \end{bmatrix} \begin{bmatrix} b_1 \\ b_2 \\ b_3 \\ b_4 \\ b_5 \end{bmatrix} = 0, \tag{4.56}$$

where $\omega = 2\pi/c$. The fundamental gap is the energy interval between the two highest of the five bands resulting from (4.56), which becomes (4.55) for $q = 0$. We now want to study the complex band structure associated with this gap, for which it proves convenient to eliminate b_2 and b_3 from (4.56). This yields

$$\begin{bmatrix} -E+\frac{1}{2}k_0^2+\frac{3}{2}q^2+V_8 & 2\sqrt{2}\,q\omega & 0 \\ 2\sqrt{(2q)}\omega & -E+\frac{1}{2}k_0^2+\frac{3}{2}q^2-V_8 & -2V_3 \\ 0 & -2V_3 & -E+\frac{1}{2}k_1^2+\frac{3}{2}q^2+\Delta \end{bmatrix}\begin{bmatrix} b_1 \\ b_4 \\ b_5 \end{bmatrix} = 0, \tag{4.57}$$

where

$$\Delta = \frac{(-E+\frac{1}{2}k_2^2+\frac{3}{2}q^2)(\sqrt{2}\,q\omega)^2}{\begin{vmatrix} -E+\frac{1}{2}k_1^2+\frac{3}{2}q^2 & \sqrt{2}\,V_3 \\ \sqrt{2}\,V_3 & -E+\frac{1}{2}k_2^2+\frac{3}{2}q^2 \end{vmatrix}}. \tag{4.58}$$

The band structure is given by the secular equation of (4.57), which is

$$(-E+\tfrac{1}{2}k_0^2+\tfrac{3}{2}q^2+V_8)\times\left(-E+\frac{1}{2}k_0^2+\tfrac{3}{2}q^2-V_8+\frac{4V_3^2}{E-\frac{1}{2}k_1^2-\frac{3}{2}q^2+\Delta}\right) = 8q^2\omega^2. \tag{4.59}$$

This is considerably simplified by noticing that Δ and $\frac{3}{2}q^2$, for energies near the gap, are neglīgible compared with $E-\frac{1}{2}k_0^2$ and $E-\frac{1}{2}k_0^2$. We then have

$$(-E+\tfrac{1}{2}k_0^2+V_8)\left(-E+\frac{1}{2}k_0^2-V_8+\frac{4V_3^2}{E-\frac{1}{2}k_1^2}\right) = 8q^2\omega^2. \tag{4.60}$$

The corresponding wavefunction is obtained by making the same simplification in (4.57) and using there the eigenvalue obtained from (4.60). The resulting band structure is shown in fig. 69. Notice in partiicular the extrema appearing at the middle point. The equation for the real energy loop joining these two extrema is given by (4.60), if we allow for imaginary values of q. The evanescent wavefunction along this loop is

$$\psi_1(\mathbf{r}) = \mathrm{e}^{-rz}\left\{\begin{bmatrix} 1 \\ 0 \\ \dfrac{V_3}{E-\frac{1}{2}k_1^2} \\ \dfrac{-V_3}{E-\frac{1}{2}k_1^2} \\ 0 \end{bmatrix} + a\begin{bmatrix} 0 \\ -1 \\ \dfrac{V_3}{E-\frac{1}{2}k_1^2} \\ \dfrac{-V_3}{E-\frac{1}{2}k_1^2} \\ 0 \end{bmatrix}\right\} \tag{4.61}$$

where

$$a=\frac{1-\mathrm{i}\sqrt{\left[\left(-E+\frac{1}{2}k_0^2-V_8+\dfrac{4V_3^2}{E-\frac{1}{2}k_1^2}\right)\Big/(E-\frac{1}{2}k_0^2-V_8)\right]}}{1+\mathrm{i}\sqrt{\left[\left(-E+\frac{1}{2}k_0^2-V_8+\dfrac{4V_3^2}{E-\frac{1}{2}k_1^2}\right)\Big/(E-\frac{1}{2}k_0^2-V_8)\right]}}, \tag{4.62a}$$

and

$$r=\frac{1}{2\sqrt{2}\omega}\sqrt{\left[(E-\tfrac{1}{2}k_0^2-V_8)\left(-E+\tfrac{1}{2}k_0^2-V_8+\frac{4V_3^2}{E-\frac{1}{2}k_1^2}\right)\right]}. \tag{4.62b}$$

Of course z is the coordinate in the direction perpendicular to the surface. The sign of r is chosen to ensure correct exponential decay inside the crystal. A reasonable approximation consists in replacing the term

$$\tfrac{1}{2}k_0^2-V_8+\frac{4V_3^2}{E-\frac{1}{2}k_1^2}$$

by its value in the conduction band, where this expression is just E_C. This yields

$$a=\frac{1-\mathrm{i}\sqrt{[(E_C-E)/(E-E_V)]}}{1+\mathrm{i}\sqrt{[(E_C-E)/(E-E_V)]}}, \tag{4.63a}$$

and

$$r=(1/2\sqrt{2}\,\omega)\sqrt{[(E-E_V)(E_C-E)]}. \tag{4.63b}$$

Compare with the one-dimensional result (4.8).

This describes the loop at the middle point. The corresponding evanescent wavefunction will contribute to the s.s. wavefunction in this gap, together with other loops appearing in fig. 69. We have insisted above that, irrespective of the model used for the bulk band structure, a correct matching requires including all the loops which yield evanescent wavefunctions for a given surface. We have seen that the total number of loops is equal to twice the number of different parallel components entering the plane-wave expansion of the wavefunction. In this case we must look at the number of independent parallel projections of the wavevectors of (4.52). These are the three vectors $\boldsymbol{\kappa}_1$, $\boldsymbol{\kappa}_2$, $\boldsymbol{\kappa}_3$ of fig. 68. Define

$$f_i=\exp\mathrm{i}(\boldsymbol{\kappa}_i\cdot\boldsymbol{\rho})\quad(i=1,2,3). \tag{4.64}$$

Then (4.61) can be rewritten as

$$\psi_1=\psi_1^1f_1+\psi_1^2f_2+\psi_1^3f_3, \tag{4.65}$$

where

$$\psi_1^1 = \exp(-rz)\left[\exp\left(\frac{i2hz}{3}\right) - \frac{V_3}{E-\frac{1}{2}k_1^2}(1+a)\exp\left(-\frac{ihz}{3}\right)\right];$$

$$\psi_1^2 = \exp(-rz)\left[-a\exp\left(-\frac{i2hz}{3}\right) + \frac{V_3}{E-\frac{1}{2}k_1^2}(1+a)\exp\left(\frac{ihz}{3}\right)\right]; \qquad (4.66)$$

$$\psi_1^3 = 0,$$

and h is the modulus of the reciprocal lattice vector (1, 1, 1). The five plane waves making up the wavefunction actually involve only three different parallel projections, giving six complex momentum loops. But only three of these have the correct sign in the imaginary component to ensure exponential decay for $z \to +\infty$. We need another two evanescent wavefunctions besides (4.65) and (4.66). In order to find these we notice that on moving away from the middle point $(2/3, \bar{2}/3, 0)$ in the (1, 1, 1) direction (fig. 69) we go through the points $(1, \bar{1}/3, 1/3)$ and $(1/3, \bar{1}, \bar{1}/3)$ at which the other two loops of interest appear. Adding reciprocal lattice vectors to these points we obtain the sequences of wavevectors

$$\left(\frac{1}{3}, \bar{1}, \frac{\bar{1}}{3}\right), \left(\frac{4}{3}, 0, \frac{2}{3}\right), \left(\frac{\bar{2}}{3}, 0, \frac{2}{3}\right), \left(\frac{1}{3}, 1, \frac{\bar{1}}{3}\right), \left(\frac{\bar{2}}{3}, 0, \frac{\bar{4}}{3}\right), \ldots$$

and

$$\left(1, \frac{\bar{1}}{3}, \frac{1}{3}\right), \left(0, \frac{\bar{4}}{3}, \frac{\bar{2}}{3}\right), \left(0, \frac{2}{3}, \frac{\bar{2}}{3}\right), \left(\bar{1}, \frac{\bar{1}}{3}, \frac{1}{3}\right), \left(0, \frac{2}{3}, \frac{4}{3}\right), \ldots$$

These are precisely the vectors obtained by taking the sequence (4.52) associated with the middle point and rotating its members through $\pm 120°$ about the (1, 1, 1) axis. Thus these two loops correspond to the same eigenvalue structure and can be described by a suitable adaptation of (4.66). In particular, in the notation of (4.65), the components of the evanescent wavefunctions associated with these loops are

$$\psi_2^1 = 0; \quad \psi_2^2 = \psi_1^1; \quad \psi_2^3 = \psi_1^2, \qquad (4.67)$$

and

$$\psi_3^1 = \psi_1^2; \quad \psi_3^2 = 0; \quad \psi_3^3 = \psi_1^1. \qquad (4.68)$$

In the a.p.m. we must match these three wavefunctions at $z = z_0$ with the three plane waves

$$\psi_i^V(\boldsymbol{r}) = \exp(i\boldsymbol{\kappa}_i \cdot \boldsymbol{\rho})\exp[z\sqrt{(W-E+\tfrac{1}{2}\kappa^2)}] \quad (i = 1, 2, 3). \qquad (4.69)$$

In other words, we write down the s.s. wavefunction as

$$\psi^{\text{ss}}=\left\{\begin{array}{ll} c_1^{\text{X}}\psi_1+c_2^{\text{X}}\psi_2+c_3^{\text{X}}\psi_3 & (z>z_0),\\ c_1^{\text{V}}\psi_1^{\text{V}}+c_2^{\text{V}}\psi_2^{\text{V}}+c_3^{\text{V}}\psi_3^{\text{V}} & (z<z_0),\end{array}\right\} \tag{4.70}$$

and match at z_0. Let us write ψ^{ss} in matrix form as follows

$$\psi^{\text{ss}}=\begin{bmatrix}\psi_1^1 & \psi_2^1 & \psi_3^1\\ \psi_1^2 & \psi_2^2 & \psi_3^2\\ \psi_1^3 & \psi_2^3 & \psi_3^3\end{bmatrix}\begin{bmatrix}c_1^{\text{X}}\\ c_2^{\text{X}}\\ c_3^{\text{X}}\end{bmatrix}\quad (z>z_0), \tag{4.71a}$$

and

$$\psi^{\text{ss}}=\exp\left[z\sqrt{(W-E+\tfrac{1}{2}\kappa^2)}\right]\begin{bmatrix}1 & 0 & 0\\ 0 & 1 & 0\\ 0 & 0 & 1\end{bmatrix}\begin{bmatrix}c_1^{\text{V}}\\ c_2^{\text{V}}\\ c_3^{\text{V}}\end{bmatrix}\quad (z<z_0). \tag{4.71b}$$

Putting Λ^{X} and Λ^{V} for the matrices of (4.71a) and (4.71b) the two matching conditions read, in synthetic notation,

$$\Lambda^{\text{X}}c^{\text{X}}=\Lambda^{\text{V}}c^{\text{V}}\quad (z=z_0), \tag{4.72a}$$

and

$$\frac{\mathrm{d}\Lambda^{\text{X}}(z)}{\mathrm{d}z}c^{\text{X}}=\frac{\mathrm{d}\Lambda^{\text{V}}(z)}{\mathrm{d}z}c^{\text{V}}\quad (z=z_0). \tag{4.72b}$$

The compatibility condition is (with subindex z_0 understood)

$$\operatorname{Det}\left[\frac{\mathrm{d}\Lambda^{\text{X}}}{\mathrm{d}z}-\frac{\mathrm{d}\Lambda^{\text{V}}}{\mathrm{d}z}(\Lambda^{\text{V}})^{-1}\Lambda^{\text{X}}\right]=0. \tag{4.73}$$

By neglecting $r\ll h$ this yields the s.s. eigenvalue equation (remember (4.63a))

$$a^3\equiv\left[\frac{(1-\mathrm{i}\tfrac{2}{3}h\mathscr{L}_0)-\dfrac{V_3}{E-\frac{1}{2}k_1^2}(1+a)\left(1+\mathrm{i}\dfrac{h}{3}\mathscr{L}_0\right)\exp(-\mathrm{i}hz_0)}{(1+\mathrm{i}\tfrac{2}{3}h\mathscr{L}_0)-\dfrac{V_3}{E-\frac{1}{2}k_1^2}(1+a^*)\left(1-\mathrm{i}\dfrac{h}{3}\mathscr{L}_0\right)\exp(\mathrm{i}hz_0)}\right]^3\exp(\mathrm{i}4hz_0), \tag{4.74}$$

where

$$\mathscr{L}_0=1/\sqrt{[2(W-E+\tfrac{1}{2}\kappa^2)]}. \tag{4.75}$$

It is interesting to compare this with the s.s. eigenvalue equation (4.9) for the two-band one-dimensional model. The essential difference comes from the terms containing the factor $V_3/(E-k_1^2/2)$. The point is that we have used a five-band approximation. Let us discuss what would happen with a two-band approximation for the three-dimensional problem. The evanescent wavefunctions are then simply obtained by setting $V_3=0$ in (4.65)–(4.58) and the s.s. eigenvalue equation results from setting $V_3=0$ in (4.74). This yields three separate equations, namely

$$a=\frac{1-\mathrm{i}\frac{2}{3}h\mathcal{L}_0}{1+\mathrm{i}\frac{2}{3}h\mathcal{L}_0}\exp\left(\mathrm{i}\tfrac{4}{3}hz_0\right), \tag{4.76}$$

and

$$a=\frac{1-\mathrm{i}\frac{2}{3}h\mathcal{L}_0}{1+\mathrm{i}\frac{2}{3}h\mathcal{L}_0}\exp\left[\mathrm{i}\left(\tfrac{4}{3}hz_0\pm\tfrac{2}{3}\pi\right)\right]. \tag{4.77}$$

Equation (4.76) is just like (4.9) of the one-dimensional case with the changes

$$g\to\tfrac{4}{3}h;\quad \sqrt{[2(W-E)]}\to\sqrt{[2(W-E+\tfrac{1}{2}\kappa^2)]}.$$

Furthermore, the two equations (4.77) are just like (4.76) with an extra phase $\pm\frac{2}{3}\pi$ in the complex exponential. We can think of (4.76) as effectively a one-dimensional problem with a potential

$$V_{8,\mathrm{eff}}\exp\left(\mathrm{i}\tfrac{4}{3}hz\right)+\mathrm{c.c.}, \tag{4.78a}$$

while (4.77) corresponds to the 'rotated' effective potentials

$$V_{8,\mathrm{eff}}\exp\left[\mathrm{i}\tfrac{4}{3}hz\pm\tfrac{2}{3}\pi)\right]. \tag{4.78b}$$

The value of $V_{8,\mathrm{eff}}$ should be adjusted to reproduce the width of the fundamental gap in the one-dimensional model. To illustrate this viewpoint compare (4.74) with (4.76) and (4.77). Both cases involve a, which by (4.63a) depends on E_C and E_V, which in turn depend on the pseudopotential components. Strictly speaking, if we set $V_3=0$ in (4.74) then we should allow for the consequent changes in E_C and E_V. The effective two-band approximation implied here consists in setting $V_3=0$ in the r.h.s. of (4.74) but using in the evaluation of a the values of E_C and E_V obtained in a more elaborate model. It will also mean putting $V_3=0$ in (4.66), while using a more accurate value of a everywhere else. We can then expect to have a decent approximation to the wavefunction for a fairly realistic gap. This is the meaning of $V_{8,\mathrm{eff}}$ in (4.78). For example, in the five-band model described above, the fundamental gap is defined by the energy levels

$$E_\mathrm{C}=\tfrac{1}{2}k_0^2-V_8+\frac{4V_3^2}{E_\mathrm{C}-\frac{1}{2}k_1^1};\quad E_\mathrm{V}=\tfrac{1}{2}k_0^2+V_8.$$

Then

$$V_{8,\mathrm{eff}}=V_8-\frac{2V_3^2}{E_\mathrm{C}-\frac{1}{2}k_1^2}\approx V_8-\frac{4V_3^2}{k_0^2-k_1^2}.$$

This is equivalent to a perturbative calculation of E_C and E_V to second order in V_3. The idea of the effective two-band approximation is to combine the simplicity of the two-band model with a more accurate estimate for the width of the gap of interest. In practice this device is not unreasonable, yielding s.s. eigenvalues whose errors compared with the results of higher-order calculations, are of the order of one tenth of the width of the gap.

Summing up the results of the matching exercise carried out for the $\bar{J}$ point of the (1, 1, 1) covalent surface we find: (i) complete matching at all points of the matching plane can be achieved by detailed analysis of the complex band structure, (ii) the eigenvalue equation for s.s. is given in the five-band approximation by (4.74), (iii) in the effective two-band approximation the three-dimensional problem can be factorised into three effectively 'rotated' one-dimensional problems. This fact will be used later for a simplified analysis of the general properties of (1, 1, 1) covalent surfaces.

Let us now consider a rather different case, namely, a (1, 1, 0) ionic surface (fig. 70). Now we have two atoms – anion and cation – in one surface unit cell. Figure 71 shows the surface projection of the three-dimensional J.z. together with the two-dimensional B.z., obtained here from the surface projections of the vectors (2, 0, 0) and (1, 1, 1). We want to study s.s. at the point $\bar{X}$. This means studying the complex band structure for fixed $\boldsymbol{\kappa}$ – corresponding to $\bar{X}$ – and the perpendicular

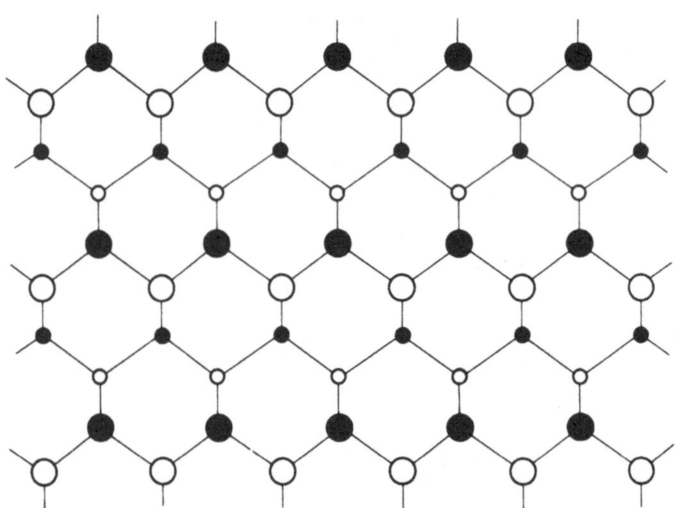

Fig. 70. Zincblende lattice projected on the (1, 1, 0) face. Open and solid circles indicate the two species in the crystal. Smaller circles indicate deeper atomic planes.

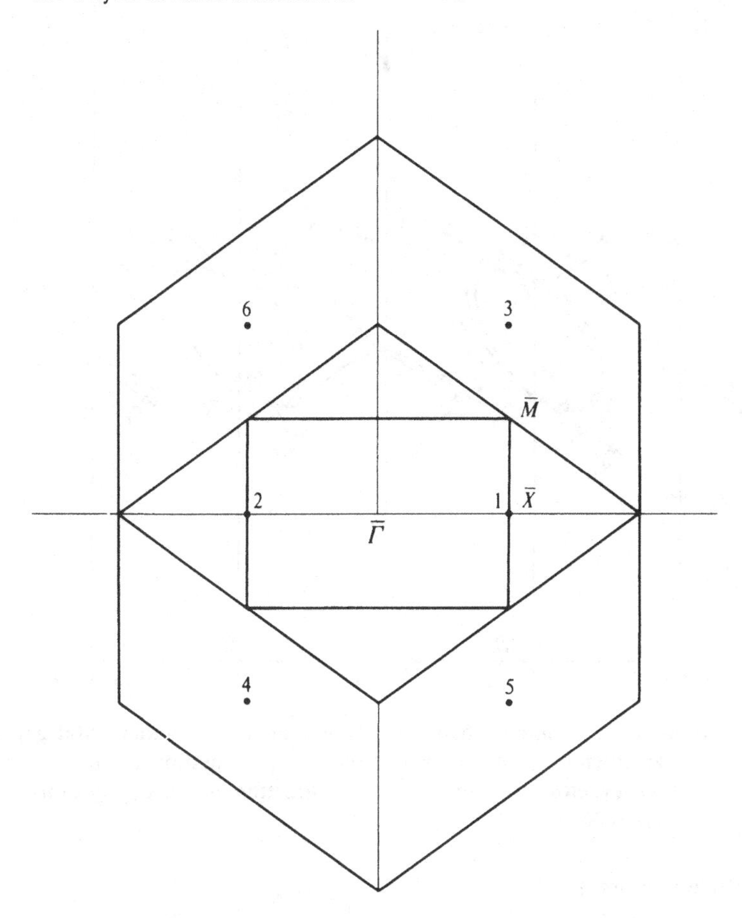

Fig. 71. The inner rectangle is the two-dimensional B.z. for the (1, 1, 0) surface of the zincblende, or diamond, lattice. The outer hexagon is the corresponding surface projection of the J.z. The projections of other faces of this zone are also shown. The figure shows the six points in this zone, labelled 1 to 6, which are equivalent to $\bar{X}$.

component of $\boldsymbol{k}$ varying in the (1, 1, 0) direction and going through $\bar{X}$. We define again the surface prism formed by the two-dimensional B.z. and the perpendicular vector (2, 2, 0) with the centre of the prism at the origin in $\boldsymbol{k}$-space. Again we study the band structure for $\boldsymbol{k}$ spanning an edge of the surface prism with middle point (fig. 72) at $(\bar{1}/2, 1/2, 0)$ and extremes at $(1/2, 3/2, 0)$ and $(\bar{3}/2, \bar{1}/2, 0)$. Also interesting in this line are the points (0, 1, 0) and $(\bar{1}, 0, 0)$ at which band extrema appear.

Consider the point $(\bar{1}, 0, 0)$: the point X of the three-dimensional band structure. The sequence of wavevectors formed by adding reciprocal

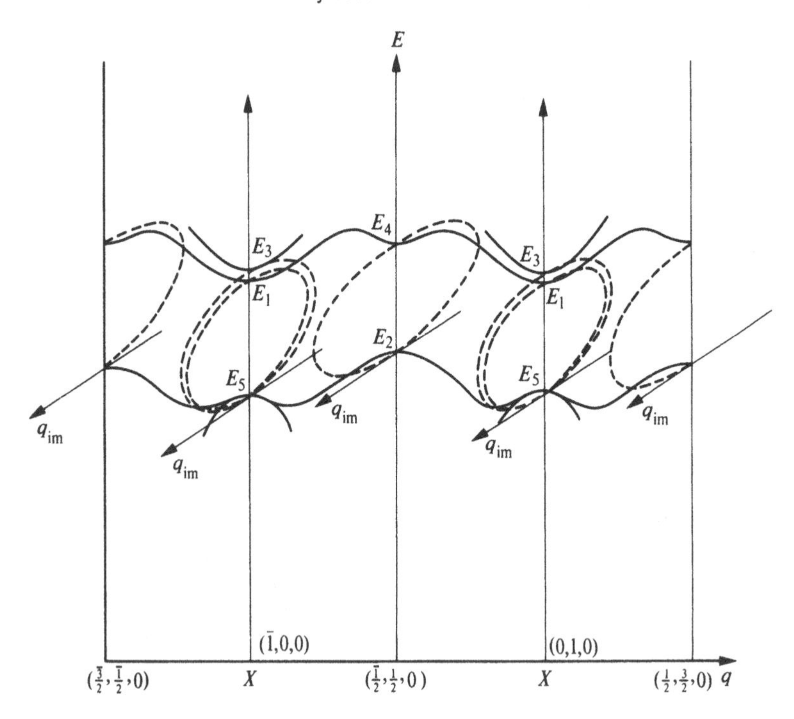

Fig. 72. Complex band structure around the fundamental gap for a zincblende crystal in a five-band approximation as in fig. 69. The wavevector grows in the (1, 1, 0) direction and its tip goes through the point $\bar{X}$.

lattice vectors is

$$(\bar{1},0,0),(1,0,0),(0,1,1),(0,\bar{1},\bar{1}),(0,1,\bar{1}),$$
$$(0,\bar{1},1),(1,0,\bar{2}),(1,2,0),(\bar{1},0,\bar{2}),\ldots \tag{4.79}$$

A reasonable approximation is obtained by using only the first six vectors of this sequence, as the rest are well past the J.z. boundary. The analysis of the eigenstates at this point shows that the highest valence-band eigenvalue, E_5 in fig. 72, is twofold degenerate, while in the conduction band there are two rather close levels E_1 and E_3. In a covalent semiconductor these two levels are degenerate. The difference between the two types of crystals comes from the difference in pseudopotential components. In general in an ionic semiconductor with zincblende structure the crystalline pseudopotential components are

$$V(\boldsymbol{G}) \equiv V^{S}(\boldsymbol{G})+\mathrm{i}V^{A}(\boldsymbol{G}) = v^{S}(\boldsymbol{G})\cos[(l+m+n)\pi/4]$$
$$+\mathrm{i}v^{A}(\boldsymbol{G})\sin[(l+m+n)\pi/4], \tag{4.80}$$

where

$$v^S(\boldsymbol{G}) = v_1(\boldsymbol{G}) + v_2(\boldsymbol{G}); \quad v^A(\boldsymbol{G}) = v_1(\boldsymbol{G}) = v_1(\boldsymbol{G}) - v_2(\boldsymbol{G}). \tag{4.81}$$

Here 1, 2 label the two species of atoms. The essential difference is the imaginary antisymmetric part, which vanishes in the covalent case, when v_1 and v_2 are equal. This is essentially the cause of the qualitative differences between ionic and covalent semiconductors. The study of the band structure about the point X is carried out in much the same way as in the (1, 1, 1) surface, this time with the addition of the vector $q(1, 1, 0)$ to the vectors of (4.79). We find two complex momentum loops joining the twofold degenerate level E_5 (fig. 72) to the conduction-band minima E_1 and E_3. In order to study the wavefunctions along these loops we define the parallel components (fig. 71)

$\boldsymbol{\kappa}_1$,	$\boldsymbol{\kappa}_2$,	$\boldsymbol{\kappa}_3$	$\boldsymbol{\kappa}_4$,	$\boldsymbol{\kappa}_5$,	$\boldsymbol{\kappa}_6$,
$(\bar{1}, 0, 0)$,	$(1, 0, 0)$,	$(0, 1, 1)$,	$(0, \bar{1}, \bar{1})$,	$(0, 1, \bar{1})$,	$(0, \bar{1}, 1)$

and the corresponding two-dimensional plane waves as in (4.64), with $i = 1, \ldots, 6$. After a similar analysis we find for the wavefunction ψ_1 of the loop linking E_5 and E_1, *in the notation of* (4.65),

$$\left.\begin{aligned}
\psi_1^1 &= \tfrac{1}{2}\exp(-s_1 z)(1-\mathrm{i})\alpha_4 \exp(-\tfrac{1}{2}\mathrm{i}h_1 z)(1+a_1);\\
\psi_1^2 &= \tfrac{1}{2}\exp(-s_1 z)(1+\mathrm{i})\alpha_4 \exp(\tfrac{1}{2}\mathrm{i}h_1 z)(1+a_1);\\
\psi_1^3 &= \mathrm{i}\exp(-s_1 z)\exp(\tfrac{1}{2}\mathrm{i}h_1 z);\\
\psi_1^4 &= -\mathrm{i}\exp(-s_1 z)a_1 \exp(-\tfrac{1}{2}\mathrm{i}h_1 z);\\
\psi_1^5 &= \exp(-s_1 z)\exp(\tfrac{1}{2}\mathrm{i}h_1 z);\\
\psi_1^6 &= \exp(-s_1 z)a_1 \exp(-\tfrac{1}{2}\mathrm{i}h_1 z),
\end{aligned}\right\} \tag{4.82}$$

where h_1 is the modulus of the vector (1, 1, 0),

$$s_1 = \frac{\sqrt{[(E_1 - E)(E - E_5)]}}{2h_1}; \quad \alpha_4 = \frac{V_3^S + V_3^A}{E_1 - \frac{1}{2}V_4^A - \frac{1}{2}h_1^2},$$

a_1 is (4.63*a*) with $E_C = E_1$, $E_V = E_5$ and V_4 is $V(2, 0, 0)$. For the wavefunction ψ_2 of the loop linking E_5 and E_3 we find

$$\left.\begin{aligned}
\psi_2^1 &= \tfrac{1}{2}\exp(-s_2 z)(1+\mathrm{i})\alpha_3 \exp(-\tfrac{1}{2}\mathrm{i}h_1 z)(1+a_2);\\
\psi_2^2 &= \tfrac{1}{2}\exp(-s_2 z)(1-\mathrm{i})\alpha_3 \exp(\tfrac{1}{2}\mathrm{i}h_1 z)(1+a_2);\\
\psi_2^3 &= -\mathrm{i}\exp(-s_2 z)\exp(\tfrac{1}{2}\mathrm{i}h_1 z);\\
\psi_2^4 &= \mathrm{i}a_2 \exp(-s_2 z)\exp(-\tfrac{1}{2}\mathrm{i}h_1 z);\\
\psi_2^5 &= \exp(-s_2 z)\exp(\tfrac{1}{2}\mathrm{i}h_1 z);\\
\psi_2^6 &= \exp(-s_2 z)a_2 \exp(\tfrac{1}{2}\mathrm{i}h_1 z),
\end{aligned}\right\} \tag{4.83}$$

where s_2 and a_2 are s_1 and a_1 with E_1 replaced by E_3, and

$$\alpha_3 = \frac{V_3^{\mathrm{S}} - V_3^{\mathrm{A}}}{E_1 + \frac{1}{2}V_4^{\mathrm{A}} - \frac{1}{2}h_1^2}.$$

Since the number of independent parallel components in this case is six, we still need another four loops and associated evanescent wavefunctions. Two of these are immediately found by noticing that if we start from the point (0, 1, 0) and add reciprocal lattice vectors we obtain a sequence of wavevectors which are (in a different order) precisely those of (4.79) with the (x, y) coordinates interchanged. Thus the band structure at this point is the same as at the point $(\bar{1}, 0, 0)$ and the wavefunctions in the two loops are constructed from the same components. Notice, however, that these two points are not equivalent, i.e. they are not joined by a reciprocal lattice vector of the f.c.c. lattice. The band structure is shown in fig. 72 and for the wavefunctions we find, after suitable rearrangement,

$$\psi_3^1 = \psi_1^2, \quad \psi_3^2 = \psi_1^1, \quad \psi_3^3 = \psi_1^6, \quad \psi_3^4 = \psi_1^5, \quad \psi_3^5 = \psi_1^4, \quad \psi_3^6 = \psi_1^3, \tag{4.84}$$

and

$$\psi_4^1 = \psi_2^2, \quad \psi_4^2 = \psi_2^1, \quad \psi_4^3 = \psi_2^6, \quad \psi_4^4 = \psi_2^5, \quad \psi_4^5 = \psi_2^4, \quad \psi_4^6 = \psi_2^3. \tag{4.85}$$

Consider now the loop at $(\bar{1}/2, 1/2, 0)$. Adding reciprocal lattice points we find the sequence of wavevectors

$$\left(\frac{\bar{1}}{2}, \frac{1}{2}, 0\right), \left(\frac{3}{2}, \frac{1}{2}, 0\right), \left(\frac{\bar{1}}{2}, \frac{\bar{3}}{2}, 0\right), \left(\frac{1}{2}, \frac{\bar{1}}{2}, 1\right), \left(\frac{1}{2}, \frac{\bar{1}}{2}, \bar{1}\right), \left(\frac{3}{2}, \frac{\bar{3}}{2}, 0\right), \ldots, \tag{4.86}$$

of which the sixth is already outside the J.z. Keeping the first five terms we obtain two non-degenerate levels, E_4 and E_2 in fig. 72, which give the fundamental gap at this point, with a complex momentum loop linking them and associated evanescent wavefunction given by

$$\left.\begin{aligned}
\psi_5^1 &= \exp(-s_3 z)[\exp(\mathrm{i}h_1 z) - a_3 \exp(-\mathrm{i}h_1 z)];\\
\psi_5^2 &= 0;\\
\psi_5^3 &= \tfrac{1}{2}\exp(-s_3 z)\mathrm{i}\gamma_1(1 + a_3);\\
\psi_5^4 &= 0;\\
\psi_5^5 &= \tfrac{1}{2}\exp(-s_3 z)\mathrm{i}\gamma_1^*(1 + a_3);\\
\psi_5^6 &= 0.
\end{aligned}\right\} \tag{4.87}$$

Here s_3 and a_3 are s_1 and a_1 of (4.82) with (E_1, E_5) replaced by (E_4, E_2), and

$$\gamma_1 = (V_3^{\mathrm{A}} + \mathrm{i}V_3^{\mathrm{S}})/(E_4 - \tfrac{3}{4}h_1^2). \tag{4.88}$$

Furthermore, starting from the point (1/2, 3/2, 0) and adding reciprocal lattice vectors yields the same wavevectors as those of (4.86), in a different order and with the (x, y) coordinates interchanged. Thus the sixth evanescent wavefunction is immediately obtained by a suitable adaptation of (4.87). This yields

$$\psi_6^1 = 0, \quad \psi_6^2 = \psi_5^1, \quad \psi_6^3 = 0, \quad \psi_6^4 = \psi_5^5, \quad \psi_6^5 = 0, \quad \psi_6^6 = \psi_5^3. \tag{4.89}$$

We now have all the elements necessary to study surface states in this gap. As before, we take the a.p.m. with the barrier at z_0 and match the six evanescent wavefunctions to the six evanescent plane waves on the vacuum side defined by (4.69) with the six two-dimensional plane waves and the $\boldsymbol{\kappa}_i$s of the (1, 1, 0) surface. The s.s. eigenvalue equation is again (4.73) with the matrices defined as follows:

$$[\Lambda^{X}]_{ij} = \psi_j^i; \quad [\Lambda^{V}]_{ij} = \exp[z\sqrt{(W - E + \tfrac{1}{2}\kappa_i^2)}]\delta_{ij}. \tag{4.90}$$

Using (4.90) in (4.73) the eigenvalue equation factorises in the two determinantal equations

$$\mathrm{Det}[D \pm Z V_3^A / V_3^S] = 0. \tag{4.91}$$

The matrices D (diamond) and Z (zincblende) are defined by

$$D = \begin{bmatrix} d_1 + id_2 & -d_1 + id_2 & d_5 \\ d_3 - id_4 & 0 & d_6 \\ 0 & d_3 + id_4 & -d_6 \end{bmatrix}; \qquad Z = \begin{bmatrix} d_1 + id_2 & d_1 - id_2 & 0 \\ 0 & 0 & d_6 \\ 0 & 0 & d_6 \end{bmatrix}, \tag{4.92}$$

where

$$d_1 = -\frac{V_3^S \exp(\tfrac{1}{2}ih_1 z_0)(1 + a_1)(1 + \tfrac{1}{2}ih_1 \mathscr{L}_1)}{2(E_1 - \tfrac{1}{2}h_1^2)}; \tag{4.93}$$

$$d_2 = \frac{V_3^S \exp(-\tfrac{1}{2}ih_1 z_0)(1 + a_1)(1 - \tfrac{1}{2}ih_1 \mathscr{L}_1)}{2(E_1 - \tfrac{1}{2}h_1^2)}; \tag{4.94}$$

$$d_3 = (1 - \tfrac{1}{2}ih_1 \mathscr{L}_3)\exp(\tfrac{1}{2}ih_1 z_0); \tag{4.95}$$

$$d_4 = -a_1(1 + \tfrac{1}{2}ih_1 \mathscr{L}_3)\exp(-\tfrac{1}{2}ih_1 z_0); \tag{4.96}$$

$$\begin{aligned} d_5 = {} & [\exp(ih_1 z_0) - a_3 \exp(-ih_1 z_0)] \\ & -[ih_1 \mathscr{L}_1 \exp(ih_1 z_0) + a_3 \exp(-ih_1 z_0)]; \end{aligned} \tag{4.97}$$

$$d_6 = \frac{V_3^S(1 + a_3)}{E_4 - \tfrac{3}{4}h_1^2}, \tag{4.98}$$

and $\mathscr{L}_i$ is (4.75) with the corresponding κ_i. The above results entail the following approximations. (i) We neglect the small difference existing in the ionic case between E_1 and E_3. This amounts to taking $a_1 = a_2$. (ii) We have taken $V_4^A \approx 0$, which is not unreasonable in practice. (iii) We have also neglected s_1, s_2 and s_3, which are very much smaller than h_1. Notice that (4.91) illustrates rather neatly the difference between covalent and ionic semiconductors. Energy eigenvalues which are different in the ionic case coalesce into degenerate levels when $V_3^A = 0$. We shall insist further on this point, but now we shall discuss the lowest-order approximation.

Consider again the six evanescent wavefunctions used to construct the s.s. eigenfunction. In the lowest-order approximation we retain, in the plane-wave expansion of these functions, only those plane waves whose wavevectors are at the J.z. boundary. The evanescent wavefunctions in this case are obtained by putting $V_3 = 0$ in the ψ_1 and ψ_2 as given in (4.82)–(4.89). With these wavefunctions, and by using the approximation $a_2 = a_1$, the s.s. eigenvalue equation is factorised into three equations, each one being twofold degenerate. These equations are:

$$a_3 = \frac{1 - \mathrm{i}h_1\mathcal{L}_1}{1 + \mathrm{i}h_1\mathcal{L}_1} \exp(\mathrm{i}2h_1z_0); \qquad (4.99)$$

$$a_1 = \mathrm{i}\frac{1 - \frac{1}{2}\mathrm{i}h_1\mathcal{L}_3}{1 + \frac{1}{2}\mathrm{i}h_1\mathcal{L}_3} \exp(\mathrm{i}h_1z_0); \qquad (4.100)$$

$$a_1 = -\mathrm{i}\frac{1 - \frac{1}{2}\mathrm{i}h_1\mathcal{L}_3}{1 + \frac{1}{2}\mathrm{i}h_1\mathcal{L}_3} \exp(\mathrm{i}h_1z_0). \qquad (4.101)$$

Notice that (4.99) is a one-dimensional equation for the loop linking E_2 with E_4, while (4.100) and (4.101) are the one-dimensional equations for the loops linking E_5 with E_1 and E_3, assumed degenerate in the approximation used here. Comparing these with the results of § 4.1 we can regard (4.99)–(4.101) as effectively one-dimensional equations for the one-dimensional potentials

$$V_{1,\mathrm{eff}} \exp(\mathrm{i}2h_1z) + \mathrm{c.c.}; \qquad (4.102)$$

$$V'_{1,\mathrm{eff}} \exp[\mathrm{i}(h_1z + \tfrac{1}{2}\pi)] + \mathrm{c.c.}; \qquad (4.103)$$

$$V'_{1,\mathrm{eff}} \exp[\mathrm{i}(h_1z - \tfrac{1}{2}\pi)] + \mathrm{c.c.} \qquad (4.104)$$

The parameters $V_{1,\mathrm{eff}}$ and $V'_{1,\mathrm{eff}}$ are real and adjusted to fit the actual gaps $E_4 - E_2$ and $E_1 - E_5$ obtained in a better approximation, in the spirit of the effective two-band model discussed for the (1, 1, 1) surface. Just as we did for $\bar{J}$, we reduce in this way the study of the s.s. at $\bar{X}$ to the study of three effectively one-dimensional problems while using an at least qualitatively reasonable approximation for the bulk band structure with which we can see how the general principles of matching work.

4.2.3. Surface states, resonances and related concepts

The results obtained in the model calculation for $\bar{X}$ can also be used to discuss another important basic point. We have been thinking all the time about distinct states with eigenvalues in the gap and evanescent wavefunctions localised near the surface, i.e. true stationary – surface – states. But a careful study shows that we may find other types of solutions and they are all part of the spectrum of the surface system, as well as the distortion of the continuum of the bulk bands, whose importance for the study of the density of states and associated surface charge distribution was stressed in § 4.1. Let us concentrate on the particular case of the point $\bar{X}$. Remember (§ 4.1) that the position of the s.s. level in the a.p.m. depends on where the surface barrier is located, i.e. on the value of z_0 (fig. 54). Suppose then that we take for z_0 a value such that the s.s. level associated with the loop of the gap $E_1 - E_5$ is near E_5. Here we face an essential difference between one-dimensional models and three-dimensional crystals. By virtue of the approximaton used, the calculation of the s.s. eigenvalue in this gap is reduced effectively to a one-dimensional problem, but we are studying an actual three-dimensional system and it suffices to look at fig. 72 in order to see that a s.s. with energy near E_5 is degenerate in energy with other states of the valence band *with the same* $\boldsymbol{\kappa}$. What happens now?

In the lowest approximation the different loops are decoupled and, *to this order*, the state near E_5 is a distinct stationary s.s. which just happens to overlap in energy with the valence band. In order to see that the approximation which decouples the loops also decouples this state from band states with the same $\boldsymbol{\kappa}$ (which is not at all obvious) we can take the wavefunctions that result from setting $V_3 = 0$ and $a_1 = a_2$ in (4.82)–(4.89) and form the following combinations of them

$$\chi_1 = [0, 0, 0, 0, 1, \mathrm{i}] \exp(-s_1 z)[\exp(\tfrac{1}{2}\mathrm{i}h_1 z) - \mathrm{i}a_1 \exp(-\tfrac{1}{2}\mathrm{i}h_1 z)]; \quad (4.105)$$

$$\chi_2 = [0, 0, 0, 0, 1, -\mathrm{i}] \exp(-s_1 z)[\exp(\tfrac{1}{2}\mathrm{i}h_1 z) + \mathrm{i}a_1 \exp(-\tfrac{1}{2}\mathrm{i}h_1 z)]; \quad (4.106)$$

$$\chi_3 = [0, 0, 1, \mathrm{i}, 0, 0] \exp(-s_1 z)[\exp(\tfrac{1}{2}\mathrm{i}h_1 z) - \mathrm{i}a_1 \exp(-\tfrac{1}{2}\mathrm{i}h_1 z)]; \quad (4.107)$$

$$\chi_4 = [0, 0, 1, -\mathrm{i}, 0, 0] \exp(-s_1 z)[\exp(\tfrac{1}{2}\mathrm{i}h_1 z) + \mathrm{i}a_1 \exp(-\tfrac{1}{2}\mathrm{i}h_1 z)]; \quad (4.108)$$

$$\chi_5 = [1, 0, 0, 0, 0, 0] \exp(-s_3 z)[\exp(\mathrm{i}h_1 z) - a_3 \exp(-\mathrm{i}h_1 z)]; \quad (4.109)$$

$$\chi_6 = [0, 1, 0, 0, 0, 0] \exp(-s_3 z)[\exp(\mathrm{i}h_1 z) - a_3 \exp(-\mathrm{i}h_1 z)]. \quad (4.110)$$

The six figures given in the square brackets indicate the amplitudes of the six two-dimensional plane waves f_i (4.64) for the $\boldsymbol{\kappa}_i$s of this case. Suppose now we take z_0 such that χ_1 and χ_3 give a s.s. in the gap E_1–E_5 whose eigenvalue level is a little below E_2 (fig. 72). The valence-band wavefunctions of the states overlapping in energy with the s.s. are then (4.109) and (4.110) with real instead of complex momentum. To be specific, they are of the following form

$$\chi_5^{\pm} = [1, 0, 0, 0, 0, 0] \exp(\pm iqz)[\exp(ih_1 z) - a_3 \exp(-ih_1 z)], \quad (4.111)$$

and

$$\chi_6^{\pm} = [0, 1, 0, 0, 0, 0] \exp(\pm iqz)[\exp(ih_1 z) - a_3 \exp(-ih_1 z)]. \quad (4.112)$$

As in § 4.1, the boundary conditions due to the surface combine $\chi_5^{\pm}$ in the phase-shifted stationary wave,

$$[1, 0, 0, 0, 0, 0]\{\sin[(q + h_1)z + \eta] - a_3 \sin[(q - h_1)z + \eta]\}, \quad (4.113)$$

and $\chi_6^{\pm}$ in

$$[0, 1, 0, 0, 0, 0]\{\sin[(q + h_1)z + \eta] - a_3 \sin[(q - h_1)z + \eta]\}. \quad (4.114)$$

With these wavefunctions one could study the density of states in the valence band as in § 4.1. The point to notice is that the wavefunctions (4.113) and (4.114) are decoupled from (4.105) and (4.107) because, although they have the same reduced $\boldsymbol{\kappa}$, they have different momenta parallel to the surface, as can be seen by looking at the $\boldsymbol{\kappa}_i$s. This is illustrated in fig. 73 which shows the z-dependence of one of the components of the wavefunctions (4.114) and (4.105). The valence wavefunction (fig. 73*a*) tends to a Bloch wave in the bulk and decays outside the surface barrier. This is the distortion of the continuum, discussed at length in § 4.1.2, where we saw that it contributes to a change in the density of states in the region near the band edge (fig. 74*a*, curve). The surface state (fig. 73*b*) is localised near the surface and decays into the bulk. Under these circumstances this state contributes a discrete δ-function term to the change in the density of states, which overlaps (fig. 74*a*) with the distortion of the continuum. But this decoupling is only an artifact of the simple model used and disappears when a higher-order approximation is used. Let us see how this comes about.

Compare the more accurate wavefunctions ψ_i given in (4.82)–(4.89) with the lower-order ones ψ_i^0 which result from these by putting $V_3 = 0$. The main point is that the more accurate wavefunctions have more non-vanishing parallel components and it is this fact that brings about the coupling, because in doing the complete matching analysis we must include in the s.s. wavefunction also the non-evanescent wavefunctions which have some parallel components in common with the wavefunction

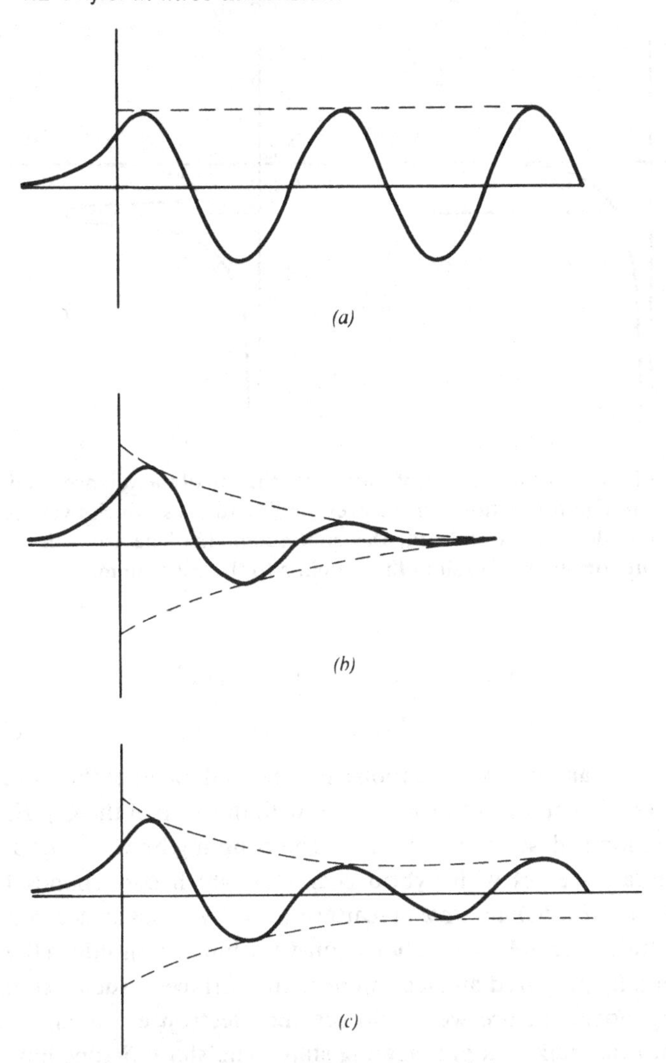

Fig. 73. (*a*) A stationary bulk state wavefunction which adapts itself to the surface boundary condition. (*b*) A stationary surface state evanescent wavefunction whose amplitude is localised near the surface. (*c*) A waveform associated with a quasistationary surface state which resonates with the bulk continuum.

we are seeking. In particular, instead of (4.105) we now have a combination of ψ_1 (4.82), ψ_2 (4.83), ψ_3 (4.84) and ψ_4 (4.85), as well as of the non-evanescent bulk wavefunctions,

$$\chi_5^{\pm} = \exp(\pm iqz)\{[\exp(ih_1 z), 0, \tfrac{1}{2}i\gamma_1, 0, \tfrac{1}{2}i\gamma_1^*, 0] + a_3[-\exp(-ih_1 z), 0, \tfrac{1}{2}i\gamma_1, 0, \tfrac{1}{2}i\gamma_1^*, 0]\} \quad (4.115)$$

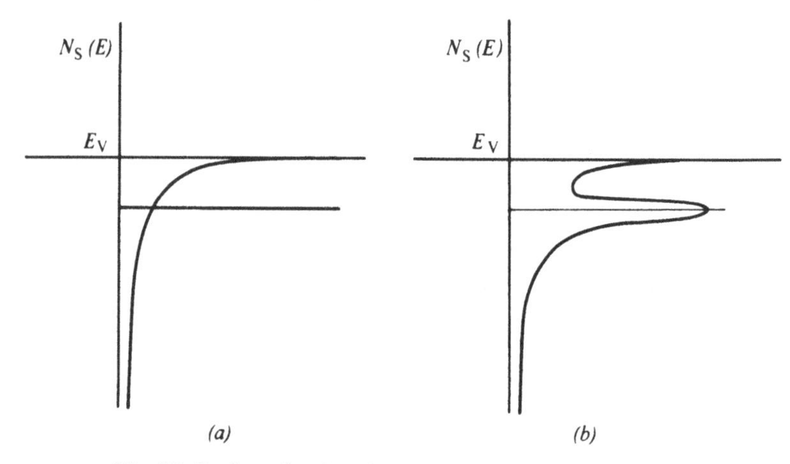

Fig. 74. Surface density of states near the top of the valence band. (*a*) A curve (a δ-function term) corresponding to a s.s. which overlaps with, but does not couple to, the continuum. (*b*) The curve is typically broadened as a result of the coupling to the continuum.

and

$$\chi_6^{\pm} = \exp(\pm iqz)\{[0, \exp(ih_1 z), 0, \tfrac{1}{2}i\gamma_1^*, 0, \tfrac{1}{2}i\gamma_1] + a_3[0, -\exp(-ih_1 z), 0, \tfrac{1}{2}i\gamma_1^*, 0, \tfrac{1}{2}i\gamma_1]\}, \tag{4.116}$$

which are the analytic continuations into the bulk band of the wavefunctions ψ_5 (4.87) and ψ_6 (4.89) associated with the loop of the gap $E_4 - E_2$. The situation is described in fig. 73*c*. The coupling between (*a*) and (*b*) produces a state (*c*) with hybrid character which corresponds to the notion of *quasistationary state* in scattering theory. It has an accumulation of amplitude near the surface, but a non-vanishing amplitude in the bulk. If we initially prepared an electron near the surface in such a state, in a time-dependent picture we would see the electron eventually leaking away into the bulk in a propagating state. The sharp δ-function of the uncoupled case now broadens out into a hump in the density of states (fig. 74*b*). We have found a *resonance* between a s.s. and the valence band.

Summing up, in a careful study of the eigensolutions of a three-dimensional crystal with a surface we may find the following situations:

(i) Distinct, true, stationary *surface states.* Their wavefunction is evanescent and localised near the surface. They contribute distinct δ-function terms to the density of states, but of course when we scan the two-dimensional values of $\boldsymbol{\kappa}$ with continuity they form continuous bands of s.s.

(ii) *Distortions of the continuum.* These occur because the bulk wavefunctions must also adapt themselves to the surface boundary condition.

The distortion of the wavefunction is also localised near the surface within distances similar to those characterising the localisation of the s.s. wavefunctions. They contribute changes in the density of states in the range of the bulk bands.

(iii) *Resonances* or quasistationary s.s. These overlap in energy with bulk band states having the same parallel momentum. The wavefunction behaves as in fig. 73*b* and one such state, for given $\boldsymbol{\kappa}$, contributes a hump in the density of states superimposed on the distortion of the continuum. Either because of special circumstances or, more often, because of a low-order approximation, such a state may not couple to bulk states, in which case it is just a distinct s.s. which happens to overlap in energy with a bulk band.

It is clear that even a very accurate calculation that is concerned only with s.s. may amount to an incomplete study of the surface problem. The analysis of some spectroscopic data, for example, may well require the study of the spectrum in its full complexity, involving the various types of solutions just discussed.

4.2.4. Selfconsistency and surface-state occupation in covalent surfaces

So far we have used simple models to study the general principles involved in a matching calculation, and the nature of the different types of solutions, but now we must go a step farther. The main limitation of the a.p.m. is that in itself it cannot be selfconsistent. In principle we ought to study the surface potential barrier and the wavefunctions for all $\boldsymbol{\kappa}$ self-consistently, but carrying out this program literally would entail a very large amount of numerical computation. However, our purpose is only to try and confirm certain basic features, even at the price of using some simplifications, provided these can furnish a reasonable picture. To begin with therefore, we shall study only the points $\bar{J}$ and $\bar{X}$ which we take as suitably representative of the (1, 1, 1) and (1, 1, 0) faces, respectively. The justification of this procedure will be discussed later. Furthermore, we have just seen that although higher-order approximations bring about minor quantitative improvements, the picture furnished by the effective two-band approximation is basically reasonable. The results of § 4.1 can then be adapted to the study of three-dimensional crystals via the factorisation into effectively one-dimensional problems, which results from this approximation.

Let us now start with $\bar{J}$. Since the study of the surface charge involves also the distortions of the continuum it is convenient to once more obtain the factorisation into one-dimensional problems in a different way; this will yield the effective one-dimensional potentials and emphasise that the distortions of the continuum can also be studied by a superposition of effectively one-dimensional problems. In general the crystal potential near a (1, 1, 1) surface can be written in the form (remember (4.48))

$$V(\boldsymbol{\rho}, z) = V_0(z) + \sum_{\boldsymbol{G}} V_{\bar{G}}(z) \exp(\mathrm{i}\bar{G} \cdot \boldsymbol{\rho}). \tag{4.117}$$

The wavefunction of a state with given $\boldsymbol{\kappa}$ satisfies the Schrödinger equation,

$$-\tfrac{1}{2}\nabla^2 \psi_{\boldsymbol{\kappa}}(\boldsymbol{\rho}, z) + V(\boldsymbol{\rho}, z)\psi_{\boldsymbol{\kappa}}(\boldsymbol{\rho}, z) = E\psi_{\boldsymbol{\kappa}}(\boldsymbol{\rho}, z). \tag{4.118}$$

Within the J.z. approximation the potential can be written as

$$V(\boldsymbol{\rho}, z) = V_0(z) + \{V_1(z)[\exp(\mathrm{i}\bar{\boldsymbol{G}}_1 \cdot \boldsymbol{\rho}) + \exp(\mathrm{i}\bar{\boldsymbol{G}}_2 \cdot \rho) + \exp(\mathrm{i}\bar{\boldsymbol{G}}_3 \cdot \boldsymbol{\rho})] + \text{c.c.}\}. \tag{4.119}$$

The vectors $\bar{\boldsymbol{G}}_1$, $\bar{\boldsymbol{G}}_2$, $\bar{\boldsymbol{G}}_3$ are shown in fig. 68. The associated values of $\boldsymbol{\kappa}_\mathrm{i}$ define the two-dimensional wavefunctions f_i (4.64). Using these as a basis the symmetry at $\bar{J}$ allows us to write for the three solutions of (4.118) at this point

$$\left.\begin{aligned} &\psi_1(\boldsymbol{r}) = [1, 1, 1]\phi_1(z); \quad \psi_2(\boldsymbol{r}) = [1, \omega, \omega^2]\phi_2(z); \\ &\psi_3(\boldsymbol{r}) = [1, \omega^2, \omega]\phi_3(z), \end{aligned}\right\} \tag{4.120}$$

where $\omega = \exp(\mathrm{i}\frac{2}{3}\pi)$ and the ϕ_is satisfy the one-dimensional equations,

$$\left[-\frac{1}{2}\frac{\mathrm{d}^2}{\mathrm{d}z^2} + \frac{1}{2}\kappa^2 + V^{(i)}(z)\right]\phi_i(z) = E^{(i)}\phi_i(z), \tag{4.121}$$

with the one-dimensional potentials

$$V^{(1)}(z) = V_0(z) + V_1(z) + V_1^*(z); \tag{4.122a}$$

$$V^{(2)}(z) = V_0(z) + V_1(z)\omega^2 + V_1^*(z)\omega; \tag{4.122b}$$

$$V^{(3)}(z) = V_0(z) + V_1(z)\omega + V_1^*(z)\omega^2. \tag{4.122c}$$

For z large and positive, well inside the crystal, we have, in the J.z. approximation,

$$V_0(z) = -V_3 \exp(\mathrm{i}hz) + \text{c.c.}; \tag{4.123a}$$

$$V_1(z) = [V_3 \exp(\tfrac{1}{3}\mathrm{i}hz) + V_8 \exp(\tfrac{1}{3}\mathrm{i}4hz)] + \text{c.c.} \tag{4.123b}$$

Then, for z large and positive,

$$V^{(1)}(z) = [-V_3 \exp(\mathrm{i}hz) + V_3 \exp(\tfrac{1}{3}\mathrm{i}hz) + V_8 \exp(\tfrac{1}{3}\mathrm{i}4hz)] + \text{c.c.}; \tag{4.124a}$$

$$V^{(2)}(z) = \{-V_3 \exp(ihz) + V_3 \exp[\tfrac{1}{3}i(hz + 2\pi)] + V_8 \exp[\tfrac{1}{3}i(4hz + 2\pi)]\} + \text{c.c.}; \quad (4.124b)$$

$$V^{(3)}(z) = \{-V_3 \exp(ihz) + V_3 \exp[\tfrac{1}{3}i(hz - 2\pi)] + V_8 \exp[\tfrac{1}{3}i(4hz - 2\pi)]\} + \text{c.c.} \quad (4.124c)$$

The various Fourier components of the potential create gaps at $\frac{2}{3}h$, $\frac{1}{2}h$ and $\frac{1}{6}h$ in the extended zone scheme. The fundamental gap is at $\frac{2}{3}h$ and, as discussed above, is due to V_8 to first order and also to V_3 to second order. A similar situation appears at $\frac{1}{3}h$, where the gap is due, to second order, to $V_3 \exp(\frac{1}{3}ihz)$. The type of band structure resulting from (4.121) and (4.124) is shown in fig. 75 in the reduced-zone scheme with the boundary of the first B.z. at $\frac{1}{6}h$. The information displayed in fig. 75 is similar to that given in fig. 69, where the band structure is shown by using a different zone scheme, namely, the surface prism defined in § 4.2.2. The one-dimensional equations (4.121) correspond to the equations (4.77) which

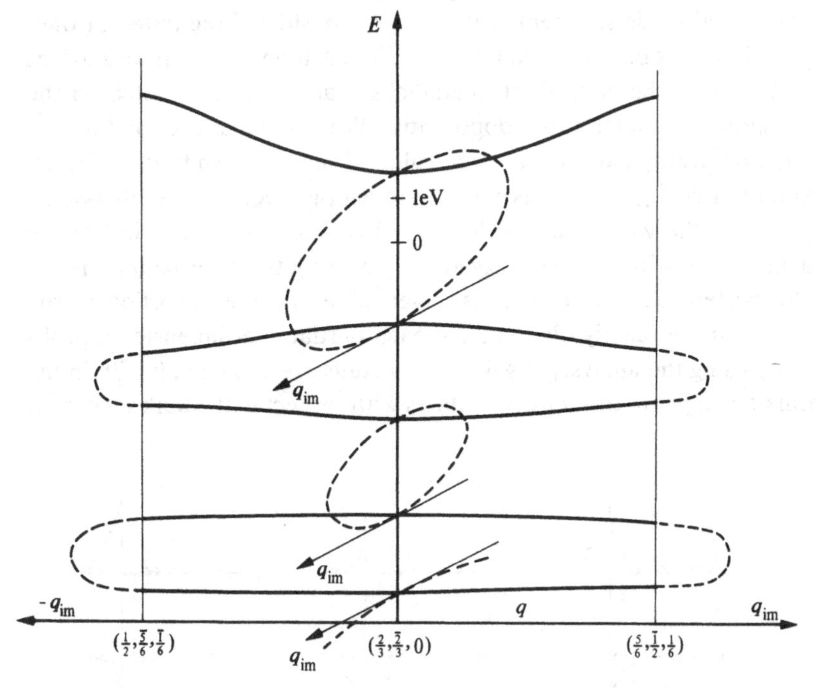

Fig. 75. Complex band structure for Si. The wavevector grows in the (1, 1, 1) direction and its tip goes through $\bar{J}$, a corner of the corresponding *two-dimensional* B.z. But the band structure is here plotted inside the first *three-dimensional* B.z. Compare with the five-band approximation of fig. 69.

give the s.s. eigenvalues in the fundamental gap in the two-band approximation. In this approximation the pseudopotentials of (4.124) are replaced by a single effective pseudopotential component adjusted to fit the width of the gap under consideration. This device can be adapted to concentrate on the study of each separate gap, if desired. For the surface problem the interest lies actually in the fundamental gap which separates occupied from unoccupied states. We have seen (§ 4.1.3) that the effects associated with the lower (occupied) gaps cancel out. In this case the potentials of (4.124) are replaced by those of (4.78), so that the effective total potentials to use are

$$V^{(i)}(z) = V_0(z) + \{V_{8,\text{eff}}(z) \exp[\tfrac{1}{3}\mathrm{i}(4hz + n_i 2\pi)] + \text{c.c.}\}, \tag{4.125}$$

where $n_1 = 0$, $n_2 = +1$, $n_3 = -1$ and $V_{8,\text{eff}}(z)$ tends to the constant value of $V_{8,\text{eff}}$ for $z \to +\infty$.

If we now accept that $\bar{J}$ can be taken as a representative point of the (1, 1, 1) surface, then we can study the problems of occupation and selfconsistency of this surface by using the results of § 4.1 for one-dimensional models, except that we must consider three different one-dimensional problems simultaneously. For each one of them we start, as in § 4.1, from the equivalent 'metallic' surface and then switch on the corresponding effective pseudopotential. We know that the condition of charge neutrality fixes the average value of the phase shift at the Fermi level, i.e. that $\langle\eta_F\rangle = \frac{1}{4}\pi$. Taking $\bar{J}$ as an average point, we shall assume that this is the value that results from $V_0(z)$ for the phase shift of the wavefunction with Fermi energy at $\bar{J}$. By making this approximation, the selfconsistent position of the s.s. eigenvalue and the distortion of the continuum are readily obtained for each partial one-dimensional problem by using the analysis of § 4.1. The results are shown in fig. 76 in the form of changes in the density of states with respect to the bulk spectrum.

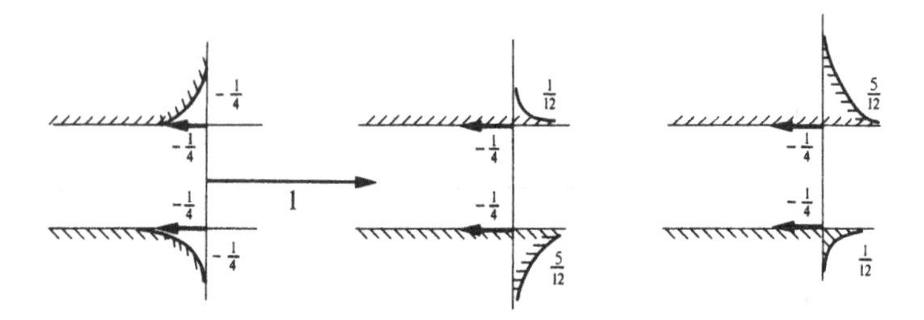

Fig. 76. Surface density of states for the three one-dimensional problems into which the three-dimensional problem associated with $\bar{J}$ factorises. After F. Flores & C. Tejedor, *J. Physique* **38**, 949 (1977).

It is seen that the potential $V^{(1)}$ of (4.125) produces a s.s. at the midgap and subtracts $\frac{1}{2}$ state from each band, while with $V^{(2)}$ and $V^{(3)}$ there are no s.s. levels in the gap; only distortions of the continuum appear in these two cases. But the results from $V^{(2)}$ and $V^{(3)}$ have a very important feature, namely, that the distortion of the conduction/valence band of one of them is equal to the distortion of the valence/conduction band of the other. This follows immediately from the discussion after (4.36), since the phase shifts corresponding to the two 'rotated' potentials are $\frac{1}{4}\pi \pm \frac{3}{4}\pi$ and add up to $\frac{1}{2}\pi$. This has an important consequence, since on combining the three partial one-dimensional problems the total changes in the density of states are symmetric about the midgap and we are back to the situation of § 4.1.3. In this approximation the charge redistributions associated with the s.s. and with the distortion of the valence band again cancel out and we obtain the following approximate picture: (i) The selfconsistent potential profile and charge distribution are very similar to those of an equivalent metal with the same electronic density. (ii) The average position of the band of s.s. is fixed by the charge neutrality condition. (iii) The occupation of this band is $\frac{1}{2}$.

Since an accurate selfconsistent calculation would be a formidable task, it is interesting to see what sort of s.s. bands one can obtain by using this simple picture in practice. A drastic simplification is effected by introducing a suitably parametrised a.p.m. in which the position of the barrier, z_0, is fixed to achieve selfconsistency within the present approximation. The key lies in the dependence of the phase shifts on z_0. What we can do is to choose z_0 so that charge neutrality is ensured in the reference 'metallic' system. This fixes η_F for the wavefunction at $\bar{J}$ at its correct value $\frac{1}{4}\pi$ and from then on, the s.s. band can be calculated by carrying out for different $\boldsymbol{\kappa}$s a matching calculation of the type outlined in § 4.2.2. Fig. 77 shows the resulting bands of s.s. obtained in this way in the fundamental gap of Si and Ge. The calculations were actually performed with the method of surface Green-function matching (chapter 2) by using an expansion in fifty-four plane waves. The figure shows also the two-dimensional projection of the bulk bands, so that the optical gap can be seen for each $\boldsymbol{\kappa}$. The s.s. band stays fairly reasonably at about the middle of the two bulk bands. In fact these results turn out to be fairly good, agreeing quite well with the results of more elaborate computations, where available.

Similar arguments can be applied to the (1, 1, 0) surface. Here we take $\bar{X}$ as an average point (fig. 71). We have seen how in the effective two-band approximation the problem again factorises into three one-dimensional problems. From (4.99)–(4.104) we find that the effective

(a)

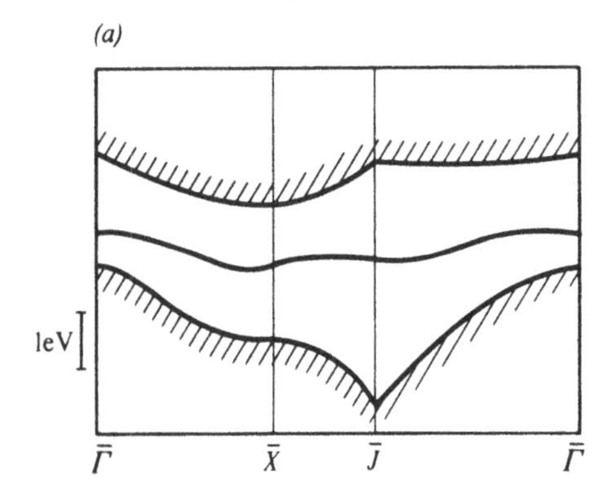

(b)

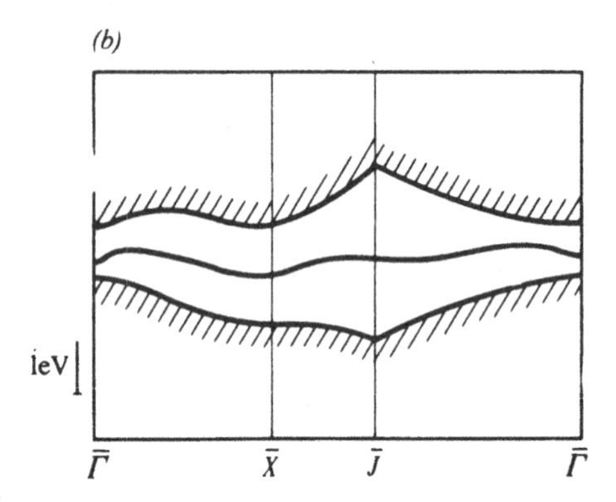

Fig. 77. Band of s.s. for (*a*) Si and (*b*) Ge; see text. The shaded areas are the two-dimensional projections of the three-dimensional bulk bands. The s.s. bands are adapted from F. Flores, F. García-Moliner, E. Louis & C. Tejedor, *J. Phys.* C: *Solid St. Phys.* **9**, L429 (1976).

potentials for the three – twofold degenerated – one-dimensional problems are

$$V_0(z)+[V_{1,\text{eff}}(z)\exp(\mathrm{i}2h_1z)+\text{c.c.}], \tag{4.126a}$$

and

$$V_0(z)+[V'_{1,\text{eff}}(z)\exp(\mathrm{i}h_1z\pm\tfrac{1}{2}\pi)+\text{c.c.}], \tag{4.126b}$$

where $V_{1,\text{eff}}(z)$ and $V'_{1,\text{eff}}(z)$, for $z\to+\infty$, tend to the constant values $V_{1,\text{eff}}$ and $V'_{1,\text{eff}}$, adjusted to fit the gaps E_4–E_2 and E_1–E_5 (fig. 72). Repeating the process applied to $\bar{J}$ in the (1, 1, 1) face, we start with the equivalent 'metallic' surface and then switch on the effective pseudopotential. The changes in the spectrum in and about the gap are shown in fig. 78 for each one-dimensional problem. We find the same situation

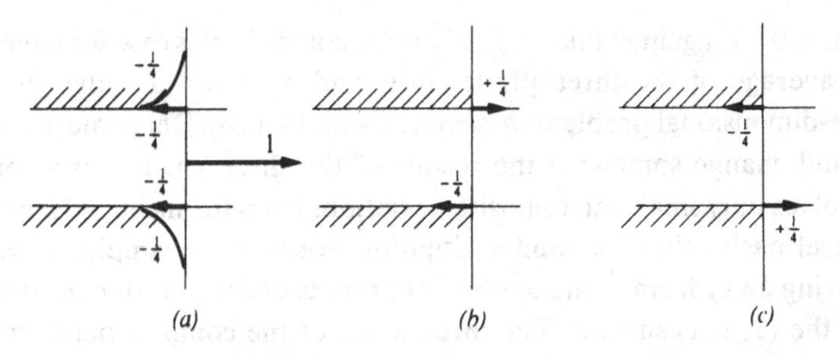

Fig. 78. Surface density of states for the three one-dimensional problems of the point $\bar{X}$.

as in the (1, 1, 1) case, i.e. the total changes in the spectrum add up to a symmetric distribution about the midgap and the net surface charge redistribution is again zero. The selfconsistent potential profile in this approximation is again very close to the equivalent metallic potential and, remembering the twofold degeneracy of the one-dimensional problems, we find two bands of s.s., with the average energy-level position fixed by charge neutrality and occupation $\frac{1}{2}$.

Now, this approximate picture of the state of affairs at the surface hinges on (i) taking $\bar{J}$ or $\bar{X}$ as suitable representative points of the corresponding surfaces, and (ii) attributing to the phase shift of the Fermi wavefunctions of each one-dimensional problem a value $\frac{1}{4}\pi$, the average metallic value. We know these approximations will result in some quantitative inaccuracy, but we must now investigate whether they constitute a plausible simplification. Consider, for example, the point $\bar{J}$. Since in the two-dimensional projection of the J.z. this point is at an intermediate position between the centre and the boundary, it is not unreasonable to expect the phase shift for this point to be close to the average value $\frac{1}{4}\pi$. This can be illustrated with a simple example. Take an abrupt infinite barrier displaced a distance l from the origin to ensure surface charge neutrality. Then the phase shift at $\bar{J}$ is

$$\eta_0(\bar{J}) = \tfrac{2}{3}hl, \tag{4.127a}$$

where $\frac{2}{3}h$ is one half of the momentum $\frac{4}{3}h$ associated with the effective potentials (4.125) of the one-dimensional problems. But we know l from (3.57). By remembering the factor $2\pi/c$ for h and the units we are using this yields $\eta_0(\bar{J}) = 0.87$, which is within 10% of $\frac{1}{4}\pi$. For $\bar{X}$ we have the potentials (4.126) which have different associated momenta. The *average* of the phase shifts produced by these three potentials with an abrupt barrier at $z = l$ is

$$\langle\eta_0(\bar{X})\rangle = \tfrac{1}{3}(h_1 d + \tfrac{1}{2}h_1 d + \tfrac{1}{2}h_1 d) = \tfrac{2}{3}h_1 \mathrm{d}. \tag{4.127b}$$

This is 0.71, again within 10% of $\frac{1}{4}\pi$. Of course in this case we have taken an average of the three phase shifts and used this to solve the three one-dimensional problems associated with (4.126). Being more accurate would change somewhat the results of the three partial cases, but this involves only small and opposite deviations from the mean, which tend to cancel each other. A similar situation arises, for example, if we start moving away from $\bar{J}$ and study other points of the two-dimensional B.z. for the (1, 1, 1) surface. The three loops of the complex band structure cease to appear in the same energy interval (fig. 69), but the departures from this also have mutually opposite effects. By using a displaced abrupt barrier as an indication it turns out that the mean phase shift is again (4.127a) and essentially the same situation is recovered.

The (1, 1, 0) surface can be discussed in a similar way. One would think that the point to take as an average one for this face should be the corner, as we did in the (1, 1, 1) surface. The results for $\bar{X}$ demonstrate how, on moving away from the corner, one obtains again an average phase shift characterising the surface in question, and how using this averaged phase shift for each one-dimensional case leads again to cancellation of the net charge redistribution.

It is important to notice that the above arguments, to the extent that one can take a given point of the two-dimensional B.z. as a representative one for an entire surface, hold only for the study of the approximate global cancellation of the charge redistribution. Of course the study of more detailed properties, such as the s.s. band structure, requires the separate analysis of different points, as emphasised above in connection with the results presented in fig. 77. This figure illustrates that the overall picture emerging from this analysis is not only qualitatively reliable, accounting for the key features, but is also quantitatively quite reasonable.

The main point of this approximate description of the covalent surfaces is that, as long as the narrow-gap approximation describes the bulk crystal fairly well, the surface potential and associated surface charge distribution are very close to those of the equivalent metallic surface, and this can be somewhat improved by treating the metallic problem as in § 3.3, where we discussed the inclusion of lattice effects on the metallic potential near the surface. Fig. 79 shows the $\bar{G}=0$ component of the crystal potential as a function of z near a (1, 1, 1) Si surface obtained in this way. The wavefunction at $\bar{J}$ can then be calculated for this potential by solving (4.121). The charge density distribution of this wavefunction is shown in fig. 80 as an example of the spatial localisation of a representative surface state.

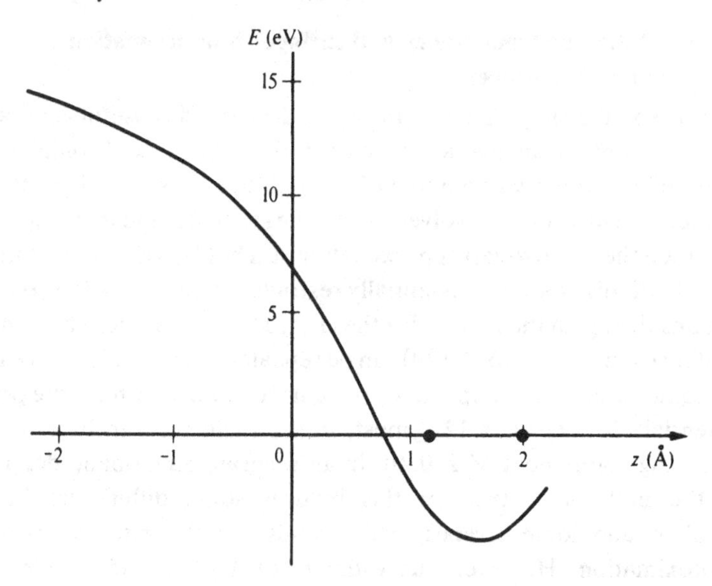

Fig. 79. Variation of the potential in the (1, 1, 1) direction, near the Si (1, 1, 1) surface, after averaging in the plane parallel to the surface. The points indicate the last two atomic planes. F. Flores & C. Tejedor, *J. Physique* **38**, 949 (1977).

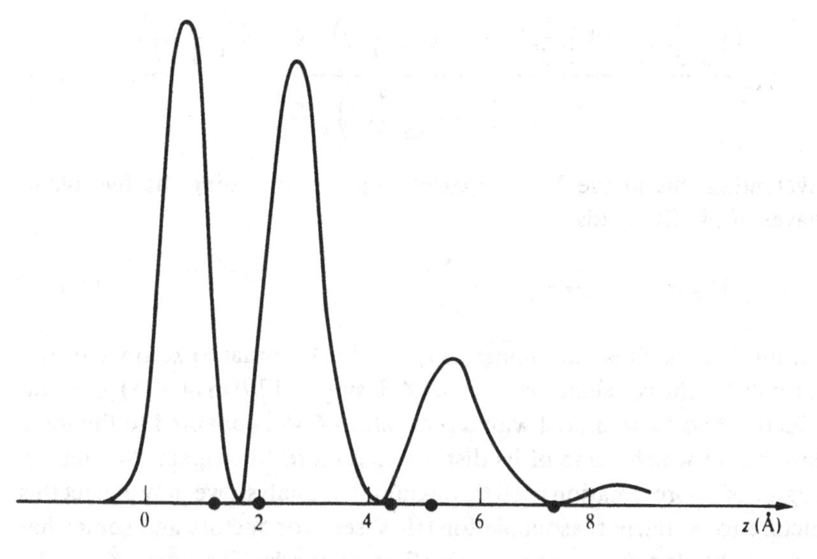

Fig. 80. Charge density associated with the s.s. at $\bar{J}$ plotted along the line which goes through an atom in the first atomic plane. F. Flores & C. Tejedor, *J. Physique* **38**, 949 (1977).

4.2.5. Selfconsistency and surface-state occupation in ionic surfaces

The reason the simplified picture so far developed is useful is because it serves to obtain an overall view of both covalent and ionic surfaces without involving a vast amount of computation. The situation with ionic surfaces is a little more involved but can be studied in much the same way, provided the narrow-gap approximation still holds, which means that the validity of this analysis is essentially restricted to relatively low ionicities.

Consider again the point $\bar{J}$ for the (1, 1, 1) ionic surface. The argument leading from (4.121) to (4.124) can be repeated with the difference that in the ionic case, V_3 is complex; V_8 remains real. Furthermore the pseudopotentials $V^{(i)}(z)$ of (4.124) must, in the ionic case, include the non-vanishing component $V(2, 0, 0)$. In an accurate many-band description of the bulk-band structure this induces some differences between covalent and ionic crystals, which vanish in the effective two-band approximation. However, the estimate of $V_{8,\text{eff}}$ is different precisely because of the new values of V_3 and V_4, as can be seen with a Rayleigh–Schrödinger perturbation analysis. At the point $\bar{J}$, $V_{8,\text{eff}}$ connects the plane waves (2/3, 4/3, 0) and $(\bar{4}/3, \bar{2}/3, 0)$. If we put $E(a, b, c)$ for the kinetic energy of the plane wave (a, b, c) the result to second order is

$$V_{8,\text{eff}} = V_8 + \sum_{\boldsymbol{G}} \frac{\left\langle\left(\frac{2}{3}, \frac{4}{3}, 0\right)\middle| V\middle|\left(\frac{2}{3}, \frac{4}{3}, 0\right) - \boldsymbol{G}\right\rangle\left\langle\left(\frac{2}{3}, \frac{4}{3}, 0\right) - \boldsymbol{G}\middle| V\middle|\left(\frac{\bar{4}}{3}, \frac{\bar{2}}{3}, 0\right)\right\rangle}{E\left(\frac{2}{3}, \frac{4}{3}, 0\right) - E\left[\left(\frac{2}{3}, \frac{4}{3}, 0\right) - \boldsymbol{G}\right]}. \tag{4.128}$$

Evaluating this in the J.z. approximation, i.e. including the five plane waves of (4.52), yields

$$V_{8,\text{eff}} = V_8 - \frac{4|V_3|^2}{k_0^2 - k_1^2} - \frac{2|V_4|^2}{k_0^2 - k_2^2}. \tag{4.129}$$

Setting V_4, which is pure imaginary, and Im V_3 equal to zero yields the formula for the covalent case of § 4.2.4. Using (4.129) in (4.125) yields an effective two-band model with a real value of $V_{8,\text{eff}}$ adapted to the ionic case but in which some of its distinctive features disappear. Within the degree of approximation used throughout our analysis we may expect this picture to be fairly reasonable for III–V semiconductors and somewhat less reliable for the more ionic II–VI compounds. The point $\bar{X}$ for the (1, 1, 0) surface can be studied in a similar manner, with an appropriate effective potential which, inserted in (4.102), (4.103), (4.104), can be used to analyse again the independent effectively one-dimensional model. Here, for example, the small splitting of the twofold degenerate s.s. eigenvalue disappears on making the effective two-band approximation. This point will be further discussed in § 4.3.

It is important to notice that while this simplified picture of the bulk-band structure loses some of the detailed features (for both bulk and surface states) which distinguish an ionic from a covalent semiconductor, it does pick out essential differences concerning the selfconsistent description of charge neutrality and s.s. occupation, because these stem basically from the difference in the charge of the two atoms in the unit cell. Our purpose is just to discuss this problem for the ionic case, emphasising the differences with the covalent case. For this we start by looking at the electrostatic potential created by the ionic and electronic charges which, after averaging in the (x, y) plane, we study as a function of z, in the direction perpendicular to the surface of interest. Fig. 81 shows this potential for the direction perpendicular to the (1, 1, 1) surface in a

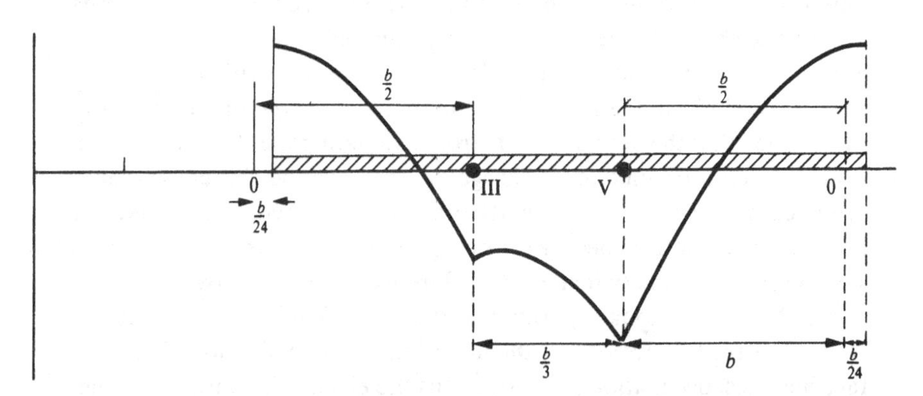

Fig. 81. Electrostatic potential created by the ionic and electronic charges of a III–V compound in the (1, 1, 1) direction, after averaging in the plane perpendicular to it. The distance between the position of the potential maxima and the midpoint between atoms is $b/24$, where b is the bond length. F. Flores, C. Tejedor & A. Martín-Rodero, *Phys. St. Solidi* (b) **88**, 591 (1978).

III–V compound. The length of the bond between neighbours is denoted by b. The potential is created by the ionic charges and by an electronic charge density assumed to be uniformly distributed in the unit cell, which is consistent with the weak crystalline perturbation of the weak pseudopotential model used in the calculation of the bulk wavefunctions. We then define a unit cell whose boundaries are at the two maxima of the electrostatic potential shown in the figure, and imagine a crystal, indefinitely extended in the (x, y) plane, but containing N such unit cells in the z-direction. It has two surfaces, as far apart as necessary, at which the normal gradient of the electrostatic potential is zero. This is then constant outside, in the vacuum, and the potential difference between the two opposite faces is zero, as a consequence of our choice of unit cell; otherwise we would have successive potential drops at the successive cells, and a potential difference between the two faces which would grow with their distance apart. Of course, this is only the electrostatic potential we start from when we create the surfaces at two points defined as ends of unit cells. We must then allow for electronic charge redistribution at the surface regions. An abrupt drop of the electronic charge density to zero would be utterly unrealistic; the charge adapts itself to the surface condition and in the final selfconsistent rearrangement decreases smoothly to zero. But it is important to realise that this rearrangement does not change the total electronic charge in the surface layer, so that there is no net electronic charge transfer between the two opposite faces; thus the potential difference between them remains finite. In other words, the surface charge in the ionic case, as compared with the covalent case, has an extra contribution which is obtained by evaluating the charge contained within the length $\frac{1}{24}b$ in the uniform charge distribution (fig. 81). Notice that this length results from evaluating the distance through which the uniform electronic charge density must be shifted so that its centre coincides with the weighted centre of the ionic charges III and V. It is this shift that determines the s.s. occupation. Take the covalent case as a starting situation and switch on the difference between the two ions in the unit cell. We must look separately at the $(1, 1, 1)$ and $(\bar{1}, \bar{1}, \bar{1})$ faces. Let $(\bar{1}, \bar{1}, \bar{1})$ be the face formed by the ions of the column V-anions. Then this face has a s.s. occupation given by $\frac{1}{2}$ plus the electronic charge contained in the distance $\frac{1}{24}b$. Considering that eight electronic charges are distributed in a distance $\frac{4}{3}b$ with a cross section equal to the surface area per atom we find an extra $\frac{1}{4}$ electronic charge per surface atom, but this includes the spin factor. Thus the s.s. occupation per spin for this surface is $(\frac{1}{2})+(\frac{1}{8})$. By the same argument the occupation of the opposite face is $(\frac{1}{2})-(\frac{1}{8})$ and both, of course, add up to unity.

Knowing the s.s. occupation we can discuss selfconsistency proceeding as in the covalent case, i.e. we first start with an equivalent jellium model from which we evaluate the phase shifts of the wavefunctions and then switch on the weak crystal pseudopotential whose effects we have described in the narrow-gap effective two-band approximation. For the first problem we have the results of the metallic case and, again, we take for the phase shifts of the Fermi wavefunctions the average value $\frac{1}{4}\pi$ in studying the factorised one-dimensional problems. For the point $\bar{J}$ of the (1, 1, 1) surface these are described by the three equations (4.121). The difference is that now the phase shift, being referred to the jellium edge in the equivalent jellium problem, must be corrected by the displacement of length $\frac{1}{24}b$ discussed above. Thus the phase shift referred to the origin of z – i.e. the midpoint of the broken bond – is, for the (1, 1, 1) ionic surfaces,

$$\eta_0 = \tfrac{1}{4}\pi \pm \eta_1; \quad \eta_1 = \tfrac{1}{24}b \times \tfrac{2}{3}h = \tfrac{1}{24}\pi.$$

The sign is +/− for the V/III face. Knowing η_0 we can estimate the position of the s.s. for each one-dimensional problem. The eigenvalue equations are

$$\sqrt{\left(\frac{E_C - E}{E - E_V}\right)} = \tan \eta_0; \quad \sqrt{\left(\frac{E_C - E}{E - E_V}\right)} = \tan (\eta_0 \pm \tfrac{2}{3}\pi). \quad (4.130a,b)$$

The s.s. eigenvalues can be readily obtained by using the above values of η_0. It is interesting to plot the general relationship between the s.s. eigenvalue position in the gap and its occupation Q_s, related to η_0 by

$$\eta_0 = \tfrac{1}{4}\pi + \tfrac{1}{3}\pi(Q_s - \tfrac{1}{2}). \quad (4.131)$$

For example, if $Q_s = \frac{1}{2}$, then $\eta_0 = \frac{1}{4}\pi$ (covalent case), and if $Q_s = (\frac{1}{2}) \pm (\frac{1}{8})$, then $\eta_0 = \frac{1}{4}\pi \pm \frac{1}{24}\pi$ ((1, 1, 1) and $(\bar{1}, \bar{1}, \bar{1})$ ionic faces). The results obtained by using (4.131) in (4.130) are shown (dashed line) in fig. 82. The s.s. levels for the two opposite ionic faces are no longer at the midgap, as in the covalent case, but they are at $\pm 0.27|V_{8,\mathrm{eff}}|$, symmetrically located with respect to the midgap.

At this stage we must study the charge redistribution due to the combined effects of the – partly occupied – surface states and the distortions of the valence band. For this we can use the analysis carried out in § 4.1, where we calculated the integrated density of states (4.38*b*) associated with the valence band in a one-dimensional system. Differentiating this result with respect to the distance d we have the local density of states in the valence band in the form

$$n_s^V(z, \eta_0) = -\frac{V}{\pi g}\int_0^\infty \frac{da}{a^2}\,\frac{1}{1+a^2+2a\cos(2\eta_0)}$$
$$\times\{-\cos(2qz)[(a^3+a)\cos(2\eta_0)+2a^2]$$
$$+\sin(2qz)(a^3-a)\sin(2\eta_0)\}. \quad (4.132)$$

As in § 4.1, q is $V(a^2-1)/ag^2$ and V and g are the effective potential and reciprocal lattice vector associated with the gap in question. The total charge distribution near the surface is then obtained by combining the results of the three one-dimensional problems with phase shifts η_0,

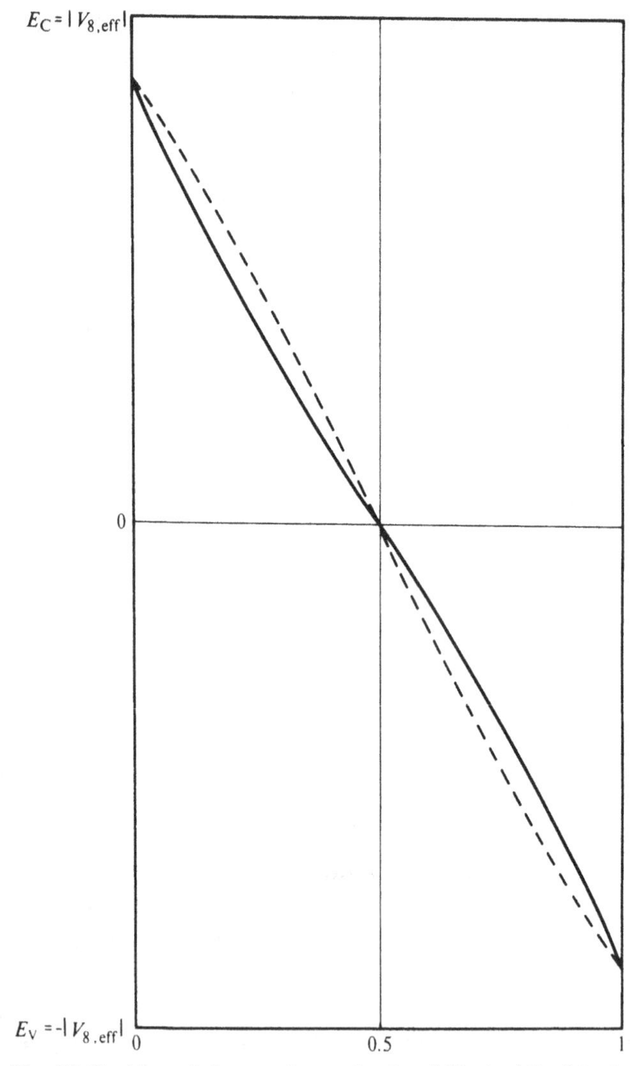

Fig. 82. Position of the s.s. eigenvalue level (dashed line) in the gap as a function of the occupation degree of the said state. The solid line gives the corresponding 'dipole neutrality' level for the same occupation degree. F. Flores, C. Tejedor & Martín-Rodero, *Phys. St. Solidi* (b) **88**, 591 (1978).

$\eta_0 \pm \frac{2}{3}\pi$. By again taking $\bar{J}$ as a representative point of a two-dimensional B.z. of area A_B in reciprocal space, the charge density profile due to the valence-band states is

$$n_s^V(z) = \frac{A_B}{(2\pi)^2}[n_s^V(z, \eta_0) + n_s^V(z, \eta_0 + \tfrac{2}{3}\pi) + n_s^V(z, \eta_0 - \tfrac{2}{3}\pi)]. \qquad (4.133)$$

On the other hand a fully occupied band of s.s. in this approximation produces a charge density

$$n^{ss}(z)=\frac{A_B}{(2\pi)^2}\frac{4|V\sin(2\eta_0)|}{g}\exp\left[-z\frac{4|V\sin(2\eta_0)|}{g}\right]. \quad (4.134)$$

As in § 4.1, we describe only the envelope and omit the comparatively fast oscillating factor $[1-\cos(gz+2\eta_0)]$. The total charge redistribution resulting from the partly occupied s.s. and from the distortion of the valence band – counted as electronic charge – has the envelope,

$$n_s^V(z)+Q_s n^{ss}(z). \quad (4.135)$$

Using the value of η_0 found for the initial equivalent 'metal' in (4.135) we can now study the effects of the s.s. and of the valence-band distortions. It is actually more convenient to study the surface dipole D_s of the charge distribution (4.135), i.e.

$$D_s=4\pi\int_0^\infty z[n_s^V(z)+Q_s n^{ss}(z)]\,dz. \quad (4.136)$$

Strictly speaking the integrand is only well defined for $z\geqslant 2\pi/g$ because of the approximation $q\ll g$ for all the states of interest, as was discussed in § 4.1, but the extension of the lower integration limit down to $z=0$ introduces only negligible errors in D_s. Using (4.132)–(4.136) yields

$$D_s=\frac{A_B}{(2\pi)^2}\left[D^V(\eta_0)+D^V(\eta_0+\tfrac{2}{3}\pi)+D^V(\eta_0-\tfrac{2}{3}\pi)+\frac{2\pi g}{V}\frac{Q_s}{\sin(2\eta_0)}\right], \quad (4.137a)$$

where

$$D^V(\eta_0)=\frac{\pi g}{2V}\frac{\sin(2\eta_0)}{1+\cos(2\eta_0)}+\frac{g}{V}\frac{2\eta_0}{\sin(2\eta_0)}. \quad (4.137b)$$

Remember that for the (1, 1, 1) surface $g=\frac{2}{3}h$ and $V=V_{8,\mathrm{eff}}$. For given s.s. occupation, (4.137) can be used to estimate the value of η_0 for which D_s vanishes. Since the s.s. eigenvalue is also determined by η_0, we can estimate, for a given Q_s, the value of it for which D_s vanishes. This is shown (full line) in fig. 82. For example, for $Q_s=(\frac{1}{2})\pm(\frac{1}{8})$ this requires s.s. levels at $\pm 0.20|V|$. The question is, what does $D_s=0$ mean in terms of charge cancellation?

It is possible to see that $D_s=0$ amounts to a fairly good cancellation of all the effects involved in surface charge redistribution. This is illustrated with a particular example in fig. 83, which shows the two terms of (4.135) for $\eta_0=0.69$. This is the value of η_0 for which $D_s=0$ with $Q_s=\frac{3}{8}$. It is seen

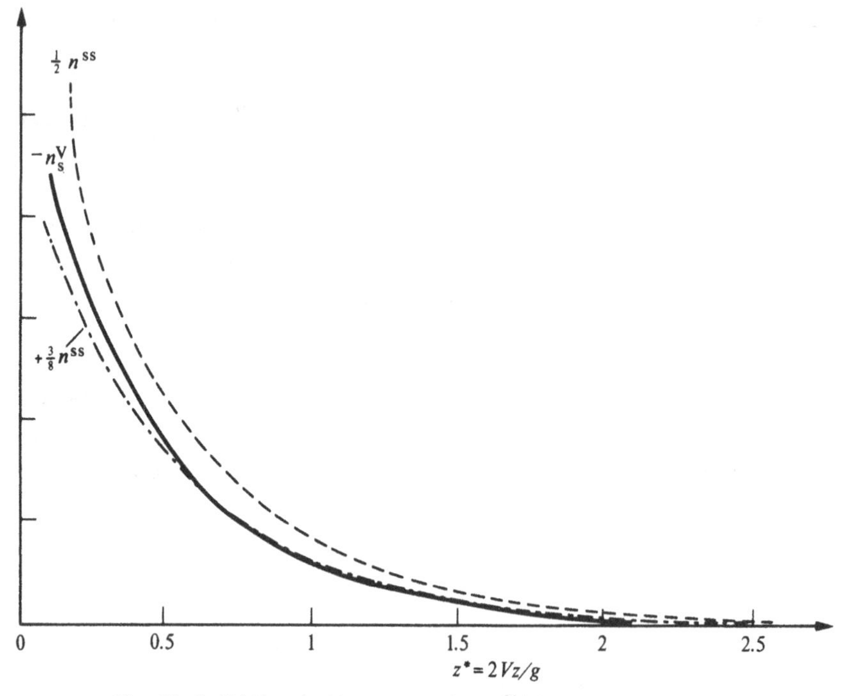

Fig. 83. Solid line (arbitrary units): $-n_s^V(z)$ as a function of the dimensionless variable $z^* = 2Vz/g$ for $\eta_0 = 0.69$. Dashed line: $\frac{1}{2}n^{ss}(z)$. Dash-dotted line: $\frac{3}{8}n^{ss}(z)$. Notice that n_s^V and $\frac{3}{8}n^{ss}$ nearly cancel out.

that the vanishing of the surface dipole corresponds to a rather good cancellation of the surface charge redistribution, except for distances very close to $z = 0$, where the approximations we have used to describe local charge become more inaccurate. It is also possible to see that if the s.s. level moves away from this 'dipole neutrality' level, then a dipole moment is induced which tends to shift the s.s. eigenvalue back to the vanishing dipole level. Notice, in the example of fig. 82, that this actually differs very little from the eigenvalue levels at $\pm 0.27|V|$ obtained from the unredistributed charge density which corresponds to the equivalent 'metallic' case, and that the difference vanishes in the covalent limit. These considerations give an approximate semiquantitative estimate of the extent to which the ionic surfaces differ from the quasimetallic character of the covalent surface layer. The indications here are that this departure is fairly small. Summing up, on the basis of this approximate analysis we may expect to find for the (1, 1, 1) faces of III–V compounds, (i) surface states with occupations $\frac{3}{8}$ and $\frac{5}{8}$ for the cation and anion faces, respectively; (ii) the bands of s.s. of the two opposite faces are nearly symmetric about a hypothetical intermediate band into which they coalesce in the covalent limit; (iii) the band with occupation $\frac{5}{8}$ is the lower one, reflecting the fact that the states associated with more electronic atoms have lower levels.

Fig. 84 describes an estimate of the s.s. bands for the (1, 1, 1) and $(\bar{1}, \bar{1}, \bar{1})$ faces of GaAs obtained by starting from the above results for the point $\bar{J}$ and assuming that the general features of these bands are similar to those found for a covalent crystal like Ge (fig. 77).

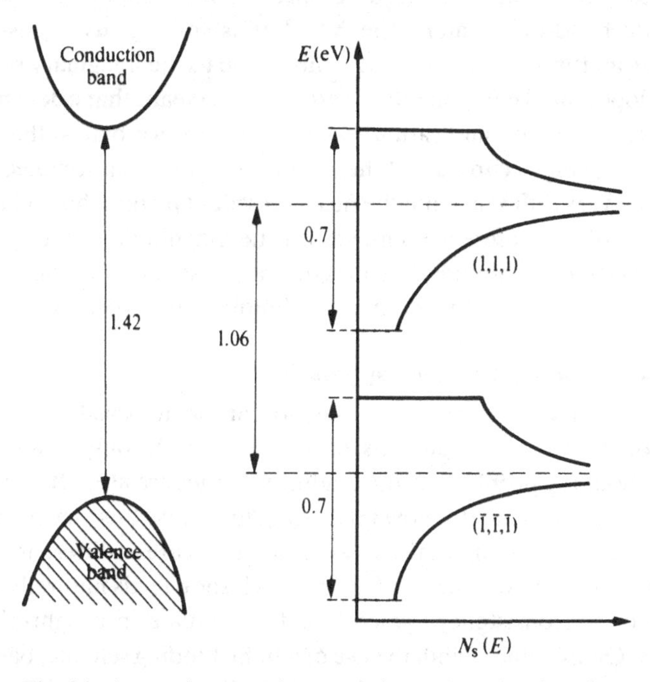

Fig. 84. Approximate estimate of the surface density of states for the polar GaAs faces, calculated by starting from the bulk band structure given in M. L. Cohen & T. K. Bergstresser, *Phys. Rev.* **141**, 789 (1966).

Although, as stressed above, the approximations we have made would be much too unreliable for II–VI compounds, it is interesting to extrapolate to this case in order to get a rough indication of what the picture would be. This would suggest that (i) the occupation of the s.s. bands is $(\frac{1}{2}) \pm (\frac{1}{4})$, and (ii) these bands are, for the two opposite faces, located about the mean gap with an average separation of about $\pm 0.41|V|$. For ZnS, for example, this would amount to 3.1 eV, compared with 1.1 eV for GaAs. This suggests that with ZnS s.s. bands stick very close to the bulk band edges.

Finally, the ionic (1, 1, 0) surfaces, unlike the (1, 1, 1) faces, are heteropolar, i.e. they consist of atoms of the two species. This is a very important difference. On averaging in the surface plane the properties of an ionic (1, 1, 0) surface are very similar to those of the covalent (1, 1, 0) surface. Furthermore, the effective two-band approximation eliminates small differences between the two cases, so that the picture of the covalent surface is essentially valid as a first approximation to the ionic case. Of course, a higher-order approximation for the bulk band structure introduces some characteristic differences. For example, the twofold degenerate band of s.s. along the $\bar{X}\bar{M}$ line is split up, as discussed in § 4.2.2. From the n.f.e. point of view this is due to the imaginary parts of the pseudopotential components. Physically this means that a degenerate band formed of anion and cation states splits into two bands, the lower one (cation) being occupied and the upper one (anion) unoccupied. This seems to be very different from the nearly covalent picture, but in fact the importance of the difference depends on the magnitude of the splitting which, for III–V compounds, is in practice rather small. This point will be further discussed in § 4.3 from the tight-binding point of view.

4.3. The tight-binding approach

Our discussion of surface states has so far been based on a n.f.e. pseudopotential approach and this, of course, is not the only possible way of going about the problem. Tight-binding schemes are also often used to study surface states and it is now interesting to discuss this viewpoint and to try and see whether or not these two approaches are as different as they at first appear. For example, a n.f.e. approach soon leads naturally to the question of selfconsistency, which is not immediate in a tight-binding approach. On the other hand, the use of a tight-binding scheme, based on the idea of atomic orbitals, naturally lends itself to a consideration of, say, the dangling bond s.s. from a rather more chemical point of view, in harmony with chemical notions like those of free or unsaturated radicals. It is also very likely that, as perfect surfaces become better understood and interest shifts to the study of various types of imperfections, an approach based on localised orbitals may prove more fruitful, and may also provide solutions to the problems of surface phonons and chemisorption. Thus it is important to try and understand the foundations of this method. The issues we are raising here are not affected by the form in which tight-binding calculations often appear in practice. It is a matter of how one uses this method. By its very nature the tight-binding approach lends itself to the construction of interpolation schemes in which by fitting parameters we may obtain a description of the band structure, but we shall not be concerned with this here. Our aim is to try and understand the theoretical basis of the tight-binding method and to stress the elements that the two approaches may have in common. We shall start with a simple one-dimensional model which serves to introduce the main theoretical issues, and then we shall discuss three-dimensional crystals.

4.3.1. One-dimensional covalent model

Let us study the linear monatomic chain shown in fig. 85. Each atom, labelled $i(i=1,\ldots,N)$, is the site of two electronic localised orbital states $|ia\rangle$ and $|ib\rangle$ and has two valence electrons. Bonds can be formed out of these orbitals and, with two valence electrons per atom, we have an elementary one-dimensional version of a possible covalent system. The model is not fully defined without specifying the interactions, and we shall define a very simple one with two parameters, namely, V_1 for the interaction between $|ia\rangle$ and $|ib\rangle$, and V_2 for the interaction between nearest neighbouring orbitals on different atoms. Furthermore, we shall neglect overlap between all orbitals. This simplifies the calculation of the total charge on each atom without diminishing the usefulness of this simple system as a model for the study of the main issues of interest. All this can be described in terms of a simple model Hamiltonian which reads as follows:

$$
\begin{aligned}
H = {} & \sum_{i=-\infty}^{\infty} E_0 |ia\rangle\langle ia| + \sum_{i=-\infty}^{\infty} E_0 |ib\rangle\langle ib| + \sum_{i=-\infty}^{\infty} V_1 |ia\rangle\langle ib| \\
& + \sum_{i=-\infty}^{\infty} V_1 |ib\rangle\langle ia| + \sum_{i=-\infty}^{\infty} V_2 |ib\rangle\langle i+1, a| \\
& + \sum_{i=-\infty}^{\infty} V_2 |i+1, a\rangle\langle i, b|. \qquad (4.138)
\end{aligned}
$$

Here E_0 is the mean energy level of the orbitals of the isolated atoms. In the following discussion we shall also treat this as a parameter.

The eigenstates of (4.138) are of the form

$$
|\psi\rangle = \alpha \sum_i e^{iqx_i} |ia\rangle + \beta \sum_i e^{iqx_i} |ib\rangle, \qquad (4.139)
$$

thus

$$
\begin{aligned}
H|\psi\rangle = {} & (\alpha E_0 + V_1\beta + V_2\beta\, e^{-iqc}) \sum_i e^{iqx_i} |ia\rangle \\
& + (\beta E_0 + V_1\alpha + V_2\alpha\, e^{+iqc}) \sum_i e^{iqx_i} |ib\rangle. \qquad (4.140)
\end{aligned}
$$

Fig. 85. Qualitative diagram describing a linear monatomic chain with two orbitals per atom.

If $|\psi\rangle$ is to be an eigenstate of H, then this must be $E|\psi\rangle$, whence the eigenvalue E must satisfy the equations

$$\left.\begin{aligned}(E_0-E)\alpha+(V_1+V_2\,\mathrm{e}^{-\mathrm{i}qc})\beta=0;\\ (V_1+V_2\,\mathrm{e}^{\mathrm{i}qc})\alpha+\beta(E_0-E)=0.\end{aligned}\right\}\qquad(4.141)$$

This yields the eigenvalues

$$E_q=E_0\pm\surd(V_1^2+V_2^2+2V_1V_2\cos qc).\qquad(4.142)$$

Using this back in (1.141) we find the ratio

$$\begin{aligned}&\beta/\alpha=\pm\omega(q)\exp(\tfrac{1}{2}\mathrm{i}qc);\\ &\omega(q)=\left[\frac{V_1\exp(-\mathrm{i}qc/2)+V_2\exp(\mathrm{i}qc/2)}{V_1\exp(\mathrm{i}qc/2)+V_2\exp(-\mathrm{i}qc/2)}\right]^{1/2}.\end{aligned}\qquad(4.143)$$

This yields the two (unnormalised) eigenstates

$$|\psi\rangle=\sum_i\exp(\mathrm{i}qx_i)|ia\rangle\pm\sum_i\omega(q)\exp[\mathrm{i}q(x_i+\tfrac{1}{2}c)]|ib\rangle.\qquad(4.144)$$

The two signs for E and $|\psi\rangle$ correspond to the two bands shown in fig. 86. These are the valence and conduction bands of our simple covalent model and, in this case, they are symmetrical about the mean free atom orbital level E_0. Furthermore, the two eigenstates for a given q differ only in the

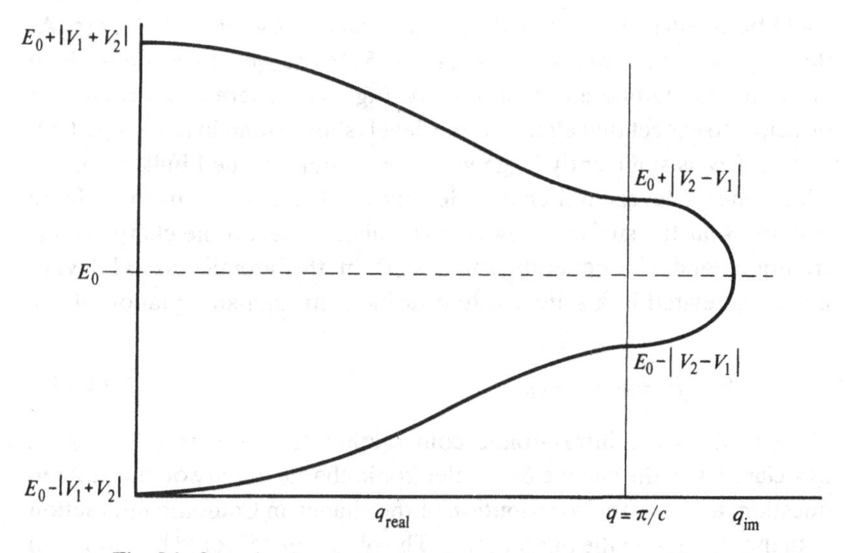

Fig. 86. Complex band structure for the linear monatomic chain of fig. 85.

sign in front of the factor $\omega(q)$ and, with the assumption of vanishing overlap between $|ia\rangle$ and $|ib\rangle$ it is easy to see that both wavefunctions have the same associated charge, and this charge, owing to the fact that ω is a complex number of unit modulus, is equally distributed among all the orbitals of the 'crystal'.

Now that we have the bulk band structure we can proceed as we did in the n.f.e. case, i.e. study the complex band structure corresponding to the real energy loop in fig. 86. This figure has been drawn for the case in which V_1 and V_2 have the same sign. Then the top of the valence band and the bottom of the conduction band are at the edge of the Brillouin zone, and the dispersion relation (4.142) can be rewritten as

$$\cos\frac{qc}{2} = \pm\left[\frac{(E-E_0)^2-(V_1-V_2)^2}{4V_1V_2}\right]^{1/2}. \tag{4.145}$$

Since we are interested in surface-state energies in the gap, we take $|E-E_0|<|V_1-V_2|$. Then (4.145) yields

$$\frac{qc}{2} = \frac{\pi}{2} \pm \mathrm{i}\sinh^{-1}\left[\frac{(V_1-V_2)^2-(E-E_0)^2}{4V_1V_2}\right]^{1/2}. \tag{4.146}$$

The real part yields just the value of q corresponding to the edge of the Brillouin zone, and the imaginary part determines the complex momentum corresponding to real energies in the gap.

We thus have the standard framework with which we could start the study of surface states much as we did in § 4.1. Evanescent wavefunctions could be constructed with appropriate choice of sign in (4.146), etc. At this stage we face the following problem. Suppose we make a finite chain of N atoms, numbered from 1 to N (fig. 85). There is no reason in principle to expect that all the atomic levels should remain unchanged. Of course if N is sufficiently large we have an unperturbed bulk region in which energy levels and charge density are the same as in the infinite system. Near the surfaces however we might expect some charge redistribution, and consequently some shift in the atomic orbital levels, mutually related in a selfconsistent fashion through an equation of the form

$$E_i = E_0 + \mathscr{E}\,\delta n_i + v_i. \tag{4.147}$$

Here $\mathscr{E}\,\delta n_i$ is the intra-atomic contribution to the energy-level shift associated with the change δn_i in electronic charge density on the atom in question, and v_i is the contribution of the change in Coulomb interaction with the charges on the other atoms. The parameter $\mathscr{E}$ could be estimated from free atom spectroscopic terms. For example, $\mathscr{E}$ could be taken as the difference between the electro-affinity and the first ionization level of the atom if $\delta n_i > 0$, or as the difference between the first and second ionization levels if $\delta n_i < 0$. On the other hand v_i is a typical parameter of the nature of a Madelung constant and can be evaluated by standard methods. We would expect E_i to be different from E_0 at least (in the simplest model) for $i = 1$ and $i = N$. Moreover, in physical terms, we should look at the interaction parameters, and see whether anything makes them change. An obvious mechanism is surface relaxation, which would clearly require at least a change in V_2 for the surface atoms. However, we shall not study this situation here, and throughout our analysis we shall assume that V_1 and V_2 are not modified. Our purpose is only to discuss how the basic ideas of a selfconsistent description can find a natural place in a tight-binding scheme. Then, consistent with (4.147), we could start from the following model Hamiltonian for the surface system:

$$H_s = \sum_{i=1}^{N} E_i |ia\rangle\langle ia| + \sum_{i=1}^{N} E_i |ib\rangle\langle ib| + \sum_{i=1}^{N} V_1 |ia\rangle\langle ib| + \sum_{i=1}^{N} V_1 |ib\rangle\langle ia| + \sum_{i=1}^{N-1} V_2 |ib\rangle\langle i+1, a| + \sum_{i=1}^{N-1} V_2 |i+1, a\rangle\langle ib|. \quad (4.148)$$

In the simplest conceivable case we would have $E_i = E_0$ for all atoms. We shall later discuss the possible relevance of this situation. For the time being let us see how the s.s. of the finite linear chain come about in this case. The corresponding eigenstates can be obtained from (4.144) by using the values of q given by (4.146) for energies in the gap. Suppose we consider the surface at $x_i = 0$. Then we can take $N \to \infty$ and we must take the + sign in (4.146) in order to ensure exponential decay of the wavefunction into the bulk. Thus we seek surface eigenstates of the form

$$|\psi_{ss}\rangle = \sum_{i=1}^{\infty} \exp(iqx_i)|ia\rangle \pm \omega(q) \sum_{i=1}^{\infty} \exp[iq(x_i + \tfrac{1}{2}c)]|ib\rangle, \quad (4.149)$$

with the above specification for the sign used in evaluating q. The $\pm$ sign appearing in (4.149) itself depends on whether the corresponding eigenvalue is above or below the midgap. Acting on (4.149) with the Hamiltonian (4.148) in which $N \to \infty$ we have

$$H_s|\psi_{ss}\rangle = E_q|\psi_{ss}\rangle \mp V_2 \exp(-\tfrac{1}{2}iqc)\omega(q)|1a\rangle. \quad (4.150)$$

Thus the condition for $|\psi_{ss}\rangle$ to be an eigenstate is

$$\omega(q) = 0: \quad V_1 \exp(-\tfrac{1}{2}iqc) + V_2 \exp(\tfrac{1}{2}iqc) = 0. \quad (4.151)$$

Using (4.140) yields the s.s. eigenvalue equation

$$(V_1+V_2)\surd[(V_1-V_2)^2-(E-E_0)^2] \\ +(V_1-V_2)\surd[(V_1+V_2)^2-(E-E_0)^2]=0. \tag{4.152}$$

The sign convention for the square roots is already explicitly expressed in this equation, so that the positive root is taken in both terms. Thus (4.152) has a real root for E only if $|V_1|<|V_2|$, and then the s.s. eigenvalue is $E=E_0$.

Suppose, on the contrary, that not all the atomic orbital levels remain unchanged. For example, imagine that the level for atom 1 has a value E_1, while all the others have the level E_0. Then if we try (4.149) for a s.s. we find

$$H_s|\psi_{ss}\rangle = E_q|\psi_{ss}\rangle \mp V_2 \exp(-\tfrac{1}{2}iqc)\omega(q)|1a\rangle \\ +(E_1-E_0)|1a\rangle \pm (E_1-E_0)\exp(\tfrac{1}{2}iqc)\omega(q)|1b\rangle. \tag{4.153}$$

Therefore (4.149) cannot be an eigenstate of H_s. In order to achieve the simultaneous vanishing of the coefficients of $|1a\rangle$ and $|1b\rangle$ we must try a combination of the form

$$|\phi_{ss}\rangle = \alpha|(4.149)\rangle + \beta|1a\rangle.$$

Then

$$H_s|\phi_{ss}\rangle = E_q|\phi_{ss}\rangle + \beta[(E_1-E_q)|1a\rangle + V_1|1b\rangle] \\ +\alpha[\mp V_2 \exp(-\tfrac{1}{2}iqc)\times\omega(q)|1a\rangle + (E_1-E_0)|1a\rangle \\ \pm(E_1-E_0)\exp(\tfrac{1}{2}iqc)\omega(q)|1b\rangle] \tag{4.154}$$

and the simultaneous vanishing of the coefficients of $|1a\rangle$ and $|1b\rangle$ yields the two equations

$$\left.\begin{aligned} &[\mp V_2 \exp(-\tfrac{1}{2}iqc)\omega(q)+(E_1-E_0)]\alpha+(E_1-E_q)\beta=0 \\ &[\pm(E_1-E_0)\exp(\tfrac{1}{2}iqc)\omega(q)]\alpha+V_1\beta=0, \end{aligned}\right\} \tag{4.155}$$

which then yield the secular equation for the s.s. eigenvalue. We shall not delve any further into this. The above suffices to see how further orbital terms must be added to (4.149) as the surface perturbation penetrates into the bulk. In fact this involves the orbitals of all perturbed atoms, except the b orbital of the last one.

In order to discuss selfconsistency we must also study, besides stationary s.s., the changes in the extended bulk states. We saw in chapter 2 and §§ 4.1, 4.2 that this contributes to the total change in the density of states, i.e. to the surface density of states and therefore to the surface charge. Thus we must study the – occupied – valence band in the presence of the surface. Consider again the simple case in which $E_i = E_0$ for all i. The eigenstates of the bulk band can now be written in the form

$$|\psi\rangle = A|\psi_q\rangle + B|\psi_{-q}\rangle, \tag{4.156}$$

where $|\psi_q\rangle$ is the state of (4.142) specified for the valence band of the finite linear chain. Notice the difference with (4.149), which describes an evanescent s.s. Owing to the requirement of exponential decay we only have terms in $+q$ in (4.149) because the terms in $-q$ would increase to infinity for $x > 0$. Here, in (4.156), with q real both $+q$ and $-q$ go into the making of an eigenstate of the bulk band. Physically this corresponds to waves incident on, and reflected from, a given surface. Operating with (4.148) on (4.156) yields, after some algebra,

$$\begin{aligned} H_s|(4.156)\rangle = E_q|(4.156)\rangle \\ - A V_2[\exp(iNqc)|Nb\rangle \pm \omega(q)\exp(-\tfrac{1}{2}iqc)|1a\rangle] \\ - B V_2[-\exp(-iNqc)|Nb\rangle \pm \omega^*(q)\exp(\tfrac{1}{2}iqc)|1a\rangle]. \end{aligned} \tag{4.157}$$

Thus the conditions for (4.156) to be an eigenstate of H_s are

$$\left.\begin{aligned} \exp(iNqc)A + \exp(-\tfrac{1}{2}iqc)B = 0, \\ \omega(q)\exp(-\tfrac{1}{2}iqc)A + \omega^*(q)\exp(\tfrac{1}{2}iqc)B = 0. \end{aligned}\right\} \tag{4.158}$$

Then the eigenvalue, by (4.157), is E_q as given in (4.142), i.e. the same as for the bulk eigenstate with the same q, while the compatibility condition of equations (4.158) yields the allowed values of q for the finite linear chain. By (4.143) this is

$$V_2 \sin(Nqc) + V_1 \sin[(N+1)qc] = 0. \tag{4.159}$$

The graphical solution of this equation is shown in fig. 87 for $N = 3$ and both V_1 and V_2 negative, in two different situations: one in which $|V_1| < |V_2|$ and another in which $|V_1| > |V_2|$. The results obtained with $N = 3$ demonstrate the following general rules:

(i) If $|V_1| < |V_2|$, then the number of solutions of (4.159), for $0 < q < \pi/c$, is $(N-1)$.

(ii) If $|V_1| > |V_2|$, then this number is N.

Notice that the values $q = 0$, π/c do not yield bulk eigenstates of H_s. For example, for $q = 0$ the combination (4.156) is simply reduced to $|\psi(q=0)\rangle$, which is not an eigenstate of H_s, as can be easily checked by acting with H_s on $|\psi(q=0)\rangle$. On the other hand, the same analysis applies

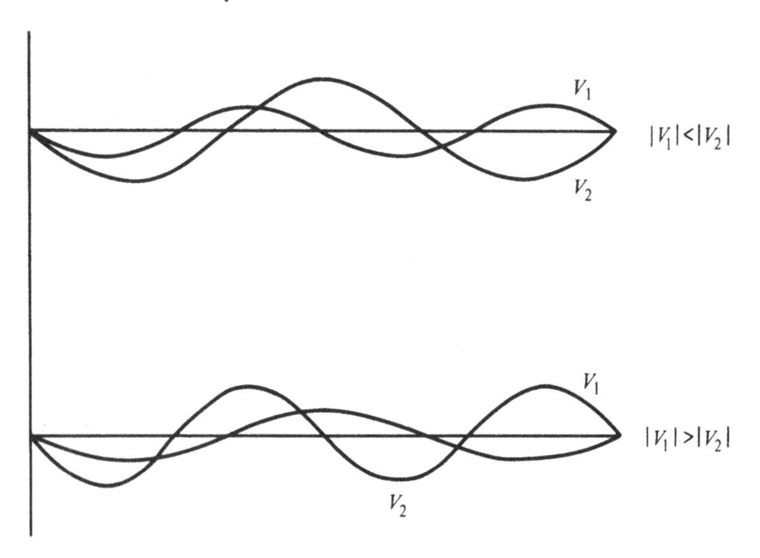

Fig. 87. Graphical solution of (4.159). See text.

equally to the valence and conduction bands, and putting all this together we have the following rules for the total number of bulk states, both bands included:

(i) If $|V_1|<|V_2|$, then there are $2(N-1)$ bulk states, i.e. $(N-1)$ in each band.

(ii) If $|V_1|>|V_2|$, then there are $2N$ bulk states, i.e. $(N-1)$ in each band.

If we recall the results obtained for the s.s. in (4.152), then we find that in both cases the *total* number of states – bulk and surface – is $2N$. Thus in this simple example we see an important conservation rule verified, while the discussion is very similar to the analysis carried out in § 4.1, including the study both of the new states which may appear in the gap, and the changes in the bulk continuum.

Let us also briefly consider how one could go about this problem when the atoms near the surface are perturbed. Consider again the case in which only the end atoms of the finite chain are perturbed. Proceeding as in the case of s.s. we find that now (4.156) cannot give a bulk eigenstate of H_s. We can then see that a combination of the form $A|\psi_q\rangle+B|\psi_{-q}\rangle+C|1a\rangle+D|Nb\rangle$ must be used, involving the two end orbital states. The condition for this to be an eigenstate of H_s now yields a determinant from which we can see how many bulk states there are. It is then easy to verify the conservation rule again and, indeed, this can be extended to the case in which more atoms near the surface are perturbed.

We are now ready to discuss the problem of selfconsistency. As in § 4.1, we shall start by studying a 'zero-order approximation'. The approach is similar in this respect although the details are different. In the n.f.e. analysis of §§ 4.1, 4.2 we started from a hypothetical model of a metal with the same number of valence electrons, solved this problem selfconsistently, and then switched on the crystalline pseudopotential. Here our zero-order approximation is a linear chain in which we assume that the surface atoms are unperturbed, i.e. all $E_i = E_0$. It is not obvious that this can be a selfconsistent solution, so we must study this problem.

Remember that we have found that the rule of conservation of the total number of states is satisfied and also that in the case $|V_1| < |V_2|$ the s.s. appearing on each face is just at the midgap. Let us in particular consider the case $V_2 < V_1 < 0$. Then the orbitals on neighbouring atoms tend to form a bonding configuration. This is the analogue of the bonding case in the pseudopotential model, when the potential is attractive in the region between atoms. Notice that if V_2 is negative, then V_1 must also be negative for the gap to appear at the edge of the Brillouin zone, as in fig. 86. With the model and the solution it is easy to see that the occupation of the s.s. is $\frac{1}{2}$. Indeed, with $V_2 < V_1 < 0$, the $2N$ electronic charges of the N atoms are distributed as follows: $2(N-1)$ electrons – spin included – in the $(N-1)$ states of the occupied valence band and two electronic charges in the two s.s., one at each surface. This leaves one electronic charge for one s.s. and, if we account for spin, this means occupation $\frac{1}{2}$. Now we must investigate whether this surface charge induces any charge redistribution in the atoms. For this it is convenient to first prove a theorem concerning the total density of states associated with each atom for the Hamiltonian H_s of (4.148). We have seen that the total number of states of the finite linear chain is $2N$, not counting spin. The theorem says, moreover, that *the total number of states on each one of the atoms, no matter which one it is, is always two* if the system is described by the Hamiltonian (4.148). Notice that in this argument the E_is in principle need not all be equal to E_0. The proof runs as follows. Since H_s is not only Hermitian but also symmetric, the eigenfunctions can be written in the form

$$\psi_q^\nu(x) = \sum_i A_i^\nu(q) f_{ia}(x) - \sum_i B_i^\nu(q) f_{ib}(x), \tag{4.160}$$

where $f_{ia}(x)$, $f_{ib}(x)$ are the normalised wavefunctions of the corresponding atomic orbitals, ν is a band index, and the coefficients $A_i^\nu(q)$, $B_i^\nu(q)$ are real. With the assumption of negligible overlap between orbitals we have

$$|\psi_q^\nu(x)|^2 = \sum_i |A_i^\nu(q)|^2 |f_{ia}(x)|^2 + \sum_i |B_i^\nu(q)|^2 |f_{ib}(x)|^2.$$

The normalisation of the wavefunction requires that

$$\sum_i |A_i^\nu(q)|^2 + \sum_i |B_i^\nu(q)|^2 = 1. \tag{4.161}$$

Now, with the coefficients $A_i^\nu(q)$, $B_i^\nu(q)$ we can form a square matrix with $2N$ rows, labelled by (Λ, i) (where $\Lambda = A, B$) and $2N$ columns, labelled by (ν, q) (where ν indicates valence, conduction). Notice that (ν, q) may refer not only to a band state but also to a s.s. when q is complex. The important point is that this matrix is orthogonal, hence besides (4.161) we also have, for the elements of all rows,

$$\sum_{\nu,q} |A_i^\nu(q)|^2 = \sum_{\nu,q} |B_i^\nu(q)|^2. \tag{4.162}$$

This shows that taking account of the two orbitals on each atom – but not accounting for spin – the total number of states on each atom is just two.

Let us now return to the simple case in which all $E_i = E_0$. Then the bulk band eigenstates can all be written in the form (4.156) and the difference between the eigenstates of the valence and conduction band with the same q lies only in the sign in front of $\omega(q)$ in (4.144). Therefore the wavefunctions of both bands localise the same number of states on any given atom. Since the valence-band states are all occupied and those of the conduction band are all empty, on that account the total amount of charge on each atom is given by the occupation of exactly half the number of band states. But we have also seen that the occupation of the s.s. is $\frac{1}{2}$ just when all $E_i = E_0$. Thus we have on each atom the charge which results from the occupation of half the *total* number of states which, by virtue of the theorem just proved, is exactly two. If we take account of spin degeneracy this means there are two electronic charges per atom, and this applies to all atoms even when there are surface states. In other words, the surface charge associated with s.s. does not modify the total charge associated with the surface atoms. Since this is all based on the assumption that $E_i = E_0$, it is clear, by (4.147) that we have a selfconsistent solution. Thus the model Hamiltonian (4.148), with $E_i = E_0$, represents an extremely simple model indeed, but it is selfconsistent. This is its main feature and it is not an immediately obvious one. However, this property, even for the simple model Hamiltonian, is a relatively unimportant peculiarity of the one-dimensional case. The really important features are (i) that the s.s. is located at the midgap, and (ii) that its occupation is $\frac{1}{2}$. It is interesting that we have reached, via a tight-binding approach, the same conclusions as we reached for the one-dimensional n.f.e. model.

As indicated above, this simplicity might disappear, for example, if we had surface relaxation. This would change the values of some of the parameters and complicate the analysis. It could be done, but we would get involved in a lengthy argument. Our purpose here is simply to see how one can go about obtaining selfconsistency within the framework of a tight-binding approach. Moreover, we shall look at the zero-order approximation in an even simpler case, namely that in which we take $V_1 = 0$ and $V_2 < 0$. This is the so-called *molecular model*, in which we neglect the interaction between orbitals on the same atom and consider only a bonding interaction between orbitals on neighbouring atoms. This model, being extremely simple, has the advantage that it is very easy to study, and for this reason it is useful as a flexible device for drawing qualitative conclusions in the more difficult three-dimensional case, as we shall see in § 4.3.2.

To begin with, the picture of the bulk system is extremely simplified. In the one-dimensional case the two bands become flat and each one coalesces into the multiply-degenerate energy level $E_0 \pm V_2$. This would be the limit $V_1 \to 0$ in fig. 86. Now consider the finite chain with N atoms. The $(N-1)$ bonds give $2(N-1)$ energy levels which might even be different if a surface perturbation penetrated into the bulk. Besides these states we have two uncoupled orbitals at the ends of the chain. These are the dangling bonds of our model, amounting in this case to two s.s. with energy levels $E_1 = E_N$. From the previous analysis it is easy to see that setting all $E_i = E_0$ we have a selfconsistent solution, with the bulk bands at $E_0 \pm V_2$ and the s.s. at E_0, the valence band fully occupied, and the s.s. with occupation $\frac{1}{2}$. We shall use this simplified molecular model to discuss qualitatively selfconsistency in three dimensions.

4.3.2. Three-dimensional models

We now discuss three-dimensional covalent crystals again using simple model Hamiltonians. The well-known structure involving tetrahedral coordination is shown in fig. 88. The four orbitals on each atom correspond to the sp^3 hybrid states obtained from $|s\rangle$, $|p_x\rangle$, $|p_y\rangle$ and $|p_z\rangle$ by means of the tetrahedral combinations

$$|1\rangle = |s\rangle + |p_x\rangle + |p_y\rangle + |p_z\rangle; \quad |2\rangle = |s\rangle + |p_x\rangle - |p_y\rangle - |p_z\rangle;$$

$$|3\rangle = |s\rangle - |p_x\rangle + |p_y\rangle - |p_z\rangle; \quad |4\rangle = |s\rangle - |p_x\rangle - |p_y\rangle + |p_z\rangle.$$

In one unit cell we have two atoms and therefore eight such orbitals. The bulk band structure, to begin with, is again determined by the interactions between these orbitals and we start once again from the model Hamiltonian

$$H = \sum_{i,j} E_0 |i\boldsymbol{R}_j\rangle\langle i\boldsymbol{R}_j| + \sum_j \sum_{i \neq i'} V_1 |i\boldsymbol{R}_j\rangle\langle i'\boldsymbol{R}_j| + \sum_{\text{bonds}} V_2 |i\boldsymbol{R}_j\rangle\langle i'\boldsymbol{R}_{j'}|. \qquad (4.163)$$

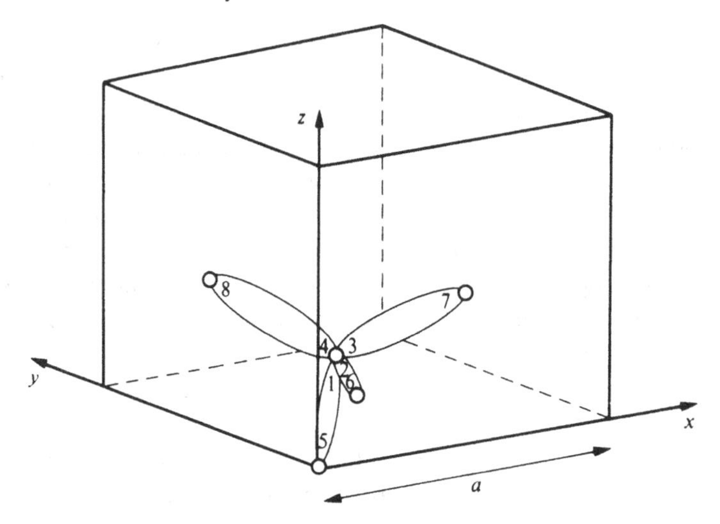

Fig. 88. Orbitals and bonds in the unit cell of the diamond lattice.

The second term describes the interaction between different orbitals on the same atom, while the third term represents the interactions between neighbouring orbitals which form the bonds between atoms. In order to seek the eigenstates of this Hamiltonian we consider the states obtained by forming the Bloch sum

$$|i, \boldsymbol{k}\rangle = \sum_{\boldsymbol{R}} \mathrm{e}^{\mathrm{i}\boldsymbol{k}\cdot\boldsymbol{R}} |\psi_i(\boldsymbol{R})\rangle,$$

where the $\boldsymbol{R}$ are the vectors of the f.c.c. lattice and the index i ($i = 1, \ldots, 8$) labels for each cell the eight orbital states of the two atoms in each cell. The eigenstate of (4.163) with momentum $\boldsymbol{k}$ is then sought in the form

$$|\psi_{\boldsymbol{k}}\rangle = \sum_i A_i(\boldsymbol{k})|i, \boldsymbol{k}\rangle. \tag{4.164}$$

Let us define the vector column formed by the eight coefficients A_i and the 4×4 matrices M_1, M_2 and M_3 as follows: $[M_1]_{ii} = E_0 - E$, $[M_1]_{ij} = V_1$ for $i \neq j$, $[M_2]_{ij} = V_2\delta_{ij}$, and

$$M_3 =$$

$$\begin{bmatrix} E - E_0 & V_1 \exp(\tfrac{1}{2}\mathrm{i}\boldsymbol{k}\cdot\boldsymbol{R}_{101}) & V_1 \exp(\tfrac{1}{2}\mathrm{i}\boldsymbol{k}\cdot\boldsymbol{R}_{110}) & V_1 \exp(\tfrac{1}{2}\mathrm{i}\boldsymbol{k}\cdot\boldsymbol{R}_{001}) \\ V_1 \exp(-\tfrac{1}{2}\mathrm{i}\boldsymbol{k}\cdot\boldsymbol{R}_{101}) & E - E_0 & V_1 \exp(\tfrac{1}{2}\mathrm{i}\boldsymbol{k}\cdot\boldsymbol{R}_{01\bar{1}}) & V_1 \exp(\tfrac{1}{2}\mathrm{i}\boldsymbol{k}\cdot\boldsymbol{R}_{\bar{1}10}) \\ V_1 \exp(-\tfrac{1}{2}\mathrm{i}\boldsymbol{k}\cdot\boldsymbol{R}_{110}) & V_1 \exp(-\tfrac{1}{2}\mathrm{i}\boldsymbol{k}\cdot\boldsymbol{R}_{01\bar{1}}) & E - E_0 & V_1 \exp(\tfrac{1}{2}\mathrm{i}\boldsymbol{k}\cdot\boldsymbol{R}_{\bar{1}01}) \\ V_1 \exp(-\tfrac{1}{2}\mathrm{i}\boldsymbol{k}\cdot\boldsymbol{R}_{001}) & V_1 \exp(-\tfrac{1}{2}\mathrm{i}\boldsymbol{k}\cdot\boldsymbol{R}_{\bar{1}10}) & V_1 \exp(-\tfrac{1}{2}\mathrm{i}\boldsymbol{k}\cdot\boldsymbol{R}_{\bar{1}01}) & E - E_0 \end{bmatrix}$$

The Schrödinger equation, obtained by acting with (4.163) on (4.164) then yields eight linear homogeneous equations which can be written in concise matrix form as

$$\begin{bmatrix} M_1 & M_2 \\ M_2 & M_3 \end{bmatrix}[A]=0. \tag{4.165}$$

The matrix multiplying the vector column A is the 8×8 one formed by arranging the 4×4 matrices as indicated. The secular equation, by means of appropriate linear combinations of the basis states, admits the following solutions (energy eigenvalues):

$$E_1=-V_1+V_2 \quad \text{(doubly degenerate);} \tag{4.166a}$$

$$E_2=-V_1-V_2 \quad \text{(doubly degenerate);} \tag{4.166b}$$

$$E_3=V_1\pm[4V_1^2+V_2^2\pm 2V_1V_2(1+\mu)^{1/2}]^{1/2}, \tag{4.166c}$$

where

$$\mu=\cos(\tfrac{1}{2}k_xc)\cos(\tfrac{1}{2}k_yc)+\cos(\tfrac{1}{2}k_yc)\cos(\tfrac{1}{2}k_zc)+\cos(\tfrac{1}{2}k_zc)\cos(\tfrac{1}{2}k_xc).$$

This yields the band structure consisting of two flat (non-dispersive) doubly degenerate bands and the four bands of (4.166c). These are shown in fig. 89 for $V_1=-2$ eV, $V_2=-6$ eV and k varying along the line

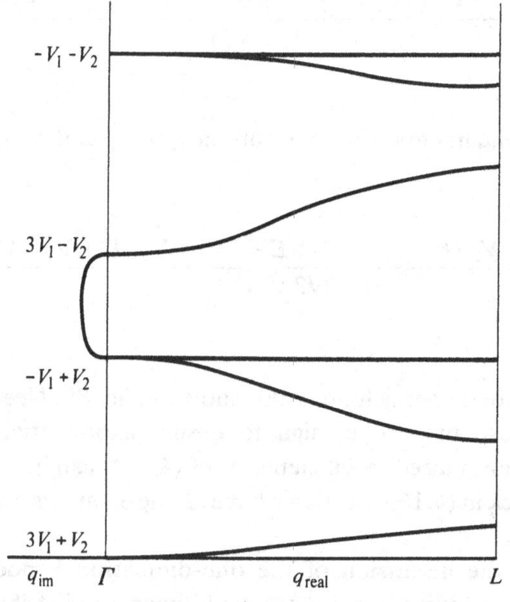

Fig. 89. Band structure – including the real energy loop spanning the principal gap – corresponding to (4.166) for a covalent semiconductor with $V_1=-2$ eV, $V_2=-6$ eV.

ΓL. Of course this is an extremely crude model, but it is the model system we shall use for a qualitative discussion of the basic theoretical issues concerning the surface problem in a tight-binding approach. The main feature is that we have a total of eight bands, distributed in two sets of four separated by an energy gap. Below this gap we have the four valence bands occupied by the eight valence electrons per unit cell. This is the fundamental gap in this model system. We are now ready to study the surface problem and, as before, we shall start by looking at the main features of the complex-band structure.

We take the (1, 1, 1) surface, whose two-dimensional Brillouin zone is shown in fig. 68. Thus, for a given (real) $\boldsymbol{\kappa}$ in this zone, we are interested in q, the component of k perpendicular to the surface, which can be real or complex. In particular we concentrate on the centre of the zone. Thus we study the complex band structure for $\boldsymbol{\kappa}=0$ and q varying along the line ΓL (fig. 89). In this direction we can write $k=q(1, 1, 1)$ and using this in (4.166c) we have

$$E_q = V_1 \pm [4V_1^2 + V_2^2 \pm 2V_1V_2|1+3\cos^2(\tfrac{1}{2}qc)|^{1/2}]^{1/2}, \qquad (4.167)$$

or, equivalently,

$$\sin(\tfrac{1}{2}qc) = \pm \frac{[(E+V_1+V_2)(3V_1+V_2-E)(E-3V_1+V_2)(E+V_1-V_2)]^{1/2}}{2\sqrt{3}\,V_1V_2} \qquad (4.168)$$

For real E in the fundamental gap, q is pure imaginary and we have

$$\tfrac{1}{2}qc = \pm \mathrm{i}\sinh^{-1} \times \left\{\frac{[(E+V_1+V_2)(E-3V_1-V_2)(E-3V_1+V_2)(E+V_1-V_2)]^{1/2}}{2\sqrt{3}\,V_1V_2}\right\}. \qquad (4.169)$$

The corresponding real energy loop is also shown in fig. 89. Needless to say, we must choose the proper sign to ensure exponential decay. Moreover, the (unnormalised) coefficients A_i of (4.164) can be obtained by using (4.169) back in (4.165). Thus we have all the elements needed to study surface states.

As indicated in the discussion of the one-dimensional model, the surface region may be perturbed, i.e. we could have a redistribution of charge and, consequently, a change in atomic levels. Thus in general the selfconsistent description of the surface may require solving simultaneously (4.147) and the eigenvalue problem of the Hamiltonian,

$$H_s = \sum_{i,j} E_i |i\boldsymbol{R}_j\rangle\langle i\boldsymbol{R}_j| + \sum_j \sum_{i\neq i'} V_1 |i\boldsymbol{R}_j\rangle\langle i'\boldsymbol{R}_j| + \sum_{\text{bonds}} V_2 |i\boldsymbol{R}_j\rangle\langle i'\boldsymbol{R}_{j'}|. \tag{4.170}$$

Remember that the assumption that V_1 and V_2 do not change means that we are not studying relaxed surfaces. The s.s. of this Hamiltonian can be studied by proceeding in a manner similar to the analysis of § 4.3.1. Let us specifically consider the (1, 1, 1) surface and study the case in which all $E_i = E_0$. Putting $\boldsymbol{k} = q(1, 1, 1)$ in (4.164), we look for a s.s. of the form

$$|\psi_{ss}\rangle = \sum_i A_i \sum_{\boldsymbol{R}} \exp\left[-q(R_x + R_y + R_z)\right] |\psi_i(\boldsymbol{R})\rangle, \tag{4.171}$$

in which $\boldsymbol{R}$ runs through the inside of the half crystal and q is given by (4.169). Putting $E_i = E_0$ in (4.170) and acting with H_s on (4.171) yields

$$H_s|\psi_{ss}\rangle = E_q|\psi_{ss}\rangle - V_2 A_5 \sum_{\boldsymbol{R}_s} |\psi_1(\boldsymbol{R}_s)\rangle. \tag{4.172}$$

Here E_q is given by (4.161) and $\boldsymbol{R}_s$ refers to lattice points of the last atomic plane. Notice that on forming the (1, 1, 1) surface we cut (fig. 88) the bond formed by orbitals 1 and 5 with coupling constant V_2. Thus the condition for (4.171) to be an eigenstate is

$$A_5(q) = 0. \tag{4.173}$$

Notice the similarity with the situation encountered in § 4.3.1. Likewise it is easy to see that, if the last atomic plane were perturbed, then the mean level of the orbitals 1, 2, 3, 4 would change from E_0 to another value, E_1. In this instance (4.171) could not be an eigenstate of (4.170) and it would then be necessary to form the combination

$$\alpha|(4.171)\rangle + \sum_{\boldsymbol{R}_s} \sum_{i\neq 5} B_i |\psi_i(\boldsymbol{R}_s)\rangle. \tag{4.174}$$

The orbital labelled 5 is precisely the one which is removed from each unit cell when the (111) surface is formed (fig. 88). This is the three-dimensional extension of the combination formed for the one-dimensional system, where one orbital is removed from the surface atom and only one remains. From the discussion of § 4.3.1 it is easy to see how one should proceed here. We shall omit further details but it is interesting to notice that the same combination (4.174) can be used if the surface perturbation affects not one but two atomic planes. From there on, when the number of perturbed atomic planes increases by two the number of orbitals to be included in the linear combination increases by eight.

Let us now return to the zero-order approximation ($E_i = E_0$) and discuss the problem of selfconsistency for the (1, 1, 1) surface. We want to see whether the creation of the surface produces a charge redistribution, i.e. in the tight-binding formulation, a change in atomic charges, which would then require a change in atomic orbital levels. The result found for the one-dimensional system depended crucially on the complete symmetry between valence and conduction bands. The situation now is different. In the band structure of the model Hamiltonian shown in fig. 89 there are only two of the four dispersive bands which are symmetric with respect to the fundamental gap. The rest of the bands are not. Because of this asymmetry the solution with all $E_i = E_0$ is no longer selfconsistent, i.e. it is necessary to effect some charge redistribution in the atoms near the surface. The study of the selfconsistent solution then becomes rather lengthy, even with a simple model Hamiltonian. In order to obtain some insight while yet avoiding an elaborate analysis we shall resort to the even simpler *molecular model*, briefly discussed at the end of § 4.3.1. With $V_1 = 0$ the bulk band structure is simplified to the utmost. As in the one-dimensional case, all the energy levels coalesce into the flat degenerate bands with energy $E_0 \pm V_2$, corresponding to the bonding and antibonding levels of the interatomic couplings. The problem of selfconsistency then becomes extremely simple, although of course this can only serve to obtain a qualitative picture. When the surface is formed the energy level of the resulting dangling bonds is E_0. This is the energy eigenvalue of the covalent s.s. in this model, with occupation again $\frac{1}{2}$. All atoms – bulk or surface – have then the same total charge and there is no need for surface correction. In the molecular model, the zero-order approximation is selfconsistent.

Although this model is extremely crude, it is interesting to note that the picture emerging from it is qualitatively identical to the outcome of the n.f.e. pseudopotential analysis, as regards the two key features, i.e. we have a band – flat in this case – of s.s. centred at the midgap, and selfconsistency is achieved with $\frac{1}{2}$ occupation of this band. If the tight-binding analysis is carried out for a better model than the molecular one, the results of course differ greatly in detail – for example, the E_is are no longer all equal to E_0 – and appear rather different, but in fact the key features remain very nearly unaltered. Fig. 90 shows the results obtained by using the Hamiltonian H_s of (4.170). The s.s. now form an actual band with dispersion. The interesting point is that the mean of the s.s. band – defined as the average value of the energy weighted with the density of states of the band – is very nearly midway between the mean levels of the valence and conduction bands, in qualitative agreement with the simple result of the molecular model.

The picture provided by this tight-binding model is also qualitatively rather similar to the picture of the same surface provided by the n.f.e.

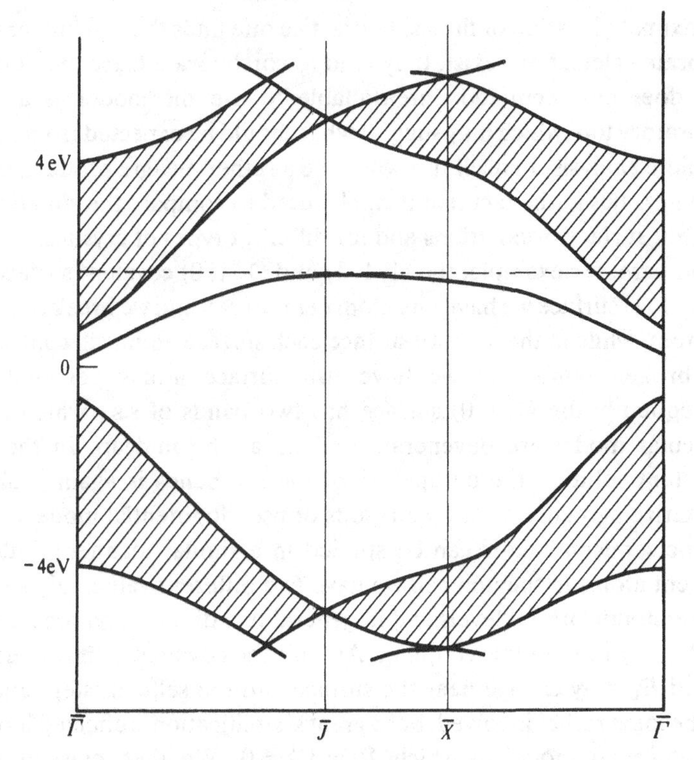

Fig. 90. Surface band structure for (1, 1, 1) Ge obtained from a self-consistent calculation by using the Hamiltonian of (4.170) with $V_1 = -1.8$ eV, $V_2 = -4.3$ eV and $E_0 = -9.45$ eV. The dot-dashed line gives the mean level position of the bulk bands. Computed by B. Djafari-Rouhani, L. Dobrzynski & M. Lannoo, current work.

pseudopotential model (fig. 77), although there is a conspicuous difference in that the tight-binding bands are on the whole very visibly shifted upwards. This is due to the initial bulk band structure (fig. 89). The weight of the degenerate flat bands, which are rather asymmetrically distributed with respect to the fundamental gap, pulls the mean levels upwards, as reflected in fig. 90. In fact an improvement of the tight-binding model Hamiltonian, including interactions beyond nearest-neighbour orbitals, tends to pull all energy levels downwards. We thus see how a picture is emerging from the tight-binding approach which tends to agree qualitatively with the results of the pseudopotential models, and in fact a substantial measure of agreement between the two viewpoints is beginning to materialise. Moreover, crude as it is, the molecular model is not altogether too absurd as a simple system from which one can obtain qualitatively indicative results concerning the occupation and mean approximate location of the s.s. In practice one finds that, wherever more elaborate calculations exist, they tend to corroborate these two features so it does not seem too unreasonable to use the model as a quick exploratory tool which may suggest what should be expected from a more complete analysis. In itself this would be a rather meagre product, but its usefulness lies in the fact that it can be used to compare the situation for different surface orientations and for different types of crystals.

Compare, for example, the (1, 1, 1) and (1, 1, 0) covalent surfaces. In the (1, 1, 1) surface we have one atom per unit cell and we break one bond per atom, while in the (1, 1, 0) surface each surface atom still contributes one broken bond, but we have *two* surface atoms per unit cell. Consequently the (1, 1, 0) surface has two bands of s.s., which in the molecular model are degenerate and lie at the midgap. In the self-consistent solution the occupation of the s.s. bands is again $\frac{1}{2}$, also in qualitative agreement with the results of pseudopotential models.

Ionic semiconductors can be studied in a similar manner. With two different atoms in the unit cell we have two different values, E_0 and E_0', for the atomic orbital levels and in general two different parameters, V_1 and V_1', for intraatomic coupling. As with the covalent semiconductors, E_0 and E_0' may change near the surface and the selfconsistent analysis may become rather involved, but a great simplification is effected by using the molecular model in which $V_1 = V_1' = 0$. We then have only the interaction measured by V_2 between neighbouring orbitals whose levels we shall indicate by $\pm E_1$. The electronic structure of the bulk is then obtained by simply combining a representative pair of neighbouring orbital states, say $|a\rangle$ and $|b\rangle$, in a state of the form $\alpha|a\rangle + \beta|b\rangle$, which yields the eigenvalue equations

$$\left.\begin{aligned} &(-E_1 - E)\beta + V_2\alpha = 0; \\ &V_2\beta + (E_1 - E)\alpha = 0, \end{aligned}\right\} \tag{4.175}$$

whence the (bonding and antibonding) eigenvalues

$$E = \pm\sqrt{(V_2^2 + E_1^2)}, \tag{4.176}$$

which constitute the (flat) degenerate conduction and valence bands of this model. The normalised eigenstates are, for the valence and conduction bands,

$$|\psi_{\mp}\rangle = \frac{V_2}{2^{1/2}[(E_1 + V_2)^2 \mp E_1(E_1^2 + V_2^2)^{1/2}]^{1/2}} \times \left\{\left[\frac{E_1}{V^2} \mp \frac{(E_1^2 + V_2^2)^{1/2}}{V^2}\right]|a\rangle + |b\rangle\right\}. \tag{4.177}$$

By once more assuming negligible overlap between orbitals it is now easy to evaluate the electronic charge on each occupied orbital of the valence band. If we account for spin this yields

$$Q_a = \frac{2E_1^2 + V_2^2 - 2E_1(E_1^2 + V_2^2)^{1/2}}{(E_1^2 + V_2^2) - E_1(E_1^2 + V_2^2)^{1/2}}; \tag{4.178}$$

$$Q_b = \frac{V_2^2}{(E_1^2 + V_2^2) - E_1(E_1^2 + V_2^2)^{1/2}}, \tag{4.179}$$

so that that $Q_a + Q_b = 2$. Notice the following limits. If $E_1/V_2 \to 0$, then $Q_a = Q_b = 1$; this is the covalent case. If $E_1/V_2 \to \infty$, then $Q_a = 0$, $Q_b = 2$; this is the extreme ionic case. Thus the picture of the bulk depends on two parameters, E_1 and V_2, which yield the energy levels of the flat bands and the charge transfer between the orbitals forming the bonds. We could think of these two parameters as obtained by fitting the mean value of the optical gap between valence and conduction bands and the difference in ionic charges, or some better estimate of the charge transfer between ions. In this way the molecular model is admittedly very crude but preserves some effective resemblance to a more realistic picture.

Now we create the surface and have dangling bonds again. We know these dangling bonds are the s.s. of this model so that their occupation is the first thing we must know in order to study the problem of self-consistency. Fig. 91 shows the distribution of electronic charges on the

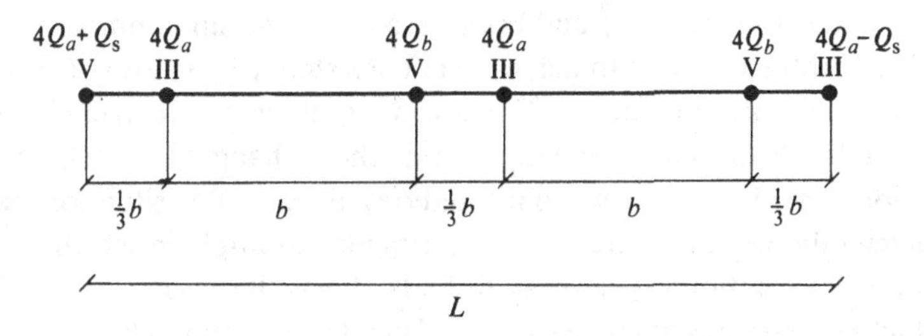

Fig. 91. Electronic charges along a chain of atoms in the (1, 1, 1) direction, after averaging in the (surface) (1, 1, 1) plane, in the molecular approximation. The net charges in the bulk atoms are $5-4Q_b$ and $3-4Q_a$.

atoms of a linear chain in the (1, 1, 1) direction of a III–V ionic compound after averaging over the perpendicular plane, i.e. the plane of the surface now under consideration. The atoms in successive atomic planes are then of types III and V, alternately, while their distance is alternately b or $\frac{1}{3}b$, where b is the length of the interatomic bond. In the bulk, each atom of type V has a charge $4Q_b$, associated with the four b orbitals, while each atom of type III has a charge $4Q_a$. The charge on the surface atoms depends on the occupation of the dangling bond. If the dangling bond for the surface atom of type V has a charge Q_b+Q_s, then the electronic charge on this surface atom is $4Q_b+Q_s$, which means that, because of charge neutrality, the surface atom on the opposite face, of type III, has an electronic charge $4Q_a-Q_s$. The existence of the charge transfer Q_s per atom between the opposite faces is related to the field inside the crystal. Indeed, a physical requirement to be imposed is that this field must remain finite no matter how large the separation L between the two faces. Now, with the charge distribution of fig. 91 the electrostatic potential difference between the opposite faces is

$$\Delta\Phi=(5-4Q_b)\tfrac{1}{4}L+Q_sL. \tag{4.180}$$

The first term is the potential created by the alternate sequence of total – electronic plus ionic – charges $(5-4Q_b)$ and $(3-4Q_a)$, while the second term is the potential difference due to the electronic charge transfer Q_s between opposite dangling bonds. Hence, if $\Delta\Phi$ of (4.180) is to remain finite we must have

$$Q_s=\tfrac{5}{4}-Q_b. \tag{4.181}$$

This means that the electronic charge on the dangling bond of the surface atom of type V is $\frac{5}{4}$, i.e. the occupation of the s.s. band for this face is $\frac{5}{8}$ and not $\frac{1}{2}$ as in the covalent case. For the opposite face the occupation of the s.s. band is found to be $\frac{3}{8}$ and both, of course, add up to unity.

The problem now is to achieve selfconsistency by starting from these occupation rules for the s.s. This means (i) allowing the average atomic orbital levels to change, and (ii) relating these changes to charge redistribution via (4.147). If we omit tedious details, the situation can be sufficiently well illustrated with a particular example in which the self-consistent solution is very easy to find. This is the case in which $Q_b=\frac{5}{4}$, which corresponds to $E_1/V_2=0.35$. Then the electronic charge on a bulk atom is $4Q_b=5$ for type V and 3 for type III. The electronic charge on the surface dangling bond for an atom of type V, by (4.181), is also $\frac{5}{4}$, as for a bulk orbital and therefore the surface does not induce any atomic charge redistribution. Thus the atomic orbital levels remain unchanged and we have a selfconsistent solution. If we associate the (1, 1, 1) face with type III atoms and the $(\bar{1}, \bar{1}, \bar{1})$ face with type V atoms, then the picture which emerges from this simple model can be summarised as follows. (i) The occupation of the s.s. is $\frac{3}{8}$ for the (1, 1, 1) face and $\frac{5}{8}$ for the $(\bar{1}, \bar{1}, \bar{1})$ face. (ii) The two (flat) bands of s.s. are symmetrically located about the middle of the fundamental gap of width $2(E_1^2+V_2^2)^{1/2}$. The lower band is associated with the more electronegative atoms. For example, when $E_1/V_2 = 0.35$ the two s.s. bands are located at energies $\pm E_1$ with respect to the midgap, and the ratio of their separation to the width of the gap is 0.33. (iii) Selfconsistency in general may require charge redistribution and shifts of atomic orbital levels. In the particular example just considered, the unperturbed solution – i.e. with all atomic charges and orbital levels unchanged – is selfconsistent.

A similar analysis can be carried out for II–VI compounds. The main results then are as follows. (i) The (flat) band of s.s. associated with the $(1, 1, 1)$ face has occupation $\frac{1}{4}$, while that associated with the $(\bar{1}, \bar{1}, \bar{1})$ face has occupation $\frac{3}{4}$. (ii) These two bands are again symmetrically located with respect to the midgap. (iii) Selfconsistency again may in general require changes in some atomic charges and orbital levels, except in the particular case $Q_b = \frac{3}{2}$. There the ratio of the separation between s.s. and the width of the band is 0.52.

For both the III–V and II–VI compounds the key features are again in general agreement with the results found in the pseudopotential models so that the simple molecular model yields a reliable qualitative picture. Moreover, the same happens if the tight-binding calculation is selfconsistently carried beyond the molecular model. This is the case, for example, with GaAs if the Hamiltonian of (4.163) is used. Fig. 92 shows the s.s. bands obtained in this way for the $(1, 1, 1)$ and $(\bar{1}, \bar{1}, \bar{1})$ faces. The main features of interest are as follows. (i) The occupations of the s.s. turn out to be the same as obtained with the molecular model, which suggests definite occupation rules for the s.s. which must be confirmed by all calculations, irrespective of the model. (ii) The mean level of the two bands of s.s. is very close to the mean level of the bulk bands. These results also compare fairly well with the results obtained with pseudopotential models, with the typical difference that all the bands tend to be higher up in energy, although as in the covalent case, an improvement of the tight-binding Hamiltonian going beyond nearest-neighbour interactions tends to correct this, pulling all the bands somewhat downwards.

4.4. A general overview of ideal semiconductor surfaces

The study of real semiconductor surfaces may encounter rather formidable difficulties due to surface relaxation, reconstruction, existence of steps and various structural imperfections. No doubt there is here vast scope for research, but first of all it is important to have a clear picture of ideal surfaces, stressing the main theoretical issues and the basic principles involved. Matters such as occupation rules for the s.s., surface charge and selfconsistency must be firmly established. A picture of this

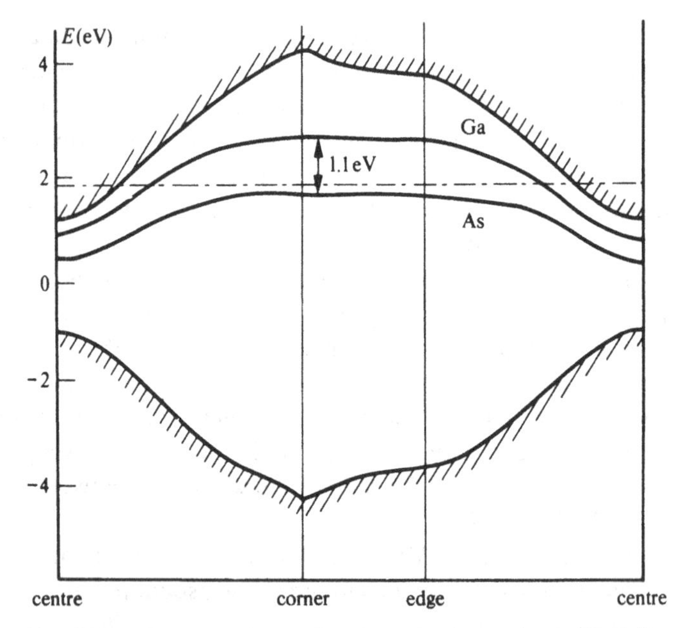

Fig. 92. Surface-state bands for the polar (1, 1, 1) and $(\bar{1}, \bar{1}, \bar{1})$ faces of GaAs, calculated with the tight-binding model described in the text. (E_0(As) = −11.3 eV, E_0(Ga) = −7.6 eV, V_1 = −1.8 eV, V_2 = −4.3 eV.) The shaded areas represent the surface projection of the bulk bands. The dot-dashed line gives the mean level position of the bulk bands. Computed by B. Djafari-Rouhani, L. Dobrzynski & M. Lannoo, current work.

kind can be obtained by careful study of the simple models considered in this chapter and an important question concerns the connection between the tight-binding and n.f.e. approaches which, as we have just discussed, can be made to converge. It may be in order at this stage to make a quick survey of the broad picture thus obtained for semiconductor surfaces.

If we begin with covalent semiconductors, the (1, 1, 1) face has already been considered in some detail and needs no further discussion. As regards the (1, 1, 0) face, fig. 93 shows some results obtained for Si with the pseudopotential model of § 4.2, by using an effective abrupt potential barrier suitably displaced to achieve charge neutrality in the zero-order 'metallic approximation' as described in § 4.2. The main feature is the degeneracy of the two bands of s.s. along the line $\bar{X}\bar{M}$ of the two-dimensional Brillouin zone (fig. 71). Remember we already encountered the degeneracy at the $\bar{X}$ point in § 4.2. The existence of two bands of s.s. for this surface is related to the fact that here we break one bond per surface atom but we have two atoms per unit cell, as already remarked in

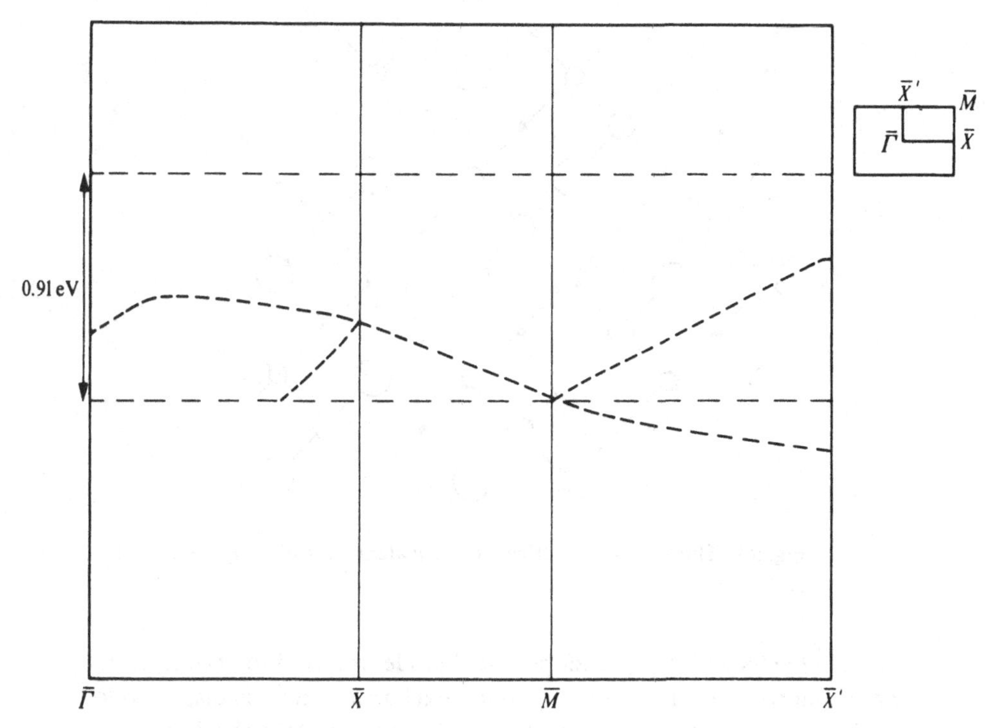

Fig. 93. Surface-state bands for (1, 1, 0) Si calculated with the pseudo-potential model indicated in the text. The figure shows also the thermal gap and the two-dimensional B.z.

the discussion of the tight-binding approach (§ 4.3.2). It is also interesting to enquire as to what we should expect to find for the (1, 0, 0) surface. Fig. 94 shows the different atomic planes for this. The point to notice is that there is one atom per unit cell but, on forming the surface, two bonds are broken per atom. Thus we expect to find two bands of s.s. The separation between these bands is roughly given by the interaction energy between the two orbitals of the same atom. Thus the mean of the energy levels of the two bands of s.s. is at about the middle of the optical gap. The occupation rule can be obtained by an analysis similar to that of § 4.2 and the result is again that half of the s.s. are occupied.

With ionic semiconductors the occupation numbers for s.s. change. Important differences also exist between homopolar – e.g. (1, 1, 1) – and heteropolar – e.g. (1, 1, 0) – surfaces. In general the occupation number of the s.s. on heteropolar surfaces is $\frac{1}{2}$, as in the covalent case, while this changes for homopolar faces. As a result the general features found for covalent surfaces apply also to heteropolar ionic surfaces, but an important detail changes. Fig. 95 shows the s.s. bands obtained for

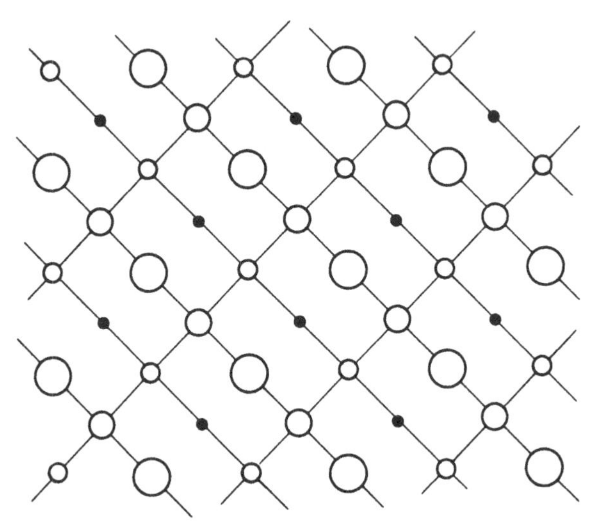

Fig. 94. The diamond lattice of a covalent crystal projected on the (1, 0, 0) surface.

(1, 1, 0) GaAs with the pseudopotential model referred to above, i.e. in particular with an effective abrupt potential barrier, suitably displaced to achieve 'metallic' charge neutrality. The degeneracy along the $\bar{X}\bar{M}$ line is now lifted, as it should be because now the two bands of s.s. represent states mainly localised about either As or Ga atoms. This is in line with the discussion carried out in § 4.2 for the $\bar{X}$ point where we found (equation (4.91)) that the splitting of this degeneracy is measured by the imaginary part of the pseudopotential component V_3. This is the main feature differentiating the (1, 1, 0) covalent and ionic surfaces. The situation is more complicated with the homopolar surfaces. As we have seen, the main result for this case is a change in the s.s. occupation which brings about charge redistribution in the surface region. The change in s.s. occupation is such that the states associated with the face of the more electronegative atoms have a higher occupation number, as was to be expected on intuitive grounds. In all cases we find that (i) the occupation numbers for both (opposite) faces add up to 1, and (ii) the s.s. of the more electronegative faces are lower. The average of the two bands found for the two opposite faces behaves more or less like the single band of the covalent case and is about the middle of the optical gap. This is illustrated in fig. 84, which shows the density of states for the s.s. bands of the (1, 1, 1) and $(\bar{1}, \bar{1}, \bar{1})$ GaAs faces.

Finally let us briefly consider the (1, 0, 0) ionic surfaces. An analysis carried out with the molecular model of § 4.3.2 yields the following

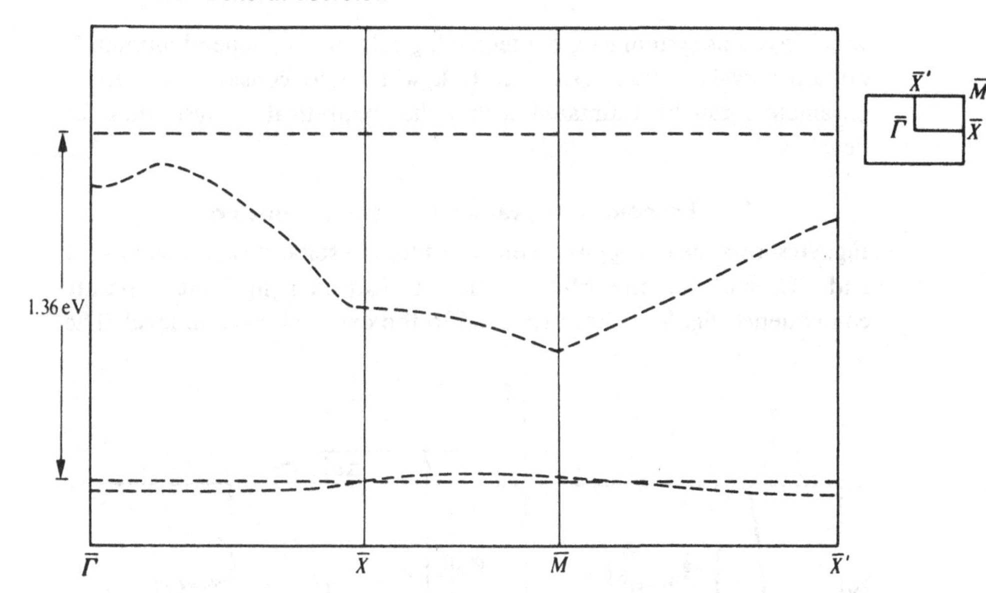

Fig. 95. Surface-state bands for (1, 1, 0) GaAs calculated with the pseudopotential model indicated in the text. The figure shows also the thermal gap and the two-dimensional B.z.

picture. (i) For III–V compounds the occupation number for the s.s. is $\frac{5}{8}$ for the $(\bar{1}, 0, 0)$ face, and $\frac{3}{8}$ for the (1, 0, 0) face. (ii) For the II–VI compounds it is $\frac{3}{4}$ and $\frac{1}{4}$ respectively. Notice what this means. Take for example the $(\bar{1}, 0, 0)$ face of a III–V compound. Between the two bands of s.s. we should expect a total occupation of $2 \times \frac{5}{8}$, i.e. $\frac{5}{4}$. This means one fully occupied band and $\frac{1}{4}$ occupation of the other one. On the basis of the various cases studied in §§ 4.2 and 4.3, with coincident conclusions, we would expect to also find these occupations if we used more elaborate selfconsistent calculations. This is the sort of guidance we can obtain from the simple analysis expounded here. On the other hand, since these occupation numbers are the same as we found for the (1, 1, 1) faces, we expect that the s.s. bands of the (1, 0, 0) and $(\bar{1}, 0, 0)$ faces of these compounds will be displaced in energy with respect to the covalent case as in the (1, 1, 1) surfaces. For example, (fig. 84) for GaAs these displacements should be of order 0.5 eV.

4.5. The metal–semiconductor interface

Having studied metal surfaces and semiconductor surfaces, we find that the study of the metal–semiconductor (M–SC) interface constitutes a natural extention of the same sort of analysis. The practical properties of the M–SC junction, such as V–I characteristics and rectifying power, which give this system its great technological interest, depend ultimately on some basic parameters. Our task will be to consider how these parameters can be estimated within the theoretical analysis thus far developed.

4.5.1. Phenomenological features and parameters

Fig. 96 shows the energy-level diagrams for the separate components, M and SC, and for the M–SC junction. Consider first the separate components (fig. 96*a*) both referred to the external vacuum level. The

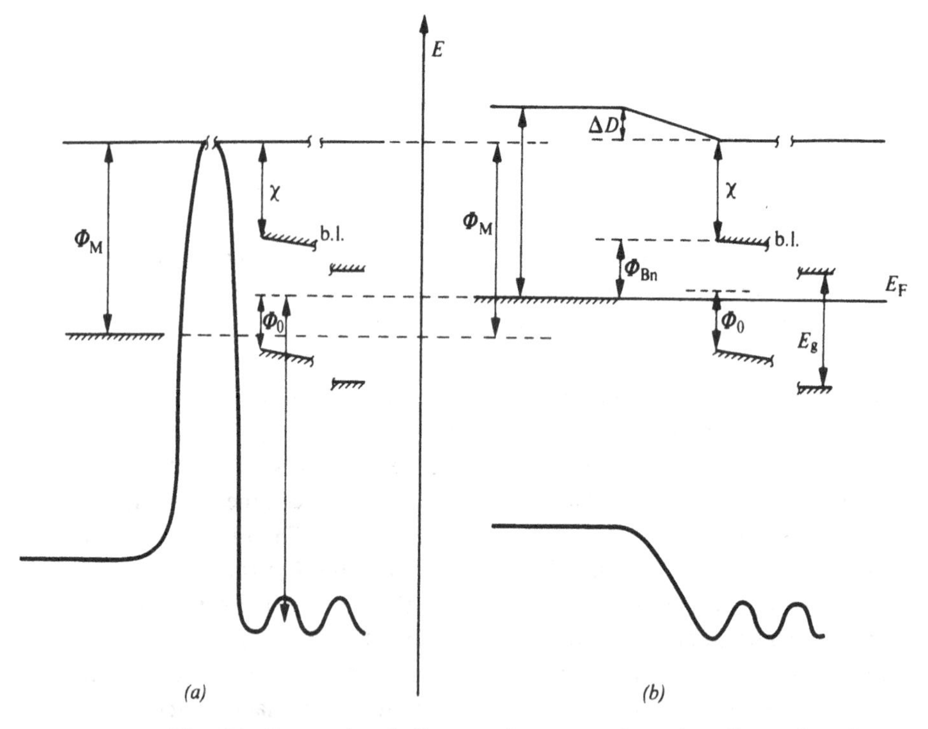

Fig. 96. Energy-level diagram for a metal–semiconductor junction (*a*) before, and (*b*) after allowing for electrical contact. E_g is the thermal energy gap and b.l. indicates the space charge boundary layer, not explicitly shown in the figure. C. Tejedor, F. Flores & E Louis, *J. Phys.* C: *Solid St. Phys.*, **10**, 2163 (1977).

work function Φ_M of the metal was discussed in (3.106) as the sum of a volume term and a surface dipole term, now called D^M. This is the term whose dependence on surface orientation produces the differences in Φ_M for the different crystal faces. We shall write

$$D^M = D_0^M + D_L^M, \tag{4.182}$$

where D_0^M is the electrostatic dipole calculated in the jellium model and D_L^M is the lattice correction. We saw in § 3.3.3 that this correction entails two effects, namely, the variation in the surface charge itself and the contribution to the height of the surface barrier created by the bulk pseudopotential, partly screened by the surface charge.

Fig. 96 shows also two different regions associated with the semiconductor surface. One is the microscopic region (typically a few angstroms) in which the surface potential barrier referred to is located. The other one is the much thicker space-charge boundary layer (typically a few hundred angstroms) of the kind considered in § 1.7, which is associated with the presence of a surface charge, classically screened by the semiconductor crystal within a length of the Debye type. We saw, for example, that for an n-type material with conduction electron concentration n_e this length is, in a.u.,

$$L_D = \left(\frac{\varepsilon_c k_B T}{8\pi n_e}\right)^{1/2}.$$

In other words, at this stage we recognise that the study of a semiconductor surface involves two problems. The purely quantum mechanical analysis, with which we have been concerned throughout this chapter, applies to an intrinsic semiconductor at zero temperature – i.e. actually to an insulator – and leads to, among other results, the selfconsistent determination of the Fermi-level position. The classical electrostatic and statistical finite-temperature analysis describes the type of semiconductor material leading to a bending of the bands and, classically, to a shift of the s.s. eigenvalues by an amount equal to the change in the band-edge levels across the space-charge boundary layer. Of course the Fermi level – i.e. the electrochemical potential when there is a space charge – remains constant if there is thermal equilibrium between bulk and surface, but it is customary to take, at every depth, the local positions of the band edges as a reference, and to speak of changes in the Fermi level *relative to this local reference.* Thus the quantum mechanical analysis fixes the Fermi level *at the surface,* in this conventional language. It follows that, strictly speaking, the whole question of s.s. occupation, charge distribution and overall neutrality should be selfconsistently reanalysed to take account of possible charge transfer between the s.s. and the boundary layer. The question is, how important are these corrections in practice? The density of surface states is of order 10^{14} electrons cm^{-2} eV^{-1}, while for ordinary dopings the conduction electron concentration is usually of order not more than 10^{18} electrons cm^{-3}. From tabulated solutions of the Poisson equation for the boundary layer one can estimate that even the *total* surface excess of charge in the boundary layer (§ 1.7) corresponds typically to something of the order of 10^{12} electrons $cm^{-2}\,eV^{-1}$. Thus, except for strong dopings, we may expect the boundary-layer effects to be negligible. The Fermi level *at the surface* and the s.s. occupation are rather effectively fixed by the high density of s.s. Under these circumstances we can study the quantum mechanical surface problem as a separate issue.

On this basis we now start the analysis of the (microscopic) interface at the M–SC junction. We learnt in § 4.2 that the surface potential and charge distribution at the semiconductor surface can be obtained to a good approximation when we study this as if it were a metal with the same valence electron concentration. Thus the electronic affinity χ of the semiconductor (fig. 96*a*) depends also on the surface dipole

$$D^s = D_0^s + D_L^s, \qquad (4.183)$$

as in (4.182). Now we put the two media in contact and create the M–SC junction (fig. 96*b*). The parameter of practical interest – for n-type semiconductor – is the barrier height Φ_{Bn}, the energy needed to take one electron from the Fermi level of the metal to the bottom of the conduction band of the semiconductor at the surface, if we disregard possible tunnelling across the barrier. Our aim is the theoretical evaluation of Φ_{Bn}. The physical picture of the process involved in the creation of the junction is described as follows. Before putting the two media in contact their Fermi levels are in general different. Thus when contact is established there will be some electronic charge flow from one material to the other, e.g. from the semiconductor to the metal in the example of fig. 96. This charge flow produces a change ΔD in the dipole at the interface, so that

$$\Phi_{Bn} = \Phi_M - \chi - \Delta D. \qquad (4.184)$$

There are two extreme situations. In the *Schottky limit*, $\Delta D = 0$: the height of the Schottky barrier, for a given semiconductor, varies linearly with Φ_M. This ignores the possibility of interface states. The opposite model is the *Bardeen limit*, in which one would say essentially that owing to a high density of interface states the Fermi level is very effectively fixed, and the barrier height for given semiconductor is a constant independent of Φ_M. It is clear that the situation requires a study of the possible interface states, which so far are still hypothetical. We have studied surface states at the surface–vacuum interface, but not at the interface between two different media. The effects of the boundary layer are still sufficiently small that the quantum mechanical problem can be studied separately, but this must be studied anew. We need a theory for the density of interface states.

4.5.2. The density of interface states

We start again with a one-dimensional problem (fig. 97) and use it, in the narrow-gap approximation, as a model system to introduce the basic concepts, of which the central one is the *charge neutrality level.* We know that for a bounded one-dimensional semiconductor this is just the midgap. We shall prove that if the Fermi level of the metal coincides with the charge neutrality level, then to a good approximation the M–SC interface can be described as an equivalent metal–metal interface. This will allow us to use concepts and results previously obtained for metal surfaces.

We start by analysing the density of states for the model of fig. 97. For this we take a matching plane at $z = z_0$, coinciding with the first minimum in the periodic semiconductor potential to occur beyond the interface. That is to say, across the interface we smooth out the first potential oscillation of the semiconductor side and describe this part only in terms of the much stronger variation of the potential barrier which goes down from the bulk-metal level to the bulk-semiconductor level, as we did in § 4.1 for the free semiconductor surface. Now we imagine three infinite barriers or hard walls, at $z = z_0$ and $z = z_0 \pm L$, where the length $L = Nc$ contains an integral number of semiconductor unit cells. The idea is to proceed as in § 4.1, i.e. to start with an infinite barrier at the surface of interest and then to lower it until the actual barrier is constructed, while leaving the two infinite barriers at $z_0 \pm L$. Now we find an important difference. Not only do we have a metal instead of a vacuum on the other side of the matching plane – this would only mean shifting the constant potential – but we have also left all the interface outside this plane. Thus on the metal side we do not have simple metallic wavefunctions: these are now perturbed by the potential of the interface. Although the same conclusions, presently to be described, can be reached with a more accurate analysis, we shall here illustrate the situation in a simple way by assuming the interface can be described in a W.K.B. approximation, which in practice can be justified for most cases of real interest. Then we write the wavefunction for $z < z_0$ as

$$\psi^{\mathrm{M}} \propto p^{-1/2} \sin\left[\int_{z_0-L}^{z} p(z')\,\mathrm{d}z'\right], \qquad (4.185)$$

where

$$p(z) = \sqrt{\{2[E - V(z)]\}}.$$

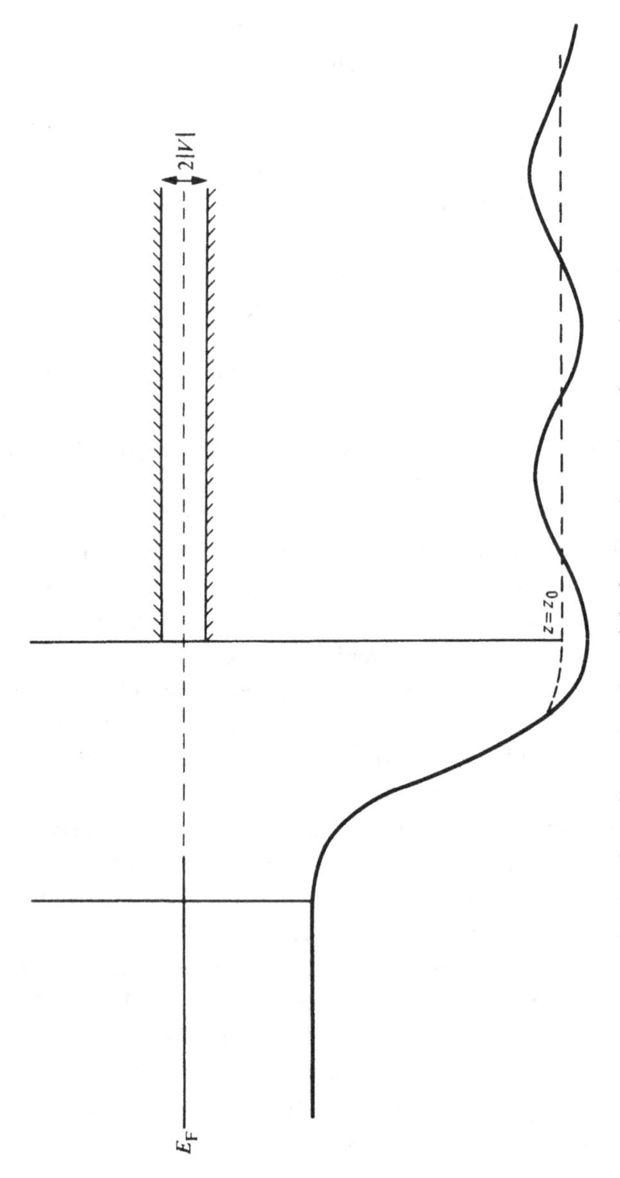

Fig. 97. One-dimensional model of a metal–semiconductor junction.

The eigenvalue condition with the two infinite barriers is

$$\int_{z_0-L}^{z_0} p(z')\,\mathrm{d}z' = m\pi. \tag{4.186}$$

For $z > z_0$ the wavefunction of a valence-band state for arbitrary surface barrier would be, from (4.16),

$$\psi^{\mathrm{SC}} \propto \sin\left[(q - \tfrac{1}{2}g)(z - z_0) + \eta\right] + a \sin\left[(q + \tfrac{1}{2}gz)(z - z_0) + \eta\right]. \tag{4.187}$$

With the infinite barriers at z_0 and $z_0 + L$ we have $\eta = 0$ and the eigenvalue condition

$$qL = m'\pi, \quad 0 < m' < N. \tag{4.188}$$

We saw in § 4.1 that this leads to the surface density of states (4.22) in the valence band. In fact we concentrate on the energy range near the fundamental gap and need only recall the loss of $\frac{1}{4}$ state at the valence band edge. We also saw that (4.188) corresponds to $(N-1)$ states in the valence band.

Now we lower the infinite barrier at $z = z_0$ and study the effect of this on the density of states in (i) the energy range near and below the valence-band edge, and (ii) the range of the gap. In the first case we must study the matching of (4.185) and (4.187), just as in § 4.1 we matched (4.12) and (4.16). (The difference is that in § 4.1 we had only the interface region on the other side of the matching plane and now we have the interface and the bulk metal extending to $z_0 - L$.) The point of using this device is that here we can apply directly some of the basic results of § 4.1, notably the quantisation rule for the number of states in the valence band of the semiconductor. We saw that this number is the sum of the number of band states with the hard wall at the interface, and the number of times the phase shift η_0 of (4.12) goes through a multiple of π. Here

$$\eta_0 = \int_{z_0-L}^{z_0} p(z')\,\mathrm{d}z', \tag{4.189}$$

as can be readily seen by comparing (4.185) and (4.12). Then, by (4.186), the number of times that η_0 goes through a multiple of π is equal to the number of eigenstates we now have in the metal. Thus the total number of states of the interface system with finite barrier is equal to the total number of states of the same system with an infinite barrier at $z = z_0$. On the other hand, in the energy range of the semiconductor gap the wavefunction of the interface system can be obtained by matching (4.185) to an evanescent wavefunction like (4.8), which we now write as

$$\psi^{\mathrm{SC}} \propto \exp\left[-q(z - z_0)\right] \cos\left[\tfrac{1}{2}g(z - z_0) + \tfrac{1}{2}\phi\right], \tag{4.190}$$

where

$$\exp(\mathrm{i}\phi) = [\mathscr{E} + \mathrm{i}\sqrt{(V^2 - \mathscr{E}^2)}]/V. \tag{4.191}$$

The point to notice (fig. 97) is that $p(z \approx 0) \approx \frac{1}{2}g$. Thus the matching of (4.190) and (4.185) is in practical terms reduced to equating the phases of the wavefunctions. Since, in the narrow-gap model, $q \ll g$ for the range of interest – the gap in this case – the matching condition becomes

$$\int_{z_0 - L}^{z_0} p(z')\,\mathrm{d}z' = \eta_0 = \frac{1}{2}\pi + \frac{1}{2}\phi + m\pi. \tag{4.192}$$

Compare with (4.186), which holds when the barrier at the interface is infinite. From the change in the value of η_0, i.e. $\frac{1}{2}\pi + \frac{1}{2}\phi$, we can obtain the change in the density of states of the interface system on passing from an infinite to a finite barrier at the interface. Thus we have, for the surface density of states of the M–SC system in the semiconductor gap range,

$$N_{\text{M-SC}}(E) = N_{\text{M}}(E) + \frac{1}{2\pi}\frac{\mathrm{d}\phi}{\mathrm{d}E}. \tag{4.193}$$

Notice the characteristic structure of this formula in which the terms other than the hard-wall contribution take the form of the derivative of a phase function, in line with the general argument of § 2.3 and of the particular results of §§ 2.7, 3.1, 4.1 and 4.2. In this case, from (4.191), we have

$$\frac{1}{2\pi}\frac{\mathrm{d}\phi}{\mathrm{d}E} = \frac{1}{2\pi}(V^2 - \mathscr{E}^2)^{-1/2}, \tag{4.194}$$

which is shown in fig. 98. Notice that this is no longer the sharp δ-function of a distinct s.s. Rather, what we find for the M–SC interface is a continuous distribution spread across the gap: in this case symmetrically about the midgap, but this need not always be so. But this distribution has a very important property, which can be readily seen by noticing that ϕ changes by π when $\mathscr{E}$ spans the gap. Thus the *total spectral strength* of (4.194) integrates up to exactly $\frac{1}{2}$.

Of course we could not have sharp surface states if the semiconductor gap overlapped with the metallic band. From the semiconductor side, a s.s. with given energy resonates with the continuum of the metal. The result (4.194) says that this broad resonance contributes $\frac{1}{2}$ state altogether. From the metal side an infinite barrier at a metal surface would require the vanishing of the wavefunction, but a finite barrier allows these wavefunctions to leak out into the semiconductor side (fig. 99). The tails of the metallic wavefunctions can hold electronic charge

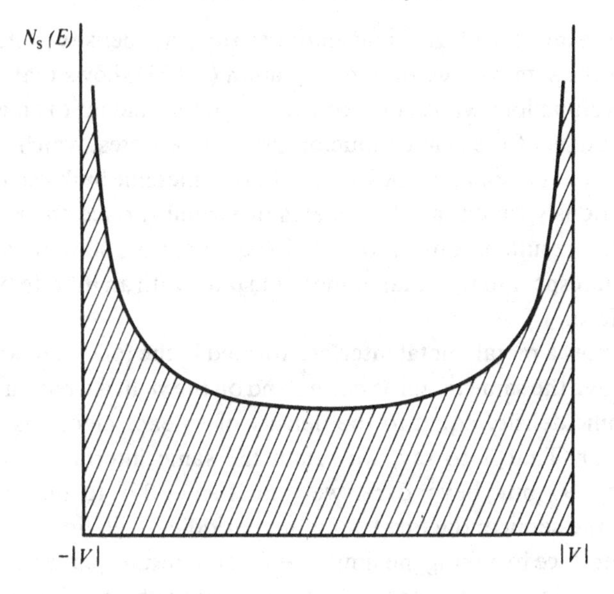

Fig. 98. Density of interface states for a metal–semiconductor junction in the energy range of the semiconductor gap.

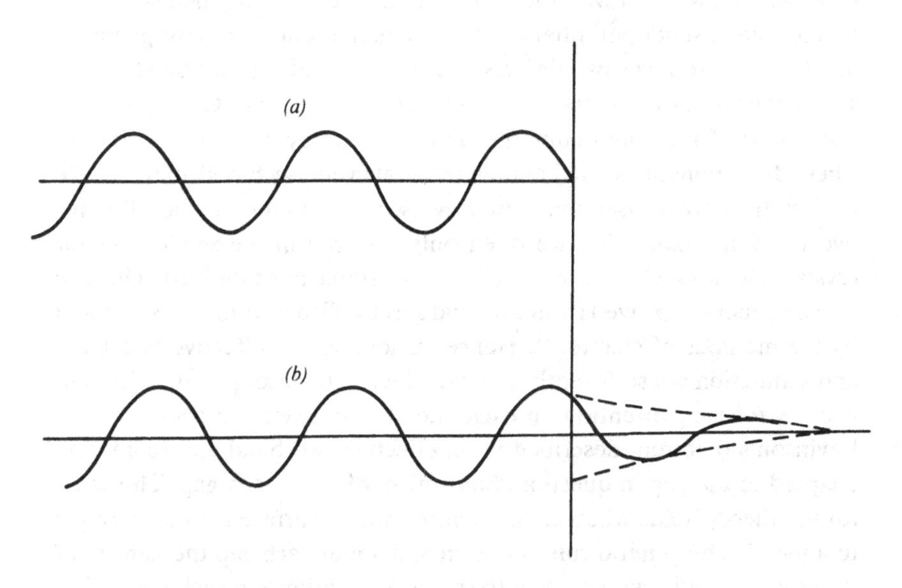

Fig. 99. Behaviour, at a metal–semiconductor interface, of the metallic wavefunctions corresponding to energies in the semiconductor gap, (*a*) when an infinite barrier is kept at $z = z_0$; (*b*) after this barrier has been eliminated and matching between the two media has been permitted.

localised in the interface region and amount to an extra density of states in the range of the semiconductor gap. Equation (4.193) shows that all the metallic wavefunctions whose tails contribute to this add up to an extra $\frac{1}{2}$ state in the range of the semiconductor gap. These states, which can be looked at either as resonances of s.s. or as tails of metallic bulk wavefunctions are variously called interface states or virtual surface states. With the symmetric results in our model we have an extra $\frac{1}{4}$ state below the midgap, with respect to the total number of states with an infinite barrier at the interface.

Consider now a metal–metal interface formed in the following way: on the left we have the same metal as before and on the right we have a metal with electronic density equal to the density of valence electrons of the semiconductor. That is to say, we retain the same interface potential barrier of fig. 97 but, beyond z_0, we join it smoothly to the average potential in the semiconductor, losing the periodic oscillations. We can study this interface by taking the limit $V \to 0$ in the results just derived for the M–SC junction. This process does not change the total density of states contained below the midgap. In particular the deficiency of $\frac{1}{4}$ state at the valence-band edge cancels out the extra $\frac{1}{4}$ state in the lower half of the gap. Turning the argument around we can imagine the following process. First we start with the metal–metal interface and then we switch on the weak pseudopotential which creates the semiconductor gap as in fig. 97. This produces two defects of $\frac{1}{4}$ state each at the two band-edges, which cancel out the extra $\frac{1}{2}$ state appearing in the gap. On top of this, distortions of the continuum appear in the energy range of the bands. These distortions are such that their integral in energy turns out to vanish. In fact this is a consequence of Levinson's theorem. In the effective two-band approximation we open only one gap in the semiconductor crystal. The total change in the number of states must be zero. This can also be generally proved from a formal surface Green-function argument by the methods of chapter 2. Hence, as long as the effective two-band approximation is a sufficiently accurate device to 'decouple' the different gaps by focusing attention on each one at a time, we can conclude that Levinson's theorem, described in an effective two-band approximation adapted to the gap in question, holds piecewise for each gap. The same formal theory holds whether the semiconductor surface is free or joined to a metal. The general rule is that in and about each gap the density of states is changed – with respect to the bulk situation – in such a way that the total change associated with each gap is zero.

The point of the above digression is to stress that, if we return to the M–SC junction, we see that, like the free semiconductor surface, it too can be studied by starting with the equivalent hypothetical 'metal', and as in § 4.1, one can see that the cancellation is not only total but also local in the sense of § 4.1. We shall not repeat the arguments but only indicate the main conclusion of this analysis. First, if we integrate in energy the local density of states we find that as soon as we integrate up to an energy level well above the gap the integral vanishes locally, i.e. at all depths z. Secondly, because of the symmetry about the midgap, which characterises the narrow-gap approximation, there is also local cancellation for all the occupied states below the midgap; i.e. the opening of the semiconductor gap does not require surface charge redistribution (if we ignore, as in § 4.1, the small effect of the periodic charge-density oscillations of the bulk semiconductor). This holds if we assume that the gap in the semiconductor forming the junction opens up with its middle level – the charge neutrality level in the one-dimensional system – coinciding with Fermi level of the metal. This point will be discussed in § 4.5.3.

The three-dimensional case can now be studied as in § 4.2, by taking special symmetry points as representative ones of the corresponding two-dimensional B.z. This leads to factorisation into effectively one-dimensional problems, for which we can use the above results. There is one technical point concerning the charge neutrality level. The three-dimensional system is studied by combining different one-dimensional problems with different associated complex momentum loops and different middle points for the different κs. Thus the charge neutrality point must be defined as an average. This is indicated in fig. 96 by the level Φ_0 referred to the valence-band edge. The implication of the result found in the one-dimensional case is that if the Fermi level of the metal coincides with the average charge neutrality level Φ_0, then the M–SC interface can be described by the same surface-charge distribution as that of the equivalent metal–metal interface.

4.5.3. The dipole at the interface

We can now study the dipole moment of the charge distribution at the interface in two stages, as follows.

(i) We assume at the start that E_F (metal) and Φ_0 (semiconductor) coincide. Then the analysis of § 4.5.2 holds. Moreover, we can also introduce lattice corrections as in chapter 3 and write

$$D^{sM} = D_0^{sM} + D_L^{sM} \quad (E_F = \Phi_0), \tag{4.195}$$

as in (4.182) and (4.183).

(ii) However, in general the final selfconsistent situation need not correspond to $E_F = \Phi_0$ (fig. 96*b*). We must then estimate a further contribution to D^{sM} which is due to the extra surface charge in the interface or virtual surface states; this extra charge, of course, can be positive or negative. Let D_{vs} be this contribution. We shall discuss presently how to estimate this term. If we suppose we know it, (4.195) becomes

$$D^{sM} = D_0^{sM} + D_L^{sM} + D_{vs} \quad (E_F \neq \Phi_0). \tag{4.196}$$

The barrier height we want is then (fig. 96)

$$\Phi_{Bn} = \Phi_M - \chi + D_J + D_{vs}, \tag{4.197}$$

where

$$D_J = (D_0^{sM} - D_0^s + D_0^M) + (D_L^{sM} - D_L^s + D_L^M) \equiv D_e + D_L. \tag{4.198}$$

The values of Φ_M and χ must correspond to the crystal faces forming the junction under study. Both D_J and D_{vs} can be estimated on the basis of a simplified model for the junction, similar to the model used in § 3.2.5. We postulate a charge distribution of the form (compare with (3.204))

$$n(z) = \left\{ \begin{array}{ll} n_M - \frac{1}{2}(n_M - n_s) \exp(\beta_0 z) & (z < 0), \\ n_s + \frac{1}{2}(n_M - n_s) \exp(-\beta_0 z) & (z > 0). \end{array} \right\} \tag{4.199}$$

The variational parameter β_0 can be determined by the method of § 3.2.5, i.e. by minimising the electronic energy of the interface. If we really had simply a metal–metal junction, β_0 would be uniquely fixed by the condition of equality between the two Fermi levels. Now, however, we are accounting for charge transfer between the metal and the interface states. In a metal–metal junction, when contact is established charge flows from one metal to the other until both Fermi levels are equal. In a M–SC junction the charge flow also tends to push E_F and Φ_0 close together, but the charge transfer to, or from, the interface states may cause a final selfconsistent situation in which E_F and Φ_0 are not exactly equal. The difficulty with the M–SC junction lies precisely in how to give a good description of this effect. Before studying this problem let us see how D_J can be estimated from (4.199).

First, suppose we have studied separately the surfaces of the metal and the semiconductor with the model of (3.204). Let β_M and β_s be the corresponding values found for the parameter of (3.204) and β_0 for (4.199). Then the term D_e of (4.198) is

$$D_e = 4\pi \left[\frac{n_M - n_s}{\beta_0^2} - \frac{n_M}{\beta_M^2} + \frac{n_s}{\beta_s^2} \right]. \tag{4.200}$$

In practice this contribution turns out to be very small. Now look at the terms of (4.198) giving D_L. In first approximation we saw in § 3.3 that the crystalline corrections to the surface dipole come from two effects, namely, the variation of the surface charge, and the potential jump at the surface partly screened by the surface charge (§ 3.3.3). The first effect is such that its contributions to D_L^{sM}, D_L^s and D_L^M nearly cancel out on forming the combination of (4.198). The second effect is mainly responsible for the contribution D_L to D_J. This justifies the use of an approximation which would be poor for estimating D_L^{sM}, D_L^s and D_L^M separately, but is quite reasonable for estimating the combination of (4.198). Then the calculation of D_J can be reduced, effectively, to the screening of a surface dipole within the charge model of (4.199). The same calculation, by setting $n_M = 0$ or $n_s = 0$, can be used to estimate D_L^{sM}, D_L^s and D_L^M with which we finally estimate D_J.

The next question is the extra surface charge due to charge transfer between the metal and the interface states. The problem is to find the extra charge $\delta n(z)$ corresponding to the energy range between E_F and Φ_0 (fig. 96*b*). This is very sensitive to the particular properties of the semiconductor in question. Well inside the metal, $\delta n(z)$ must tend to the bulk metallic charge with energies between E_F and Φ_0. On the semiconductor side the behaviour of the charge density is described by the evanescent wavefunction (4.190). It we neglect the much faster variation of the oscillations of the cosine term we have a charge that decays as $\exp(-2qz)$; again, the three-dimensional case must be described with a mean value q_m for the different wavefunctions. We can then imagine that in the semiconductor side δn is of the form $\delta n_1 \exp(-2q_m z)$. The factor δn_1 is determined by the condition

$$\int_0^\infty \delta n_1 \exp(-2q_m z)\,dz = \int_{\Phi_0}^{E_F} N_{vs}(E)\,dE. \tag{4.201}$$

Like Φ_0 and q_m, the density of interface states $N_{vs}(E)$ can be obtained by averaging over the two-dimensional B.z. Now, the difference between E_F and Φ_0 is in practice small and we can effectively take $N_{vs}(E)$ in (4.201) as constant. This yields

$$\delta n_1 = 2q_m N_{vs}(E_g - \Phi_{Bn} - \Phi_0), \tag{4.202}$$

where E_g is the thermal energy gap and we must remember that Φ_0 is referred to the valence-band edge. But now the extra charge $\delta n(z)$ interacts with the rest of the system. This interaction can be divided in two parts: (i) A long range interaction, through which the semiconductor screens classically the surface charge of $\delta n(z)$ with its dielectric constant ε_0. At long distances this reduces $\delta n(z)$ to $(\delta n_1/\varepsilon_0)\exp(-2q_m z)$. At the same time an extra charge concentration accumulates at the interface, analogous to the charge concentration induced classically at the surface of a dielectric. (ii) The extra charge induced at the interface with the total surface charge. Let us see how we can construct a model to estimate this interaction.

Notice that the effect of the long-range interaction is simply to redistribute the charge $\delta n_1 \exp(-2q_m z)$ in such a way that its total amount is conserved. Consistent with the model of the interface of (4.199), classical long-range screening effectively starts beyond distances of order $1/\beta_0$. Thus the charge transfer $\delta n(z)$, whose total amount remains constant and is given by (4.201), is redistributed via its interaction with the system. After the above considerations we postulate for this redistribution the form

$$\rho_{vs}(z) = (\delta n_1/\varepsilon_0)\exp(-2q_m z) + \delta n_1\,\gamma_0 z\exp(-\beta z). \qquad (4.203)$$

The parameter γ_0 is fixed by the condition that (4.203) must contain the total amount of charge given by (4.201). This yields

$$\gamma_0 = \frac{\beta^2}{2q_m}\left(1 - \frac{1}{\varepsilon_0}\right). \qquad (4.204)$$

Now we can give explicitly our model for the *total* electronic charge density at the interface. This can be thought of as consisting of two terms. One would be the total charge in the equivalent metal–metal interface below Φ_0. If m_0 is the bulk metallic charge between E_F and Φ_0, then in view of the preceding discussion we write this term in the form

$$n_1(z) = \begin{cases} (n_M - m_0) - \frac{1}{2}(n_M - m_0 - n_s)\exp[\beta(z-a)] & (z \leqslant a), \\ n_s + \frac{1}{2}(n_M - m_0 - n_s)\exp[-\beta(z-a)] & (z \geqslant a). \end{cases} \qquad (4.205)$$

The other term is the total charge associated with the energy range between E_F and Φ_0, which we write for the same reasons as

$$n_2(z) = \begin{cases} m_0 + \gamma_1 \exp[\beta(z-a)] & (z \leqslant a), \\ \gamma_2 \exp[-\beta(z-a)] \\ \quad + \delta n_1\left\{\dfrac{\exp[-2q_m(z-a)]}{\varepsilon_0} + \gamma_0 z \exp[-\beta(z-a)]\right\} & (z \geqslant a). \end{cases} \qquad (4.206)$$

The requirements of continuity of the charge density and its derivative fix the parameters γ_1 and γ_2, and the parameter a is fixed by total charge neutrality, while γ_0 is fixed as indicated above. Note that the minimisation process that yields β now automatically includes the interaction between the metallic-type charge and the charge transferred to the interface states, thus the value of β will be slightly different β_0 of (4.199). In practice δn_1 is rather small, since the difference between E_F and Φ_0 is small, and δn_1 can effectively be treated linearly. The value of β finally determines the charge distributions (4.205) and (4.206). With this we can estimate the change in surface dipole on passing from $E_F = \Phi_0$ to $E_F - \Phi_0$, i.e. the term D_{vs} in (4.197). The result is of the form

$$D_{vs} = \alpha N_{vs}(E_g - \Phi_{Bn} - \Phi_0), \tag{4.207}$$

where α is a parameter which depends in a complicated way on (4.205) and (4.206). This form for D_{vs} results from the assumption that E_F and Φ_0, although different, are fairly close to each other.

4.5.4. The barrier height

The last step in the estimate of Φ_{Bn} consists in using (4.207) in (4.197), which yields

$$\Phi_{Bn} = \frac{1}{1+\alpha N_{vs}}[\Phi_M - \chi + D_J + \alpha N_{vs}(E_g - \Phi_0)]. \tag{4.208}$$

All the parameters in this formula can be determined either from experiment or from the theoretical analysis of this section. The form of this result is quite instructive. The crucial feature is the factor αN_{vs}, which gives an effective measure of the importance of the interface states. The limit $\alpha N_{vs} \gg 1$ is the *Bardeen limit*, $\Phi_{Bn} = E_g - \Phi_0$, in which Φ_{Bn} does not depend on Φ_M because the Fermi level at the interface is very efficiently anchored by a high effective density of states. The opposite extreme, $\alpha N_{vs} \ll 1$, is the *Schottky Limit*, $\Phi_{Br} = \Phi_M - \chi + D_J$, in which Φ_{Bn} varies linearly with Φ_M. Strictly speaking this is not the literal Schottky model. There is a small difference due to lattice effects, which is measured by D_J.

Two distinct features can be appreciated in (4.208). One is the role of the semiconductor. It turns out that the parameter α depends very little on the metal that forms the junction. In fact, the whole term αN_{vs} is almost a distinctive feature of the semiconductor; or, in a more detailed analysis, of its band structure. Notice (table 9) how αN_{vs} is large for Si (covalent) and GaAs (slightly ionic), and small for ZnS (rather more ionic). For Si, naturally, the behaviour of the barrier height is of the Bardeen type. For example, for the Si–Al junction, taking $E_g = 1.1$ eV we find $\Phi_{Bn} \approx 0.62$ eV – actually very close to the experimental value – while $\Phi_0 \approx 0.59$ eV, which indicates how effective the pegging of the Fermi level is (fig. 96*b*).

The other distinctive feature of (4.208) is the role of the metal that forms the junction, which manifests itself through the parameters Φ_M and

TABLE 9. *Values of* αN_{vs} *and* S_0 *calculated, as explained in the text, for* Si, GaAs *and* ZnS

	Si	GaAs†	ZnS†
αN_{vs}	6.14	7.33	2.83
S_0	0.14	0.12	0.30

† Source: C. Tejedor, F. Flores & E. Louis, *J. Phys. C: Solid St. Phys.* **10**, 2163 (1977).

D_J. Incidentally, an important caveat is in order here: our analysis assumes that the bottom of the metallic band is always below the semiconductor gap. Then the semiconductor surface states must necessarily overlap with metallic bulk states, as explained, and we can only have the situation studied here. It may happen with some light metals, e.g. Na, that the bottom of the metallic band – at least for some domains of κ – is higher up, in the gap range. Then we may find true stationary surface states whose amplitude decays in both directions. The analysis must then be modified to account for the extra terms (s.s.) which may appear in the density of states of the system. Returning to the standard case, the role of the metal is reflected in the values of Φ_M and D_J. The latter turns out to vary, among different metals, much less than Φ_M. Thus the dependence on the metal is essentially determined by Φ_M. Hence we can write (4.208) approximately in the form

$$\Phi_{Bn} = S_0 \Phi_M + S_1, \tag{4.209}$$

where S_0 and S_1 are parameters which in practical terms characterise the semiconductor, irrespective of the metal. The practical interest is centred on the slope S_0, which is also the parameter often sought experimentally. Some representative values of it, estimated in the approximation discussed here, are shown in table 9. Notice the increasing trend on going towards more-ionic semiconductors. This is, of course, what one would expect on general grounds, although existing experimental evidence appears to suggest a sharper rise of S_0, i.e. a sharper transition towards Schottky-type behaviour. It may be, for example, that in the analysis expounded here the narrow-gap approximation has been pushed too far, but it is also clear that further experimental work is necessary. In this respect it is interesting that the results of this simple model are consistent with those of more elaborate and accurate computations. In any case a thorough discussion of the problem is a matter for further detailed study, and is really outside the scope of this basic theoretical survey.

5

The dynamical response of simple metal surfaces

5.1. Ground-state energy and non-local response

In chapter 1 we discussed the normal modes of surface systems, while in chapter 3 we studied the static properties of the electron gas metallic surfaces. We shall now study the dynamical, i.e. frequency dependent, properties of the bounded electron gas as a model of simple metal surfaces. This will give a different view of metallic surfaces, which connects with concepts seen in chapters 1, 2 and 3.

A useful object, which plays a central part in this analysis, is the non-local, ω-dependent polarisability or response function. We shall now study some general relations which involve the non-local dynamic polarisability and are applicable quite generally to inhomogeneous systems. We consider the ground-state value of the electron–electron interaction energy

$$E_{e\text{-}e}=\left\langle 0\left|\frac{1}{2}\sum_{i\neq j}\frac{e^2}{|\boldsymbol{r}_i-\boldsymbol{r}_j|}\right|0\right\rangle. \tag{5.1}$$

For reasons which will be seen later we still put $\hbar=m=1$, but write e explicitly. It is convenient to rewrite (5.1) in the form

$$E_{e\text{-}e}=\frac{e^2}{2}\int\frac{1}{|\boldsymbol{r}-\boldsymbol{r}'|}\left\langle 0\left|\sum_{i\neq j}\delta(\boldsymbol{r}-\boldsymbol{r}_i)\delta(\boldsymbol{r}'-\boldsymbol{r}_j)\right|0\right\rangle \mathbf{d}\boldsymbol{r}\,\mathbf{d}\boldsymbol{r}'. \tag{5.2}$$

Notice that the ground-state wavefunctions depend on the electron coordinates $\boldsymbol{r}_i$, $\boldsymbol{r}_j$, but not on $\boldsymbol{r}$, $\boldsymbol{r}'$. We introduce the density operator

$$\rho(\boldsymbol{r})=\sum_i\delta(\boldsymbol{r}-\boldsymbol{r}_i), \tag{5.3}$$

whose expectation value is the ground-state average particle density $n_0(\boldsymbol{r})$. Writing out the self-energy term explicitly we then rewrite (5.2) as

$$\begin{aligned}E_{e\text{-}e}&=\frac{e^2}{2}\int\frac{1}{|\boldsymbol{r}-\boldsymbol{r}'|}\\&\quad\times\left\langle 0\left|\sum_{i,j}\delta(\boldsymbol{r}-\boldsymbol{r}_i)\delta(\boldsymbol{r}'-\boldsymbol{r}_j)-\sum_i\delta(\boldsymbol{r}-\boldsymbol{r}_i)\delta(\boldsymbol{r}'-\boldsymbol{r}_i)\right|0\right\rangle \mathbf{d}\boldsymbol{r}\,\mathbf{d}\boldsymbol{r}'\\&=\frac{e^2}{2}\int\frac{1}{|\boldsymbol{r}-\boldsymbol{r}'|}\{\langle 0|\rho(\boldsymbol{r})\rho(\boldsymbol{r}')|0\rangle-\delta(\boldsymbol{r}-\boldsymbol{r}')n_0(\boldsymbol{r})\},\end{aligned} \tag{5.4}$$

which in turn, by suitable additions and identical subtractions can be written as

$$E_{\text{e-e}} = \int \frac{e^2}{2|\boldsymbol{r}-\boldsymbol{r}'|} \times \{\langle 0|[\rho(\boldsymbol{r}) - n_0(\boldsymbol{r})][\rho(\boldsymbol{r}') - n_0(\boldsymbol{r}')]|0\rangle - \delta(\boldsymbol{r}-\boldsymbol{r}')n_0(\boldsymbol{r}')\}\,\mathbf{d}\boldsymbol{r}\,\mathbf{d}\boldsymbol{r}' + \int \frac{e^2}{2|\boldsymbol{r}-\boldsymbol{r}'|} n_0(\boldsymbol{r})n_0(\boldsymbol{r}')\,\mathbf{d}\boldsymbol{r}\,\mathbf{d}\boldsymbol{r}'. \tag{5.5}$$

What we have done is to (i) separate the Hartree energy from short-range interactions, (ii) subtract explicitly the self-energy term, and (iii) introduce the density fluctuations about the average value. Using this notion we can now start from (5.5) and express $E_{\text{e-e}}$ in term of the non-local response function.

We define the polarisability $\chi(\boldsymbol{r}, \boldsymbol{r}'; \omega)$ so that an external potential $V^{\text{e}}(\boldsymbol{r}', \omega)$ induces a particle density fluctuation $\rho^{\text{i}}(\boldsymbol{r}) = \rho(\boldsymbol{r}) - n_0(\boldsymbol{r})$ given by

$$e\rho^{\text{i}}(\boldsymbol{r}, \omega) = \int \chi(\boldsymbol{r}, \boldsymbol{r}'; \omega) V^{\text{e}}(\boldsymbol{r}', \omega)\,\mathbf{d}\boldsymbol{r}'. \tag{5.6}$$

Note that this is not exactly the susceptibility χ defined in chapter 1, which relates the polarisation to the electric field, although they are of course intimately related. We shall work in terms of χ as defined in (5.6). The frequency ω will be explicitly given or not, according to convenience, but in any case the ω-dependence is to be understood throughout. Now, it is well known from linear response theory that χ is given by

$$\chi(\boldsymbol{r}, \boldsymbol{r}'; t) = \begin{cases} -\text{i}e^2\langle 0|[\rho^{\text{i}}(\boldsymbol{r}, t), \rho^{\text{i}}(\boldsymbol{r}', 0)]|0\rangle & (t > 0) \\ 0 \quad (t < 0). \end{cases} \tag{5.7}$$

This relates the polarisability to the density fluctuations and involves the commutator of the density operators in Heisenberg representation. In practice it is more convenient to Fourier transform t into ω, and then, in order to comply formally with the retarded causal response expressed by (5.7), we must add the usual infinitesimal imaginary part. Thus one finds

$$\chi(\boldsymbol{r}, \boldsymbol{r}'; \omega^+) = e^2 \sum_n \left\{ \frac{\langle 0|\rho^{\text{i}}(\boldsymbol{r})|n\rangle\langle n|\rho^{\text{i}}(\boldsymbol{r}')|0\rangle}{\omega - \omega_{n0} + \text{i}\eta} - \frac{\langle 0|\rho^{\text{i}}(\boldsymbol{r}')\rangle\langle n|\rho^{\text{i}}(\boldsymbol{r})|0\rangle}{\omega + \omega_{n0} + \text{i}\eta} \right\}. \tag{5.8}$$

Here $|n\rangle$ is an eigenstate of the interacting electron system, with excitation frequency ω_{n0}. Notice that this formula involves matrix elements of density operators which are independent of the external (small) perturbation, which has been divided away. Using the well-known symbolic formula for $(\omega \pm \omega_{n0} + \mathrm{i}\eta)^{-1}$ we have

$$-\frac{1}{\pi}\int_0^\infty \mathrm{Im}\,\chi(\boldsymbol{r},\boldsymbol{r}';\omega^+)\,\mathrm{d}\omega$$
$$= e^2 \sum_n \langle 0|\rho^{\mathrm{i}}(\boldsymbol{r})|n\rangle\langle n|\rho^{\mathrm{i}}(\boldsymbol{r}')|0\rangle = e^2\langle 0|\rho^{\mathrm{i}}(\boldsymbol{r})\rho^{\mathrm{i}}(\boldsymbol{r}')|0\rangle, \tag{5.9}$$

which used in (5.5) yields

$$E_{\mathrm{e\text{-}e}} = \int \frac{e^2}{2|\boldsymbol{r}-\boldsymbol{r}'|} n_0(\boldsymbol{r}) n_0(\boldsymbol{r}')$$
$$+ \int \frac{1}{2|\boldsymbol{r}-\boldsymbol{r}'|}\left\{-\int_0^\infty \mathrm{Im}\,\chi(\boldsymbol{r},\boldsymbol{r}';\omega^+)\frac{\mathrm{d}\omega}{\pi} - e^2\delta(\boldsymbol{r}-\boldsymbol{r}')n_0(\boldsymbol{r}')\right\}\mathbf{d}\boldsymbol{r}\,\mathbf{d}\boldsymbol{r}'. \tag{5.10}$$

We could also think of V^{e} as created by an external charge, i.e.

$$V^{\mathrm{e}}(\boldsymbol{r}') = \int \frac{e}{|\boldsymbol{r}'-\boldsymbol{r}''|}\rho^{\mathrm{e}}(\boldsymbol{r}'')\,\mathbf{d}\boldsymbol{r}'', \tag{5.11}$$

so that

$$e\rho^{\mathrm{i}}(\boldsymbol{r}) = \int \chi(\boldsymbol{r},\boldsymbol{r}')\frac{e}{|\boldsymbol{r}'-\boldsymbol{r}''|}\rho^{\mathrm{e}}(\boldsymbol{r}'')\,\mathbf{d}\boldsymbol{r}'\,\mathbf{d}\boldsymbol{r}'', \tag{5.12}$$

whence the functional derivative

$$\frac{\delta\rho^{\mathrm{i}}(\boldsymbol{r})}{\delta\rho^{\mathrm{e}}(\boldsymbol{r}'')} = \int \chi(\boldsymbol{r},\boldsymbol{r}')\frac{1}{|\boldsymbol{r}'-\boldsymbol{r}''|}\,\mathbf{d}\boldsymbol{r}'. \tag{5.13}$$

This allows us to rewrite (5.10) as

$$E_{\mathrm{e\text{-}e}} = \int \frac{e^2}{2|\boldsymbol{r}-\boldsymbol{r}'|} n_0(\boldsymbol{r}) n_0(\boldsymbol{r}')\,\mathbf{d}\boldsymbol{r}\,\mathbf{d}\boldsymbol{r}'$$
$$-\int_0^\infty \frac{\mathrm{d}\omega}{2\pi}\int \mathbf{d}\boldsymbol{r}\,\mathrm{Im}\,\frac{\delta\rho^{\mathrm{i}}(\boldsymbol{r})}{\delta\rho^{\mathrm{e}}(\boldsymbol{r})} - \int \frac{e^2}{2|\boldsymbol{r}-\boldsymbol{r}'|}\delta(\boldsymbol{r}-\boldsymbol{r}')n_0(\boldsymbol{r}')\,\mathbf{d}\boldsymbol{r}\,\mathbf{d}\boldsymbol{r}'. \tag{5.14}$$

We now have three equivalent ways of writing $E_{\mathrm{e\text{-}e}}$, namely, (5.5), (5.10) and (5.14). Each one is most useful in a different context. For our immediate purpose it will be more convenient to use the last one, which involves the response function $\delta\rho^{\mathrm{i}}/\delta\rho^{\mathrm{e}}$. So far we have only studied the electron–electron interaction energy, but we can now see that this enables us to express the total ground state energy – kinetic energy included – in terms of the response function. This applies generally to any system, homogeneous or inhomogeneous, and in particular to a surface system.

For this we make the hypothetical assumption that the strength of the electron–electron interaction is gradually increased from zero to its full value. This can be achieved by replacing e^2 by λe^2, where $0 \leqslant \lambda \leqslant 1$. If we have a Hamiltonian which depends on the parameter λ, then the *Hellmann–Feynman theorem* says that

$$\frac{\mathrm{d}E}{\mathrm{d}\lambda} = \left\langle 0, \lambda \left| \frac{\mathrm{d}H(\lambda)}{\mathrm{d}\lambda} \right| 0, \lambda \right\rangle, \tag{5.15}$$

where $|0, \lambda\rangle$ is the ground state of the Hamiltonian $H(\lambda)$. The problem is how to define $H(\lambda)$, and this can be done in three stages as follows.

(i) For $\lambda = 1$ we clearly have

$$H = H(\lambda = 1) = -\frac{1}{2}\sum_i \nabla_i^2 + \frac{e^2}{2}\sum_{i \neq j} \frac{1}{|\boldsymbol{r}_i - \boldsymbol{r}_j|} + \sum_i V_1(\boldsymbol{r}_i). \tag{5.16}$$

In principle $V_1(\boldsymbol{r}_i)$ is the potential of the ions in the crystal lattice. In practice we shall spread this out into a positive background, but the formulation at this stage is still general.

(ii) For $\lambda = 0$ we choose a reference Hamiltonian H_0 of the form

$$H_0 = -\frac{1}{2}\sum_i \nabla_i^2 + \sum_i V_a(\boldsymbol{r}_i). \tag{5.17}$$

Here V_a is a hypothetical applied potential which must be such that the particle density corresponding to (5.17) is the same as the actual density $n_0(\boldsymbol{r})$ of the fully interacting Hamiltonian (5.16). We know from the *Hohenberg–Kohn theorem* (§ 3.2) that V_a is uniquely determined by n_0.

(iii) For $0 \neq \lambda \neq 1$ the Hamiltonian will undoubtedly contain, besides the kinetic energy T, the term

$$\sum_{i \neq j} \frac{\lambda e^2}{2|\boldsymbol{r}_i - \boldsymbol{r}_j|},$$

but we also add an extra potential $V(\boldsymbol{r}, \lambda)$, such that the density, for every value of λ, remains equal to $n_0(\boldsymbol{r})$. This is also uniquely defined, though not necessarily in a straightforward fashion. The point is that $V(\boldsymbol{r}, \lambda)$ exists and is uniquely defined. Since T is independent of λ, we can then write

$$\frac{\mathrm{d}H(\lambda)}{\mathrm{d}\lambda} = \sum_{i \neq j} \frac{e^2}{2|\boldsymbol{r}_i - \boldsymbol{r}_j|} + \sum_i \frac{\mathrm{d}V(\boldsymbol{r}_i, \lambda)}{\mathrm{d}\lambda}. \tag{5.18}$$

Hence, from (5.15) and (5.18):

$$\frac{dE}{d\lambda}=\left\langle 0,\lambda\left|\sum_{i\neq j}\frac{e^2}{2|\boldsymbol{r}_i-\boldsymbol{r}_j|}\right|0,\lambda\right\rangle+\left\langle 0,\lambda\left|\sum_i\frac{dV(\boldsymbol{r}_i,\lambda)}{d\lambda}\right|0,\lambda\right\rangle$$

$$=\frac{1}{\lambda}E_{e\text{-}e}(\lambda)+\int\frac{dV(\boldsymbol{r},\lambda)}{d\lambda}n_0(\boldsymbol{r})\,d\boldsymbol{r}, \qquad (5.19)$$

where we stress that $n_0(\boldsymbol{r})$ is independent of λ. We now have a first-order differential equation which, in terms of the kinetic energy T_0 of the non-interacting system (5.17), has the boundary condition

$$E(\lambda=0)=T_0+\int V_a(\boldsymbol{r})n_0(\boldsymbol{r})\,d\boldsymbol{r}.$$

With this we can integrate (5.19). This yields

$$E=T_0+\int V_a(\boldsymbol{r})n_0(\boldsymbol{r})\,d\boldsymbol{r}+\int_0^1\frac{1}{\lambda}E_{e\text{-}e}(\lambda)\,d\lambda$$

$$+\int_0^1 d\lambda\int\frac{dV(\boldsymbol{r},\lambda)}{d\lambda}n_0(\boldsymbol{r})\,d\boldsymbol{r}$$

$$=T_0+\int\left[V_a(\boldsymbol{r})+\int_0^1\frac{dV(\boldsymbol{r},\lambda)}{d\lambda}d\lambda\right]n_0(\boldsymbol{r})\,d\boldsymbol{r}+\int_0^1\frac{1}{\lambda}E_{e\text{-}e}(\lambda)\,d\lambda. \qquad (5.20)$$

From (5.17), (5.16) and (5.18), the square bracket in the integrand of the second term of (5.20) is just $V_1(\boldsymbol{r})$. Thus

$$E=T_0+\int V_1(\boldsymbol{r})n_0(\boldsymbol{r})\,d\boldsymbol{r}+\int_0^1\frac{1}{\lambda}E_{e\text{-}e}(\lambda)\,d\lambda. \qquad (5.21)$$

But $E_{e\text{-}e}(\lambda)$ is just what we have studied in (5.5)–(5.14), replacing e^2 everywhere by λe^2. Thus, from (5.5),

$$\int_0^1\frac{d\lambda}{\lambda}E_{e\text{-}e}(\lambda)=\int\frac{e^2}{2|\boldsymbol{r}-\boldsymbol{r}'|}n_0(\boldsymbol{r})n_0(\boldsymbol{r}')\,d\boldsymbol{r}\,d\boldsymbol{r}'+\int_0^1\frac{d\lambda}{\lambda}$$

$$\times\int\frac{\lambda e^2}{2|\boldsymbol{r}-\boldsymbol{r}'|}\{\langle 0,\lambda\,|\rho_i(\boldsymbol{r})\rho_i(\boldsymbol{r}')|0,\lambda\rangle-\delta(\boldsymbol{r}-\boldsymbol{r}')n_0(\boldsymbol{r}')\}\,d\boldsymbol{r}\,d\boldsymbol{r}'.$$

Using this now in (5.21) and subtracting the Hartree and kinetic energies we have the exchange and correlation energy

$$\int_0^1\frac{d\lambda}{\lambda}\int\frac{\lambda e^2}{2|\boldsymbol{r}-\boldsymbol{r}'|}$$

$$\times\{\langle 0,\lambda\,|\rho^{i}(\boldsymbol{r})\rho^{i}(\boldsymbol{r}')|0,\lambda\rangle-\delta(\boldsymbol{r}-\boldsymbol{r}')n_0(\boldsymbol{r}')\}\,d\boldsymbol{r}\,d\boldsymbol{r}'. \qquad (5.22)$$

Alternatively, by using (5.10):

$$E_{xc}=\int_0^1 \frac{d\lambda}{\lambda}\int \frac{1}{2|\boldsymbol{r}-\boldsymbol{r}'|}$$
$$\times\left\{-\int_0^\infty \frac{d\omega}{\pi}\,\mathrm{Im}\,\chi(\boldsymbol{r},\boldsymbol{r}';\omega^+;\lambda)-\lambda e^2\delta(\boldsymbol{r}-\boldsymbol{r}')n_0(\boldsymbol{r}')\right\}\mathbf{d}\boldsymbol{r}\,\mathbf{d}\boldsymbol{r}', \tag{5.23}$$

where $\chi(\boldsymbol{r},\boldsymbol{r}';\omega^+;\lambda)$ is the polarisability of an electron gas interacting with coupling constant λe^2. Finally, by using (5.14):

$$E_{xc}=-\int_0^1 \frac{d\lambda}{\lambda}\int_0^\infty \frac{d\omega}{2\pi}\int \mathbf{d}\boldsymbol{r}\,\mathrm{Im}\,\frac{\delta\rho^{\mathrm{i}}(\boldsymbol{r},\lambda)}{\delta\rho^{\mathrm{e}}(\boldsymbol{r})}$$
$$-\int \frac{e^2}{2|\boldsymbol{r}-\boldsymbol{r}'|}\delta(\boldsymbol{r}-\boldsymbol{r}')n_0(\boldsymbol{r}')\,\mathbf{d}\boldsymbol{r}\,\mathbf{d}\boldsymbol{r}'. \tag{5.24}$$

So far, this is general formal theory. We shall now use this as a framework to study some of the basic problems associated with surface systems. Our purpose is to discuss elementary points of theory, for which we shall use simple and tractable models which may not always be the most adequate for accurate quantitative calculations, but are perfectly suitable for bringing out the essential physics of the various surface problems presently to be discussed. We shall first (§ 5.2) analyse the semiclassical approximation, in which specific quantum mechanical interference effects are ignored. Quantum theory will be considered in § 5.3 and finally (§ 5.3) we shall briefly discuss the relationship between the approach to the surface energy problem which emerges from this picture and the approach of chapter 3.

5.2. The semiclassical infinite-barrier model

We now concentrate on the following model: we assume a jellium model of a metal with an infinite barrier which is treated in the semiclassical approximation. We further assume the surface to be specular, which corresponds to the case $p=1$ of chapter 1, and we work in the quasistatic limit. Other effects like, say, non specularity could be included by direct application of the methods developed in chapter 1. However this would complicate the analysis quite considerably and we shall here stick to the simple model just defined. In spite of the many simplifications, it admits of an analysis which emphasises quite clearly the role of general concepts.

We start from (5.24), for which we must first study the non-local response function of our model surface system. Let the metal, with dielectric function $\varepsilon(\boldsymbol{k};\omega)$, be contained in $z>0$. We want to study, at every point $\boldsymbol{r}=(\boldsymbol{\rho}, z)$, the response to a small external test charge $Q\delta(\boldsymbol{r}')$, where $\boldsymbol{r}'=(\boldsymbol{\rho}', z'>0)$ is a point inside the metal. The reason for putting Q inside is that in (5.24) we need (5.13) with $\boldsymbol{r}''=\boldsymbol{r}$. Thus we shall eventually take $\boldsymbol{r}'=\boldsymbol{r}$ and it suffices for the time being to study the case in which the test charge is inside the metal. The case of a real external charge outside will be discussed later. We can solve this problem by using the method of extended pseudomedia as in chapter 1. Since we assume a specular model, the extended pseudometal (M) contains the charge $+Q$ at $+z'$, its image $+Q$ at $-z'$ and a fictitious surface charge σ^{M} whose dependence on $(\boldsymbol{\kappa}, \omega)$ is understood throughout. For the extended pseudovacuum it suffices to introduce a surface charge σ^{V}. Both σ^{M} and σ^{V} are to be eliminated from the matching conditions for the electrostatic potential ϕ and its normal derivative. But we learnt from our experience in chapter 1 that the second matching condition yields $\sigma^{\mathrm{V}}=-\sigma^{\mathrm{M}}$. Thus it suffices to use one single parameter σ and to simply match ϕ. Now, ϕ^{M} is created by the *three* charges as explained. Thus

$$\phi^{\mathrm{M}}(\boldsymbol{k})=2Q\cos(qz')\frac{4\pi}{k^2\varepsilon(\boldsymbol{k},\omega)}+\frac{4\pi\sigma}{k^2\varepsilon(\boldsymbol{k},\omega)}, \tag{5.25}$$

while

$$\phi^{\mathrm{V}}(\boldsymbol{k})=-4\pi\sigma/k^2. \tag{5.26}$$

Remember the real potential is ϕ^{M} for $z>0$ and ϕ^{V} for $z<0$. Matching at $z=0$, we have

$$\int\frac{2Q\cos(qz')}{(\kappa^2+q^2)\varepsilon(\boldsymbol{\kappa},q)}\,\mathrm{d}q+\int\frac{\sigma}{(\kappa^2+q^2)\varepsilon(\boldsymbol{\kappa},q)}\,\mathrm{d}q$$
$$=-\int\frac{\sigma}{(\kappa^2+q^2)}\,\mathrm{d}q, \tag{5.27}$$

whence (ω-dependence understood)

$$\sigma=-\frac{1}{\mathscr{B}}\int\frac{2Q\cos(qz')}{(\kappa^2+q^2)\varepsilon(\boldsymbol{\kappa},q)}\,\mathrm{d}q, \tag{5.28}$$

where we have defined

$$\mathscr{B}(\boldsymbol{\kappa},\omega)\equiv\int\frac{\mathrm{d}q}{(\kappa^2+q^2)}+\int\frac{\mathrm{d}q}{(\kappa^2+q^2)\varepsilon(\boldsymbol{\kappa},q,\omega)}\equiv I[1]+I[\varepsilon]. \tag{5.29}$$

This is a familiar object. For example we saw in (1.208) that the vanishing of $\mathscr{B}$ yields the s.m.d.r. for the specular surface model in the quasistatic limit. More generally we would have the sum of the surface impedances – of vacuum and metal – which is what $\mathscr{B}$ is for the particular model studied here. We saw in chapter 1 how the surface impedances can be obtained for $p \neq 1$. Formally, in the analysis presently to be developed we could think of $\mathscr{B}$ as the more general object obtained in chapter 1. We shall see that $\mathscr{B}$ will play a central part in bringing the normal modes of the surface system into the theory. However, the more general formula for $\mathscr{B}$ is considerably more complicated and we shall restrict ourselves to the explicit form of (5.29). Using (5.28) now in (5.25) we have

$$\phi^{\mathrm{M}}(\boldsymbol{k}) = 2Q\cos(qz')\frac{4\pi}{k^2\varepsilon(\boldsymbol{k})} - \frac{4\pi}{k^2\varepsilon(\boldsymbol{k})}\frac{1}{\mathscr{B}}\int 2Q\cos(qz')\frac{\mathrm{d}q}{(\kappa^2+q^2)\varepsilon(\boldsymbol{\kappa}, q)}. \tag{5.30}$$

Thus

$$e\rho^{\mathrm{M}}(\boldsymbol{k}) = (k^2/4\pi)\times(5.30). \tag{5.31}$$

The Fourier transform of this yields a charge distribution $e\rho^{\mathrm{M}}(\boldsymbol{r})$. In order to obtain from this the real induced charge that we are interested in, we must proceed with some care. First we are still studying the extended pseudomedium (M), in which $e\rho^{\mathrm{M}}$ includes the three external charges discussed above, and the induced charge. Thus we must first subtract the three external charges and *then* we have an induced charge which, for $z > 0$, is equal to the real induced charge inside the medium under consideration. Thus we need

$$\rho^{\mathrm{M,i}}(\boldsymbol{r}) = \int \frac{\mathrm{d}^3k}{(2\pi)^3}\rho^{\mathrm{M,i}}(\boldsymbol{k})\,\mathrm{e}^{\mathrm{i}\boldsymbol{\kappa}\cdot\boldsymbol{\rho}}\,\mathrm{e}^{\mathrm{i}qz}, \tag{5.32}$$

where

$$e\rho^{\mathrm{M,i}}(\boldsymbol{k}) = \frac{2Q\cos(qz')}{(2\pi)^3}\left[\frac{1}{\varepsilon(\boldsymbol{k})} - 1\right] - \frac{1}{\mathscr{B}}\left[\frac{1}{\varepsilon(\boldsymbol{k})} - 1\right]\int \frac{2Q\cos(qz')}{(2\pi)^3}\frac{\mathrm{d}q}{(\kappa^2+q^2)\varepsilon(\kappa, q)}. \tag{5.33}$$

Notice, incidentally, that the induced charge, at $\boldsymbol{r}$, depends of course on the position of the *real* external charge at $\boldsymbol{r}'$. Now we can say that $e\times(5.32)$ is the real induced charge for $z > 0$, whence, for $z > 0$ and $z' > 0$,

$$\frac{\delta\rho^{\mathrm{i}}(\boldsymbol{r})}{\delta\rho^{\mathrm{e}}(\boldsymbol{r}')} = \int \frac{\mathrm{d}^3k}{(2\pi)^3}\left[\frac{1}{\varepsilon(\boldsymbol{k})} - 1\right]\mathrm{e}^{\mathrm{i}\boldsymbol{\kappa}\cdot(\boldsymbol{\rho}-\boldsymbol{\rho}')}[\mathrm{e}^{\mathrm{i}q(z-z')} + \mathrm{e}^{\mathrm{i}q(z+z')}] - \int \frac{\mathrm{d}^3k}{(2\pi)^3}\left[\frac{1}{\varepsilon(\boldsymbol{k})} - 1\right]\mathrm{e}^{\mathrm{i}\boldsymbol{\kappa}\cdot(\boldsymbol{\rho}-\boldsymbol{\rho}')}\frac{\mathrm{e}^{\mathrm{i}qz}}{\mathscr{B}(\boldsymbol{\kappa})}\int \mathrm{d}p\frac{\mathrm{e}^{\mathrm{i}pz'} + \mathrm{e}^{-\mathrm{i}pz'}}{(\kappa^2+p^2)\varepsilon(\boldsymbol{\kappa}, p)}. \tag{5.34}$$

For future reference, we indicate that a similar analysis yields, for $z>0$ and $z'<0$,

$$\frac{\delta\rho^{\mathrm{i}}(\boldsymbol{r})}{\delta\rho^{\mathrm{e}}(\boldsymbol{r}')}=\int\frac{\mathrm{d}^3k}{(2\pi)^3}\left[\frac{1}{\varepsilon(\boldsymbol{k})}-1\right]\mathrm{e}^{\mathrm{i}\boldsymbol{\kappa}\cdot(\boldsymbol{\rho}-\boldsymbol{\rho}')}\frac{\mathrm{e}^{\mathrm{i}qz}}{\mathscr{B}(\boldsymbol{\kappa})}\int\mathrm{d}p\frac{\mathrm{e}^{\mathrm{i}pz'}+\mathrm{e}^{-\mathrm{i}pz'}}{(\kappa^2+p^2)\varepsilon(\boldsymbol{\kappa},p)}. \tag{5.35}$$

Obviously, ρ^{i} is always zero for $z<0$, irrespective of sgn z. Thus

$$\delta\rho^{\mathrm{i}}(\boldsymbol{r})/\delta\rho^{\mathrm{e}}(\boldsymbol{r}')=0 \quad (z<0,\ z'\gtrless 0). \tag{5.36}$$

Equations (5.34)–(5.36) describe completely the response of our model surface system but, for the application of (5.24), we need the case $\boldsymbol{r}'=\boldsymbol{r}$ and, on integrating over $\boldsymbol{r}$, the contribution of (5.36) vanishes identically. Thus, it suffices to use (5.34). Notice the role of the normal modes in the zeros of ε (bulk system) and of $\mathscr{B}$ (surface system). This will soon be studied in detail, but first we shall clarify a point concerning the application of (5.24).

What we want to do is to separate out bulk and surface terms, for which we shall use a procedure which is somewhat informal but yields the correct results. The idea is to see how the different terms scale according to factors proportional to the volume V or area A. Strictly speaking we ought to take a finite volume, evaluate discrete sums and then take the limit of infinite volume, but in order to see the relative scaling of the terms arising from the integration of (5.34) we can evaluate the integrals and formally introduce the factors V or A as they would arise on evaluating discrete sums for a system of finite size. Looking at (5.34), the first term on the r.h.s. gives two contributions due to the two exponentials. The first one is simply equal to unity on setting $z=z'$. This contributes a scale factor V on integrating over $\boldsymbol{r}$. The second contribution involves (remember (5.36)) the term

$$\int\mathrm{d}^2\rho\int_0^\infty\mathrm{d}z\ \mathrm{e}^{(2\mathrm{i}q-\eta)z}. \tag{5.37}$$

We have added the infinitesimal convergence factor, as usual. Thus we have a scale factor A from the two-dimensional integral and also a factor

$$\frac{1}{2\mathrm{i}(q+\frac{1}{2}\mathrm{i}\eta)}=\frac{1}{2\mathrm{i}}\left[\mathscr{P}\left(\frac{1}{q}\right)-\mathrm{i}\pi\delta(q)\right]. \tag{5.38}$$

Since $\varepsilon(\boldsymbol{\kappa},q)$ in (5.34) is an even function of q, on integrating over q the only non-vanishing contribution comes from $\delta(q)$. The second term on the r.h.s. of (5.34) gives other contributions to the integral over $\boldsymbol{r}$ of the form

$$\int\mathrm{d}^2\rho\int_0^\infty\mathrm{d}z\ \mathrm{e}^{[\mathrm{i}(q\pm p)-\eta]z}, \tag{5.39}$$

which again introduces a scale factor A and a factor $[\mathrm{i}(q \pm p) - \eta]^{-1}$. These in turn produce two δ-function contributions to the integral over q in (5.34). Putting all this together we find

$$\int \frac{\delta\rho^{\mathrm{i}}(\boldsymbol{r})}{\delta\rho^{\mathrm{e}}(\boldsymbol{r})}\,\mathrm{d}\boldsymbol{r} = V\int \frac{\mathrm{d}^3k}{(2\pi)^3}\left[\frac{1}{\varepsilon(\boldsymbol{k})} - 1\right] + \frac{A}{4}\int \frac{\mathrm{d}^2\kappa}{(2\pi)^2}\left[\frac{1}{\varepsilon(\boldsymbol{\kappa}, 0)} - 1\right]$$
$$- A\int \frac{\mathrm{d}^2\kappa}{(2\pi)^2}\,\frac{1}{\mathscr{B}(\boldsymbol{\kappa})}\int \mathrm{d}q\left[\frac{1}{\varepsilon(\boldsymbol{\kappa}, q)} - 1\right]\frac{1}{(\kappa^2 + q^2)\varepsilon(\boldsymbol{\kappa}, q)}. \qquad (5.40)$$

The main thing is to notice the different scaling factors, which signal the bulk and surface terms. For the integration over the strength parameter λ in (5.24) we must note that (5.40) has been evaluated with the full strength of the coupling constant, i.e. with λe^2 for $\lambda = 1$. The dependence of (5.40) on λ is contained in the dielectric function, which depends on λ in the form $\varepsilon = 1 + \lambda a$. This happens, for example, in both the r.p.a. and the hydrodynamic models, as well as in the simplified dielectric functions used in chapter 1. Thus

$$\int_0^1 \frac{\mathrm{d}\lambda}{\lambda}\left[\frac{1}{\varepsilon} - 1\right] = -\int_0^1 \frac{\lambda a}{1 + \lambda a}\,\frac{\mathrm{d}\lambda}{\lambda} = -[\ln(1 + \lambda a)]_0^1 = -\ln \varepsilon. \qquad (5.41)$$

This holds both for $\varepsilon(\boldsymbol{\kappa}, q)$ and for $\varepsilon(\boldsymbol{\kappa}, 0)$. In the third term on the r.h.s. of (5.40) we must note that $\mathscr{B}(\boldsymbol{\kappa})$ also depends on λ through the dielectric function, according to (5.29). We then have

$$\mathscr{B}(\kappa) = \int \frac{\mathrm{d}q}{k^2} + \int \frac{\mathrm{d}q}{k^2(1 + \lambda a)},$$

and

$$\int \mathrm{d}q\left(\frac{1}{\varepsilon} - 1\right)\frac{1}{k^2\varepsilon} = -\int \mathrm{d}q\,\frac{\lambda a}{k^2(1 + \lambda a)^2} = \lambda\frac{\mathrm{d}\mathscr{B}}{\mathrm{d}\lambda}.$$

Thus from the third term on the r.h.s. of (5.40) we obtain, before integrating over $\boldsymbol{\kappa}$,

$$\int_0^1 \mathrm{d}\lambda\,\frac{\mathrm{d}}{\mathrm{d}\lambda}\ln\left\{\int \frac{\mathrm{d}q}{(\kappa^2 + q^2)} + \int \frac{\mathrm{d}q}{(\kappa^2 + q^2)[1 + \lambda a(\boldsymbol{\kappa}, q)]}\right\}$$
$$= \ln\left[\frac{1}{2} + \frac{\kappa}{2\pi}\int \frac{\mathrm{d}q}{k^2\varepsilon(\boldsymbol{\kappa}, q)}\right]. \qquad (5.42)$$

Using (5.41) and (5.42) in (5.40) and (5.24), separating out bulk and surface terms and expressing the results as energies per unit volume or area, we finally obtain (remember (5.24))

$$E^{\mathrm{b}}_{\mathrm{xc}} = \int \frac{\mathrm{d}^3k}{(2\pi)^3} \int_0^\infty \frac{\mathrm{d}\omega}{2\pi} \operatorname{Im} \ln \varepsilon(\boldsymbol{\kappa}, q; \omega^+)$$
$$- \frac{1}{2V} \int \frac{e^2}{|\boldsymbol{r}-\boldsymbol{r}'|} \delta(\boldsymbol{r}-\boldsymbol{r}') n_0(\boldsymbol{r}') \, \mathbf{d}\boldsymbol{r} \, \mathbf{d}\boldsymbol{r}', \tag{5.43}$$

and

$$E^{\mathrm{s}}_{\mathrm{xc}} = \frac{1}{4} \int \frac{\mathrm{d}^2\kappa}{(2\pi)^2} \int_0^\infty \frac{\mathrm{d}\omega}{2\pi} \operatorname{Im} \ln \varepsilon(\boldsymbol{\kappa}, 0; \omega^+)$$
$$+ \int \frac{\mathrm{d}^2\kappa}{(2\pi)^2} \int_0^\infty \frac{\mathrm{d}\omega}{2\pi} \operatorname{Im} \ln \left[\frac{1}{2} + \frac{\kappa}{2\pi} \int \frac{\mathrm{d}q}{(\kappa^2+q^2)\varepsilon(\boldsymbol{\kappa}, q; \omega^+)} \right]. \tag{5.44}$$

In the formula for the bulk energy we have explicitly separated out the self-energy term. For later reference this term can be re-expressed as

$$- \frac{e^2}{2V} \int \frac{\mathrm{d}^3k}{(2\pi)^3} \frac{4\pi}{k^2} \int \mathbf{d}\boldsymbol{r} \, \mathbf{d}\boldsymbol{r}' \, \mathrm{e}^{\mathrm{i}\boldsymbol{k}\cdot(\boldsymbol{r}-\boldsymbol{r}')} \delta(\boldsymbol{r}-\boldsymbol{r}') n_0(\boldsymbol{r}')$$
$$= - \frac{e^2}{2V} \int \frac{\mathrm{d}^3k}{(2\pi)^3} \frac{4\pi}{k^2} n_0(\boldsymbol{r}) \, \mathbf{d}\boldsymbol{r} = - \frac{Ne^2}{2V} \int \frac{\mathrm{d}^3k}{(2\pi)^3} \frac{4\pi}{k^2}. \tag{5.45}$$

Ignoring this term for the moment, notice that the terms appearing in (5.43) and (5.44) have the form

$$\int_0^\infty \operatorname{Im} f(\omega^+) \, \mathrm{d}\omega. \tag{5.46}$$

Since

$$\operatorname{Im} f(\omega^+) = (1/2\mathrm{i})[f(\omega^+) - f(\omega^-)],$$

we have

$$\int_0^\infty \operatorname{Im} f(\omega^+) \frac{\mathrm{d}\omega}{2\pi} = \frac{1}{4\pi\mathrm{i}} \int_C f(\omega) \, \mathrm{d}\omega, \tag{5.47}$$

where the integration contour is shown in fig. 100. This allows for an identical rewriting of (5.43) and (5.44). For example,

$$E^{\mathrm{s}}_{\mathrm{xc}} = \frac{1}{4\pi\mathrm{i}} \int \frac{\mathrm{d}^2\kappa}{(2\pi)^2} \int_C \mathrm{d}\omega$$
$$\times \left\{ \frac{1}{4} \ln \varepsilon(\kappa, 0; \omega) + \ln \left[\frac{1}{2} + \frac{\kappa}{2\pi} \int \frac{\mathrm{d}q}{(\kappa^2+q^2)\varepsilon(\boldsymbol{\kappa}, q; \omega)} \right] \right\}.$$

The important point is that the integrand is a sum of logarithms of functions of ω. In general this could have poles and branch cuts and the

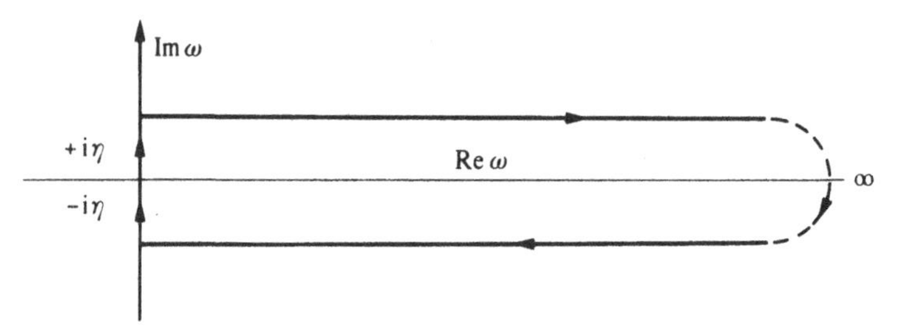

Fig. 100. Integration contour for the integral of (4.47).

evaluation of the integral can be rather cumbersome. This can be very neatly circumvented by means of a formal device. It all hinges on the quantisation scheme one uses. If we do the formal analysis for a system of finite size, then $\boldsymbol{\kappa}$ and q take discrete values and the functions of ω that we are concerned with have only discrete poles. If we took the limit of infinite size first, a sequence of discrete poles would coalesce into a branch cut, but instead we can do all the formal analysis first with a finite size, and carry on to obtain the desired formulae. These will have the form of discrete sums, and *then* when it comes to evaluating them, we can proceed to the continuum limit if convenient. The point of this formal procedure is that if the function $f(\omega)$ of (5.47) has the form $\ln u(\omega)$ and if $u(\omega)$ has poles and zeros only on the real axis and is analytic otherwise, then we can use a well-known theorem which says that

$$\int_C \ln u(\omega)\,\mathrm{d}\omega = -2\pi\mathrm{i}\Big(\sum_i \omega_i - \sum_j \omega_j\Big), \tag{5.48}$$

where ω_j are the zeros and ω_i are the poles of $u(\omega)$. Alternatively, for direct application to (5.43) and (5.44) we can write

$$\int_0^\infty \frac{\mathrm{d}\omega}{2\pi} \operatorname{Im} f(\omega^+) = \frac{1}{2}\Big(\sum_j \omega_j - \sum_i \omega_i\Big). \tag{5.49}$$

From now on we shall use (5.49) applying it directly to the integrals which will appear presently, on the understanding that these may be regarded as discrete sums according to the formal procedure just described. With this we have a mathematical method which is ideally suited to elucidate (5.43) and (5.44) in a way which has a clear physical meaning.

5.2.1. Surface energy and the normal modes of a surface system

The physical interpretation of (5.49) will lead us obviously to a normal-mode analysis, but before we consider this more fully it is very instructive to look at the results of some simple models. At this stage we must discuss a point which was overlooked in chapter 1. Suppose we are interested in the normal modes – plasmons – of a bulk system, with dispersion relation $\omega(\boldsymbol{k})$. We know from the theory of the electron gas that the plasmons are not well-defined elementary excitations for very high k. In fact they are stable modes only up to a critical value k_c, at which the collective modes decay into incoherent single-particle or electron-hole excitations. Suppose we want to make a very simple model. We must at least introduce *ad hoc* the above fact. Thus the simplest conceivable model complying with this is

$$\varepsilon = \begin{cases} 1 - \omega_p^2/\omega^2 = \varepsilon(\omega) & (k < k_c), \\ 1 & (k > k_c). \end{cases} \tag{5.50}$$

Notice (i) this function *is* dispersive, since it depends on k, and (ii) it has no single-particle structure, but the effect of the electron-hole pairs is actually present in the artificial cutoff. Thus the main feature is, so to speak, fudged in, so that it is not missing from the model. For example, suppose we study the first term of (5.43) and we concentrate on the integration over ω, before doing the integration over κ. On applying (5.49) we notice that the dielectric function of (5.50) has a pole at $\omega = 0$ and a zero at $\omega = \omega_p$, the bulk plasma frequency. Thus, for each value of κ the first term of (5.34) amount to $\frac{1}{2}\omega_p$. The physical meaning is clear: with this model dielectric function there is no single-particle structure. Then we are evaluating just ground-state correlation energy and, remembering the factor $\hbar$, we see that this appears as zero-point energy of the collective excitations of the system, which are the only normal-mode excitations existing in this model. This notion of the zero-point energy of the normal modes will recur throughout this chapter.

Let us now look, within the same model (5.50), at the first term of (5.44) and again concentrate first on the ω integration. Applying (5.49) we have, for fixed $\boldsymbol{\kappa}$,

$$\frac{1}{4}\int_0^\infty \frac{d\omega}{2\pi} \operatorname{Im} \ln \varepsilon = \frac{\omega_p}{8}. \tag{5.51}$$

For the second term of (5.44) the function of interest, apart from an irrelevant factor $\kappa/2\pi$, is (remember (5.29)) precisely $\ln \mathcal{B}(\boldsymbol{\kappa}, \omega)$. It is actually more convenient to define

$$\mathcal{B}'(\boldsymbol{\kappa}, \omega) = (\kappa/2\pi)\mathcal{B}(\boldsymbol{\kappa}, \omega)$$

and to study $\mathcal{B}'$. We must first study $I[\varepsilon]$ which, because of the cutoff introduced in (5.50), is

$$\int_{-\infty}^{\infty} \frac{dq}{(\kappa^2+q^2)\varepsilon} = 2\int_0^{q_c} \frac{dq}{(\kappa^2+q^2)\varepsilon(\omega)} + 2\int_{q_c}^{\infty} \frac{dq}{(\kappa^2+q^2)},$$

where

$$q_c = \sqrt{(k_c^2-\kappa^2)} < k_c.$$

The integrals are immediate and we find

$$\mathscr{B}'(\boldsymbol{\kappa}) = 1 + \frac{1}{\pi}\left[\frac{1}{\varepsilon(\omega)} - 1\right] \tan^{-1}(q_c/\kappa). \tag{5.52}$$

We know from (1.208) that the vanishing of this expression yields the s.m.d.r. for the specular surface model. If we remove the cutoff to infinity, then $q_c \to \infty$, $\tan^{-1}(q_c/\kappa) \to \frac{1}{2}\pi$ and the s.m.d.r. is simply

$$\frac{1}{2}\left[\frac{1}{\varepsilon(\omega)} - 1\right] = 0; \quad \omega = \frac{\omega_p}{\sqrt{2}}.$$

Otherwise we have

$$\mathscr{B}'(\kappa) = [\omega^2 - \omega_s^2(\boldsymbol{\kappa})]/(\omega^2 - \omega_p^2). \tag{5.53}$$

The pole is at $\omega = \omega_p$, the bulk plasmon frequency. The zero is at

$$\omega = \omega_s(\boldsymbol{\kappa}) = \omega_p \sqrt{\left[1 - \frac{1}{\pi}\tan^{-1}(q_c/\kappa)\right]}, \tag{5.54}$$

the surface plasmon frequency. For $\kappa \to 0$ we have

$$q_c/\kappa \to \infty, \quad \tan^{-1}(q_c/\kappa) \to \tfrac{1}{2}\pi, \quad \omega_s \to \omega_p/\sqrt{2}, \tag{5.55}$$

while $\kappa = k_c$ yields $q_c = 0$ and $\omega_s = \omega_p$. The corresponding s.m.d.r. is shown in fig. 101. This figure displays a result which arises very simply in this crude model, but is quite significant and general. The existence of a cutoff in the bulk induces also the same cutoff for the surface modes. We shall come back to this point in the r.p.a. analysis of § 5.3. For the time being we notice that by using (5.53) in (5.44) the contribution of the second term, before integrating over $\boldsymbol{\kappa}$, is

$$\tfrac{1}{2}[\omega_s(\kappa) - \omega_p]. \tag{5.56}$$

Adding (5.56) and (5.51) we obtain, in this model,

$$E_c^s = \sum_{\kappa<k_c} \frac{1}{2}\left[\omega_s(\boldsymbol{\kappa}) - \frac{3}{4}\omega_p\right]. \tag{5.57}$$

In fact the same formula would hold formally if we had no cutoff, with two differences, namely (i) $\omega_s(\boldsymbol{\kappa})$ would not depend on κ; it would be simply

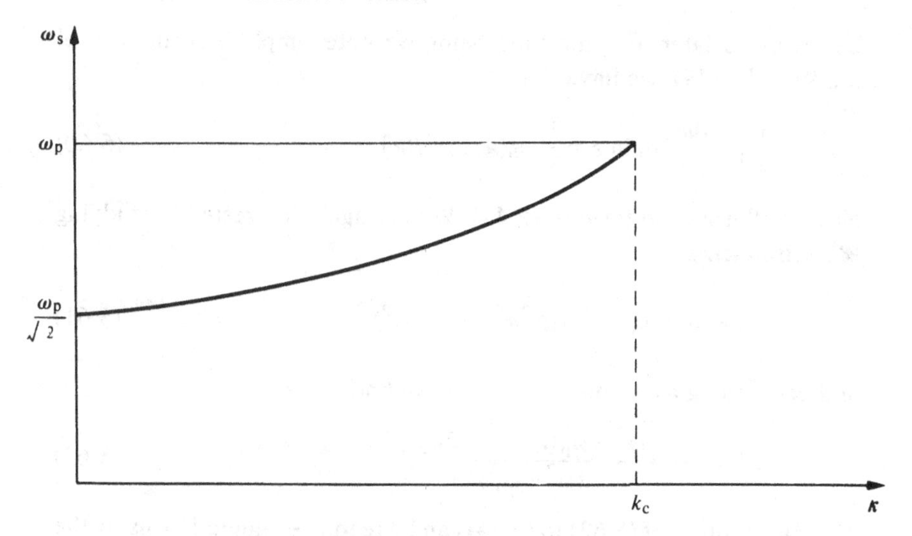

Fig. 101. S.m.d.r. (5.54) for the model of (5.50), showing (i) that there is some dispersion; (ii) the effect of the cutoff.

$\omega_p/\sqrt{2}$; and, more important, (ii) the sum over $\boldsymbol{\kappa}$ would diverge. This is why the cutoff is necessary for the present analysis, although it did not arise in chapter 1.

Notice a curious feature of (5.57). The term $\frac{1}{2}\omega_s$ is easy to interpret as zero-point energy of the surface mode, but we see that it is as if $\frac{3}{4}$ of a bulk plasmon were subtracted. Of course we intuitively expect that in creating a surface we create surface modes, and some spectral strength must be removed from bulk modes. In order to see the origin of the factor $\frac{3}{4}$ it is convenient to interpret these results in terms of the changes in the density of states of the system due to the introduction of the surface. But before doing this it is instructive to consider another, somewhat more sophisticated, model, which exhibits some dispersion explicitly; it is (if we forget about cutoff for time being) given by

$$\varepsilon = \varepsilon(\boldsymbol{k}, \omega) = 1 - \frac{\omega_p^2}{\omega^2 - \beta^2 k^2}; \quad \beta = \sqrt{\frac{3}{5}} V_F. \tag{5.58}$$

This is the dielectric function of the *hydrodynamic approximation*, and is also the longitudinal dielectric function that we used in § 1.6. We have

$$\varepsilon = \frac{\omega^2 - \omega_b^2(\boldsymbol{k})}{\omega^2 - \beta^2 k^2}; \quad \omega_b^2(\boldsymbol{k}) = \omega_p^2 + \beta^2 k^2. \tag{5.59}$$

The point about this function is not only that it explicitly exhibits some dispersion, but also that it contains some single-particle structure, as will be discussed later. For the time being we note simply that, in view of (5.59) and (5.49), we have

$$\frac{1}{4}\int_0^\infty \frac{\mathrm{d}\omega}{2\pi}\,\mathrm{Im}\ln\varepsilon = \frac{1}{8}[\omega_b(\boldsymbol{\kappa},0)-\beta\kappa]. \tag{5.60}$$

Now for the second term of (5.44). We are again interested in studying $\mathcal{B}'$. Introducing

$$\kappa_p(\kappa,\omega)=\kappa_p=\frac{1}{\beta}\sqrt{(\beta^2\kappa^2+\omega_p^2-\omega^2)} \tag{5.61}$$

and performing a contour integration we find

$$\mathcal{B}'(\kappa,\omega)=\frac{\omega^2-\omega_p^2(\kappa+\kappa_p)/2\kappa_p}{(\omega^2-\omega_p^2)}. \tag{5.62}$$

The contribution of (5.62) to (5.44) can be readily evaluated by using the theorem (5.49) again, but it is more instructive, for reasons which will become apparent, to proceed otherwise. By partial integration we can write, for fixed $\boldsymbol{\kappa}$,

$$\int_0^\infty \frac{\mathrm{d}\omega}{2\pi}\,\mathrm{Im}\ln\mathcal{B}'(\boldsymbol{\kappa};\omega)=-\int_0^\infty\frac{\mathrm{d}\omega}{2\pi}\,\omega\frac{\mathrm{d}\Phi_{\boldsymbol{\kappa}}(\omega)}{\mathrm{d}\omega}, \tag{5.63}$$

where we have introduced the phase function

$$\Phi_{\boldsymbol{\kappa}}(\omega)=\arg\mathcal{B}'(\boldsymbol{\kappa},\omega). \tag{5.64}$$

We want to study (5.64), for fixed $\boldsymbol{\kappa}$, as a function of ω. Notice a subtle point. Although the numerator of $\mathcal{B}'(\kappa)$ in (5.62), formally vanishes for $\omega=\omega_p$ (remember (5.61)) this does not mean that ω_p is the frequency of the surface mode, i.e. a zero of $\mathcal{B}'(\boldsymbol{\kappa},\omega)$, because the denominator in (5.62) also vanishes for $\omega=\omega_p$ and the limit of (5.62) is not zero for $\omega\to\omega_p$. In fact it is obvious that the frequency of the surface mode, whose dispersion relation is given by the solution of the implicit equation,

$$\omega^2=\frac{\omega_p^2}{2}\left[1+\frac{\kappa}{\kappa_p(\kappa,\omega)}\right], \tag{5.65}$$

tends to $\omega_p/\sqrt{2}$, as usual, when $\kappa\to 0$. It is when (5.65) is satisfied that $\mathcal{B}'(\boldsymbol{\kappa},\omega)$ changes sign and its argument jumps by π. At $\omega=\omega_b$, $\Phi_{\boldsymbol{\kappa}}(\omega)$ also has a discontinuity of magnitude $\frac{1}{2}\pi$, which does not correspond to a change in the sign of $\mathcal{B}'(\kappa,\omega)$. The behaviour of the phase function (5.64), for the model (5.59), is summarised in fig. 102, where $\omega_s(\boldsymbol{\kappa})$ is the solution of (5.65). Notice how the different terms arise. Using this form of $\Phi_{\boldsymbol{\kappa}}(\omega)$, we find the following contributions to the derivative appearing in

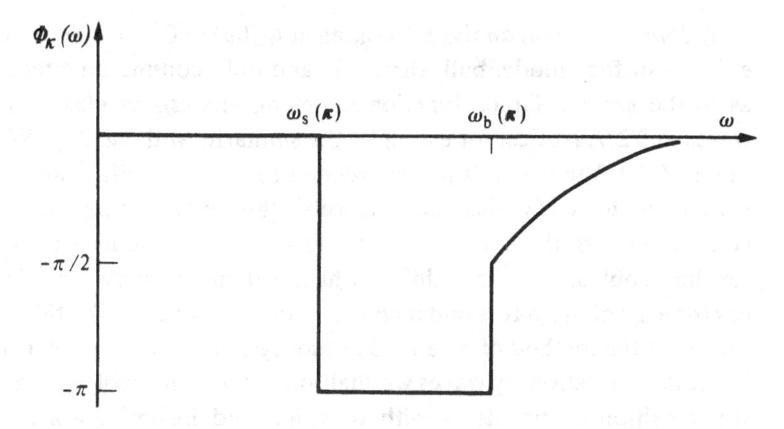

Fig. 102. Qualitative behaviour of the phase function (5.64) evaluated for the model (5.58).

(5.63): (i) a term $-\pi\delta[\omega-\omega_s(\boldsymbol{\kappa})]$, which contributes $+\frac{1}{2}\omega_s(\kappa)$ to (5.63) and corresponds to the appearance of *one* surface plasmon; (ii) a term $+(\frac{1}{2}\pi)\delta(\omega-\omega_b)$, which contributes $-\frac{1}{4}\omega_b$ and corresponds to a loss of $\frac{1}{2}$ bulk plasmon at the threshold of the bulk continuum; (iii) finally, a third term which corresponds to a total change of $+(\frac{1}{2}\pi)$ in the phase function. The change in $\Phi_{\kappa}(\omega)$ tends to be accumulated near the bottom of the bulk $\boldsymbol{\kappa}$-sub-band, in the language of chapter 4, and its spread reflects the dispersion contained in the dielectric function, i.e. in the hydrodynamic model, on the value of β. It corresponds to a loss of $\frac{1}{2}$ bulk plasmon from the (low energy frequency range of the) continuum. In the limit of very small β this contributes $-\frac{1}{4}\omega_p$ to (5.63) and adding these contributions to (5.60) in the same limit we recover precisely (5.57) in such a way that the origin of the different contributions to the term in ω_p is clearly traced. The argument is exactly the same if we introduce a cutoff, as needed for convergence in the sum – or integration – over $\boldsymbol{\kappa}$, with the difference that the s.m.d.r. is then explicitly modified by the cutoff, as we saw with the model of (5.50).

Let us pause to see where we stand. We have anticipated that the mathematical analysis developed above will lead to a normal-mode analysis, but we have also seen, on the basis of simple models, that the change in the spectrum can be described in terms of the derivative of a phase function, which is very strongly reminiscent of the examples of analysis seen in chapters 2 and 3; in fact this was the point of doing the above exercise based on (5.64). We shall now tackle these two separate questions, namely: (i) phase-function analysis, and (ii) normal mode analysis.

(i) *Phase-function analysis.* Looking at $\Phi_{\kappa}(\omega)$ in fig. 102 we see that it exhibits surface-mode, bulk-threshold and bulk-continuum effects, just as in the surface Green-function matching analysis of elastic surface waves in § 2.7. Notice, for example, the similarity with fig. 26*a*. We also saw in § 3.1.3 that this is in turn related to a phase-shift analysis. It is instructive to study this more closely, thus establishing an explicit connection with the concepts of chapters 2 and 3. The idea is to study for this problem the phase shift η which will enter the asymptotic form $\cos(qz+\eta)$ of the waves under consideration, for fixed $\boldsymbol{\kappa}$. For this we can use again the method of extended pseudosystems as in chapter 1. Being interested in stationary waves we shall study the real problem as a linear superposition of problems with outgoing and incoming waves, with respect to the surface. Thus, we shall superimpose the solutions for extended pseudomedia $\mathrm{M}^{\pm}$, with surface stimuli $\sigma^{\pm}$ and dielectric functions $\varepsilon^{\pm}$, with outgoing/incoming waves, and match the solution to an extended pseudovacuum with surface stimulus $-(\sigma^{+}+\sigma^{-})$. Then

$$\phi^{\mathrm{M}}(\boldsymbol{k})=\frac{4\pi\sigma^{+}}{k^{2}\varepsilon(\boldsymbol{k},\omega^{+})}+\frac{4\pi\sigma^{-}}{k^{2}\varepsilon(\boldsymbol{k},\omega^{-})},\tag{5.66}$$

and

$$\phi^{\mathrm{V}}(\boldsymbol{k})=-\frac{4\pi(\sigma^{+}+\sigma^{-})}{k^{2}}.\tag{5.67}$$

Matching the electrostatic potential at $z=0$ yields the ratio

$$\frac{\sigma^{+}}{\sigma^{-}}=-\frac{I[1]+I[\varepsilon(\omega^{-})]}{I[1]+I[\varepsilon(\omega^{+})]}.\tag{5.68}$$

The integrals $I[1]$, $I[\varepsilon]$ were defined in (5.29). This is all we need to find the phase shift. Apart from an irrelevant amplitude factor we have

$$\begin{aligned}\phi^{\mathrm{M}}(z)\propto&\int\frac{\mathrm{d}q}{(\kappa^{2}+q^{2})}\left[1+\frac{1}{\varepsilon(\boldsymbol{\kappa},q;\omega^{-})}\right]\times\int\frac{\mathrm{d}q}{(\kappa^{2}+q^{2})\varepsilon(\boldsymbol{\kappa},q;\omega^{+})}\,\mathrm{e}^{\mathrm{i}qz}\\&-\int\frac{\mathrm{d}q}{(\kappa^{2}+q^{2})}\left[1+\frac{1}{\varepsilon(\boldsymbol{\kappa},q;\omega^{+})}\right]\times\int\frac{\mathrm{d}q}{(\kappa^{2}+q^{2})\varepsilon(\boldsymbol{\kappa},q;\omega^{-})}\,\mathrm{e}^{\mathrm{i}qz}.\end{aligned}\tag{5.69}$$

On evaluating these integrals by contour integration the contributions from the zeros of $(\kappa^{2}+q^{2})$ decay with z and we need only worry about the contributions from the zeros of $\varepsilon(\kappa,q;\omega^{\pm})$. These take place at $q=\pm q_{\mathrm{p}}$, where q_{p} is the zero of $\varepsilon(\kappa,q;\omega)$. The signs correspond to outgoing/incoming waves and we are left with

$$\phi^{\mathrm{M}}(z) \propto \int \mathrm{d}q \left[\frac{1}{(\kappa^2+q^2)} + \frac{1}{(\kappa^2+q^2)\varepsilon(\boldsymbol{\kappa}, q; \omega^+)}\right] \mathrm{e}^{-\mathrm{i}q_{\mathrm{p}}z}$$

$$+ \int \mathrm{d}q \left[\frac{1}{(\kappa^2+q^2)} + \frac{1}{(\kappa^2+q^2)\varepsilon(\boldsymbol{\kappa}, q; \omega^-)}\right] \mathrm{e}^{\mathrm{i}q_{\mathrm{p}}z}. \tag{5.70}$$

The point is that the factors in front of the exponentials depend only on $(\boldsymbol{\kappa}, \omega)$ and the sum of the two terms in (5.70) can be expressed in the form $\cos(q_{\mathrm{p}}z + \eta)$ that we were seeking, with

$$\eta(\boldsymbol{\kappa}, \omega) = -\arg \int \left[\frac{1}{(\kappa^2+q^2)} + \frac{1}{(\kappa^2+q^2)\varepsilon(\boldsymbol{\kappa}, q; \omega^+)}\right] \mathrm{d}q.$$

Since $I[1]$ is real, from (5.29) and (5.64) we have

$$\eta(\boldsymbol{\kappa}, \omega) = -\arg \mathscr{B}'(\kappa, \omega) = -\phi_{\boldsymbol{\kappa}}(\omega).$$

Thus the phase shift is just the phase function shown in fig. 102 with a change of sign.

Now, in view of the discussion of § 3.1.2, and by recalling the argument after (3.26), the surface density of modes is

$$N_{\boldsymbol{\kappa},\mathrm{s}}(\omega) = \frac{1}{\pi}\frac{\mathrm{d}\eta}{\mathrm{d}\omega} + \text{density for } \eta = 0.$$

The problem is to find the mode density for $\eta = 0$. We can repeat the discussion leading to (3.54) with one important difference, namely, that there the wavefunction was of the form $\sin(qz)$ for $\eta = 0$, while here it is of the form $\cos(qz)$ for $\eta = 0$. This changes the sign of the δ-function term appearing at the bulk threshold and we obtain $+\frac{1}{4}$ instead of $-\frac{1}{4}$. The two situations are complementary. Their threshold contributions add up to zero, thereby restoring the bulk density of modes. Thus, in this case,

$$N_{\boldsymbol{\kappa},\mathrm{s}}(\omega) = \tfrac{1}{4}\delta[\omega - \omega_{\mathrm{b}}(\boldsymbol{\kappa}, 0)] + \frac{1}{\pi}\,\mathrm{d}\eta(\boldsymbol{\kappa}, \omega)/\mathrm{d}\omega.$$

Let us now return to the general discussion of the surface energy. Having obtained the surface density of modes, in view of the results so far obtained, particularly (5.44) and (5.49), we could straight away get a formula of the type (for fixed $\boldsymbol{\kappa}$)

$$E^{\mathrm{s}}_{\boldsymbol{\kappa}} = \frac{1}{2}\int_0^\infty \left\{\frac{1}{4}\delta[\omega - \omega_{\mathrm{b}}(\boldsymbol{\kappa}, 0)] + \frac{1}{\pi}\frac{\mathrm{d}\eta(\boldsymbol{\kappa}, \omega)}{\mathrm{d}\omega}\right\}\omega\,\mathrm{d}\omega. \tag{5.71}$$

Indeed, it is easy to evaluate this for the model (5.50) and this yields precisely (5.57). However, there is a fallacy in this argument. We need only try the model (5.58) to see that (5.71) is not generally valid. It suffices to consider the contribution of the bulk threshold, which here is simply

$$\tfrac{1}{8}\omega_b(\boldsymbol{\kappa}, 0),$$

whereas in (5.60) we obtained a different result by explicit evaluation. The disagreement is in the term in $\beta\kappa$. In order to understand the meaning of this difference we rewrite (5.60) by partial integration as

$$\frac{1}{4}\int_0^{\infty} \frac{d\omega}{2\pi} \operatorname{Im} \ln \varepsilon = -\frac{1}{8\pi}\int_0^{\infty} d\omega \frac{\omega \, d \arg \varepsilon(\omega)}{d\omega},$$

and recall (5.59), which yields

$$-\frac{d \arg \varepsilon(\omega)}{d\omega} = +\pi\delta[\omega - \omega_b(\boldsymbol{\kappa}, 0)] - \pi\delta[\omega - \beta\kappa]. \tag{5.72}$$

The answer to the puzzle is now clear. In guessing the formula (5.71) we attributed all the surface energy to changes in the spectrum of *collective modes*, and forgot about *single-particle modes*, i.e. electron-hole excitations. The reason why (5.71) works for the model (5.50) is that this model has no single-particle structure, but on evalutating (5.72) by using (5.59), we find two contributions. The first one, which is due to the zero of (5.59), corresponds to a bulk mode. Its positive sign in (5.72) reflects the appearance of a surface mode on creating the surface. The second term clearly signals the disappearance of some spectral strength connected with electron-hole pairs, since it comes from the pole of the dielectric function (5.59). On general grounds – conservation of the number of modes – this must correspond to the removal of precisely one single-particle mode. In this particular model this can be intuitively understood on the basis of the following crude argument. Consider the electron-hole excitation energy with momentum transfer k, i.e.

$$\tfrac{1}{2}|\boldsymbol{p} + \boldsymbol{k}|^2 - \tfrac{1}{2}p^2 = \boldsymbol{p} \cdot \boldsymbol{k} + \tfrac{1}{2}k^2.$$

Since we are looking at a dielectric function which entails essentially a long-wave approximation – the cutoff k_c has still to be fudged in – we can neglect k compared with the average momentum of the Fermi distribution $\langle p \rangle = \frac{3}{5}p_F$. Using this average value and remembering that we took $m = 1$, we see that the electron-hole excitation energy is just $\frac{3}{5}V_F k$, i.e. βk. The hydrodynamic model or, alternatively, the model obtained by interpolation of the Lindhard formula could be viewed as a sort of 'electron-hole pole' approximation to ε_{rpa}. Thus, if we start from the fact that the appearance of a collective mode entails the disappearance of a single-particle mode, the formula to use is not (5.65) but

$$E^{s}_{\kappa}=\frac{1}{2}\int_0^{\infty}\left\{\frac{1}{4}\delta[\omega-\omega_b(\boldsymbol{\kappa},0)]+\frac{1}{\pi}\frac{\mathrm{d}\eta(\boldsymbol{\kappa},\omega)}{\mathrm{d}\omega}\right\}(\omega-\beta\kappa)\,\mathrm{d}\omega. \tag{5.73}$$

In view of the result shown in fig. 102, it is convenient to separate out the bulk threshold and bulk continuum contributions. Indicating by η_b the phase shift in the bulk continuum *only*, we write

$$\frac{1}{\pi}\frac{\mathrm{d}\eta(\boldsymbol{\kappa},\omega)}{\mathrm{d}\omega}=\delta[\omega-\omega_s(\boldsymbol{\kappa})]-\frac{1}{2}\delta[\omega-\omega_b(\boldsymbol{\kappa},0)]+\frac{1}{\pi}\frac{\mathrm{d}\eta_b(\boldsymbol{\kappa},0)}{\mathrm{d}\omega},$$

which added to the first term in the integrand of (5.68) yields

$$E^{s}_{xc}=\sum_{\boldsymbol{\kappa}}\frac{1}{2}\left\{[\omega_s(\boldsymbol{\kappa})-\beta\kappa]-\frac{1}{4}[\omega_b(\boldsymbol{\kappa},0)-\beta\kappa]\right.$$
$$\left.+\frac{1}{\pi}\int\frac{\mathrm{d}\eta_b(\boldsymbol{\kappa},\omega)}{\mathrm{d}\omega}(\omega-\beta\kappa)\,\mathrm{d}\omega\right\}, \tag{5.74}$$

which shows the typical band-edge defect of strength $\frac{1}{4}$.

(ii) *Normal-mode analysis.* We saw that equations (5.44)–(5.49) have a form which strongly suggests using a normal-mode analysis. A rudimentary example of this is provided by the simple model leading to (5.74). Let us now formulate the problem more generally.

For the sake of completeness we first return to (5.43) and discuss again the bulk energy. Using (5.45) we can define an energy density in position and momentum space,

$$e^{b}_{xc}(\boldsymbol{k})=\int_0^{\infty}\frac{\mathrm{d}\omega}{2\pi}\,\mathrm{Im}\ln\varepsilon(\boldsymbol{k};\omega^{+})-\frac{2\pi Ne^2}{Vk^2}. \tag{5.75}$$

For the application of the theorem (5.49) we shall be interested in the zeros and poles of ε. Let us summarise some general facts in the theory of the electron gas. We recall the well-known formula

$$\varepsilon_{\mathrm{rpa}}(\boldsymbol{k};\omega^{+})=1+\frac{4\pi e^2}{Vk^2}\sum_{\substack{p<k_F\\ |\boldsymbol{p}+\boldsymbol{k}|>k_F}}\left[\frac{1}{\omega+\omega(\boldsymbol{k},\boldsymbol{p})+\mathrm{i}\eta}-\frac{1}{\omega-\omega(\boldsymbol{k},\boldsymbol{p})+\mathrm{i}\eta}\right], \tag{5.76a}$$

where

$$\omega(\boldsymbol{k},\boldsymbol{p})=\tfrac{1}{2}k^2+\boldsymbol{k}\cdot\boldsymbol{p}. \tag{5.76b}$$

This is the formula for the dielectric function in the *random phase approximation*, into which we add, to a given Fourier component of the external field, the same Fourier component of the field induced inside the system. As a result of this the response function is, in the r.p.a.,

$$\frac{\delta\rho^{\mathrm{i}}}{\delta\rho^{\mathrm{e}}}=\frac{1}{\varepsilon_{\mathrm{rpa}}(\boldsymbol{k};\omega)}-1. \tag{5.77}$$

Now the poles of the response function give the normal modes of the system, i.e. the elementary excitations which can exist with vanishing external field. These poles are the zeros of the dielectric function, as is well known. But suppose we neglect the effect of the induced field, i.e. we treat the electrons as independent. The response function would then simply be the susceptibility

$$\frac{\delta\rho^{\mathrm{i}}}{\delta\rho^{\mathrm{e}}}=\frac{\varepsilon_{\mathrm{rpa}}(\boldsymbol{k};\omega)-1}{4\pi}. \tag{5.78}$$

The normal modes of the system are then, *within this approximation*, the poles of (5.78), i.e. the zeros of the denominators of (5.76*a*). Obviously, in the absence of interactions capable of organising the electron gas into a collective mode the only possible normal modes of the system are the bare single-particle or electron-hole excitations, whose frequencies are given in (5.76*b*).

It is now convenient to use again a box of side L in which the values of $\boldsymbol{k}$ and $\boldsymbol{p}$ are quantised. In particular we are interested in looking at a fixed $\boldsymbol{k}$ and considering the quantised values

$$p_m=(2\pi/L)(m_1, m_2, m_3).$$

The behaviour of the dielectric function as a function of ω for given $\boldsymbol{k}$ is shown in fig. 103. The poles of (5.76*a*) appear at the frequencies ω_m, given by the values of (5.76*b*) for $\boldsymbol{p}=\boldsymbol{p}_m$. As we have just seen, these are the normal-mode frequencies of the non-interacting electron gas. Now switch on the interaction up to the r.p.a. The new normal modes are the zeros of (5.76*a*), indicated by ω'_m in fig. 103. What do they mean? We know that if we switch off the electron–electron interaction, the frequencies ω'_m become the bare single-particle excitation frequencies ω_m. Thus with each bare frequency ω_m we associate its renormalised normal-mode frequency ω'_m, which is thus viewed as a shift of the renormalised electron-hole pair excitation. There is a special case. The largest of all ω_m, which we shall call $\omega_{\max}$, moves off to a separate frequency, $\omega_{\mathrm{b}}(\boldsymbol{k})$, which is the bulk plasmon frequency. For example, in the limit $k\to 0$ all the structure contained within the vertical asymptotes of fig. 103 collapses into the vertical coordinate axis and we are left with just the function $1-\omega_{\mathrm{p}}^2/\omega^2$ which vanishes at the distant zero $\omega=\omega_{\mathrm{p}}$. Thus the picture is as follows. Before switching on the electron–electron interactions all the

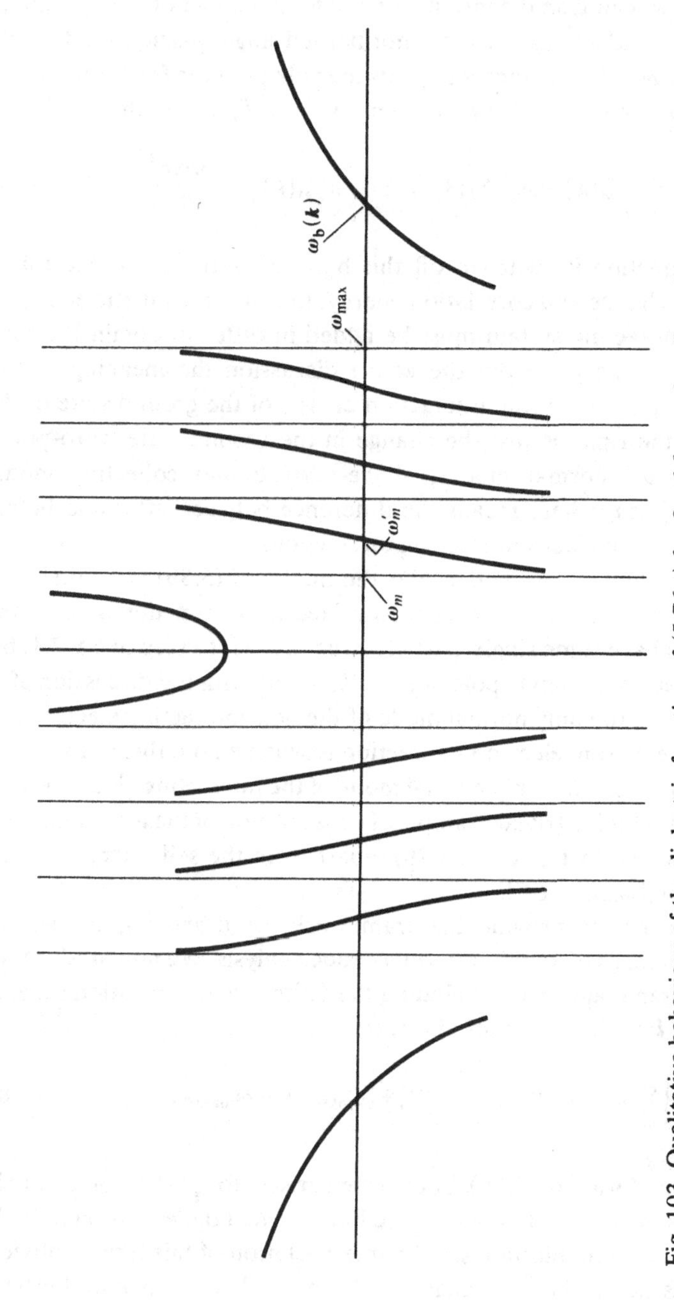

Fig. 103. Qualitative behaviour of the dielectric function of (5.76*a*) for fixed $\boldsymbol{k}$, as a function of ω, with quantised values $(\boldsymbol{p}_m)$ of $\boldsymbol{p}$.

normal modes of the system are single-particle modes with frequencies ω_m. After switching on the interaction the total number of normal modes is conserved and all but one are renormalised single-particle modes with frequencies ω'_m. The distant zero gives the bulk plasmon frequency. Now, if we apply (5.49) to (5.75) we obtain, for fixed $\boldsymbol{k}$, the result

$$e^{\mathrm{b}}_{\mathrm{xc}}(\boldsymbol{k})=\frac{1}{2}\left\{\sum_m [\omega'_m(\boldsymbol{k})-\omega_m(\boldsymbol{k})]+\omega_{\mathrm{b}}(\boldsymbol{k})-\omega_{\max}(\boldsymbol{k})\right\}-\frac{2\pi Ne^2}{Vk^2}. \tag{5.79}$$

If the interaction is switched off this becomes zero. Of course this is only the exchange and correlation energy, to which the kinetic energy of the non-interacting system must be added in order to obtain the total ground-state energy. From the above discussion the meaning of this formula is quite clear: the interaction energy of the ground state of the system in the r.p.a. is just the change in the ground-state (zero-point) energy of *all* normal modes, single-particle and collective modes included. *Change* here means the difference between after and before switching on the electron–electron interaction.

As a simple application consider the model of (5.59). This dielectric function is shown in fig. 104, again for fixed $\boldsymbol{k}$ as a function of ω. In this case we do have some single-particle structure, as stressed in § 5.2.1, but it is reduced to the single pole at $\omega=\beta k$, in line with the discussion after (5.72). This is the only normal mode of the non-interacting electron gas. When the electron–electron interaction is switched on, this becomes the bulk plasmon, i.e. the only normal mode of the interacting electron gas in this model. This is a trivial example of conservation of the total number of normal modes. In this case (5.79), apart from the self-energy term, is reduced to $\frac{1}{2}[\omega_{\mathrm{b}}(\boldsymbol{k})-\beta k]$.

These concepts provide the framework for discussing the surface energy by means of a similar normal-mode analysis. We now study (5.44) using the same approach, including the finite-size quantisation scheme. For given $\boldsymbol{k}$ the first term of (5.44) is simply

$$\frac{1}{8}\left\{\sum_m [\omega'_m(\boldsymbol{\kappa},0)-\omega_m(\boldsymbol{\kappa},0)]+[\omega_{\mathrm{b}}(\boldsymbol{\kappa},0)-\omega_{\max}(\boldsymbol{\kappa},0)]\right\}, \tag{5.80}$$

identical in form to (5.79) but particularised to $q=0$. For example, $\omega_{\mathrm{b}}(\boldsymbol{\kappa},0)$ is the threshold for what we have termed the $\boldsymbol{\kappa}$-sub-band in the continuum of bulk plasmons. (The interpretation of this term is obvious and needs no further explanation.) With the hydrodynamic dielectric function (fig. 104) this reduces to (5.60). In order to discuss the second term of (5.44) we use for q the same quantisation scheme as for the

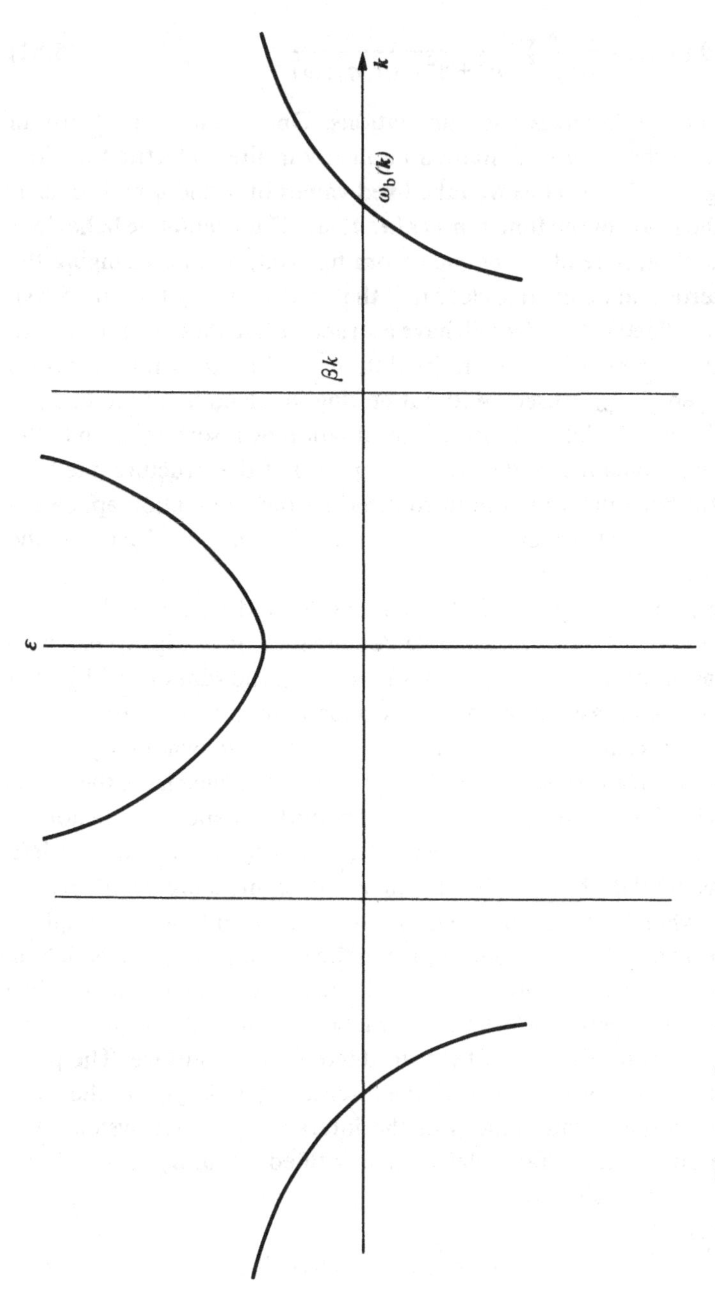

Fig. 104. Qualitative behaviour of the dielectric function (5.58), for fixed $\boldsymbol{k}$, as a function of ω.

components of $\boldsymbol{p}$ in (5.76a). Thus, for fixed $\boldsymbol{\kappa}$, we consider

$$\mathscr{B}'(\boldsymbol{\kappa}, \omega) = \frac{1}{2} + \frac{\kappa}{L} \sum_{q_n} \frac{1}{(\kappa^2 + q_n^2)\varepsilon(\boldsymbol{\kappa}, q_n; \omega)}. \tag{5.81}$$

Notice that this involves two summations. One is shown explicitly in (5.81). The other one is contained implicitly in the dielectric function, according to (5.76a). Thus we take fixed values of $\boldsymbol{\kappa}$ and q and concentrate on the study of the function $1/\varepsilon(\boldsymbol{\kappa}, q; \omega)$. The qualitative behaviour of this function is readily obtained from fig. 103, by interchanging the roles of zeros and poles. It is clear that the corresponding term in (5.81), even for *one* fixed value of q, will have a structure like that of fig. 103, with zeros and poles associated with the different values $\boldsymbol{p}_m$ with their roles interchanged if ε_{rpa} is used. Without getting involved in all the complications of this case, let us illustrate the situation by resorting again to the simpler hydrodynamic model. We then get rid of the structure due to ε, but we still have the structure introduced by the summation appearing explicitly in (5.81), which is the essential one for the physics of the problem we are studying.

Thus we consider the hydrodynamic model and take $\boldsymbol{\kappa}$ and q fixed. Then $1/\varepsilon(\boldsymbol{\kappa}, q; \omega)$ has a very simple dependence on ω (fig. 105a), with just one asymptote at $\omega = \omega_{\text{b}}(\boldsymbol{\kappa}, q)$. On forming the sum of (5.81), each value q_n gives an asymptote at the corresponding value $\omega_{\text{b}}(\boldsymbol{\kappa}, q_n)$ and $\mathscr{B}'(\boldsymbol{\kappa})$ acquires the more complicated structure shown in fig. 105b. In between the frequencies $\omega_{\text{b}}(\boldsymbol{\kappa}, q_n)$, we have the frequencies of the points indicated by $\Omega(\boldsymbol{\kappa}, q_n)$ in fig. 105b. These are the frequencies of the normal modes of the surface system. Each pole has an associated zero to its left. We stress that q_n has the literal quantitative meaning of $n2\pi/L$ in $\omega_{\text{b}}(\boldsymbol{\kappa}, q_n)$, whereas it has only the meaning of a label in $\Omega(\boldsymbol{\kappa}, q_n)$, simply to indicate that it is the zero associated with the pole $\omega_b(\boldsymbol{\kappa}, q_n)$ on its right in fig. 105b. Otherwise the situation is similar to that of fig. 103, but this time we keep the electron–electron interaction at its full strength. The frequency shift now is due to the introduction of the surface. The poles $\omega_{\text{b}}(\boldsymbol{\kappa}, q_n)$ are the normal modes of the interacting bulk system; the zeros $\Omega(\boldsymbol{\kappa}, q_n)$ are the normal modes of the interacting surface system. The distant zero is the surface plasmon, identified as $\omega_{\text{s}}(\boldsymbol{\kappa})$ in fig. 105b. Application of (5.49) now yields

$$\int_0^\infty \frac{\mathrm{d}\omega}{2\pi} \operatorname{Im} \ln \mathscr{B}'(\boldsymbol{\kappa}, \omega) = \frac{1}{2}[\omega_{\text{s}}(\boldsymbol{\kappa}) - \omega_{\text{b}}(\boldsymbol{\kappa}, 0)]$$

$$+ \frac{1}{2} \sum_n [\Omega(\boldsymbol{\kappa}, q_n) - \omega_{\text{b}}(\boldsymbol{\kappa}, q_n)]. \tag{5.82}$$

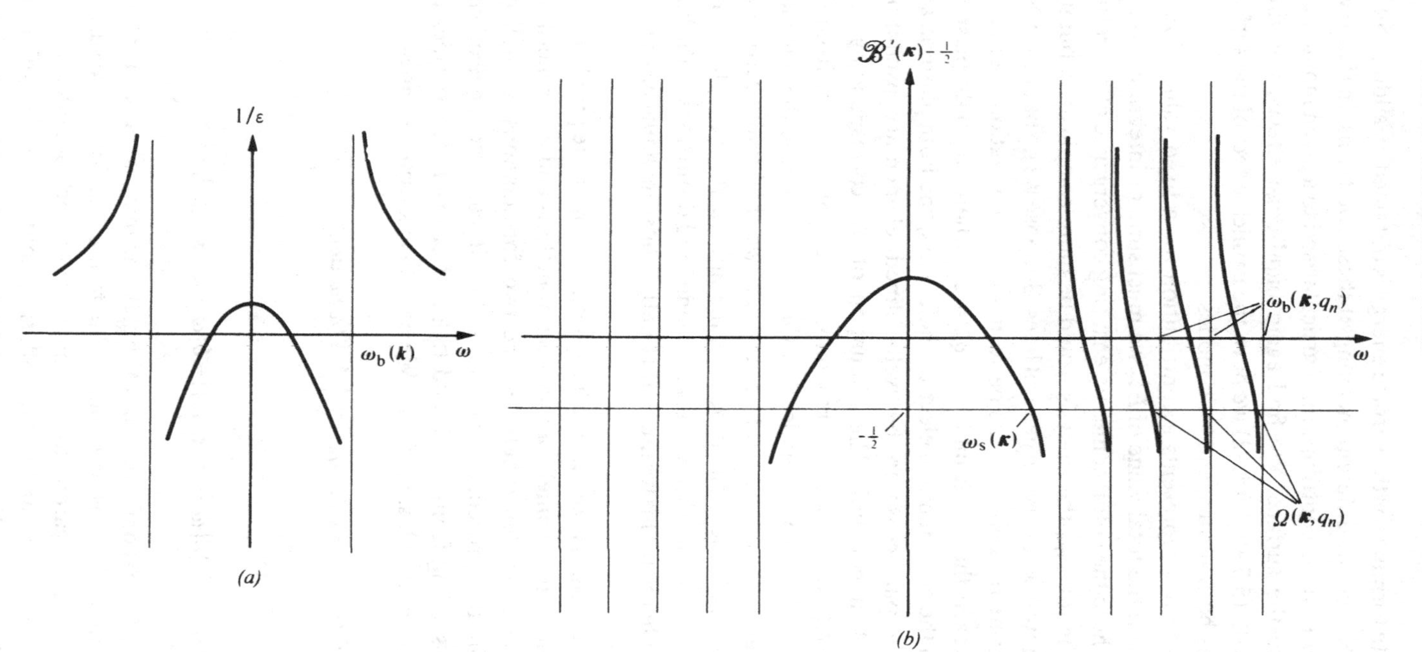

Fig. 105. Fixed $\boldsymbol{\kappa}$ and q; hydrodynamic dielectric function (5.58): (*a*) $1/\varepsilon$; (*b*) $\mathscr{B}'(\boldsymbol{\kappa})-\frac{1}{2}$. See (5.81).

The complete formula for the surface energy is obtained by adding (5.80) and (5.82). Notice the different meaning of these two terms. In (5.80) we have only bulk-mode contributions, particularised to a given momentum $\boldsymbol{\kappa}$ parallel to the surface. We find again band-edge effects, this time associated with (5.79). In (5.82) we have the explicit effect of the surface as it modifies both surface and bulk modes.

Note that (5.82) represents a contribution for a fixed value of $\boldsymbol{\kappa}$. In order to obtain the total value of E^{s}_{xc} we must sum, or integrate, over $\boldsymbol{\kappa}$, and this is where the cutoff comes in, ensuring convergence of the result. If we used the full ε_{rpa} the analysis would be more complicated, but it is easy to see how it would go on general lines. Suppose we go back to (5.81) and concentrate on a given $\boldsymbol{\kappa}$. On performing the summation over q_n we eventually reach the cutoff limit $\kappa^2+q_n^2=k_c^2$, where the bulk plasmon merges with the continuum of electron-hole pairs, and both plasmon and electron-hole pair cease to be well-defined elementary excitations because of their coupling. The evaluation of the corresponding contribution to E^{s}_{xc} will now be more cumbersome, but it is clear from the study of the hydrodynamic model that the result will have the same key features as appear in (5.82) and fig. 105, i.e. we shall find a zero-point shift of the normal modes on introducing the surface. As in chapters 3 and 4, this will contain both distinct surface modes and distortions of the bulk continuum. On top of this we must add (5.80), whose meaning has already been discussed.

A final comment is in order. We have looked at the same problem from two different angles, namely, a phase-shift analysis and a normal-mode analysis. The equivalence between the two formulations is implicitly contained in the preceding arguments, but it can also be explicitly established starting from (5.80) and (5.82). For example, consider the second term on the r.h.s. of (5.82). We can always write, identically,

$$\sum_n [\Omega(\boldsymbol{\kappa}, q_n)-\omega_b(\boldsymbol{\kappa}, q_n)]=\int \Delta\mathcal{N}(\omega)\omega\, d\omega,$$

where

$$\Delta\mathcal{N}(\omega)=\sum_n\{\delta[\omega-\Omega(\boldsymbol{\kappa}, q_n)]-\delta[\omega-\omega_b(\boldsymbol{\kappa}, q_n)]\}.$$

This is only a formal definition of $\Delta\mathcal{N}$, but its meaning is transparent. It is actually a change in density of modes due to the creation of the surface, although we must be careful to note that this only includes effects related to the bulk continuum and to the bulk threshold or 'band edge', in the language of chapters 3 and 4. This is because the effect of the distinct surface mode has been explicitly separated out in (5.82). Thus, in the light of our previous discussion of the phase-shift analysis, $\Delta\mathcal{N}$ is precisely the sum of $\pi^{-1}\,\mathrm{d}\eta_b/\mathrm{d}\omega$, i.e. the bulk-continuum term, and the bulk-threshold term. In this way we introduce the phase-shift concepts in the normal-mode results. It is an instructive exercise to see how, by following this path the sum of (5.80) and (5.82) can be identically rewritten so that we recover the results of the phase-shift analysis.

5.2.2. Polarisation interaction between two metal surfaces

The analysis of the surface energy can be extended to the case of two surfaces at a given distance, with the two half media in mutual interaction. Within the jellium model this yields a theory of the *van der Waals forces* between two metals, owing to their mutual polarisation and response. The system under consideration is defined in fig. 106*a*. In order to study it we go back to (5.24) and concentrate on the first term on the r.h.s., disregarding the self-energy term. Our problem is to find the response function $\delta\rho^{\mathrm{i}}/\delta\rho^{\mathrm{e}}$ of the system of fig. 106*a*, for which we introduce an external test charge Q (fig. 106*b*). Since we use linear response theory we can study separately the response of the system to a symmetric and an antisymmetric charge distribution (fig. 106*c*), which add up to the given test-charge distribution of fig. 106*b*. We again use the method of extended pseudosystems as in chapter 1. Let us tackle each of the two cases of fig. 106*c* separately.

(i) *Symmetric external stimulus.* An expedient way to define extended pseudomedia is described in fig. 107. Given the two half media on both sides of the vacuum gap we ought to define an extended pseudomedium, M^+, for the material on the right, and another one, M^-, for the material on the left. However, given the symmetry about the middle of the vacuum gap is suffices to define just one extended pseudomedium (M) as in fig. 107*b* and match it to the extended pseudovacuum. It is convenient to use two different reference frames, R_{M} and R_{V}, for which the origin $z=0$ is shown in fig. 107*b*, *c*. Notice, on defining M, that the charge $\frac{1}{2}Q$ that we put at $z=-z_0$ (in R_{M}) is due to the assumption of specular surface and is *not* the charge $\frac{1}{2}Q$ on the l.h.s. in fig. 107*a*, which is part of the given external charge. For example, if we used the same method to study non-specular surfaces, as in chapter 1, the charge on the l.h.s. in fig. 107*a* would still be $\frac{1}{2}Q$, while the system of fictitious stimuli in fig. 107*b* would have to be modified in the manner of § 1.6. On the other hand, on defining the extended pseudovacuum, although for reasons to be seen presently, it is convenient to use the reference frame R_{V} of fig. 107*c*, we must actually define the extension of the vacuum starting *from the surface separating*

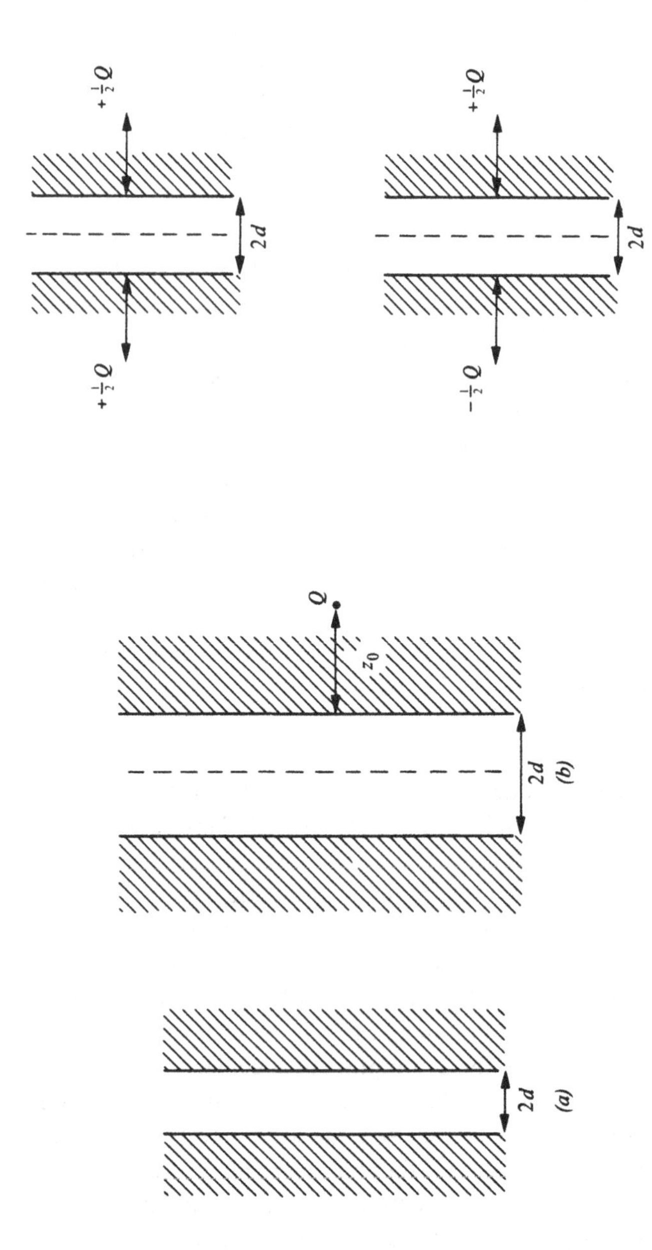

Fig. 106. (*a*) The system under study: two half media with dielectric function $\varepsilon(\boldsymbol{k}, \omega)$. (*b*) An external test charge Q to which the system responds. (*c*) The external test charge as a superposition of a symmetric and an antisymmetric charge distribution.

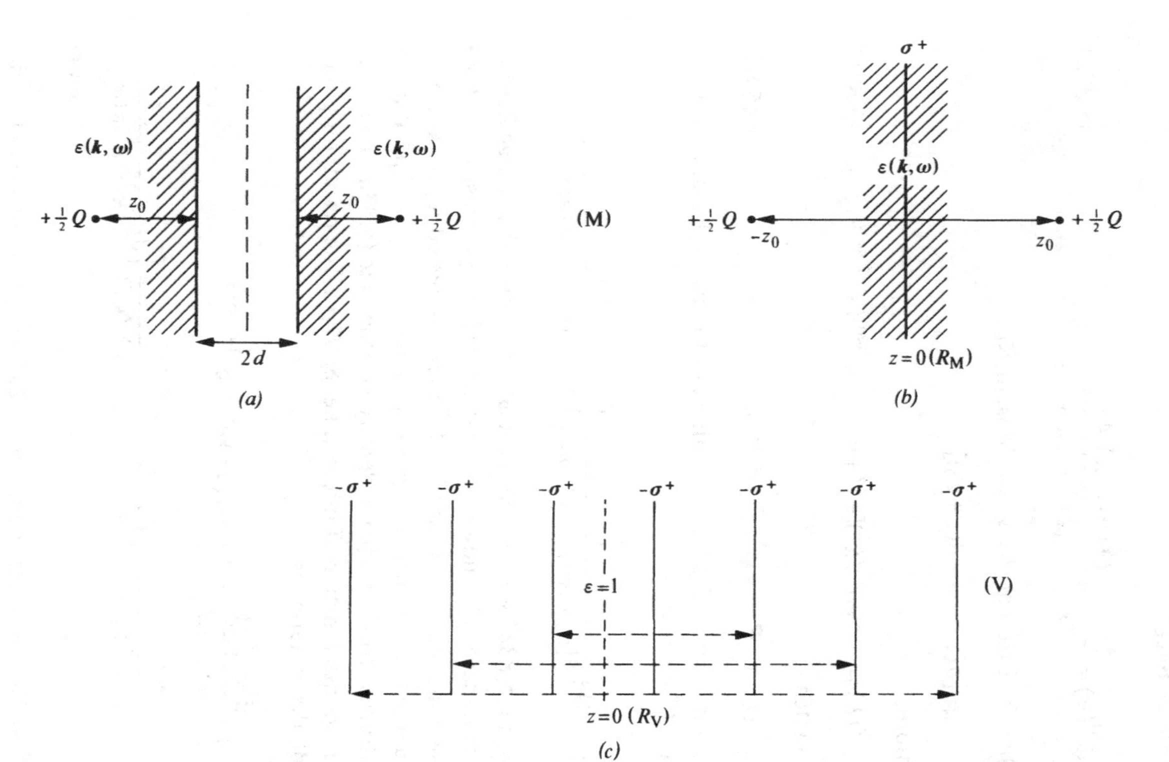

Fig. 107. The situation in (a) defines the given problem; (b) defines the extension of the half medium on the right of the vacuum gap; (c) is the extended pseudovacuum. Notice the different origins $z = 0$ taken for (b) and (c).

one half metal from the vacuum gap. This is why we have the sequence of fictitious charges $-\sigma^+$, which has specular symmetry about the surface $z=d$ (in R_V).

With this we have

$$\phi^M(\boldsymbol{k})=\frac{4\pi\sigma^+}{k^2\varepsilon}+\frac{Q\cos(qz_0)}{(2\pi)^2}\frac{4\pi}{k^2\varepsilon}. \tag{5.83a}$$

Notice that the external charge for M is, in R_M,

$$Q\delta(\boldsymbol{\rho})[\delta(z-z_0)+\delta(z+z_0)].$$

Furthermore,

$$\phi^V(\boldsymbol{k})=-(4\pi\sigma^+/k^2)[e^{iqd}+e^{-iqd}+e^{2iqd}+e^{-2iqd}+\ldots]. \tag{5.83b}$$

We now match

$$\phi^M(z=0,R_M)=\phi^V(z=d,R_V). \tag{5.84}$$

Fourier transforming (5.83*a*, *b*), writing out the two sides of (5.84) and solving for ϕ^+ we find

$$\sigma^+(\boldsymbol{\kappa})=\frac{1}{I[\varepsilon]+(\pi/\kappa)\coth(\kappa d)}\int\frac{Q\cos(qz_0)}{(2\pi)^2(\kappa^2+q^2)\varepsilon(\boldsymbol{\kappa},q)}\,dq.$$

Using this in (5.83*a*) we can construct ϕ^M and hence the charge density $\rho^M(z)$ in real space. We know that this is the real charge density $n(z)$ for a point inside the medium, which is all we are interested in, since $\rho(z)$ is zero for a point in the vacuum gap. But ρ is the sum of ρ^i and ρ^e; thus we must subtract ρ^e from the charge density associated to (5.83*a*) in order to find the response function of interest, i.e. $\delta\rho^i/\delta\rho^e$. The result, for points $\boldsymbol{r}$, $\boldsymbol{r}'$ inside the material, is

$$\frac{\delta\rho^i(\boldsymbol{r})}{\delta\rho^e(\boldsymbol{r}')}=\int\frac{d^3k}{(2\pi)^3}\left[\frac{1}{\varepsilon(k)}-1\right]\cos(qz')\,e^{-iqz}\,e^{-i\boldsymbol{\kappa}\cdot(\boldsymbol{\rho}-\boldsymbol{\rho}')}$$
$$-\int\frac{d^3k}{(2\pi)^3}\left[\frac{1}{\varepsilon(\boldsymbol{\kappa})}-1\right]\frac{e^{-iqz}\,e^{i\boldsymbol{\kappa}\cdot(\boldsymbol{\rho}-\boldsymbol{\rho}')}}{I[\varepsilon]+(\pi/\kappa)\coth(\kappa d)}\int dp\,\frac{\cos(pz')}{(\kappa^2+p^2)\varepsilon(\boldsymbol{\kappa},\boldsymbol{p})}. \tag{5.85}$$

Remember that, from the definition (5.29), $I[\varepsilon]$ is a function of $\boldsymbol{\kappa}$, the perpendicular component of $\boldsymbol{k}$ having been integrated over. This formula holds only for a symmetric external charge, as in fig. 107.

(ii) *Antisymmetric external stimulus.* We have often stressed that the definition of the extended pseudomedia is quite arbitrary, provided one half of each of the extended pseudomedia to be matched is equal to the real system in its own domain. This flexibility can be used to suit convenience. A case in point is the study of the response to an antisymmetric external charge (fig. 108*a*). We can formally define M so that it is equal to the original system in $z>0$ (in R_M) and has the specular-image charge $+\frac{1}{2}Q$ as a fictitious stimulus on the 'other side' $z<0$ (in R_M). This describes the specular surface model we are using here. This fictitious stimulus is quite a different thing from the given external charge $-\frac{1}{2}Q$ on the l.h.s. of fig. 108*a*. The antisymmetry about the middle of the vacuum gap is introduced by the system of fictitious surface stimuli of fig. 108*c*, which on the other hand is symmetric about the surfaces separating the vacuum from the medium. Then $\phi^M(\boldsymbol{k})$ is given by the same formula of (5.83a) with σ^- instead of σ^+, while

$$\phi^V(\boldsymbol{k}) = -\frac{4\pi\sigma^-}{k^2}[(e^{iqd}-e^{3iqd}+e^{5iqd}-\ldots)-(e^{-iqd}-e^{-3iqd}+e^{-5iqd}-\ldots)] \tag{5.86}$$

Matching as before we obtain $\sigma^-(\boldsymbol{\kappa})$, which is given by the same formula as $\sigma^+(\boldsymbol{\kappa})$, with $\tanh(\kappa d)$ instead of $\coth(\kappa d)$. Arguing as in the symmetric case we can find the response function of interest, which is given by the same formula as (5.85), again with $\coth(\kappa d)$ replaced by $\tanh(\kappa d)$.

Adding $\delta\rho^i/\delta\rho^e$ for the symmetric and antisymmetric cases we obtain the response function of interest for the problem of fig. 107. The result is, for $\boldsymbol{r}$ and $\boldsymbol{r}'$ both inside the metal,

$$\begin{aligned}\frac{\delta\rho^i(\boldsymbol{r})}{\delta\rho^e(\boldsymbol{r}')} &= \int \frac{d^3k}{(2\pi)^3}\left[\frac{1}{\varepsilon(\boldsymbol{k})}-1\right]e^{-i\boldsymbol{k}\cdot(\boldsymbol{r}-\boldsymbol{r}')} \\ &+\int \frac{d^3k}{(2\pi)^3}e^{-iq(z+z')}e^{-i\boldsymbol{\kappa}\cdot(\boldsymbol{\rho}-\boldsymbol{\rho}')} \\ &-\int \frac{d^3k}{(2\pi)^3}\left[\frac{1}{\varepsilon(\boldsymbol{k})}-1\right]e^{-iqz}e^{-i\boldsymbol{\kappa}\cdot(\boldsymbol{\rho}-\boldsymbol{\rho}')} \\ &\times\left\{\frac{1}{I[\varepsilon]+(\pi/\kappa)\coth(\kappa d)}+\frac{1}{I[\varepsilon]+(\pi/\kappa)\tanh(\kappa d)}\right\} \\ &\times\int dp\frac{\cos(pz')}{(\kappa^2+p^2)\varepsilon(\boldsymbol{\kappa},p)}.\end{aligned} \tag{5.87}$$

What we actually need for its use in (5.24) is the integral of (5.87) with $\boldsymbol{r}'=\boldsymbol{r}$. If we take this we single out again the volume or area scale factors. Using the same analysis which led to (5.40), but remembering that we

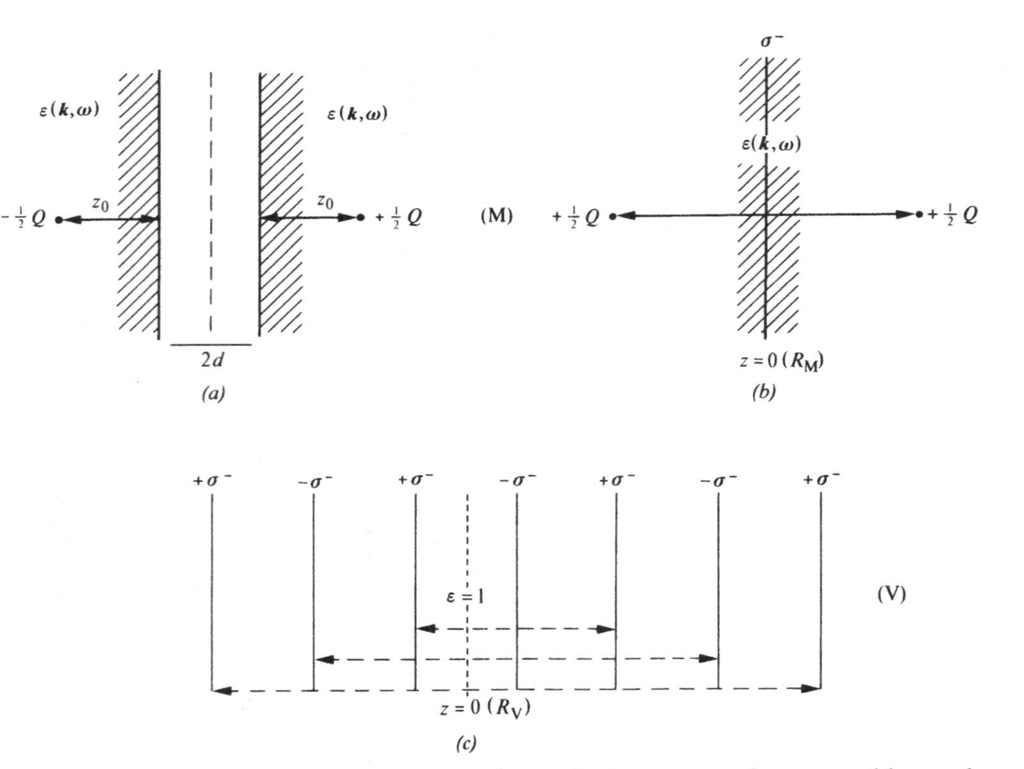

Fig. 108. Scheme of extended pseudomedia used to study the response of a system with specular surfaces to (*a*) an antisymmetric external charge. The origin $z = 0$ is explicitly shown for the two reference frames used in (*b*) and (*c*).

now have two surfaces, i.e. a factor $2A$, we find

$$\int \frac{\delta\rho^{i}(\boldsymbol{r})}{\delta\rho^{e}(\boldsymbol{r})}\mathbf{d}\boldsymbol{r} = V\int \frac{d^3k}{(2\pi)^3}\left[\frac{1}{\varepsilon(\boldsymbol{k})}-1\right]+\frac{A}{2}\int \frac{d^2\boldsymbol{\kappa}}{(2\pi)^2}\left[\frac{1}{\varepsilon(\boldsymbol{\kappa},0)}-1\right]$$

$$-A\int \frac{d^2\kappa}{(2\pi)^2}\left\{\frac{1}{I[\varepsilon]+(\pi/\kappa)\coth(\kappa d)}+\frac{1}{I[\varepsilon]+(\pi/\kappa)\tanh(\kappa d)}\right\}$$

$$\times\int dp\left[\frac{1}{\varepsilon(\boldsymbol{\kappa},p)}-1\right]\frac{1}{(\kappa^2+p^2)\varepsilon(\boldsymbol{\kappa},p)}. \tag{5.88}$$

Thus, for each surface we have, per unit area,

$$E^{s}_{xc}=\frac{1}{4}\int \frac{d^2\kappa}{(2\pi)^2}\int_0^{\infty}\frac{d\omega}{2\pi}\,\mathrm{Im}\ln\varepsilon(\boldsymbol{\kappa},0;\omega^+)+\int \frac{d^2\kappa}{(2\pi)^2}\int_0^{\infty}\frac{d\omega}{2\pi}$$

$$\times\left\{\frac{1}{2}\mathrm{Im}\ln\frac{I[\varepsilon]+(\pi/\kappa)\coth(\kappa d)}{(\pi/\kappa)[1+\coth(\kappa d)]}\right.$$

$$\left.+\frac{1}{2}\mathrm{Im}\ln\frac{I[\varepsilon]+(\pi/\kappa)\tanh(\kappa d)}{(\pi/\kappa)[1+\tanh(\kappa d)]}\right\}. \tag{5.89}$$

Notice that $\coth(\kappa d)\to\tanh(\kappa d)\to 1$ for $d\to\infty$, and then (5.89) reproduces (5.44).

We now define the (van der Waals) interaction potential

$$V(d)=2A[E^{s}_{xc}(d)-E_{xc}(\infty)], \tag{5.90}$$

whence the interaction potential between the two half media per unit area, $v(d)=V(d)/A$, is, by using (5.89),

$$v(d)=\int \frac{d^2\kappa}{(2\pi)^2}\int \frac{d\omega}{2\pi}$$

$$\times\left\{\mathrm{Im}\ln\frac{I[\varepsilon]+(\pi/\kappa)\coth(\kappa d)}{I[\varepsilon]+(\pi/\kappa)}\right.$$

$$\left.+\mathrm{Im}\ln\frac{I[\varepsilon]+(\pi/\kappa)\tanh(\kappa d)}{I[\varepsilon]+(\pi/\kappa)}\right\}. \tag{5.91}$$

The term in coth/tanh describes the role of the symmetric/antisymmetric modes. It is convenient to define

$$\left.\begin{aligned}\mathscr{B}'_+&=\frac{I[\varepsilon]+(\pi/\kappa)\coth(\kappa d)}{(\pi/\kappa)[1+\coth(\kappa d)]};\\ \mathscr{B}'_-&=\frac{I[\varepsilon]+(\pi/\kappa)\tanh(\kappa d)}{(\pi/\kappa)[1+\tanh(\kappa d)]}.\end{aligned}\right\} \tag{5.92}$$

Clearly, for $d \to \infty$ we have $\mathscr{B}'_+ \to \mathscr{B}'_- \to \mathscr{B}'$ where, remember, $\mathscr{B}'$ is $(\kappa/2\pi)\mathscr{B}$ (5.29). As a simple example consider the model of (5.50). Evaluating $I[\varepsilon]$ as in (5.52) we find

$$\mathscr{B}'_+ = \frac{\omega^2 - \omega_+^2(\kappa)}{\omega^2 - \omega_p^2}; \quad \mathscr{B}'_- = \frac{\omega^2 - \omega_-^2(\kappa)}{\omega^2 - \omega_p^2}, \tag{5.93}$$

where

$$\omega_+^2(\kappa) = \omega_p^2[1 - (1 - e^{-2\kappa d}) \frac{1}{\pi} \tan^{-1} (q_c/\kappa)] \tag{5.94}$$

is the square of the symmetric surface plasmon frequency and

$$\omega_-^2(\kappa) = \omega_p^2[1 - (1 + e^{-2\kappa d}) \frac{1}{\pi} \tan^{-1} (q_c/\kappa)] \tag{5.95}$$

is the same for the antisymmetric mode. Again, for $d \to \infty$ we have $\omega_+(\kappa) \to \omega_-(\kappa) \to \omega_s(\kappa)$ (5.54).

More generally the symmetric/antisymmetric surface modes are given by the zeros of $\mathscr{B}'_+/\mathscr{B}'_-$ (5.92). Thus we must study the extensions of (5.81) defined in (5.92). The situation is illustrated in fig. 109, which is basically fig. 105*b* with appropriate modifications, and again is based on the hydrodynamic dielectric function (5.57) and with the usual finite-size quantisation scheme for q. The surface energy (5.89) can now be studied in terms of a normal-mode analysis, as in § 5.2.1, but we are actually interested in discussing the interaction potential (5.91), from which derive the attractive polarisation forces between two metal surfaces. According to (5.90) this potential measures an energy difference between two situations, corresponding to coupled and free ($d \to \infty$) surfaces. This is reflected in the numerators and denominators appearing in (5.91). From a look at fig. 109 it is immediate to see that

$$v(d) = \int \frac{d^2\kappa}{(2\pi)^2} \frac{1}{2} \Big\{ [\omega_-(\kappa) + \omega_+(\kappa) - 2\omega_s(\kappa)] + \sum_n [\Omega_-(\boldsymbol{\kappa}, q_n) + \Omega_+(\boldsymbol{\kappa}, q_n) - 2\Omega(\boldsymbol{\kappa}, q_n)] \Big\}, \tag{5.96}$$

which has an immediate physical interpretation in terms of a change in the ground-state energy of the normal modes of a surface system on bringing the second surface system from infinity to a finite distance d, at which both half media feel their mutual interaction. For models with small dispersion, or for long distances, the dominant contribution comes from the first square bracket in (5.96). For example, consider the non-dispersive model of (5.50). It is instructive to study explicitly the effect of the cutoff.

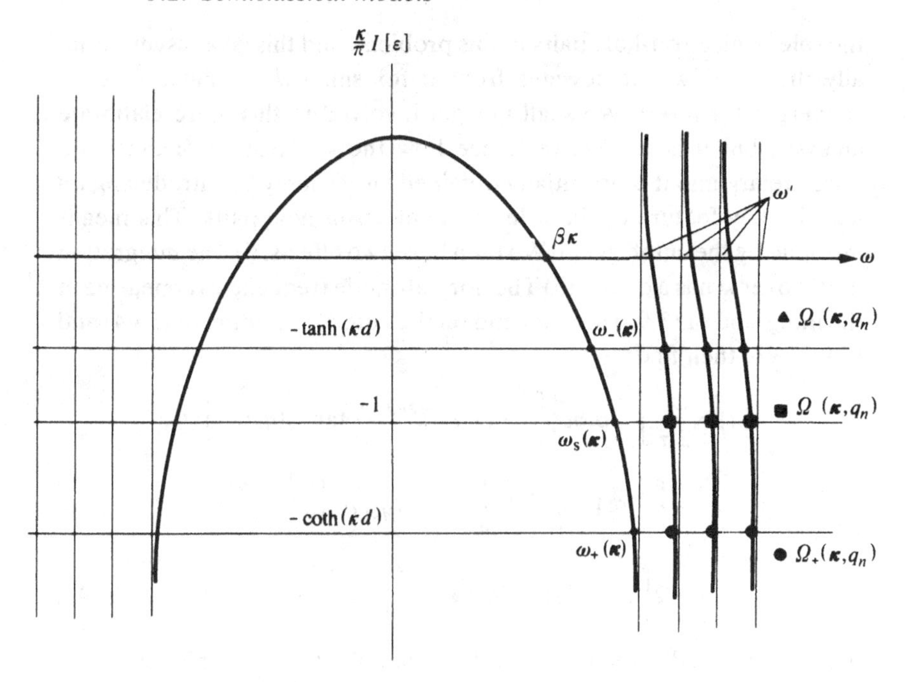

Fig. 109. Qualitative behaviour of $(\kappa/\pi)I[\varepsilon]$ for fixed κ, as a function of ω, for the hydrodynamic dielectric function model. ●, Symmetric surface modes of the two-surface system. ▲, Antisymmetric surface modes of the two-surface system. ■, Surface modes of the free surface system.

Let us first do this problem *without* cutoff. Then, using the limit $q_c \to \infty$ of (5.94) and (5.95) we find

$$v(d) = \int \frac{d^2\kappa}{(2\pi)^2} \frac{\omega_p}{2\sqrt{2}} [(1 + e^{-2\kappa d})^{1/2} + (1 - e^{-2\kappa d})^{1/2} - 2], \quad (5.97a)$$

which, by defining the auxiliary variable $\alpha = \kappa d$, can be written as

$$v(d) = \frac{\omega_p}{4\pi\sqrt{2}} \frac{1}{d^2} \int_0^\infty \alpha[(1 + e^{-2\alpha})^{1/2} + (1 - e^{-2\alpha})^{1/2} - 2]. \quad (5.97b)$$

The point is that the integral is just a number, independent of d. Thus this model predicts an interaction potential which varies as d^{-2}, a well-known result. But it is important to remember that this approximation is only valid for long distances. If we took (5.97b) literally we would find that it diverges for $d \to 0$. This would actually be a gross inconsistency because it turns out that a more careful study for short distances brings in the effects associated with the distortions of the bulk continuum and, in particular, the role of electron-hole pairs in this problem, and this gives asymptotically the d^{-2} law but deviates from it for small d so that the result converges for $d \to 0$. We shall not get involved in this more elaborate analysis, but it is instructive to see how the key feature of the more accurate argume 1t is essentially obtained very simply by introducing *ad hoc* the key feature of the role of the electron-hole pairs. This means introducing the cutoff again. We then have two effects. (i) The integration (5.96) over $\boldsymbol{\kappa}$ has a cutoff. (ii) The normal-mode frequencies appearing in the integrand of (5.96) also depend on the cutoff, according to (5.94) and (5.95). We then find

$$v(d) = \frac{\omega_{\mathrm{p}}}{2\pi} \int_0^{k_c} \mathrm{d}\kappa\kappa \left\{ \left[1 - (1 + \mathrm{e}^{-2\kappa d}) \frac{1}{\pi} \tan^{-1}(q_c/\kappa) \right]^{1/2} + \left[1 - (1 - \mathrm{e}^{-2\kappa d}) \frac{1}{\pi} \tan^{-1}(q_c/\kappa) \right]^{1/2} - 2\left[1 - \frac{1}{\pi} \tan^{-1}(q_c/\kappa) \right]^{1/2} \right\}, \tag{5.98}$$

which does not diverge for $d \to 0$. For example, for $d = 0$ this yields

$$v(d) = \frac{\omega_{\mathrm{p}}}{2\pi} \int_0^{k_c} \mathrm{d}\kappa \times \kappa \left\{ 1 + \left[1 - \frac{2}{\pi} \tan^{-1}(q_c/\kappa) \right]^{1/2} - 2\left[1 - \frac{1}{\pi} \tan^{-1}(q_c/\kappa) \right]^{1/2} \right\}, \tag{5.99}$$

which of course is finite. It is easy to see that this integral is negative, so that $v(d) < 0$, as corresponds to an attractive interaction. Thus this simple model yields a picture which is physically acceptable and altogether not too inaccurate.

It is also interesting to discuss the relationship between the energy of interaction of the two surfaces and the surface energy of a free surface system. From (5.89) we define, per unit area,

$$E^{\mathrm{s}}_{\mathrm{xc}}(d = \infty) - E^{\mathrm{s}}_{\mathrm{xc}}(d = 0) = \int \frac{\mathrm{d}^2\kappa}{(2\pi)^2} \int_0^\infty \frac{\mathrm{d}\omega}{\omega} \times \frac{1}{2} \left\{ \mathrm{Im} \ln \mathscr{B}(\boldsymbol{\kappa}, \omega) + \mathrm{Im} \ln \frac{\mathscr{B}(\boldsymbol{\kappa}, \omega)}{I[\varepsilon]} \right\}, \tag{5.100}$$

which is the energy needed to remove the two half metals infinitely far away from each other against their mutual interaction. Notice the different contributions of the symmetric and antisymmetric modes. in order to see this consider the dispersion relations

$$I[\varepsilon]+\frac{\pi}{\kappa}\coth(\kappa d)=0, \tag{5.101}$$

and

$$I[\varepsilon]+\frac{\pi}{\kappa}\tanh(\kappa d)=0. \tag{5.102}$$

These become identical for $d\to\infty$, where (5.101) and (5.102) express the vanishing of the function $\mathscr{B}(\kappa,\omega)$ appearing in the two terms in the integrand of (5.100), but they behave very differently for $d\to 0$. In this limit we reach a situation which is neither a bulk nor a free-surface situation. It corresponds to a bulk system with a hard wall at $z=0$ and the question is, how does this differ from the bulk system? It is easy to see that the bulk normal modes which are symmetric about $z=0$ remain unchanged, while the antisymmetric ones are modified so that new normal-mode frequencies appear. Consider, for example, (5.102) for $d\to 0$. Then $\tanh(\kappa d)\to 0$ and (5.102) is a new dispersion relation, whose roots are not among the eigenfrequencies of the bulk system. These are actually (antisymmetric) normal modes of the surface system formed by the infinite medium with a hard-wall barrier at $z=0$. This prevents the electronic wavefunctions from crossing the surface, where their amplitudes must vanish, but the two half media *are* coupled to each other via the electrostatic potential. Thus we obtain new normal modes of the hard-wall system which were not normal modes of the bulk medium. The position with the symmetric modes is entirely different. Taking $d\to 0$ in (5.101), since $\coth(\kappa d)\to\infty$, this dispersion relation requires the vanishing of the dielectric function, which appears in the denominator of $I[\varepsilon]$. But this gives just normal modes of the bulk medium. Thus the hard-wall system has no new symmetric modes, and no denominator appears in the first logarithm of (5.100). Notice that this is only a feature of the specular-surface model. In a symmetric mode everything – fields and electron trajectories – is symmetric about the plane $z=0$. If we introduce a surface at this plane and make a specular-surface model (§ 1.6), then we symmetrise about $z=0$, thus forming a mode which is identically the symmetric bulk mode. This would not be the case if we used, for example, a non-specular model with $p\neq 1$, as in § 1.6. Returning now to (5.100) for the specular model, we have to evaluate

$$\frac{1}{2}\int_0^\infty \frac{d\omega}{2\pi}\left\{2\,\mathrm{Im}\ln\mathscr{B}(\boldsymbol{\kappa},\omega)-\mathrm{Im}\ln I[\varepsilon]\right\}.$$

The two elements appearing in this integral have already been studied. Remembering that (i) the zeros of $\mathscr{B}(\boldsymbol{\kappa}, \omega)$ are surface modes, (ii) the zeros of $I[\varepsilon]$ are *antisymmetric* modes of the system formed by the infinite medium with a hard wall at $z=0$, (iii) the poles of $\mathscr{B}(\boldsymbol{\kappa}, \omega)$ are bulk modes, and (iv) the poles of $I[\varepsilon]$ are also bulk modes; and putting all this together, we have

$$E^{\mathrm{s}}_{\mathrm{xc}}(d=\infty)-E^{\mathrm{s}}_{\mathrm{xc}}(d=0)=\int\frac{\mathrm{d}^2\kappa}{(2\pi)^2}$$

$$\times\tfrac{1}{2}\{[\omega_{\mathrm{s}}(\boldsymbol{\kappa})-\omega_-(\boldsymbol{\kappa};d=0)]-\tfrac{1}{2}[\omega_{\mathrm{b}}(\boldsymbol{\kappa},0)-\omega_-(\boldsymbol{\kappa};d=0)]\}$$

$$+\frac{1}{2}\int\frac{\mathrm{d}^2\kappa}{(2\pi)^2}\sum_n\{[\Omega(\boldsymbol{\kappa},q_n)-\Omega_-(\boldsymbol{\kappa},q_n;d=0)]$$

$$-\tfrac{1}{2}[\omega'(\boldsymbol{\kappa},q_n)-\Omega_-(\boldsymbol{\kappa},q_n;d=0)]\}. \tag{5.103}$$

This formula displays explicitly the distinct contribution of the antisymmetric modes of the hard-wall system. As indicated above, $\omega_-(\boldsymbol{\kappa};d=0)$ is obtained from the zero of $I[\varepsilon]$ associated with $q=0$, i.e. from the pole of $\varepsilon(\boldsymbol{\kappa},q;\omega)$ associated with $q=0$. That this frequency is associated with $q=0$ is correct only for $d=0$, and is not in contradiction with the remark made about fig. 105*b*. We saw in § 5.2.1 that in the hydrodynamic model this is just $\beta\kappa$. Thus, in this model, $\omega_-(\kappa;d=0)=\beta\kappa$. In the limit of small dispersion the contribution of the sum on the r.h.s. of (5.103) is as usual very small. Taking the limit $\beta\to 0$ this yields

$$E^{\mathrm{s}}_{\mathrm{xc}}(d=\infty)-E_{\mathrm{xc}}(d=0)=\int\frac{\mathrm{d}^2\kappa}{(2\pi)^2}\frac{1}{2}\left(\omega_{\mathrm{s}}-\frac{1}{2}\omega_{\mathrm{p}}\right), \tag{5.104}$$

where of course we would have to introduce a cutoff in order to ensure convergence.

This simple form is very suggestive, but it would be erroneous to interpret this as the energy of formation of the free surface. We have indicated that (5.100) is only the energy needed to remove one half medium infinitely far away from the other, but *after* the bulk material has been cleaved in two by erecting the infinite barrier at $z=0$. Thus we must include this 'cleavage energy', i.e. from (5.89),

$$E^{\mathrm{s}}_{\mathrm{xc}}(d=0)=\frac{1}{4}\int\frac{\mathrm{d}^2\kappa}{(2\pi)^2}\int_0^\infty\frac{\mathrm{d}\omega}{2\pi}\,\mathrm{Im}\ln\cdot\varepsilon(\boldsymbol{\kappa},0;\omega)$$

$$+\int\frac{\mathrm{d}^2\kappa}{(2\pi)^2}\int_0^\infty\frac{\mathrm{d}\omega}{2\pi}\frac{1}{2}\,\mathrm{Im}\ln\left(\frac{\kappa}{\pi}I[\varepsilon]\right) \tag{5.105}$$

Take, for example, the hydrodynamic approximation. The first term of

(5.105) is then

$$\frac{1}{4}\int \frac{d^2\kappa}{2\pi}\frac{1}{2}[\omega_b(\boldsymbol{\kappa},0)-\beta\kappa]. \tag{5.106}$$

The evaluation of the second term follows the lines of the several similar examples just discussed, and combining with (5.106) we find, for the hydrodynamic model,

$$E^s_{xc}(d=0)=\int \frac{d^2\kappa}{(2\pi)^2}$$

$$\times\frac{1}{2}\left\{-\frac{1}{4}[\omega_b(\boldsymbol{\kappa},0)-\beta\kappa]+\frac{1}{2}\sum_n[\Omega_-(\boldsymbol{\kappa},q_n;d=0)-\omega'(\boldsymbol{\kappa},q_n)]\right\}. \tag{5.107}$$

To sum up, the relevant quantities studied in this analysis are: (i) the (attractive) interaction energy between two half metals whose surfaces are a finite distance apart; this is given by (5.96). (ii) The energy of separation, from $d=0$ to ∞, of the two half media against their mutual attraction; this is given by (5.103). (iii) The energy of formation of the infinite barrier, or cleavage energy; this is given by (5.107). (iv) The surface energy, i.e. the total energy of creation of a free surface; this results from the addition of (ii) ((5.103)) and (iii) ((5.107)) and reproduces the surface energy evaluated by direct analysis in § 5.2.1.

5.2.3. Polarisation interaction between metal surfaces and neutral external particles

The normal-mode analysis of §§ 5.2.1 and 5.2.2. can be extended to study the van der Waals polarisation interaction between a metal surface and an external particle which we shall call a 'molecule' but which could also be an atom, i.e. a polarisable but neutral particle. The interest of this problem lies in that this is essentially the situation in physical adsorption. We shall again use a semiclassical analysis and, when we introduce a surface, we shall also treat it as an infinite-barrier model. Before studying the actual molecule–metal system we must construct the framework by studying in general the interaction between a molecule and some other polarisable body.

The natural extension of the methods used in §§ 5.2.1 and 5.2.2 would be based again on the first term of (5.24) and would run as follows: we put a test charge $Q\,\delta(\boldsymbol{r})$ at a point inside the molecule under consideration, and we want to study the induced charge at another point $\boldsymbol{r}''$ inside this molecule because for use in (5.24) we shall set $\boldsymbol{r}''=\boldsymbol{r}$. The situation is described in fig. 110. If the molecule were alone, then $Q\delta(\boldsymbol{r})$ would create

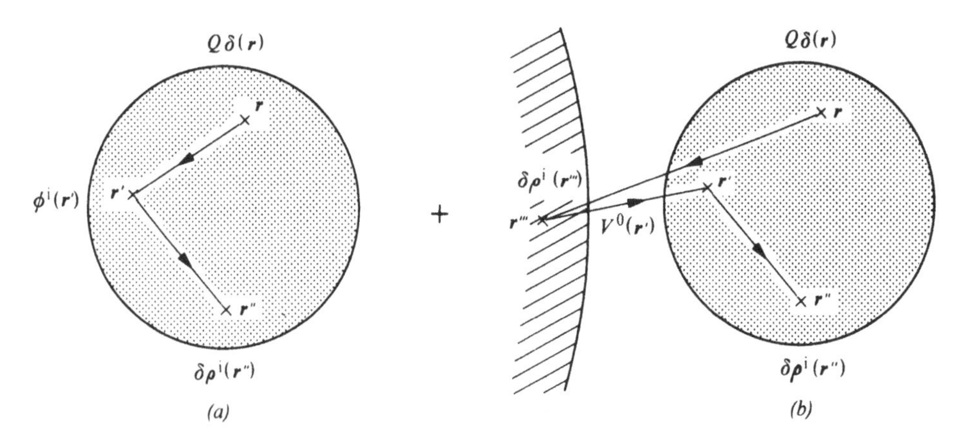

Fig. 110. Scheme for the calculation of the charge induced by an external test charge inside a molecule in the presence of some other polarisable body.

a potential $Q/|\boldsymbol{r}-\boldsymbol{r}'|$ at every point $\boldsymbol{r}'$ (fig. 110*a*), which would then produce at $\boldsymbol{r}''$ the induced charge

$$e\,\delta\rho^{\mathrm{i}}(\boldsymbol{r}'') = \int \chi(\boldsymbol{r}'', \boldsymbol{r}') \frac{Q}{|\boldsymbol{r}'-\boldsymbol{r}|}\,\mathbf{d}\boldsymbol{r}'. \tag{5.108}$$

Here $\chi(\boldsymbol{r}'', \boldsymbol{r}')$ is a polarisability function of the molecule, not exactly the susceptibility χ introduced in chapter 1($P = \chi E$) but closely related to it. Although we often omit the ω-dependence for expediency, it is clear that this is a function $\chi(\boldsymbol{r}'', \boldsymbol{r}'; \omega)$. But the molecule under study is not alone. The presence of another polarisable body adds a further contribution. Referring to fig. 110*b*, the test charge $Q\,\delta(\boldsymbol{r})$ induces a charge density $e\,\delta\rho^{\mathrm{i}}(\boldsymbol{r}''')$. If we integrate over $\boldsymbol{r}'''$, this creates some potential $V^0(\boldsymbol{r}')$, which in turn produces an extra contribution to $e\,\delta\rho^{\mathrm{i}}(\boldsymbol{r}'')$ of the form $\chi(\boldsymbol{r}'', \boldsymbol{r}')\,V^0(\boldsymbol{r}')$. Thus, adding these two terms, we have

$$e\,\delta\rho^{\mathrm{i}}(\boldsymbol{r}'') = \int \chi(\boldsymbol{r}'', \boldsymbol{r}') \frac{Q}{|\boldsymbol{r}'-\boldsymbol{r}|}\,\mathbf{d}\boldsymbol{r}' + \int \chi(\boldsymbol{r}'', \boldsymbol{r}')\,V^0(\boldsymbol{r}')\,\mathbf{d}\boldsymbol{r}'. \tag{5.109}$$

Clearly $V^0(\boldsymbol{r}')$ is in turn proportional to Q and this is the formula we need to find the response function of interest. The question is, what is $V^0(\boldsymbol{r}')$? Strictly speaking the induced charge $e\,\delta\rho^{\mathrm{i}}(\boldsymbol{r}''')$ which in turn creates $V^0(\boldsymbol{r}')$ (fig. 110*b*) is not only induced by the test charge $Q\,\delta(\boldsymbol{r})$, but also by the entire selfconsistent charge redistribution, which includes $e\,\delta\rho^{\mathrm{i}}(\boldsymbol{r}'')$ for all points $\boldsymbol{r}''$. In this simple analysis we assume that the size of the molecule is very small compared with other lengths of interest. This means that we treat the molecule as a punctual polarisable entity, which may be the site of a point dipole, for example, but whose internal structure we neglect on evaluating $V^0(\boldsymbol{r}')$. In other words, on evaluating $e\,\delta\rho^{\mathrm{i}}(\boldsymbol{r}''')$ (fig. 110*b*) we say that this charge is only induced by the external charge at $\boldsymbol{r}$ and the induced charge $e\delta\rho^{\mathrm{i}}(\boldsymbol{r}'')$, this being evaluated by neglecting the small contribution of the first integral in (5.109). Let $\boldsymbol{R}$ be the centre of the molecule and write $\boldsymbol{r}=\boldsymbol{R}+\delta\boldsymbol{r}$, $\boldsymbol{r}'=\boldsymbol{R}+\delta\boldsymbol{r}'$, etc. We then have

$$V^0(\boldsymbol{r},\boldsymbol{r}')=V^0(\boldsymbol{r},\boldsymbol{R})+[\nabla_{\boldsymbol{r}'}V^0(\boldsymbol{r},\boldsymbol{r}')]_{\boldsymbol{r}'=\boldsymbol{R}}\cdot\delta\boldsymbol{r}'+\ldots, \tag{5.110}$$

where, besides expanding the function of $\boldsymbol{r}'$ about $\boldsymbol{R}$, we emphasise that $V^0(\boldsymbol{r}')$ depends also on the point $\boldsymbol{r}$ where the external charge is located. The constant term $V^0(\boldsymbol{r},\boldsymbol{R})$ of (5.110) gives a vanishing contribution to (5.109), since a constant potential does not produce any induced charge. The gradient appearing in (5.110) is minus the electric field $\boldsymbol{E}_{\mathrm{i}}(\boldsymbol{r},\boldsymbol{R})$ encountered at the centre of the molecule as due to the induced potential shown in fig. 110*b*. We can also expand

$$\boldsymbol{E}_{\mathrm{i}}(\boldsymbol{r},\boldsymbol{R})=\boldsymbol{E}_{\mathrm{i}}(\boldsymbol{R},\boldsymbol{R})+\left[\frac{\mathrm{d}\boldsymbol{E}_{\mathrm{i}}(\boldsymbol{r},\boldsymbol{R})}{d(\delta\boldsymbol{r})}\right]_{\boldsymbol{r}=\boldsymbol{R}}\cdot\delta\boldsymbol{r}+\ldots \tag{5.111}$$

The first term of this expansion again gives a vanishing contribution when used in (5.109) and integrated as in (5.115). Thus, to first order in $\delta\boldsymbol{r}$ and $\delta\boldsymbol{r}'$, we write

$$e\,\delta\rho^{\mathrm{i}}(\boldsymbol{r}'')=\int\chi(\boldsymbol{r}'',\boldsymbol{r}')\frac{Q}{|\boldsymbol{r}-\boldsymbol{r}'|}\,\mathbf{d}\boldsymbol{r}' \\ -\int\chi(\boldsymbol{r}'',\boldsymbol{r}')\,\delta\boldsymbol{r}\cdot\left[\frac{\mathrm{d}\boldsymbol{E}(\boldsymbol{r},\boldsymbol{R})}{\mathrm{d}(\delta\boldsymbol{r})}\right]_{\boldsymbol{r}=\boldsymbol{R}}\cdot\delta\boldsymbol{r}'\,\mathbf{d}\boldsymbol{r}' \tag{5.112}$$

whence, setting $\boldsymbol{r}''=\boldsymbol{r}$, we have

$$\frac{\delta\rho^{\mathrm{i}}(\boldsymbol{r})}{\delta\rho^{\mathrm{e}}(\boldsymbol{r})}=\int\chi(\boldsymbol{r},\boldsymbol{r}')\frac{1}{|\boldsymbol{r}-\boldsymbol{r}'|}\,\mathbf{d}\boldsymbol{r}' \\ -\frac{1}{Q}\int\chi(\boldsymbol{r},\boldsymbol{r}')\,\delta\boldsymbol{r}\cdot\left[\frac{\mathrm{d}\boldsymbol{E}(\boldsymbol{r},\boldsymbol{R})}{\mathrm{d}(\delta\boldsymbol{r})}\right]_{\boldsymbol{r}=\boldsymbol{R}}\cdot\delta\boldsymbol{r}'\,\mathbf{d}\boldsymbol{r}'. \tag{5.113}$$

For the application of (5.24) we actually need the integral of (5.113) over $\boldsymbol{r}$. Define

$$\chi_{ij}^{\mathrm{P}}(\omega)=-\int\chi(\boldsymbol{r},\boldsymbol{r}',\omega)\,\delta r_i\,\delta r_j'\,\mathbf{d}\boldsymbol{r}\,\mathbf{d}\boldsymbol{r}'; \tag{5.114}$$

this is a susceptibility function as defined in chapter 1. A simple way to see this is to notice that $\chi^{\mathrm{P}}(t)$ can be expressed, by using (5.7), as

$$\chi^{\mathrm{P}}(t) = -\int \chi(\boldsymbol{r}, \boldsymbol{r}'; t)\, \delta r_i\, \delta r_j'\, \mathbf{d}\boldsymbol{r}\, \mathbf{d}\boldsymbol{r}'$$

$$= \mathrm{i} \int [e\, \delta\rho^{\mathrm{i}}(\boldsymbol{r}, t), e\, \delta\rho^{\mathrm{i}}(\boldsymbol{r}', 0)]\, \delta r_i\, \delta r_j'\, \mathbf{d}\boldsymbol{r}\, \mathbf{d}\boldsymbol{r}'$$

$$= \mathrm{i} \left[\int e\, \delta\rho^{\mathrm{i}}(\boldsymbol{r}, t)\, \delta r_i\, \mathbf{d}\boldsymbol{r}, \int e\, \delta\rho^{\mathrm{i}}(\boldsymbol{r}', 0)\, \delta r_j'\, \mathbf{d}\boldsymbol{r}' \right]$$

$$= \mathrm{i}\, [P_i(t), P_j(0)],$$

where $\boldsymbol{P}$ is the total polarisation induced in the molecule. Thus χ^{P} is, by definition, the response function which gives the induced polarisation, i.e. the object defined in chapter 1. We can then interpret (5.113) in an interesting way. Notice that $Q\,\delta\mathbf{r}$ can be regarded as an external dipole which stimulates the molecule described in the point approximation. If we use (5.114) to rewrite (5.113) as

$$\int \frac{\delta\rho^{\mathrm{i}}(\boldsymbol{r})}{\delta\rho^{\mathrm{e}}(\boldsymbol{r})}\, \mathrm{d}\boldsymbol{r} = \int \frac{\chi(\boldsymbol{r}, \boldsymbol{r}')}{|\boldsymbol{r} - \boldsymbol{r}'|}\, \mathbf{d}\boldsymbol{r}\, \mathbf{d}\boldsymbol{r}' + \sum_{ij} \chi_{ij}^{\mathrm{P}}(\omega) \left[\frac{\mathrm{d}E_j(\boldsymbol{r}, \boldsymbol{R})}{\mathrm{d}(\delta r_i)} \right]_{\boldsymbol{r}=\boldsymbol{R}}, \tag{5.115}$$

then, noticing that

$$\sum_j \chi_{ij}^{\mathrm{P}}(\omega) E_j$$

has the nature of an induced polarisation P_i, we can write the second term of (5.115) as a dipole–dipole response function. The first term is of no concern, since it would give the energy of the molecule alone and we are only interested in studying the energy of interaction between the molecule and the other polarisable body. If we ignore this term, (5.115) becomes

$$\int \frac{\delta\rho^{\mathrm{i}}(\boldsymbol{r})}{\delta\rho^{\mathrm{e}}(\boldsymbol{r})}\, \mathbf{d}\boldsymbol{r} = \sum_i \left(\frac{\mathrm{d}P_i^{\mathrm{i}}}{\mathrm{d}P_i^{\mathrm{e}}} \right)_{\boldsymbol{r}=\boldsymbol{R}}. \tag{5.116}$$

This describes an external stimulus in the form of a point-dipole distribution $\boldsymbol{P}^{\mathrm{e}}\delta(\boldsymbol{r})$ which induces a polarisation $\boldsymbol{P}^{\mathrm{i}}$ in the entire molecule. If we assume the size of the molecule is small compared with other lengths of interest, everything is described within a point dipole approximation ($\boldsymbol{r} = \boldsymbol{R}$).

Our purpose is to study the interaction between a molecule and a semi-infinite metal, but for reasons which will soon be seen it is convenient to first study the interaction between two molecules, for which we assume the external point dipole is in the molecule labelled (1) in fig. 111,

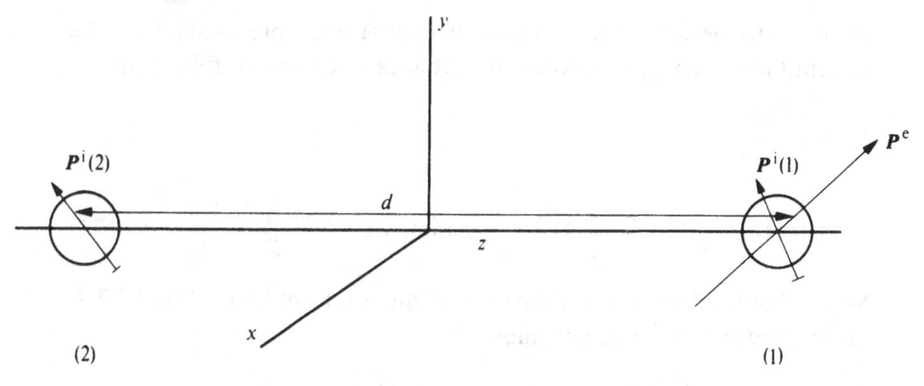

Fig. 111. Scheme for the calculation of the van der Waals interaction energy between two molecules a distance d apart, described in the point dipole approximation.

which also shows the induced dipoles in molecules (1) and (2). The standard formulae for the electric field of a dipole yield the following results. For a component perpendicular to the line joining the two molecules, e.g. E_x

$$E_x(2) = -\frac{P_x^{e}(1) + P_x^{i}(1)}{d^3}; \quad E_x(1) = -\frac{P_x^{i}(2)}{d^3}. \tag{5.117}$$

Let $\chi_1(\omega)$ and $\chi_2(\omega)$ be the corresponding susceptibilities. Then

$$P_x^{i}(2) = \chi_2 E_x(2); \quad P_x^{i}(1) = \chi_1 E_x(1),$$

whence using (5.117) we can relate $P_x^{i}(1)$ to $P_x^{e}(1)$ and find

$$\frac{\mathrm{d}P_x^{i}(1)}{\mathrm{d}P_x^{e}(1)} = \frac{\chi_1\chi_2/d^6}{1-\chi_1\chi_2/d^6}. \tag{5.118}$$

Clearly the same formula holds for $\mathrm{d}P_x^{i}(2)/\mathrm{d}P_x^{e}(2)$, and likewise for the derivative involving the y-components. For the z-components we find, proceeding in the same manner,

$$\frac{\mathrm{d}P_z^{i}(1)}{\mathrm{d}P_z^{e}(1)} = \frac{4\chi_1\chi_2/d^6}{1-4\chi_1\chi_2/d^6} = \frac{\mathrm{d}P_z^{i}(2)}{\mathrm{d}P_z^{e}(2)}. \tag{5.119}$$

We can now write the van der Waals energy for the two-molecule system, namely,

$$E_{\mathrm{xc}} = -\int_0^1 \frac{\mathrm{d}\lambda}{\lambda}\int_0^\infty \frac{\mathrm{d}\omega}{2\pi}$$
$$\times\left[\mathrm{Im}\,\frac{8\chi_1(\omega)\chi_2(\omega)/d^6}{1-4\chi_1(\omega)\chi_2(\omega)/d^6} + \mathrm{Im}\,\frac{4\chi_1(\omega)\chi_2(\omega)/d^6}{1-\chi_1(\omega)\chi_2(\omega)/d^6}\right]. \tag{5.120}$$

In order to perform the integration over λ we note again that the susceptibilities are proportional to λ. It is easy to see that this yields

$$E_{xc}=\int_0^{\infty}\frac{d\omega}{2\pi}$$

$$\times\left\{\operatorname{Im}\ln\left[1-\frac{4\chi_1(\omega)\chi_2(\omega)}{d^6}\right]+2\operatorname{Im}\ln\left[1-\frac{\chi_1(\omega)\chi_2(\omega)}{d^6}\right]\right\}. \quad (5.121)$$

Notice that this has the standard form of the results of §§ 5.2.1 and 5.2.2. To first order in d^{-6} this becomes

$$E_{xc}=-\frac{6}{d^6}\int\frac{d\omega}{2\pi}\operatorname{Im}\left[\chi_1(\omega)\chi_2(\omega)\right], \quad (5.122)$$

which is a well-known result. We stress that (5.121), based on the point-dipole approximation, is a fair physical result only to first order, as in (5.122). However the form is suggestive and we shall later see how it admits the standard interpretation in terms of normal modes.

Now the point to stress at this stage is that we have used for the two-molecule problem the same approach as we used in § 5.2.2, in the sense that we treat the two halves of the system symmetrically, and when calculating the interaction energy, we integrate the response function over the two halves of the system. This would seem a natural procedure, but it is actually not a very practical approach to use when it comes to studying the interaction between a molecule and a metal surface, because there is an inherent asymmetry between the two parts of the system. It is therefore useful to see how the correct result can also be obtained by concentrating only on the molecule under consideration. In order to see this we go back to the argument started in (5.15). Suppose we have two distinct subsystems and we put λ_i ($i=1,2$) for the coupling parameter in each subsystem. For the electron–electron interaction we can then formally write, excluding self-energy,

$$\frac{e^2}{2}\int\lambda_1\frac{\rho_1(\boldsymbol{r})\rho_1(\boldsymbol{r}')}{|\boldsymbol{r}-\boldsymbol{r}'|}\,d\boldsymbol{r}\,d\boldsymbol{r}'+\frac{e^2}{2}\int\lambda_2\frac{\rho_2(\boldsymbol{r})\rho_2(\boldsymbol{r}')}{|\boldsymbol{r}-\boldsymbol{r}'|}\,d\boldsymbol{r}\,d\boldsymbol{r}'$$

$$+e^2\int(\lambda_1\lambda_2)^{1/2}\frac{\rho_1(\boldsymbol{r})\rho_2(\boldsymbol{r}')}{|\boldsymbol{r}-\boldsymbol{r}'|}\,d\boldsymbol{r}\,d\boldsymbol{r}'.$$

Now we can formally think of a Hamiltonian $H(\lambda_1,\lambda_2)$, for which we have, independently,

$$\frac{dE}{d\lambda_i}=\left\langle 0,\lambda_1,\lambda_2\left|\frac{\partial H(\lambda_1,\lambda_2)}{\partial\lambda_i}\right|0,\lambda_1,\lambda_2\right\rangle \quad (i=1,2). \quad (5.123)$$

We can then go from $\lambda_1 = \lambda_2 = 0$ to $\lambda_1 = \lambda_2 = 1$ by first keeping, say, $\lambda_2 = 0$ and varying λ_1 from 0 to 1, and then keeping $\lambda_1 = 1$ and varying λ_2. In the first process we would obtain only the energy of subsystem 1 when isolated from any interaction with the rest, while from the second process we have, keeping $\lambda_1 = 1$ all the time,

$$\frac{\partial H}{\partial \lambda_2} = \frac{e^2}{2} \int \frac{\rho_2(\boldsymbol{r})\rho_2(\boldsymbol{r}')}{|\boldsymbol{r}-\boldsymbol{r}'|} \,\mathbf{d}\boldsymbol{r}\,\mathbf{d}\boldsymbol{r}' + \frac{e^2}{2} \lambda_2^{-1/2} \int \frac{\rho_1(\boldsymbol{r})\rho_2(\boldsymbol{r}')}{|\boldsymbol{r}-\boldsymbol{r}'|} \,\mathbf{d}\boldsymbol{r}\,\mathbf{d}\boldsymbol{r}',$$

whence

$$\left\langle 0, \lambda_2 \left| \frac{\partial H}{\partial \lambda_2} \right| 0, \lambda_2 \right\rangle = \frac{e^2}{2} \int \frac{1}{|\boldsymbol{r}-\boldsymbol{r}'|} \langle 0, \lambda_2 | \rho_2(\boldsymbol{r})\rho_2(\boldsymbol{r}') | 0, \lambda_2 \rangle \,\mathbf{d}\boldsymbol{r}\,\mathbf{d}\boldsymbol{r}'$$

$$+ \frac{e^2}{2} \lambda_2^{-1/2} \int \frac{n_{01}(\boldsymbol{r}) n_{02}(\boldsymbol{r}')}{|\boldsymbol{r}-\boldsymbol{r}'|} \,\mathbf{d}\boldsymbol{r}\,\mathbf{d}\boldsymbol{r}'.$$

Since $e n_{02}(\boldsymbol{r}')$ is the actual charge density with $\lambda_2 = 1$, independent of λ_2, integration of the second term over λ_2 yields just the average electrostatic interaction energy between the two subsystems, i.e.

$$e^2 \int \frac{n_{01}(\boldsymbol{r}) n_{02}(\boldsymbol{r}')}{|\boldsymbol{r}-\boldsymbol{r}'|} \,\mathbf{d}\boldsymbol{r}\,\mathbf{d}\boldsymbol{r}'.$$

Subtracting this term, treating the first one as in § 5.1, and adding the energy obtained in the first process (with $\lambda_2 = 0$), we obtain for the total exchange and correlation energy

$$E_{\text{xc}} = -\int \frac{\mathrm{d}\lambda_2}{\lambda_2} \int_0^\infty \frac{\mathrm{d}\omega}{2\pi} \operatorname{Im} \left[\int \frac{\delta\rho^{\mathrm{i}}(\boldsymbol{r}_2)}{\delta\rho^{\mathrm{e}}(\boldsymbol{r}_2)} \,\mathbf{d}\boldsymbol{r}_2 \right]_{\lambda_2;\, \lambda_1 = 1}$$

$$- \int \frac{\mathrm{d}\lambda_1}{\lambda_1} \int_0^\infty \frac{\mathrm{d}\omega}{2\pi} \operatorname{Im} \left[\int \frac{\delta\rho^{\mathrm{i}}(\boldsymbol{r}_1)}{\delta\rho^{\mathrm{e}}(\boldsymbol{r}_1)} \,\mathbf{d}\boldsymbol{r}_1 \right]_{\lambda_1;\, \lambda_2 = 0}. \tag{5.124}$$

We stress that the second term in this formula is only the energy of subsystem (1) without interaction with subsystem (2). Thus the interaction energy is given by the first term of (5.124). If subsystem (1) is a molecule in the point-dipole approximation, in view of the above discussion we can write for the exchange and correlation interaction energy,

$$E_{\text{xc}} = -\int_0^1 \frac{\mathrm{d}\lambda_2}{\lambda_2} \int_0^\infty \frac{\mathrm{d}\omega}{2\pi} \operatorname{Im} \left[\sum_i \frac{\mathrm{d}P_i^{\mathrm{i}}}{\mathrm{d}P_i^{\mathrm{e}}} \right]_{\lambda_2;\, \lambda_1 = 1}. \tag{5.125}$$

It is important to notice that now by definition this formula involves only the induced dipole in the *one* molecule under consideration. That this yields the correct result (5.121) – i.e. correct within the point-dipole approximation – for the two-molecule problem can be seen in the following way.

Consider again the problem of molecule (1) in interaction with molecule (2). From (5.118) and (5.119) we have

$$\sum_i \frac{\mathrm{d}P_i^{\mathrm{i}}(1)}{\mathrm{d}P_i^{\mathrm{e}}(1)} = \frac{4\chi_1\chi_2/d^6}{1-4\chi_1\chi_2/d^6} + \frac{2\chi_1\chi_2/d^6}{1-\chi_1\chi_2/d^6}, \tag{5.126}$$

but now we must keep $\lambda_1 = 1$ all the time and switch on the interaction from $\lambda_2 = 0$ to $\lambda_2 = 1$. Thus the terms to be used in (5.125) have the form

$$\int \frac{\mathrm{d}\lambda_2}{\lambda_2}\left[\frac{-4c\lambda_2}{1+4c\lambda_2} + \frac{2c\lambda_2}{1-c\lambda_2}\right].$$

It is immediate to see that this yields (5.121) again.

Thus we have seen how to pass gradually from the problem of two half metals where these are treated on the same footing, through the problem of two molecules treated also on the same footing, to the same problems solved by a different approach, in which the attention is concentrated on only one molecule, which is treated in a different way. This approach proves more practical for the molecule–metal problem, owing to its inherent asymmetry. Let us now see how this problem can be solved using this approach.

The first question is the evaluation of the response function of interest. We consider the molecule at a point where we put an external point dipole. We want to calculate the polarisation induced in this molecule in the presence of the metal with its surface. Consider first the case in which the external dipole is parallel to the surface, e.g. consider just P_x^{e} (fig. 112*a*). We use again the method of extended pseudomedia (chapter 1), as indicated in fig. 112. Notice that m indicates now an extended pseudo-vacuum which also includes the molecule under study, with the external point dipole, and their mirror images. We look for the x-component of the field met by the molecule which, for $z > 0$, is ($P_z^{\mathrm{i}} = 0$ by symmetry)

$$E_x^{\mathrm{m}} = -\frac{P_x^{\mathrm{e}} + P_x^{\mathrm{i}}(\mathrm{m})}{(2d)^3} - \int \frac{4\pi\sigma}{k^2}\,\mathrm{i}\kappa_x\,\mathrm{e}^{\mathrm{i}qd}\frac{\mathrm{d}^3k}{(2\pi)^3}. \tag{5.127}$$

If $\chi(\omega)$ is the susceptibility of the molecule, then

$$P_x^{\mathrm{i}}(\mathrm{m}) = \chi \mathrm{E}_x^{\mathrm{m}}.$$

Using this in (5.127) yields

$$P_x^{\mathrm{i}}(\mathrm{m}) = -\frac{\chi}{(2d)^3+\chi}P_x^{\mathrm{e}} - \frac{(2d)^3\chi}{(2d)^3+\chi}\int \frac{4\pi\sigma}{k^2}\,\mathrm{i}\kappa_x\mathrm{e}^{\mathrm{i}qd}\frac{\mathrm{d}^3k}{(2\pi)^3}. \tag{5.128}$$

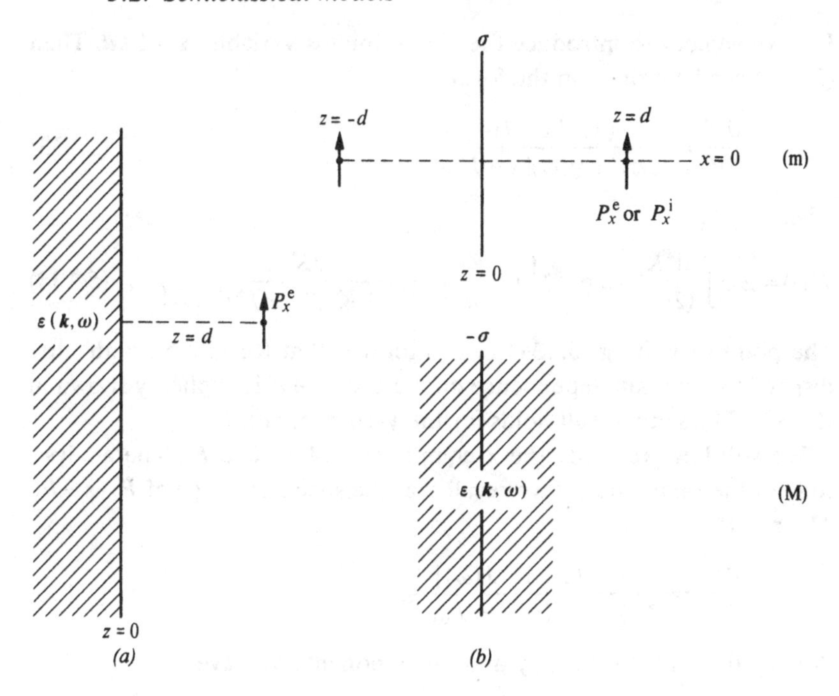

Fig. 112. (*a*) The system under study: a point molecule subject to an external point dipole in the presence of a metal surface. (*b*) Definition of the extended pseudomedia for the study of this problem.

We must find σ, for which as usual we notice that (5.127) derives from a potential which, evaluated at $z = +0$, is

$$\phi^{\mathrm{m}}(0) = \int \frac{4\pi\sigma}{k^2}\,\mathrm{d}q - \int \frac{4\pi}{k^2}\,\mathrm{i}\kappa_x[P_x^{\mathrm{e}} + P_x^{\mathrm{i}}(\mathrm{m})]2\cos(qd)\,\mathrm{d}q. \quad (5.129)$$

This is to be matched to

$$\phi^{\mathrm{M}}(0) = -\int \frac{4\pi\sigma}{k^2\varepsilon}\,\mathrm{d}q. \quad (5.130)$$

Equating (5.129) and (5.130) yields σ, which we then insert in (5.128). After some manipulation this yields, in terms of real quantities at the location of the molecule,

$$\frac{\mathrm{d}P_x^{\mathrm{i}}}{\mathrm{d}P_x^{\mathrm{e}}} = \frac{-\chi\{1 + (2d)^3 J[d;\varepsilon]\}}{(2d)^3 + \chi\{1 + (2d)^3 J[d;\varepsilon]\}}, \quad (5.131)$$

where (remember (5.29))

$$J[d;\varepsilon] = \int \frac{\mathrm{d}^3k}{(2\pi)^3}\frac{4\pi\mathrm{i}\kappa_x}{(\kappa^2+q^2)}\frac{\mathrm{e}^{\mathrm{i}qd}}{I[1]+I[\varepsilon]}\int \mathrm{d}p\,\frac{2\mathrm{i}\kappa_x\cos(pd)}{(\kappa^2+p^2)}. \quad (5.132)$$

It is convenient to introduce the dimensionless variable $\boldsymbol{K} = 2kd$. Then (5.131) can be written in the form

$$\frac{\mathrm{d}P_x^{\mathrm{i}}}{\mathrm{d}P_x^{\mathrm{e}}} = \frac{-\chi(\omega)[1-J(\omega)]}{(2d)^3 + \chi(\omega)[1-J(\omega)]}, \tag{5.133}$$

where

$$J(\omega) = 2\pi \int \frac{\mathrm{d}^2 K_\parallel}{(2\pi)^2} K_\parallel \,\mathrm{e}^{-K_\parallel} \left\{ 1 + \frac{K_\parallel}{\pi} \int \frac{\mathrm{d}K_\perp}{(K_\parallel^2 + K_\perp)^2 \varepsilon(\boldsymbol{K}/2d;\omega)} \right\}^{-1}. \tag{5.134}$$

The point of writing (5.131) in this form is that for models with little dispersion the main dependence on the distance d is explicitly shown in (5.133). The same result holds for the y-component.

We still have to study the z-components of $\dot{P}^{\mathrm{e}}$ and P^{i}. The analysis follows the same lines, if we recall that the specular image of P_z is $-P_z$. The result is

$$\frac{\mathrm{d}P_z^{\mathrm{i}}}{\mathrm{d}P_z^{\mathrm{e}}} = \frac{-2\chi(\omega)[1-J(\omega)]}{(2d)^3 + 2\chi(\omega)[1-J(\omega)]}. \tag{5.135}$$

Adding the terms of the x, y and z-components we have

$$\sum_i \frac{\mathrm{d}P_i^{\mathrm{i}}}{\mathrm{d}P_i^{\mathrm{e}}} = -\frac{2\chi(\omega)[1-J(\omega)]}{(2d)^3 + 2\chi(\omega)[1-J(\omega)]} - \frac{2\chi(\omega)[1-J(\omega)]}{(2d)^3 + \chi(\omega)[1-J(\omega)]}. \tag{5.136}$$

Now, as shown in (5.125), we only need to integrate over the coupling parameter λ of the molecule, keeping the coupling constant of the other polarisable body – the metal in this case – with its full strength. Using (5.136) in (5.125) we find for the interaction potential

$$V(d) = \int_0^\infty \frac{\mathrm{d}\omega}{2\pi} \left[\operatorname{Im} \ln \left\{ 1 + \frac{2\chi(\omega)[1-J(\omega)]}{(2d)^3} \right\} \right.$$
$$\left. + 2 \operatorname{Im} \ln \left\{ 1 + \frac{\chi(\omega)[1-J(\omega)]}{(2d)^3} \right\} \right]. \tag{5.137}$$

To lowest order,

$$V(d) = \frac{1}{d^3} \int_0^\infty \frac{\mathrm{d}\omega}{2\pi} \operatorname{Im} \left\{ \frac{1}{2} \chi(\omega)[1-J(\omega)] \right\}, \tag{5.138}$$

which yields the well-known d^{-3} law. For example, if we use a non-dispersive dielectric constant $\varepsilon(\omega)$, then

$$J(\omega) = \int \mathrm{d}K_\parallel \frac{K_\parallel^2 \,\mathrm{e}^{-K_\parallel}}{1+\varepsilon^{-1}} = \frac{2\varepsilon(\omega)}{1+\varepsilon(\omega)}, \tag{5.139}$$

and

$$V(d)=\frac{1}{2d^3}\int_0^\infty \frac{d\omega}{2\pi}\,\mathrm{Im}\left\{\chi(\omega)\frac{[1-\varepsilon(\omega)]}{[1+\varepsilon(\omega)]}\right\}, \tag{5.140}$$

which is the well-known *Lifshitz formula.*

More generally, (5.137) has the standard form of the results interpreted in §§ 5.2.2 and 5.2.3 in terms of a normal-mode analysis. For the application of (5.49) we must look at the following normal-mode frequencies: (i) the poles $\omega_i(\mathrm{m})$ of $\chi(\omega)$, of which in general there may be more than one. In the point dipole approximation the only field acting on the molecule is the external field. Thus the poles of $\chi(\omega)$ are the normal modes of the molecule by itself. (ii) The poles of $1-J(\omega)$. By (5.134), these are the zeros of $\mathscr{B}(\kappa)$, i.e. the normal modes of the semi-infinite metal with its surface. These were the frequencies called $\Omega(\boldsymbol{\kappa}, q_n)$ in § 5.2.1, e.g. in (5.80). (iii) The zeros of the equations

$$1+\frac{2\chi(\omega)[1-J(\omega)]}{(2d)^3}=0; \quad 1+\frac{\chi(\omega)[1-J(\omega)]}{(2d)^3}=0. \tag{5.141}$$

In order to see what these frequencies mean it suffices to notice that, according to (5.136), these are precisely the poles of the response function of the coupled molecule–metal system. Let these frequencies be Ω_j. Then (5.137) becomes

$$V(d)=\frac{1}{2}\left\{\sum_j \Omega_j-\left[\sum_n \Omega(\kappa, q_n)+\sum_i \omega_i(\mathrm{m})\right]\right\}. \tag{5.142}$$

This shows very neatly that the van der Waals interaction between the molecule and the semi-infinite metal is due to the renormalisation of the normal modes of the system on switching on the interaction, as it produces a shift in the zero-point energy of the corresponding normal modes. The dependence on the distance is contained in the normal frequencies Ω_j of the coupled system, and shows up explicitly when a simple model is used, as in (5.138), within the point-dipole approximation.

5.2.4. Polarisation interaction between metal surfaces and fast charged external particles

In many spectroscopic experiments the situation in essence involves the interaction between a metal surface and an external charged particle in motion. Imagine, for example, a typical experiment of fast-electron spectroscopy. This may mean electrons with energies of order 10^3–10^5 eV. Let us consider, say, an incoming electron with energy of order

10^4 eV. This means k_{el} of order 10^9 cm^{-1} and an electron velocity v_{el} of order 10^9 cm s^{-1}. Typical plasmon energies may be of order close to 10 eV. Moreover, the phase velocity of the excited plasmon, ω/k_{plas} must be of order 10^6 cm^{-1}. Thus both the energy and the momentum of the incoming electron are much larger than the energy and momentum of the excitation created, and the electron can be regarded as sufficiently fast for its recoil to be neglected. This is the meaning of 'fast' in this context.

The complete analysis of the scattering process, and of the many complicated events which can take place when an external charge collides with the surface, constitute a vast field of study which is outside our scope. We shall only discuss the prototype of the situation just described, i.e. the interaction between a metal surface and an external fast charge whose recoil we shall neglect. Thus we shall regard the kinetic energy of the charge as an infinite-energy source and will therefore take the velocity of the charge as a given constant. We want to study the potential interaction energy when the external charge is at a given position with given velocity and we shall use a classical analysis which suffices to bring out the essential physical issues. Let $\boldsymbol{v}$ be the velocity of an external charge Q approaching a metal surface and let $\boldsymbol{w}$ and u be the components of $\boldsymbol{v}$ parallel and perpendicular to the surface. Classically speaking we have an external charge distribution $Q\delta(\boldsymbol{r}-\boldsymbol{v}t)$. In order to solve the problem we use again the method of extended pseudomedia. Here the extended pseudovacuum contains the external charge and its mirror image. This is denoted as P in fig. 113. This means, as already indicated, that we restrict ourselves to the study of the interaction energy, i.e. the potential experienced by the charge when it is at $\boldsymbol{r}=\boldsymbol{v}t$ before reaching the surface. Furthermore, we define M as the symmetric extension of the semi-infinite metal, which means that we describe the *inner* metal surface as reflecting specularly the electrons of the electron gas forming the metal itself. After these preliminaries we are ready to start our analysis of the extended pseudomedia.

In order to study P consider first the charge distribution of the moving charge and its image, i.e.

$$\int Q[\delta(\boldsymbol{r}-\boldsymbol{v}t)+\delta(\boldsymbol{r}-\tilde{\boldsymbol{v}}t)]\,\mathrm{e}^{-\mathrm{i}\boldsymbol{k}\cdot\boldsymbol{r}}\,\mathrm{e}^{\mathrm{i}\omega t}\,\mathbf{d}\boldsymbol{r}\,\mathrm{d}t$$
$$=2\pi[\delta(\omega-\boldsymbol{k}\cdot\boldsymbol{v})+\delta(\omega-\boldsymbol{k}\cdot\tilde{\boldsymbol{v}})].$$

Thus, adding the effect of the fictitious surface charge and taking $Q=1$, we have

$$\phi^{\mathrm{P}}(\boldsymbol{k},\omega)=\frac{4\pi}{k^2}2\pi[\delta(\omega-\boldsymbol{k}\cdot\boldsymbol{v})+\delta(\omega-\boldsymbol{k}\cdot\tilde{\boldsymbol{v}})]+\frac{4\pi\sigma(\boldsymbol{\kappa},\omega)}{k^2}. \tag{5.143}$$

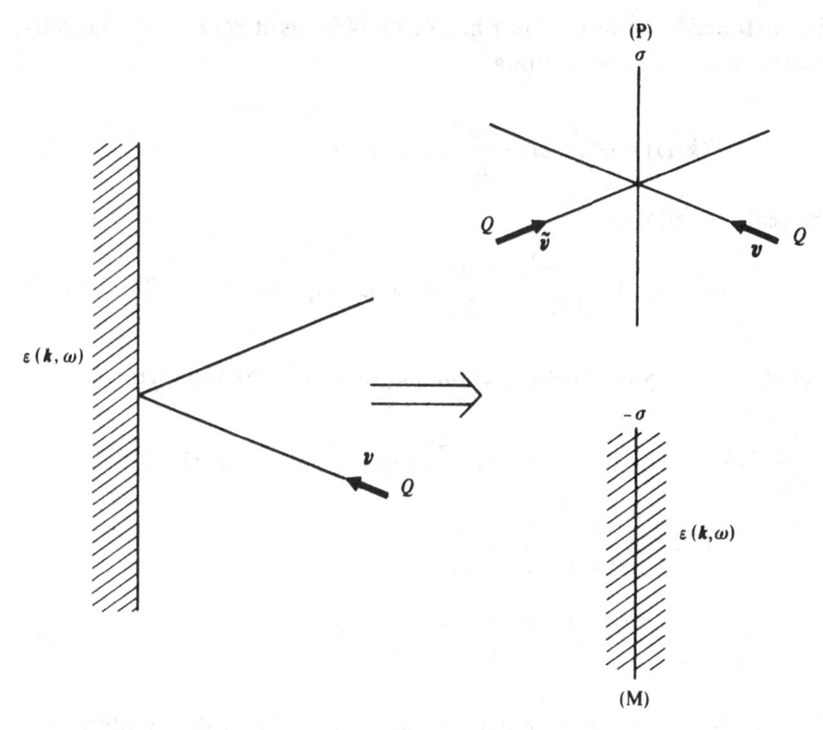

Fig. 113. Definition of extended pseudomedia used to study the interaction between a metal surface and an external (fast) moving charge. The velocity $\boldsymbol{v}_s = (\boldsymbol{w}, -u)$ is the specular image of $\boldsymbol{v} = (\boldsymbol{w}, u)$.

For M we have simply

$$\phi^{\mathrm{M}}(\boldsymbol{k}, \omega) = -4\pi\sigma(\boldsymbol{\kappa}, \omega)/k^2\varepsilon(\boldsymbol{k}, \omega). \tag{5.144}$$

Matching $\phi^{\mathrm{P}}(z = +0)$ to $\phi^{\mathrm{M}}(z = -0)$ yields σ and using this in (5.143) we find

$$\begin{aligned}\phi^{\mathrm{P}}(\boldsymbol{k}, \omega) = {} & \frac{8\pi^2}{k^2}[\delta(\omega - \boldsymbol{k}\cdot\boldsymbol{v}) + \delta(\omega - \boldsymbol{k}\cdot\tilde{\boldsymbol{v}})] \\ & - \frac{4\pi}{k^2\{I[1] + I[\varepsilon]\}}\int \frac{2\pi}{k_1^2}[\delta(\omega - \boldsymbol{k}_1\cdot\boldsymbol{v}) \\ & + \delta(\omega - \boldsymbol{k}_1\cdot\tilde{\boldsymbol{v}})]\,\mathrm{d}q_1.\end{aligned} \tag{5.145}$$

Here and henceforth we use the notation $\boldsymbol{k}_i = (\boldsymbol{\kappa}, q_i)$, where $i = 1, 2, \ldots$ Expressions involving $\boldsymbol{k}_i$ are functions of $\boldsymbol{\kappa}$, after q_i has been integrated over.

We are interested in the real potential at $\boldsymbol{r} = \boldsymbol{v}t$ $(z > 0, t < 0)$, and this is obtained from the Fourier transform of $\phi^{\mathrm{P}}(\boldsymbol{k}, \omega)$ but after removing the first δ-function term on the r.h.s. of (5.145), as it represents the self-energy of the particle. Putting

$$\psi^{\mathrm{P}}(\boldsymbol{k},\omega)=\phi^{\mathrm{P}}(\boldsymbol{k},\omega)-\frac{8\pi^2}{k^2}\delta(\omega-\boldsymbol{k}\cdot\boldsymbol{v}), \tag{5.146}$$

the real potential is

$$\phi(\boldsymbol{r}=\boldsymbol{v}t,t)=\int\frac{\mathrm{d}^3k}{(2\pi)^3}\int\frac{\mathrm{d}\omega}{2\pi}\psi^{\mathrm{P}}(\boldsymbol{k},\omega)\exp\left[\mathrm{i}(\boldsymbol{k}\cdot\boldsymbol{v}-\omega)t\right]. \tag{5.147}$$

Let us first study the ω-integration, i.e. from (5.145) and (5.146),

$$\begin{aligned}\int\frac{\mathrm{d}\omega}{2\pi}\psi^{\mathrm{P}}(\boldsymbol{k},\omega)\exp\left[\mathrm{i}(\boldsymbol{k}\cdot\boldsymbol{v}-\omega)t\right]&=\frac{4\pi}{k^2}\exp\left[\mathrm{i}(\boldsymbol{k}\cdot\boldsymbol{v}-k\cdot\tilde{\boldsymbol{v}})t\right]\\&-\frac{4\pi}{k^2}\int\frac{\mathrm{d}q_1}{k_1^2}\frac{\exp\left[\mathrm{i}(\boldsymbol{k}\cdot\boldsymbol{v}-\boldsymbol{k}_1\cdot\boldsymbol{v})\right]}{I[1]+I[\varepsilon;\boldsymbol{k}_1\cdot\boldsymbol{v}]}\\&-\frac{4\pi}{k^2}\int\frac{\mathrm{d}q_1}{k_1^2}\frac{\exp\left[\mathrm{i}(\boldsymbol{k}\cdot\boldsymbol{v}-\boldsymbol{k}_1\cdot\tilde{\boldsymbol{v}})t\right]}{I[1]+I[\varepsilon;\boldsymbol{k}_1\cdot\tilde{\boldsymbol{v}}]}.\end{aligned} \tag{5.148}$$

Here we have introduced a more explicit extension of the notation $I[\varepsilon]$ which was defined in (5.29). Since ε is a function $\varepsilon(\boldsymbol{k},\omega)$, and the perpendicular component of $\boldsymbol{k}$ is integrated away, $I[\varepsilon]$ is only a function of $(\boldsymbol{\kappa},\omega)$. This will be henceforth denoted as $I[\varepsilon;\omega]$, whereas $I[\varepsilon;\boldsymbol{a}\cdot\boldsymbol{b}]$ indicates $\omega=\boldsymbol{a}\cdot\boldsymbol{b}$ in $I[\varepsilon;\omega]$. In (5.148) $\boldsymbol{a}$ is $\boldsymbol{k}_1=(\boldsymbol{\kappa},q_1)$, so that q_1 is also integrated away, but in the integration explicitly indicated in (5.148); while $\boldsymbol{b}$ is $\boldsymbol{v}$ or $\tilde{\boldsymbol{v}}$. At this stage it is convenient to recall that ω must be understood as $\omega+\mathrm{i}\eta$ $(\eta\to 0)$. Thus we rewrite (5.148) more explicitly as

$$\begin{aligned}&\int\frac{\mathrm{d}\omega}{2\pi}\psi^{\mathrm{P}}(\boldsymbol{k},\omega)\exp\left[\mathrm{i}(\boldsymbol{k}\cdot\boldsymbol{v}-\omega)t\right]\\&\quad=\frac{4\pi}{k^2}\mathrm{e}^{\mathrm{i}2qut}-\frac{4\pi}{k^2}\mathrm{e}^{\mathrm{i}qut}\int\frac{\mathrm{d}q_1}{k_1^2}\frac{\mathrm{e}^{-\mathrm{i}q_1ut}}{I[1]+I[\varepsilon;\boldsymbol{k}_1\cdot\boldsymbol{v}+\mathrm{i}\eta]}\\&\quad-\frac{4\pi}{k^2}\mathrm{e}^{\mathrm{i}qut}\int\frac{\mathrm{d}q_1}{k_1^2}\frac{\mathrm{e}^{\mathrm{i}q_1ut}}{I[1]+I[\varepsilon;\boldsymbol{k}_1\cdot\tilde{\boldsymbol{v}}+\mathrm{i}\eta]}.\end{aligned} \tag{5.149}$$

The evaluation of the integral over q_1 depends on sgn t. We study the case of a charge approaching the surface ($u<0$) during $t<0$ and reaching it at $t=0$. Under these conditions it is easy to see that the only poles contributing to the integrations over q_1 in (5.149) are due to the vanishing of k_1^2. The result of performing these integrations can be finally expressed as

$$\int \frac{d\omega}{2\pi} \psi^{P}(\boldsymbol{k}, \omega) \exp[i(\boldsymbol{k}\cdot\boldsymbol{v}-\omega)t]$$

$$= \frac{4\pi}{k^2} e^{i2qut} - \frac{8\pi}{k^2} e^{iqut} e^{-\kappa ut} \left\{1 + \frac{\kappa}{\pi} I[\varepsilon; \boldsymbol{\kappa}\cdot\boldsymbol{w} - i\kappa u]\right\}. \qquad (5.150)$$

The vanishingly-small imaginary part $i\eta$ explicitly shown in (5.149) is necessary to make clear on which half plane the poles are on evaluating the integrations over q_1 by contour integration. Having sorted this out, $i\eta$ becomes irrelevant and can be omitted from (5.150), which however contains the finite imaginary term in $i\kappa u$.

We can now perform the last integration of (5.147) and it is convenient to do this in two steps, according to

$$\int \frac{d^3k}{(2\pi)^3} f(\boldsymbol{k}) = \int \frac{d^2\kappa}{(2\pi)^2} \int \frac{dq}{2\pi} f(\boldsymbol{\kappa}, q).$$

Applying this to (5.150) we have for the first step, i.e. q-integration,

$$\int \frac{dq}{2\pi} \int \frac{d\omega}{\pi} \psi^{(P)}(\boldsymbol{k}, \omega) \exp[i(\boldsymbol{k}\cdot\boldsymbol{v}-\omega)t]$$

$$= \frac{2\pi}{\kappa} e^{-2\kappa ut} - \frac{4\pi}{\kappa} e^{-2\kappa ut} \left\{1 + \frac{\kappa}{\pi} I[\varepsilon; \boldsymbol{\kappa}\cdot\boldsymbol{w} - i\kappa u]\right\}^{-1}. \qquad (5.151)$$

If we integrate this over $\boldsymbol{\kappa}$ we obtain the potential experienced by the unit external charge at $(\boldsymbol{r}-\boldsymbol{v}t, t)$. What we want to find is the (electrostatic potential) interaction energy of the configuration metal–external charge, i.e. the difference in potential energy of the system between the configurations $r = vt$ for finite $t<0$ and for $t \to -\infty$, i.e. $z \to \infty$. In other words, the energy spent against the electrostatic field in order to bring the charge from infinity up to a finite distance $\boldsymbol{r} = \boldsymbol{v}t$ where it experiences the potential ϕ of (5.147). From classical electrostatics this is

$$V = \frac{1}{2} \int \frac{d^2\kappa}{(2\pi)^2} \frac{2\pi\, e^{-2\kappa ut}}{\kappa} \left\{1 - \frac{2}{1 + (\kappa/\pi) I[\varepsilon; \boldsymbol{\kappa}\cdot\boldsymbol{w} - i\kappa u]}\right\}. \qquad (5.152)$$

This result takes a more suggestive form if it is rewritten so that $I[\varepsilon; \boldsymbol{\kappa}\cdot\boldsymbol{w} - i\kappa u]$ becomes $I[\varepsilon; \omega]$. It is easy to see by means of a contour integration that a formula equivalent to (5.152) is

$$V = \int_{-\infty}^{\infty} \frac{d\omega}{\pi} \int \frac{d^2\kappa}{2\pi\kappa} e^{-2\kappa ut} \frac{\kappa u}{(\omega - \boldsymbol{\kappa}\cdot\boldsymbol{w})^2 + \kappa^2 u^2} \left\{1 - \frac{2}{1 + (\kappa/\pi) I[\varepsilon; \omega]}\right\}. \qquad (5.153)$$

Changing the integration variables κ and ω into $-\kappa$ and $-\omega$ and using the property $\varepsilon(\boldsymbol{k}, \omega) = \varepsilon^*(-\boldsymbol{k}, -\boldsymbol{\omega})$, this can be rewritten as

$$V = 2\int_0^\infty \frac{d\omega}{2\pi}\int \frac{d^2\kappa}{2\pi\kappa} e^{-2\kappa ut} \frac{\kappa u}{(\omega - \boldsymbol{\kappa}\cdot\boldsymbol{w})^2 + \kappa^2 u^2} \times \mathrm{Re}\left\{1 - \frac{2}{1 + (\kappa/\pi)I[\varepsilon\,;\omega]}\right\}, \tag{5.154}$$

which shows that V is real.

Consider now the case in which the velocity of the external charge is parallel to the surface. Then, by taking the limit $u \to 0$, (5.153) becomes

$$V = \int_{-\infty}^\infty \frac{d\omega}{2\pi}\int \frac{d^2\kappa}{2\pi\kappa} e^{-2\kappa z_0}\pi\delta(\omega - \boldsymbol{\kappa}\cdot\boldsymbol{\omega})\left\{1 - \frac{2}{1 + (\kappa/\pi)I[\varepsilon\,;\omega]}\right\}. \tag{5.155}$$

This form is very suggestive indeed. Except for an irrelevant factor, the denominator appearing in (5.155) is just the function $\mathcal{B}(\boldsymbol{\kappa}, \omega)$ defined in (5.29), whose vanishing yields the normal modes of the semi-infinite metal with its surface. The δ-function factor clearly signals a matching condition between the phase velocity of the surface mode and the velocity of the external charge. In order to see this more clearly let us evaluate the work W done by the external charge against the field ϕ of (5.147). Indicating by $W_\parallel$ the case of parallel motion we have

$$W_\parallel = -\boldsymbol{E}\cdot\boldsymbol{v} = +\boldsymbol{v}\cdot\nabla\phi. \tag{5.156}$$

Following the same steps which lead to (5.152) we find

$$W_\parallel = \int \frac{d^2\kappa}{2\pi\kappa} e^{-2\kappa z_0}\, i\boldsymbol{\kappa}\cdot\boldsymbol{w}\left\{1 - \frac{2}{1 + (\kappa/\pi)I[\varepsilon\,;\boldsymbol{\kappa}\cdot\boldsymbol{w}]}\right\}, \tag{5.157}$$

which can be rewritten as

$$W_\parallel = \int_{-\infty}^\infty \frac{d\omega}{\pi}\int \frac{d^2\kappa}{2\kappa} e^{-2\kappa z_0}\,\delta(\omega - \boldsymbol{\kappa}\cdot\boldsymbol{w})\, i\omega\left\{1 - \frac{2}{1 + (\kappa/\pi)I[\varepsilon\,;\omega]}\right\}. \tag{5.158}$$

In spite of the imaginary factor it can be seen by again using the property $\varepsilon(\boldsymbol{k}\,;\omega) = \varepsilon^*(-\boldsymbol{k}, -\omega)$, that (5.158) is equal to

$$W_\parallel = -\int_0^\infty \frac{d\omega}{\pi}\int \frac{d^2\kappa}{\kappa} e^{-2\kappa z_0}\,\delta(\omega - \boldsymbol{\kappa}\cdot\boldsymbol{w})\omega\; \mathrm{Im}\left\{1 - \frac{2}{1 + (\kappa/\pi)I[\varepsilon\,;\omega]}\right\}. \tag{5.159}$$

In order to make this formula more suggestive it is convenient to reintroduce the factor $\hbar$. We then rewrite (5.159) as

$$W_\parallel = \int_0^\infty d\omega \int d^2\kappa\,\mathcal{P}(\boldsymbol{\kappa}, \omega)\hbar\omega, \tag{5.160}$$

where

$$\mathscr{P}(\boldsymbol{\kappa}, \omega) = -\frac{1}{\pi\kappa\hbar} e^{-2\kappa z_0} \delta(\omega - \boldsymbol{\kappa}\cdot\boldsymbol{\omega}) \operatorname{Im}\left\{1 - \frac{2}{1+(\kappa/\pi)I[\varepsilon;\omega]}\right\}. \quad (5.161)$$

The two factors which appeared in (5.155) appear now in (5.161) and (5.160) in a more transparent form. Notice that the imaginary part of the term in curly brackets arises precisely from the zeros of the denominators, i.e. from the normal modes in question. The physical interpretation of all this is now obvious. $\mathscr{P}(\boldsymbol{\kappa}, \omega)$ is in fact an excitation probability per unit time of the normal modes of the surface system on which the moving external charge is doing work at the rate $W_{\parallel}$, as expressed in (5.160).

Carrying further the quantum mechanical interpretation of the results of this simple classical analysis we could define a complex (potential) energy $V - \frac{1}{2}\mathrm{i}\Gamma$. The real part V is given by (5.155) while the imaginary part Γ is given by

$$\Gamma = \frac{\hbar}{\tau} = \hbar \int_0^\infty \mathrm{d}\omega \int \mathrm{d}^2\kappa\, \mathscr{P}(\boldsymbol{\kappa}, \omega). \quad (5.162)$$

We then have, in compact form, the complex energy

$$V - \mathrm{i}\frac{\Gamma}{2} = \int_0^\infty \frac{\mathrm{d}\omega}{\pi} \int \frac{\mathrm{d}^2\kappa}{2\pi\kappa} e^{-2\kappa z_0} \pi\delta(\omega - \boldsymbol{\kappa}\cdot\mathbf{w})\left\{1 - \frac{2}{1+(\kappa/\pi)I[\varepsilon;\omega]}\right\}.$$

In the standard way we can either regard Γ as measuring the level width caused by the interaction process, or τ – defined by (5.162) – as the lifetime of the state of the external charge, which decays via normal-mode excitations. This is only a crude argument, but it is very suggestive and, on the basis of a very simple model, brings out the main physical features while providing some formulae which can be used for plausible estimates.

Let us now discuss the opposite case, namely, that in which the motion of the external charge is perpendicular to the surface. The evaluation of the work done by the particle moving against the field in this case yields

$$W_{\perp} = \int \frac{\mathrm{d}^2\kappa}{2\pi\kappa} e^{-2\kappa u t}(-\kappa u)\left\{1 - \frac{2}{1+(\kappa/\pi)I[\varepsilon;-\mathrm{i}\kappa u]}\right\}. \quad (5.163)$$

If we look at (5.152) for $w = 0$, it is immediately obvious that $W_{\perp} = \mathrm{d}V/\mathrm{d}t$. It is instructive to discuss the meaning of this result. The total work done, per unit time, by the moving charge consists of two parts. One is the work done against the field $\boldsymbol{\phi}$ of (5.147), i.e. $2\,\mathrm{d}V/\mathrm{d}t$. The other is the work done by the charge in polarising the medium. This is equal to $-\mathrm{d}V/\mathrm{d}t$ and the total work involved in the motion of the charge is then $\mathrm{d}V/\mathrm{d}t$. In other words, the *dissipative* work of excitation of normal modes of the system is precisely $-W_{\perp}$ which, from (5.163), can be written as

$$-\int_0^\infty \frac{d\omega}{\pi}\int\frac{d^2\kappa}{\pi}e^{-2\kappa ut}\frac{u}{\omega^2+\kappa^2u^2}\omega\,\mathrm{Im}\left\{1-\frac{2}{1+(\kappa/\pi)I[\varepsilon;\omega]}\right\}.$$

From this we can obtain the complex potential energy $V-\frac{1}{2}i\Gamma$ which for perpendicular motion is

$$V-\tfrac{1}{2}i\Gamma=\int_0^\infty\frac{d\omega}{\pi}\int\frac{d^2\kappa}{2\pi}e^{-2\kappa ut}\frac{u}{\omega^2+\kappa^2u^2}\left\{1-\frac{2}{1+(\kappa/\pi)I[\varepsilon;\omega]}\right\}. \qquad (5.164)$$

Notice that this holds for $t<0$, when the moving charge is approaching the surface. The case $t>0$ and also the case of arbitrary velocity ($w\neq 0$, $u\neq 0$) can be studied by using the same method.

The above analysis amounts to a semiclassical discussion of the complex self energy for this problem. Knowing Γ as a function of time we could obtain the total probability of excitation of a normal mode after the charge has gone through the entire trajectory, by integrating $\Gamma(t)$. Alternatively, we can calculate directly the total work stressing once more that we only study the energy of the charge in the field (5.147) and do not get involved in the complete study of all possible interaction processes when the charge collides with the surface. We must then go back to (5.145) and notice that this time we must subtract not only the term in $\delta(\omega-\boldsymbol{\kappa}\cdot\boldsymbol{v})$, which represents the self energy of the moving charge, but also the term in $\delta(\omega-\boldsymbol{k}\cdot\tilde{\boldsymbol{v}})$ which, according to fig. 113, represents the interaction of the moving charge with the image charge moving along the image trajectory. The total work done against this interaction is identically zero after the external charge has returned to infinity. Thus, if we define

$$\Phi^{\mathrm{p}}(\boldsymbol{k},\omega)=\phi^{\mathrm{p}}(\boldsymbol{k},\omega)-(8\pi^2/k^2)[\delta(\omega-\boldsymbol{k}\cdot\boldsymbol{v})+\delta(\omega-\boldsymbol{k}\cdot\tilde{\boldsymbol{v}})], \qquad (5.165)$$

where ϕ^{p} is given by (5.145), the total work done by the charge is

$$\begin{aligned}-\int_{-\infty}^{\infty}&\boldsymbol{v}\cdot\boldsymbol{E}\,dt\\ &=\int d^3k\int d\omega\int dt\, i\boldsymbol{k}\cdot\boldsymbol{v}\Phi^{\mathrm{p}}(\boldsymbol{k},\omega)\exp[i(\boldsymbol{k}\cdot\boldsymbol{v}-\omega)t]\\ &=\int d^3k\int d\omega\, 2\pi i\boldsymbol{k}\cdot\boldsymbol{v}\Phi^{\mathrm{p}}(\boldsymbol{k},\omega)\,\delta(\omega-\boldsymbol{k}\cdot\boldsymbol{v}).\end{aligned} \qquad (5.166)$$

On inserting Φ^{p} in this expression we have an integration over q_1, according to (5.145). Doing first this integration and then the integration over q in the three-dimensional integral over $\boldsymbol{k}$ in (5.166) we finally obtain

$$-\int_{-\infty}^{\infty} \boldsymbol{v}\cdot\boldsymbol{E}\,\mathrm{d}t$$

$$= -\frac{2\mathrm{i}}{\pi^2}\int \mathrm{d}^2\kappa \int \mathrm{d}\omega \frac{u^2\kappa}{(\omega-\kappa\cdot\boldsymbol{w})^2+u^2\kappa^2}\,\frac{\omega}{1+(\kappa/\pi)I[\varepsilon\,;\omega]}$$

$$= \frac{4}{\pi^2}\int \mathrm{d}^2\kappa \int_0^{\infty} \mathrm{d}\omega\omega\ \mathrm{Im}\left\{\frac{u^2\kappa}{[(\omega-\boldsymbol{\kappa}\cdot\boldsymbol{w})^2+u^2\kappa^2]^2}\,\frac{1}{1+(\kappa/\pi)I[\varepsilon\,;\omega]}\right\}. \tag{5.167}$$

Introducing again a factor $\hbar$ explicitly we can express this in a form similar to (5.160) and (5.161), namely

$$\int_{-\infty}^{\infty} W\,\mathrm{d}t = \int \mathrm{d}^2\kappa \int_0^{\infty} \mathrm{d}\omega\,\mathcal{Q}(\kappa,\omega)\hbar\omega, \tag{5.168}$$

where

$$\mathcal{Q}(\boldsymbol{\kappa},\omega) = \frac{4}{\pi^2\hbar}\,\mathrm{Im}\left\{\frac{u^2\kappa}{[(\omega-\boldsymbol{\kappa}\cdot\boldsymbol{w})^2+u^2\kappa^2]^2}\,\frac{1}{1+(\kappa/\pi)I[\varepsilon\,;\omega]}\right\}. \tag{5.169}$$

This is now the integrated probability of excitation of the surface mode $(\boldsymbol{\kappa},\omega)$. The 'reversible' part of the work is retrieved as the external charge, originally at infinity, has receded to infinity again. The net total work done by the charge is now the total work of excitation of the normal modes, which appear as the zeros of the second denominator of (5.169) and give the imaginary part of the curly bracket.

As a simple illustration consider the now dispersive dielectric function of (5.50). Then

$$1-\frac{2}{1+(\kappa/\pi)I[\varepsilon\,;\omega]} = \frac{\frac{1}{2}\omega_\mathrm{p}^2}{(\omega+\mathrm{i}\eta)^2-\frac{1}{2}\omega_\mathrm{p}^2}. \tag{5.170}$$

The reason why in this case it is important to keep track of the infinitesimal imaginary part to be added to ω is that this tells us on which side of the real axis the poles of (5.170) are. For example, consider a unit charge at a distance $d = ut$, following a trajectory normal to the surface and approaching it with speed u. Then setting $\boldsymbol{w} = 0$ in (5.153), using (5.170) and closing the integration contour – for ω – through the upper half circle at infinity yields

$$V = -\int \frac{\omega_\mathrm{p}^2\,\mathrm{e}^{-2\kappa ut}}{2(\omega_\mathrm{p}^2+2\kappa^2u^2)}\,\mathrm{d}\kappa, \tag{5.171}$$

which describes an attractive interaction. Consider the asymptotic limit

$t \gg \omega_p^{-1}$, which means a long distance away. Then, if $\kappa u \gtrsim \omega_p$ we have $\kappa ut \gtrsim \omega_p t \gg 1$. In this range of values of κ the integrand is negligible because of the strong exponential decay. Thus it suffices for a crude estimate to make the approximation,

$$\frac{\omega_p^2}{\omega_p^2 + 2\kappa^2 u^2} \approx 1, \tag{5.172}$$

corresponding to $\kappa u \ll \omega_p$. Then (5.171) yields the classical image force result

$$V \approx -1/4d. \tag{5.173}$$

Corrections of order d^{-2}, d^{-4} etc. could be obtained by taking further terms in the expansion of the l.h.s. of (5.172) in powers of $(\kappa u/\omega_p)^2$. However, there would not be much point in further refinements of this formula because it is the model itself that would require a fair amount of improvement for an accurate and more complete study of this problem. Our purpose is only to stress the basic physical facts which underline the theory of the various types of interactions involving metal surface. In particular, with the non-dispersive model of (5.50), we have

$$\left\{1 + \frac{\pi}{\kappa} I[\varepsilon;\omega]\right\}^{-1} = \frac{\varepsilon}{1+\varepsilon} = \frac{(\omega + i\eta)^2 - \omega_p^2}{2(\omega + i\eta)^2 - \omega_p^2}, \tag{5.174}$$

whence

$$\begin{aligned} \operatorname{Im}\left\{1 + \frac{\pi}{\kappa}[\varepsilon;\omega]\right\} &= -\pi\left(\frac{\omega_p^2}{2} - \omega_p^2\right)\delta(2\omega^2\omega_p^2) \\ &= +\frac{\pi\omega_p}{4\surd 2}\,\delta\left(\omega - \frac{\omega_p}{\surd 2}\right). \end{aligned} \tag{5.175}$$

Using this in (5.169) the integrated excitation probability becomes

$$\mathcal{Q}(\kappa, \omega) = \frac{\omega}{\pi\hbar\surd 2}\,\frac{\kappa u^2}{[(\omega - \boldsymbol{\kappa}\cdot\boldsymbol{w})^2 + \kappa^2 u^2]^2}\,\delta\left(\omega - \frac{\omega_p}{\surd 2}\right). \tag{5.176}$$

The physical interpretation of this result is obvious. If we used a more elaborate dielectric function, then the denominator of (5.174) would be the product of terms of the form $[\omega + i\eta - \Omega(\boldsymbol{\kappa}, q_n)]^2$ and, after decomposition into simple fractions, we would have a result of the form

$$\left\{1 + \frac{\pi}{\kappa} I[\varepsilon;\omega]\right\}^{-1} = \sum_n \frac{N_n(\boldsymbol{\kappa};\omega)}{[\omega + i\eta - \Omega(\boldsymbol{\kappa}, q_n)]^2}, \tag{5.177}$$

whence

$$\mathrm{Im}\left\{1+\frac{\pi}{\kappa}I[\varepsilon;\omega]\right\}^{-1}$$

$$= -\sum_n \pi N_n[\boldsymbol{\kappa};\omega=\Omega(\boldsymbol{\kappa},q_n)]\delta[\omega-\Omega(\boldsymbol{\kappa},q_n)]. \qquad (5.178)$$

For example, the only term of (5.175) corresponding to the only surface normal mode of the non-dispersive model is

$$-N_n[\boldsymbol{\kappa};\omega=\Omega(\boldsymbol{\kappa},q_n)] = +\omega_p/4\sqrt{2}.$$

Notice that the sign in the end is positive, corresponding to a probability. It is clear that using (5.178) in (5.169) yields a normal-mode analysis like those of §§ 5.2.1–5.2.3.

5.3. The random-phase approximation for model surface systems

The semiclassical approximation has been profusely used in § 5.2 to study a range of typical surface problems. The essence of the semiclassical approximation can be seen graphically by reference to fig. 16*a*, which can be understood in general terms as describing the construction of any response function such as, for example, the non-local polarisability. The point is that in quantum mechanics there is an interference between the direct and reflected waves. We now want to see what happens when we go beyond the semiclassical approximation in which this interference is neglected. The r.p.a. represents a positive step in this direction. It still has some serious limitations, and we saw for example in chapter 3 that some problems necessitate going beyond this. However, for the study of dynamical problems, the r.p.a. is for the time being the only approximation which does take quantum effects into account while being amenable to general use as a systematic tool for the study of several surface problems. So far we have simply made occasional use of the r.p.a. formula for the polarisability, without further enquiry. We must now study this in greater detail, for which we shall recall some general facts of the theory of the interacting electron gas while setting out some formal aspects which are directly relevant to the analysis which will follow.

Consider, for example, the formula (5.8). This gives the polarisability in general, the degree of accuracy depending on the approximation used to construct the eigenstates $|n\rangle$. Suppose we construct some approximate one-electron Hamiltonian H_0 of the form

$$H_0 = -\sum_i \tfrac{1}{2}\nabla_i^2 + \sum_i V_0(\mathbf{r}_i).$$

For example, for the homogeneous electron gas, V_0 could be an average one-electron potential. For the bounded electron gas it could be some approximate local one-electron potential constructed by the methods of chapter 3. Suppose we then construct the eigenstates of the N-electron system as Slater determinants of the one-electron eigenstates of H_0. We are then working within the framework of the r.p.a. In the ground state all one-electron energies E_α are less than E_F, while in the excited states some electrons are excited to one-electron energy levels $E_\beta > E_F$. With these states one finds, remembering (5.3), and writing $\psi(r)$ for the one-electron eigenfunctions of H_0,

$$\langle 0|\rho(\boldsymbol{r})|n\rangle = \psi_\alpha^*(\boldsymbol{r})\psi_\beta(\boldsymbol{r}) \quad (E_\alpha < E_F < E_\beta).$$

Using this in (5.8) yields the well-known formula (remember $\hbar = 1$)

$$\chi_0(\boldsymbol{r}, \boldsymbol{r}'; \omega^+) = e^2 \sum_{\substack{E_\alpha < E_F \\ E_\beta > E_F}} \left\{ \frac{\psi_\beta(\boldsymbol{r})\psi_\alpha^*(\boldsymbol{r})\psi_\beta^*(\boldsymbol{r}')\psi_\alpha(\boldsymbol{r}')}{\omega - (E_\beta - E_\alpha) + i\eta} - \frac{\psi_\beta^*(\boldsymbol{r})\psi_\alpha(\boldsymbol{r})\psi_\beta(\boldsymbol{r}')\psi_\alpha^*(\boldsymbol{r}')}{\omega + (E_\beta - E_\alpha) + i\eta} \right\}. \tag{5.179}$$

This is the r.p.a. result for χ. We stress that this need not be restricted to a homogeneous system. The ψs could be some approximate one-electron wavefunctions of a given model surface system, as we shall see later.

Now, this is the response function in the *retarded-response formalism*, which so far we have used throughout. It is sometimes convenient to work in the *time-ordered formalism*, in which the Green function of the independent electron Hamiltonian H_0 is

$$\hat{G}_0(\boldsymbol{r}, \boldsymbol{r}'; \omega) = \sum_{E_\alpha < E_F} \frac{\psi_\alpha(\boldsymbol{r})\psi_\alpha^*(\boldsymbol{r}')}{\omega - E_\alpha - i\eta} + \sum_{E_\alpha > E_F} \frac{\psi_\alpha(\boldsymbol{r})\psi_\alpha^*(\boldsymbol{r}')}{\omega - E_\alpha + i\eta}. \tag{5.180}$$

In this formalism it is easier to prove some useful general results. For example, it is easy to see using (5.180) that

$$\chi_0(\boldsymbol{r}, \boldsymbol{r}'; \omega) = -\frac{ie^2}{2\pi} \int \hat{G}_0(\boldsymbol{r}, \boldsymbol{r}'; \omega')\hat{G}_0(\boldsymbol{r}', \boldsymbol{r}; \omega' - \omega)\, d\omega', \tag{5.181}$$

where

$$\chi_0(\boldsymbol{r}, \boldsymbol{r}'; \omega) = e^2 \sum_{\substack{E_\alpha < E_F \\ E_\beta > E_F}} \left\{ \frac{\psi_\beta(\boldsymbol{r})\psi_\alpha^*(\boldsymbol{r})\psi_\beta^*(\boldsymbol{r}')\psi_\alpha(\boldsymbol{r}')}{\omega - (E_\beta - E_\alpha) + i\eta} - \frac{\psi_\beta^*(\boldsymbol{r})\psi_\alpha(\boldsymbol{r})\psi_\beta(\boldsymbol{r}')\psi_\alpha^*(\boldsymbol{r}')}{\omega + (E_\beta - E_\alpha) - i\eta} \right\}. \tag{5.182}$$

This is the time-ordered susceptibility evaluated in the same approximation as the retarded susceptibility $\chi_0(\omega^+)$ of (5.179). The passage from one formalism to the other is very easy in practice. Comparing (5.182) and (5.179) all we need is an infinitesimal shift in the poles of the second term, so that they appear on the desired side of the real axis. With these preliminaries we can now study surface problems. In general these would require numerical computation, so we shall restrict ourselves to the infinite-barrier model which is amenable to a discussion in analytical terms.

5.3.1. Application to the infinite-barrier model

In this case, if we again assume a sufficiently large but finite size L in the z-direction, into the bulk, the one-electron wavefunctions are of the form

$$\psi_\alpha(r) = \sqrt{\frac{2}{L}}\,\frac{e^{iqz} - e^{-iqz}}{2i}\,e^{i\boldsymbol{\kappa}\cdot\boldsymbol{\rho}},$$

which used in (5.180) yield the time-ordered Green function of the surface system in the r.p.a. On doing this we have the factors

$$(e^{iqz} - e^{-iqz})(e^{iqz'} - e^{-iqz'})$$
$$= e^{iq(z+z')} - e^{iq(z'-z)} - e^{iq(z-z')} + e^{-iq(z+z')}.$$

The first and fourth terms give the same contribution to the summation indicated in (5.180), and similarly for the second and third terms. Omitting the label 0 this yields

$$\hat{G}_s(\mathbf{r}, \mathbf{r}'; \omega) = \sum_{\boldsymbol{k}} \frac{e^{i\boldsymbol{\kappa}\cdot(\boldsymbol{\rho}-\boldsymbol{\rho}')}\,e^{iq(z-z')}}{2L(\omega - \omega_{\boldsymbol{k}})} - \sum_{\boldsymbol{k}} \frac{e^{i\boldsymbol{\kappa}\cdot(\boldsymbol{\rho}-\boldsymbol{\rho}')}\,e^{iq(z+z')}}{2L(\omega - \omega_{\boldsymbol{k}})}, \tag{5.183}$$

where

$$\omega_k = \begin{Bmatrix} E_k + i\eta;\, E_k < E_F, \\ E_k - i\eta;\, E_k > E_F, \end{Bmatrix} \tag{5.184}$$

and both $\boldsymbol{r}$ and $\boldsymbol{r}'$ are inside the metal. Let us indicate by $\tilde{\boldsymbol{r}}$ the mirror-image point $(\boldsymbol{\rho}, -z)$ of a point $\boldsymbol{r}(\boldsymbol{\rho}, z)$. Then, comparing (5.183) and (5.180) we find for the time-ordered Green function of the hard-wall surface system in the r.p.a.,

$$\hat{G}_s(\boldsymbol{r}, \boldsymbol{r}'; \omega) = \hat{G}_b(\boldsymbol{r} - \boldsymbol{r}'; \omega) - \hat{G}_b(\boldsymbol{r} - \tilde{\boldsymbol{r}}'; \omega), \tag{5.185}$$

where $\hat{G}_b$ is the bulk Green function. In fact this formula need not be restricted to the r.p.a. derivation here outlined. Indeed we could go back to the surface Green-function matching formalism of chapter 2 and recall (2.39), which gives the generic matrix element between two points inside the medium. The point is that the surface projection G_s vanishes for the infinite-barrier model, since all wavefunctions must then vanish at $z=0$. The reflection operator $\mathcal{R}$ (2.74) is then exactly equal to *minus* the unit operator; note that, consistent with this $\mathcal{T}$ (2.78) vanishes. Then (2.39) directly yields (5.185), since the formalism of chapter 2 applies indifferently to advanced, retarded, or time-ordered Green functions.

At this stage we encounter the quantum mechanical interference terms. This happens on using (5.185) in (5.181), which yields

$$\chi_0^\infty(\mathbf{r},\mathbf{r}';\omega)=-\frac{ie^2}{2\pi}\int[\hat{G}_b(\mathbf{r}-\mathbf{r}';\omega)-\hat{G}_b(\mathbf{r}-\tilde{\mathbf{r}}';\omega)]$$
$$\times[\hat{G}_b(\mathbf{r}'-\mathbf{r};\omega'-\omega)-\hat{G}_b(\mathbf{r}'-\tilde{\mathbf{r}};\omega'-\omega)]\,d\omega' \qquad (5.186)$$

for the polarisability of the hard-wall surface system where we recall that $z, z'>0$. Studying the four terms which arise on developing the integrand and comparing with (5.181) again, this time applied to the bulk system, we find

$$\chi_0^\infty(\mathbf{r},\mathbf{r}';\omega)=\chi_b(\mathbf{r}-\mathbf{r}';\omega)+\chi_b(\mathbf{r}-\tilde{\mathbf{r}}';\omega)+\Delta\chi(\mathbf{r},\mathbf{r}';\omega), \qquad (5.187)$$

where

$$\Delta\chi(\mathbf{r},\mathbf{r}';\omega)=\frac{ie^2}{2\pi}\int[G_b(\mathbf{r}-\mathbf{r}';\omega)G_b(\mathbf{r}'-\tilde{\mathbf{r}};\omega'-\omega)$$
$$+G_b(\mathbf{r}-\tilde{\mathbf{r}}';\omega')G_b(\mathbf{r}-\mathbf{r}';\omega'-\omega)]\,d\omega'. \qquad (5.188)$$

If we neglect (5.188) then (5.187) leads us back to the semiclassical approximation of chapter 1 – for the specular case – and § 5.2. The interference term is precisely $\Delta\chi$.

It is obvious, on the other hand, that $\chi_0^\infty(\mathbf{r},\mathbf{r}')$ vanishes for $\mathbf{r}$ and $\mathbf{r}'$ on different sides of the surface. This follows directly from (2.40) by using it in (5.181) applied to the hard-wall surface system, since $\hat{G}_s(\mathbf{r},\mathbf{r}')$ vanishes in this case. The position is then the following: equation (5.6) *defines* χ, while the r.p.a. polarisability χ_0 is defined by

$$e\rho^i(\mathbf{r},\omega)=\int\chi_0(\mathbf{r},\mathbf{r}';\omega)V(\mathbf{r}',\omega)\,\mathbf{dr}', \qquad (5.189)$$

where

$$V(\mathbf{r}',\omega)=V^e(\mathbf{r}',\omega)+V^i(\mathbf{r}',\omega). \qquad (5.190)$$

In particular, χ_0 is χ_b for the bulk system, or χ_0^∞ (5.187) for the hard-wall surface system. If we want to evaluate $\rho^i(\boldsymbol{r}, t)$ for $z > 0$ – i.e. inside the semi-infinite metal – then from (5.189) and (5.187) we have

$$e\rho^i(\boldsymbol{\rho}, z > 0) = \int_{z'>0} [\chi_b(\boldsymbol{r}-\boldsymbol{r}') + \chi_b(\boldsymbol{r}-\tilde{\boldsymbol{r}}') + \Delta\chi(\boldsymbol{r}, \boldsymbol{r}'; \omega)] V(\boldsymbol{r}') \, \mathbf{d}\boldsymbol{r}'. \quad (5.191)$$

It is more useful to rewrite this by once more using the method of extended pseudomedia. Putting the vacuum in $z < 0$ and the metal in $z > 0$ we define the symmetrized potentials $V_+(\boldsymbol{r})$ and $V_+^i(\boldsymbol{r})$ which are equal to the real potentials V and V^i for $z > 0$, and the also symmetrised potentials $V_-(\boldsymbol{r})$ and $V_-^i(\boldsymbol{r})$ which are equal to the real potentials V and V^i in $z < 0$. We furthermore define

$$\delta\chi(\boldsymbol{r}, \boldsymbol{r}'; \omega) = \frac{ie^2}{2\pi} \int G_b(\boldsymbol{r}-\boldsymbol{r}'; \omega') \times G_b(\boldsymbol{r}'-\tilde{\boldsymbol{r}}; \omega'-\omega) \, d\omega' \quad (z, z' > 0), \quad (5.192)$$

and use this instead of $\Delta\chi$ (5.188). We can then rewrite (5.191) as

$$e\rho^i(\boldsymbol{\rho}, z > 0) = \int \chi_b(\boldsymbol{r}-\boldsymbol{r}') V_+(\boldsymbol{r}') \, \mathbf{d}\boldsymbol{r}' + \int \delta\chi(\boldsymbol{r}, \boldsymbol{r}') V_+(\boldsymbol{r}') \, \mathbf{d}\boldsymbol{r}', \quad (5.193)$$

where both integrals cover the entire space. We must now find V_+, which we do by matching the extended pseudomedia as usual. If $Q^e(\boldsymbol{r})$ is an external charge density distribution in $z > 0$ we define likewise the extended symmetrised distribution $Q_+^e(\boldsymbol{r})$. We then have, in Fourier transform and taking the ω-dependence as understood where not shown explicitly,

$$V_+(\boldsymbol{k}) = \frac{4\pi}{k^2} \chi_b(\boldsymbol{k}) V_+(k) + \frac{4\pi}{k^2} \int \delta\chi(\boldsymbol{\kappa}, q, q_2) V_+(\boldsymbol{k}_2) \, dq_2 + \frac{4\pi}{k^2} \sigma(\boldsymbol{\kappa}) + \frac{4\pi}{k^2} Q_+^e(\boldsymbol{k}). \quad (5.194)$$

The first two terms represent the potential of the induced charge (5.193) and $\delta\chi(\boldsymbol{k}, q, q_2)$ indicates the matrix element $\langle \boldsymbol{\kappa}, q | \delta\chi | \boldsymbol{\kappa}, q_2 \rangle$. For the extended pseudovacuum

$$V_-(\boldsymbol{\kappa}) = -\frac{4\pi}{k^2} \sigma(\boldsymbol{\kappa}) + \frac{4\pi}{k^2} Q_-^e(\boldsymbol{k}), \quad (5.195)$$

where we have also symmetrised whatever external charge there may be in $z<0$. Matching the real potentials at $z=0$ yields σ and using this back in (5.195) we find

$$V_+(\boldsymbol{k})=\frac{4\pi}{k^2}\chi_b(\boldsymbol{k})V_+(\boldsymbol{k})+\frac{4\pi}{k^2}\int\delta\chi(\boldsymbol{\kappa},q,q_1)V_+(q_1)\,\mathrm{d}q_1$$

$$+\frac{4\pi}{k^2}Q^e_+(\boldsymbol{k})+\frac{2\kappa}{k^2}\int\mathrm{d}q_1$$

$$\times\left[\frac{Q^e_-(\boldsymbol{k}_1)}{k_1^2}-\frac{Q^e_+(\boldsymbol{k}_1)}{k_1^2}-\frac{\chi_b(\boldsymbol{k}_1)}{k_1^2}V_+(\boldsymbol{k}_1)\right.$$

$$\left.-\frac{1}{k_1^2}\int\mathrm{d}q_2\,\delta\chi(\boldsymbol{\kappa},q_1,q_2)V_+(\boldsymbol{k}_2)\right].\tag{5.196}$$

This contains the (symmetrised) potential of the external charges

$$V^e_+(\boldsymbol{k})=\frac{4\pi}{k^2}Q^e_+(\boldsymbol{k})-\frac{2\kappa}{k^2}\int\frac{\mathrm{d}q_1}{k_1^2}[-Q^e_-(\boldsymbol{k}_1)+Q^e_+(\boldsymbol{k}_1)].\tag{5.197}$$

Separating out this term in (5.196) and recalling the well-known relation

$$\varepsilon_b(\boldsymbol{k},\omega)=1-\frac{4\pi}{k^2}\chi_b(\boldsymbol{k},\omega),\tag{5.198}$$

we can rewrite the basic integral equation (5.196) as

$$\varepsilon_b(\boldsymbol{k})V_+(\boldsymbol{k})=V^e_+(\boldsymbol{k})+\frac{4\pi}{k^2}\int\delta\chi(q,q_1)V_+(\boldsymbol{k}_1)\,\mathrm{d}q_1$$

$$-\frac{2\kappa}{k^2}\int\mathrm{d}q_1\left[\frac{\chi_b(\boldsymbol{k}_1)}{k_1^2}V_+(\boldsymbol{k}_1)\right.$$

$$\left.+\frac{1}{k_1^2}\int\mathrm{d}q_2\,\delta\chi(q_1,q_2)V_+(\boldsymbol{k}_2)\right].\tag{5.199}$$

The ω-dependence is understood throughout and $\delta\chi(q,q_1)$ is, for fixed $\boldsymbol{\kappa}$, the matrix element $\langle k|\delta\chi|\boldsymbol{k}_1\rangle$. If, on the other hand, we subtract (5.197) from both sides of (5.196), we have V^i_+ as a function of V_+, which yields

$$\frac{\delta V^i_+(\boldsymbol{k})}{\delta V_+(\boldsymbol{k}_1)}=\frac{4\pi}{k^2}\chi_b(\boldsymbol{k})\,\delta(q-q_1)+\frac{4\pi}{k^2}\,\delta\chi(q,q_1)$$

$$-\frac{2\kappa}{k^2}\frac{\chi_b(\boldsymbol{k}_1)}{k_1^2}-\frac{2\kappa}{k^2}\int\frac{\mathrm{d}q_2}{k_2^2}\,\delta\chi(q_2,q_1).\tag{5.200}$$

This formula will be used later.

5.3.2. Static dielectric response to an external charge

As a case of physical interest let us study the response of the semi-infinite metal to an external charge. In principle we must solve the integral equation (5.199), which cannot be done analytically. The difficulties are due to the interference term $\delta\chi$, which adds a non-separable term to the kernel of the basic integral equation, unlike the terms in χ_b, which are separable, i.e. in real space they depend only on the distance $|\boldsymbol{r}-\boldsymbol{r}'|$. An accurate solution of the problem would require considerable numerical computation. Our purpose is to try and see whether some reasonable approximation can be devised so that the kernel of the integral equation becomes separable and the problem can be discussed analytically. We shall stick to the infinite-barrier model. In order to obtain a clue it is useful to first discuss some general properties of $\delta\chi$.

Let us look at (5.192) for $\boldsymbol{r}=\boldsymbol{r}_s=(\boldsymbol{\rho}, z=0)$, a point on the surface. Remember that our χ_b is given by (5.181) with $\hat{G}_b$ used for $\hat{G}_0$, and remember that $\hat{G}_b$ is a function of $|\boldsymbol{r}-\boldsymbol{r}'|$. Then

$$\delta\chi(\boldsymbol{r}_s, \boldsymbol{r}'; \omega)=\frac{\mathrm{i}}{2\pi}\int \hat{G}_b(|z'|; \hat{G}_b(|z'|; \omega'-\omega)\,\mathrm{d}\omega'$$

$$= -\chi_b(\boldsymbol{r}_s, \boldsymbol{r}'; \omega). \tag{5.201}$$

Similarly

$$\delta\chi(\boldsymbol{r}, \boldsymbol{r}_s'; \omega)=\frac{\mathrm{i}}{2\pi}\int \hat{G}_b(|z|; \omega')\hat{G}_b(|z|; \omega'-\omega)\,\mathrm{d}\omega'$$

$$=-\chi_b(\boldsymbol{r}, \boldsymbol{r}_s'; \omega). \tag{5.202}$$

Expressing the l.h.s. of (5.201) and (5.202) as obtained from the evaluation of the Fourier transform for $z=0$ these imply, *for the infinite-barrier model,*

$$\left.\begin{aligned} \int_{-\infty}^{\infty}\frac{\mathrm{d}q_1}{2\pi}\,\delta\chi(q, q_1; \omega) &= -\chi_b(\boldsymbol{k}, \omega);\\ \int_{-\infty}^{\infty}\frac{\mathrm{d}q}{2\pi}\,\delta\chi(q, q_1; \omega) &= -\chi_b(\boldsymbol{k}_1; \omega). \end{aligned}\right\} \tag{5.203}$$

What does this mean? Suppose we evaluate (5.193) using (5.203). Then the induced-charge density is zero at $z=0$, as indeed it must at a hard wall. This suggests that we can try for $\delta\chi$ a separable expression of the form

$$\delta\chi(q, q_1)=-2\pi\chi_b(q)\chi_b(q_1)\Big/\int \chi_b(q_2)\,\mathrm{d}q_2. \tag{5.204}$$

We know this is only approximate, but the point is that (5.204) satisfies the key condition (5.203) and we may expect this approximation to be reasonable. The great advantage is that with this separable kernel the integral equation (5.199) can be solved analytically. In fact it suffices to try an expression of the form

$$V_+(\boldsymbol{k}) = \frac{V_+^{\mathrm{e}}(\boldsymbol{k})}{\varepsilon(\boldsymbol{k})} + A\frac{4\pi}{k^2\varepsilon(\boldsymbol{k})} + B\frac{4\pi}{k^2\varepsilon(\boldsymbol{k})}\chi_{\mathrm{b}}(\boldsymbol{k}), \tag{5.205}$$

where A and B are obtained by directly inserting (5.205) in (5.199).

As an example suppose we put an external point charge at $z = -a$, outside the metal, and we want to evaluate the potential $V_-(z)$ met by the unit external charge. All we have to do is to insert σ, obtained from the matching of (5.194) and (5.195), back into (5.195). Then we again face the difficulty of the non-separability of $\delta\chi$, but if we use the approxima-

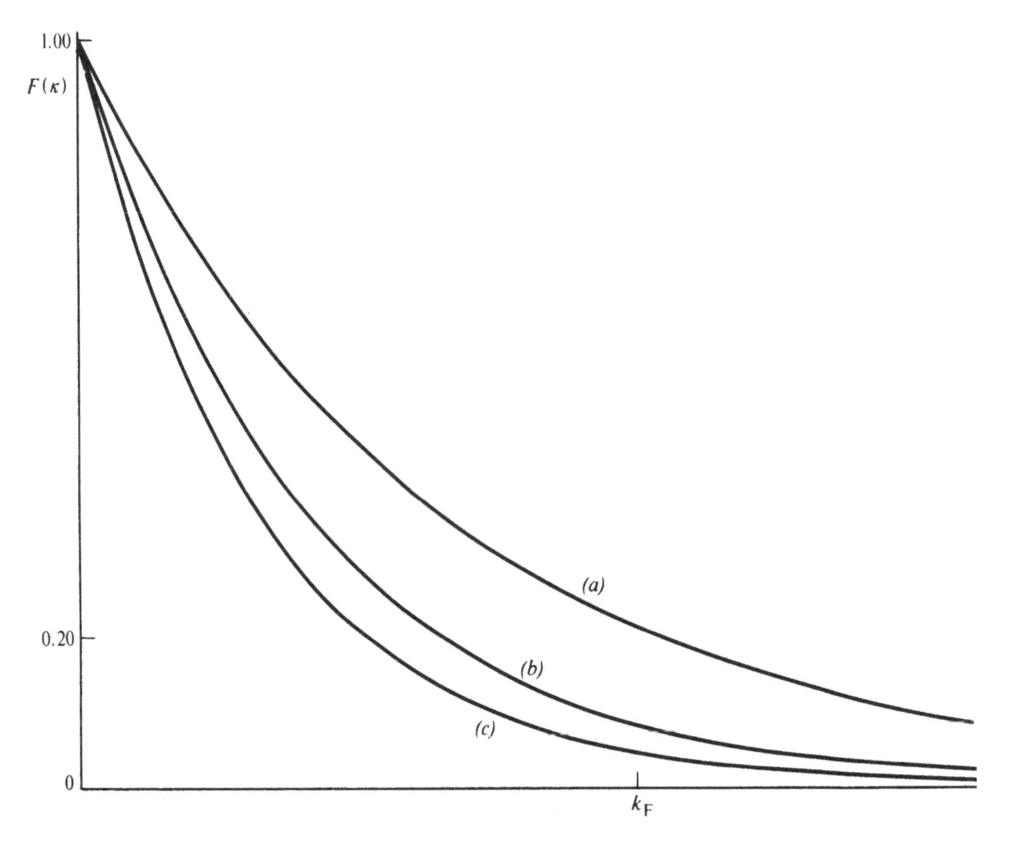

Fig. 114. The function $F(\kappa)$ of (5.206) for different approximatons: (*a*) semiclassical; (*b*) analytical, by using (5.204); (*c*) full numerical computation. D. E. Beck & V. Celli, *Phys. Rev.* B **2**, 2955 (1970).

tion (5.204) then the calculation can be done analytically. The accuracy of the results can be described in terms of the function

$$F(\kappa)=\mathrm{e}^{a\kappa}\sigma(\kappa)-1, \tag{5.206}$$

because it turns out that $\sigma(\kappa)$ varies as $\mathrm{e}^{-a\kappa}$. The results are shown in fig. 114 for the case $r_s=3.4$. It is seen that the approximate form (5.204) represents a considerable improvement upon the semiclassical model. The results are not as accurate as those of a full numerical computation, but they can be obtained by hand, which amounts to a considerable simplification. A better approximation, which works rather well in practice, can be obtained because, *a posteriori*, it turns out that the function $\delta\chi(q, q')$ falls off very sharply for $|q|$ or $|q'|$ approaching $2k_F$. This suggests another trial form, namely

$$\delta\chi(q,q')=\alpha\theta[2k_F-|q|]\theta[2k_F-|q'|]\theta[2k_F-\kappa]. \tag{5.207}$$

This does not guarantee complete vanishing of $\rho^{i}(z=0)$, but the error thus introduced is very small, while the accuracy of the approximation can be improved by adjusting α for each value of r_s. In fact curve (*b*) of fig. 114 then becomes almost equal to (*c*). The situation is further illustrated in fig. 115 which shows, again for $r_s=3.4$, the potential V met by an external unit charge when it is located at different positions outside or inside the metal. Altogether (5.207) represents a considerable improvement upon the semiclassical approximation. These examples

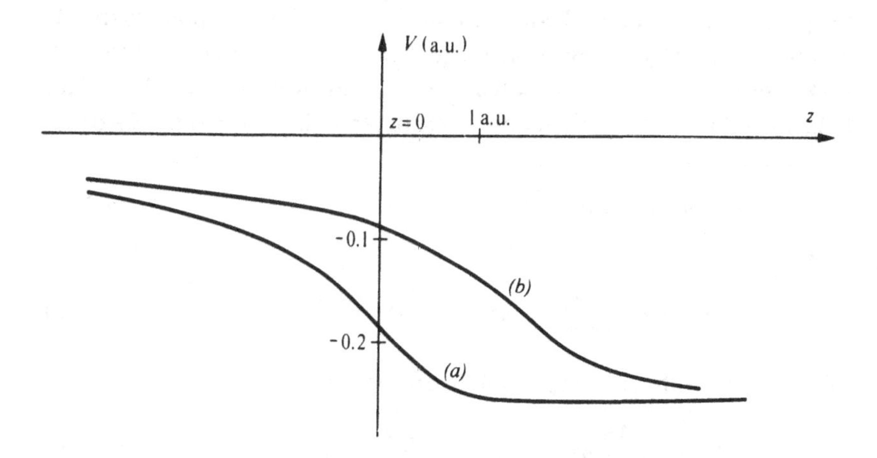

Figure 115. Potential encountered by an external unit charge as a function of its position, both outside ($z<0$) and inside ($z>0$) the metal: (*a*) semiclassical approximation; (*b*) calculated within the approximation described in the text.

suffice to illustrate the situation. A proper account of quantum interference effects requires numerical computation because we come up against an integral equation with a non-separable kernel. However, plausible analytical approximations can be devised which yield a reasonable picture. This is equivalent to having obtained a reasonable approximation to the polarisability of the surface system, which is the basic object in terms of which different physical problems can be studied.

5.3.3. Dynamical response: surface plasmons

We have discussed the static case, $\omega = 0$. Let us now briefly look at the case $\omega \neq 0$, taking the study of surface plasmons as an example. These were discussed at length in chapter 1, where the emphasis was on the model – specular or otherwise – used for the scattering of electrons off the surface. While the analysis of chapter 1 was semiclassical we want to focus now on the specific quantum effects, for which we take a specular surface model and concentrate on the effect of the interference term. We shall discuss the quasistatic limit and also the long-wave limit, which is the case of main experimental interest.

The problem now is to solve the integral equation (5.199), this time with $V_+^e = 0$ and $\omega \neq 0$ It is therefore an eigenvalue problem. The eigenvalue is the s.m.d.r. $\omega = \omega(\kappa)$. The difficulty again is that the kernel of this integral equation is non-separable, but there is something we can do in the long-wave limit. Notice that the analysis involves charges and potentials and we have chosen to eliminate the former and to formulate the problem as an integral equation for the potential. The point is that V varies much more smoothly than ρ and therefore we are better off trying a power-series expansion to solve the integral equation for V in the long-wave limit. Let us then define a separable linear operator L_0 by its matrix elements

$$\langle q|L_0|q_1\rangle = -\langle q|a\rangle\langle b|q_1\rangle, \tag{5.208}$$

where

$$\langle q|a\rangle = a(q) = \frac{2\kappa}{k^2}; \quad \langle b|q_1\rangle = b(q_1) = \frac{1-\varepsilon(\boldsymbol{k}_1, \omega)}{4\pi}. \tag{5.209}$$

We also define the non-separable operator L_1 by

$$\langle q|L_1|q_1\rangle = \frac{4\pi}{k^2}\,\delta\chi(q, q_1) - \frac{2\kappa}{\pi}\int \frac{\mathrm{d}q_2}{k_2^2}\,\delta\chi(q_2, q_1). \tag{5.210}$$

Then the integral equation can be written in operator form as

$$\varepsilon|V_+\rangle = L_0|V_+\rangle + L_1|V_+\rangle, \tag{5.211}$$

whence

$$|V_+\rangle=(1-\varepsilon^{-1}L_1)^{-1}\varepsilon^{-1}L_0|V_+\rangle. \tag{5.212}$$

Taking the scalar product with $\langle b|$ and recalling (5.208), we find that the eigenvalue equation is

$$1=\langle b|(1-\varepsilon^{-1}L_1)^{-1}\varepsilon^{-1}|a\rangle. \tag{5.213}$$

This can now be expanded as

$$1=\langle b|\varepsilon^{-1}|a\rangle+\langle b|\varepsilon^{-1}L_1\varepsilon^{-1}|a\rangle+\ \ldots, \tag{5.214}$$

where the successive approximations can be studied analytically and the numerical evaluation is very simple. The lowest approximation is

$$1=\langle b|\varepsilon^{-1}|a\rangle. \tag{5.215}$$

By using (5.209) it is trivial to see that this again yields the semiclassical approximation (1.208) for the case $p=1$, $c\to\infty$. Of course, in the lowest approximation we take no account of the interference term and it is obvious that (5.215) must yield the semiclassical result. In itself (5.215) is formally valid for arbitrary κ, but this only means that the integral equation can be solved exactly if the interference term is totally neglected. The power series expansion of (5.214) amounts in practice to an expansion in powers of κ. It is thus a device to include in an approximate manner the effects of quantum mechanical interference on the dispersion of long-wave surface plasmons. It should be stressed that even if one studied only the long-wave limit, i.e. the frequency of the surface plasmon to first order in κ, the inclusion of interference effects requires a full numerical solution of the integral equation. The second term on the r.h.s. of (5.214) can be easily evaluated by hand and it turns out to give quite good results compared with those of full numerical computations.

The picture obtained from this analysis has interesting connections with the discussion of § 1.8, and in particular with the formula (1.272), which relates the slope of the long-wave dispersion relation to the dipole moment of the charge density fluctuation of the plasma wave itself. The successive stages of approximation are described in fig. 116. In the semiclassical approximation (*a*), the jellium edge is at the surface barrier. The long-wave dispersion relation then is

$$\omega=(\omega_p/\sqrt{2})\{1+F_{sc}\kappa\},$$

where F_{sc} is a positive factor of the order of $(E_F/\hbar\omega_p)$. The first quantum correction amounts to displacing the jellium edge through a distance $3\pi/8k_F$ without changing $\delta\rho(z)$ (fig. 116*b*). The origin $z=0$ for the

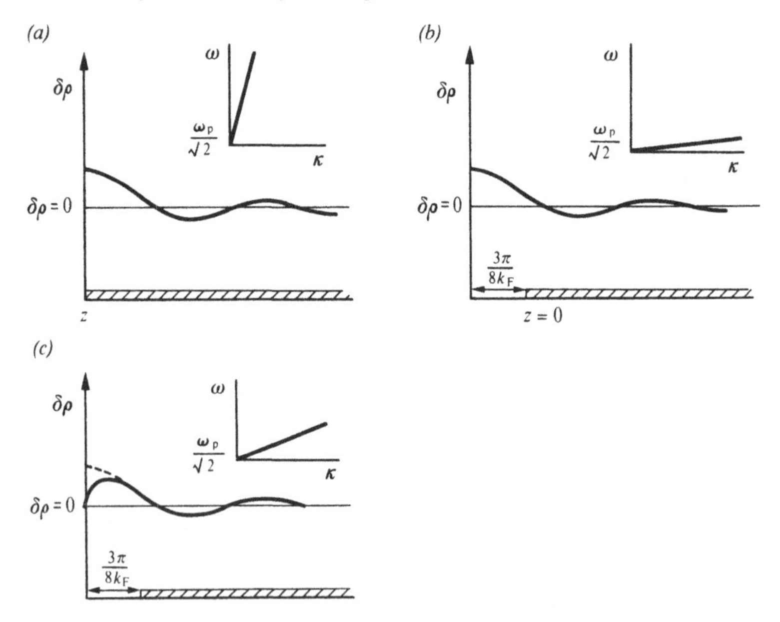

Fig. 116. Successive stages of approximation in the power expansion approach to the study of the long dispersion of surface plasmons. (*a*) Semiclassical. (*b*) Jellium edge withdrawn from hard-wall barrier. Electronic charge density still as in semiclassical approximation. (*c*) Proper quantum mechanical treatment of the electronic wavefunctions causes their amplitude, and hence $\delta\rho$, to be zero at the hard-wall barrier.

application of (1.272) follows the shift of the jellium edge and (fig. 116*b*) the slope of the long-wave dispersion relation is then considerably lowered, this relation becoming

$$\omega = \frac{\omega_p}{\sqrt{2}}\left\{1+\left(F_{sc}-\frac{1}{2}\frac{3\pi}{8k_F}\right)\kappa\right\}.$$

A smaller correction comes from a proper quantum mechanical treatment of the electronic wavefunctions. Then $\delta\rho(z)$ must vanish at the hard wall (fig. 116*c*) and this raises a little the slope according to

$$\omega = \frac{\omega_p}{\sqrt{2}}\left\{1+\left[F_{sc}-\frac{1}{2}\frac{3\pi}{8k_F}(1-f_M)\right]\kappa\right\},$$

where f_M is a numerical factor which depends on each metal and is typically – e.g. for Mg and Al – of order 0.4. Altogether, if we compare this with full numerical calculations for the same model, it results in a very considerable improvement upon the semiclassical approximation. The meaning of this is that the change in the slope reflects a change in the spectrum of the single-particle states. Incidentally, the study of the charge density fluctuation accompanying the surface plasmon itself is also interesting because different external probes may reach different depths and thus sample apparently different dipole moments. A careful study of $\rho^{\mathrm{i}}(z)$ might help us to understand the apparent discrepancies between different experimental determinations of the slope of long-wave surface plasmons. We shall not go into further details, but shall indicate only that this can be a study of some interest. We also stress that we have only discussed real dispersion here. The damping was considered in § 1.6, where we saw that it is mainly due to the non-specularity of the surface. Such effects are not brought out in a collisionless r.p.a. analysis with a specular surface; remember also the discussion of § 1.8. However, a new feature appears if the study of surface plasmons is carried out for arbitrary κ. Here one must really resort to numerical work and what one finds of course is that, like the bulk plasmons, the surface plasmons cease to be well-defined elementary excitations when their wavevector reaches the cutoff value k_{c} at which all collective modes merge into the continuum of electron-hole pair excitations. This basic feature has already appeared in § 5.2 on the basis of very simple models. It is a general fact which of course is borne out by more accurate calculations.

Finally, another interesting extension of the analysis presented here would be the study of finite barriers. The case of an arbitrary smooth barrier can only be studied numerically, but an analytical discussion is possible in theory – though very complex in practice – for an abrupt-barrier model. The idea is to return to the analysis of § 5.3.1 where the crucial step is the formula (5.185) for $\hat{G}_{\mathrm{s}}$ in terms of $\hat{G}_{\mathrm{b}}$. What we must do is to go back to chapter 2 and to use, instead of (5.185), the formulae (2.39) and (2.40), which describe reflection and transmission for the finite surface barrier according to (2.74) and (2.78). The point is that these formulae can be expressed analytically for the abrupt-barrier model, unless one uses a sophisticated model for the bulk band structure of the metal. Even under these circumstances the analysis becomes rather involved. However, it is an interesting and suggestive program. By adjusting the parameters of the abrupt barrier one may hope to obtain a reasonable simulation of a more realistic barrier and one has an analysis in terms of Green functions which lends itself to a direct physical interpretation. For example, it is possible to extract from this a fairly good estimate of the degree to which surface plasmon dispersion depends on the key features of the surface barrier. However, we shall not get involved in such a complicated analysis. We shall instead return to the infinite surface barrier as a tractable model which can serve our purpose of discussing the basic principles involved in this range of surface problems.

5.3.4. The surface energy for the infinite-barrier model

We now return to (5.23), where χ is the polarisability function defined by (5.6). This can be studied by following the *Wikborg–Inglesfield analysis*, of which the basic idea is to rewrite (5.23) in terms of the polarisability function χ_0 defined by (5.189). We recall that now χ_0 for us is the r.p.a. result expressed in (5.179). The desired transformations can be more fluently carried out in a condensed but obvious operator notation as follows.

From (5.6) and (5.189) we write

$$e\rho^{\mathrm{i}} = \chi V^{\mathrm{e}} = \chi_0 V^{\mathrm{e}} + \chi_0 V^{\mathrm{i}}. \tag{5.216}$$

We also define the operator Ω so that the equation

$$V^{\mathrm{i}}(\boldsymbol{r}) = \int \frac{e\rho^{\mathrm{i}}(\boldsymbol{r}')}{|\boldsymbol{r}-\boldsymbol{r}'|}\,\mathbf{d}\boldsymbol{r}'$$

is written in operator notation as

$$V^{\mathrm{i}} = \Omega e\rho^{\mathrm{i}}. \tag{5.217}$$

From (5.217) and (5.216) we obtain

$$\chi = \chi_0 + \chi_0\Omega\chi, \tag{5.218}$$

which can be regarded as an integral equation for χ, given χ_0. It is still convenient to introduce the new operators

$$R_0 = \Omega\chi_0; \quad R = \Omega\chi. \tag{5.219}$$

Then (5.218) becomes

$$R = R_0 + R_0 R. \tag{5.220}$$

The properties of R and R_0 can be seen in the following way. From (5.217), (5.216) and (5.219) we have

$$V^{\mathrm{i}} = RV^{\mathrm{e}}. \tag{5.221}$$

Likewise, using the second equality of (5.216) in which $V^{\mathrm{e}} + V^{\mathrm{i}}$ is re-expressed as V we have

$$V^{\mathrm{i}} = R_0 V. \tag{5.222}$$

Thus, from (5.221),

$$R = \frac{\delta V^{i}}{\delta V^{\mathrm{e}}} = \frac{\delta(V - V^{\mathrm{e}})}{\delta V^{\mathrm{e}}}, \tag{5.223}$$

i.e. more explicitly,

$$R(\boldsymbol{r}, \boldsymbol{r}'; \omega) = \varepsilon^{-1}(\boldsymbol{r}, \boldsymbol{r}'; \omega) - \delta(\boldsymbol{r} - \boldsymbol{r}'). \tag{5.224}$$

Similarly, from (5.222),

$$R_0 = \frac{\delta V^{\mathrm{i}}}{\delta V} = \frac{\delta(V - V^{\mathrm{e}})}{\delta V}, \tag{5.225}$$

i.e.

$$R_0(\boldsymbol{r}, \boldsymbol{r}'; \omega) = \delta(\boldsymbol{r} - \boldsymbol{r}') - \varepsilon(\boldsymbol{r}, \boldsymbol{r}'; \omega). \tag{5.226}$$

Omitting for the moment the explicit ω-dependence, let us concentrate on the term

$$\int \frac{1}{|\boldsymbol{r} - \boldsymbol{r}'|} \chi(\boldsymbol{r}, \boldsymbol{r}') \,\mathbf{d}\boldsymbol{r}\,\mathbf{d}\boldsymbol{r}',$$

which appears in (5.23). Remembering the definitions of Ω and R we can express this as

$$\int \frac{1}{|\boldsymbol{r}' - \boldsymbol{r}|} \chi(\boldsymbol{r}, \boldsymbol{r}') \,\mathbf{d}\boldsymbol{r}\,\mathbf{d}\boldsymbol{r}' = \int R(\boldsymbol{r}', \boldsymbol{r}') \,\mathbf{d}\boldsymbol{r}' = \mathrm{Tr}\, R. \tag{5.227}$$

Thus we can rewrite (5.23) as

$$E_{\mathrm{xc}} = -\frac{1}{2\pi} \int_0^1 \frac{\mathrm{d}\lambda}{\lambda} \int_0^\infty \mathrm{d}\omega \;\mathrm{Im}\,\mathrm{Tr}\, R(\boldsymbol{r}, \boldsymbol{r}'; \omega^+; \lambda) - \frac{1}{2} \int \frac{e^2}{|\boldsymbol{r} - \boldsymbol{r}'|} \delta(\boldsymbol{r} - \boldsymbol{r}') n_0(\boldsymbol{r}') \,\mathbf{d}\boldsymbol{r}\,\mathbf{d}\boldsymbol{r}'. \tag{5.228}$$

We can also solve (5.220) for R, which yields

$$R = (1 - R_0)^{-1} R_0, \tag{5.229}$$

and using this in (5.228), we have

$$E_{\mathrm{xc}} = -\frac{1}{2\pi} \int_0^1 \frac{\mathrm{d}\lambda}{\lambda} \int_0^\infty \mathrm{d}\omega \;\mathrm{Im}\,\mathrm{Tr}\, \{[1 - R_0(\lambda)]^{-1} R_0(\lambda)\} - \frac{1}{2} \int \frac{e^2}{|\boldsymbol{r} - \boldsymbol{r}'|} \delta(\boldsymbol{r} - \boldsymbol{r}') n_0(\boldsymbol{r}') \,\mathbf{d}\boldsymbol{r}\,\mathbf{d}\boldsymbol{r}', \tag{5.230}$$

where $R_0(\lambda)$ is to be understood as $R_0(\boldsymbol{r}, \boldsymbol{r}'; \omega^+; \lambda)$, with $\boldsymbol{r}' = \boldsymbol{r}$ on taking the trace. Now, R_0, according to (5.219), is proportional to χ_0. Thus $R_0(\lambda)$ is proportional to the coupling parameter λ. Let us put, in concise notation,

$$R_0(\lambda) = \lambda R_0(1) \equiv \lambda R_0.$$

Then

$$E_{xc} = -\frac{1}{2\pi}\int_0^1 \frac{d\lambda}{\lambda}\int_0^\infty d\omega \, \mathrm{Im}\,\mathrm{Tr}\,[(1-\lambda R_0)^{-1}\lambda R_0] \tag{5.231}$$

$$-\frac{1}{2}\int \frac{e^2}{|\boldsymbol{r}-\boldsymbol{r}'|}\delta(\boldsymbol{r}-\boldsymbol{r}')n_0(\boldsymbol{r}')\,\mathbf{d}\boldsymbol{r}\,\mathbf{d}\boldsymbol{r}'.$$

We stress that this formula for E_{xc} holds in the r.p.a. Since

$$\int_0^1 \frac{d\lambda}{\lambda}\,\mathrm{Im}\,\mathrm{Tr}\,[(1-\lambda R_0)^{-1}\lambda R_0] = -\mathrm{Im}\,\ln\,\mathrm{Det}\,(1-R_0),$$

we can write (5.231) as

$$E_{xc} = \frac{1}{2\pi}\int_0^\infty d\omega\,\mathrm{Im}\,\ln\,\mathrm{Det}\,[1-R_0(\omega^+)]$$

$$-\frac{1}{2}\int \frac{e^2}{|\boldsymbol{r}-\boldsymbol{r}'|}\delta(\boldsymbol{r}-\boldsymbol{r}')n_0(\boldsymbol{r}')\,\mathbf{d}\boldsymbol{r}\,\mathbf{d}\boldsymbol{r}'. \tag{5.232}$$

This formula has two interesting properties. The first one, which need not be spelled out in detail again, is (recall (5.226)) that writing out the contributions of the zeros and poles in the first integrand again yields a normal-mode analysis like those of § 5.2, only that now we are in the r.p.a., beyond the semiclassical approximation. The second interesting property of (5.232) arises on looking at its lowest approximation. From the power series expansion of the function $\ln(1+x)$ we find to lowest order

$$E_{xc} = \frac{1}{2\pi}\int_0^\infty d\omega\,\mathrm{Im}\,\mathrm{Tr}\,R_0(\omega^+)$$

$$-\frac{1}{2}\int \frac{e^2}{|\boldsymbol{r}-\boldsymbol{r}'|}\delta(\boldsymbol{r}-\boldsymbol{r}')n_0(\boldsymbol{r}')\,\mathbf{d}\boldsymbol{r}\,\mathbf{d}\boldsymbol{r}'. \tag{5.233}$$

Notice that this is precisely the result obtained by evaluating the integrand of (5.231), in the integral over λ, for $\lambda = 0$, i.e. for zero coupling strength of the electron–electron interaction. But this is just the notion of exchange energy, indicated for this reason as E_x in (5.233). Indeed one can evaluate (5.231) explicitly for a *homogeneous* electron gas using the formula (5.179) in terms of plane waves, and find that this yields precisely the exchange energy of the Hartree–Fock approximation. Thus (5.232) has an interesting property. Its lowest approximation (5.233) can be taken as the definition of the exchange energy of the semi-infinite electron gas. The difference between (5.232) and (5.233) then yields the correlation energy for the surface system. This allows us to study E_x and E_c separately.

For this we shall need (5.225) in Fourier transform. From (5.222), expressed explicitly as

$$V^i(\boldsymbol{k}) = \int R_0(\boldsymbol{k}, \boldsymbol{k}_1) V(\boldsymbol{k}_1) \frac{d^3k_1}{(2\pi)^3},$$

we have

$$R_0(\boldsymbol{k}, \boldsymbol{k}_1) = (2\pi)^3 \frac{\delta V^i(\boldsymbol{k})}{\delta V(\boldsymbol{k}_1)}. \tag{5.234}$$

This holds generally, for any $\boldsymbol{k} = (\boldsymbol{\kappa}, q)$ and $\boldsymbol{k}' = (\boldsymbol{\kappa}', q')$, but remember we are studying surface problems, for which we work always at a fixed $\boldsymbol{\kappa}$. Thus it suffices to consider $\boldsymbol{k}$ and all other vectors $\boldsymbol{k}_i = (\boldsymbol{\kappa}, q_i)$ with the same $\boldsymbol{\kappa}$. This simplifies the notation when integrating over q_i. Now remember (5.200) which is already symmetrised. Making explicit use of the symmetry, (5.200) can be cast in the equivalent form

$$\begin{aligned}\frac{\delta V^i_+(\boldsymbol{k})}{\delta V_+(\boldsymbol{k}_1)} = {} & \frac{2\pi}{k^2}\chi_b(\boldsymbol{k})[\delta(q-q_1)+\delta(q+q_1)] \\ & +\frac{2\pi}{k^2}[\delta\chi(q, q_1)+\delta\chi(q, -q_1)] - \frac{2\kappa}{k^2}\frac{\chi_b(\boldsymbol{k}_1)}{k_1^2} \\ & -\frac{\kappa}{k^2}[\delta\chi(q_2, q_1)+\delta\chi(q_2, -q_1)].\end{aligned} \tag{5.235}$$

Using the symmetrisation scheme in this explicit form is mathematically equivalent to using one-sided Fourier transforms, as appropriate for a semi-infinite medium. Suppose we Fourier transform back from $(\boldsymbol{\kappa}, q)$ to $(\boldsymbol{\kappa}, z)$. Since (5.235) is a symmetrised response function, if we introduce an external stimulus at $z_0 > 0$, so that the total potential at z_0 is $V(\kappa)\,\delta(z-z_0)$, the symmetrised response function (5.235) then automatically selects only the symmetrical part of the stimulus and thus gives the response to

$$\tfrac{1}{2}V(\boldsymbol{\kappa})[\delta(z-z_0)+(z+z_0)].$$

Then (5.235) gives, to a real stimulus at $z = z_0 > 0$, only one half of the real response. This reduction by a factor 2 is compensated for by the fact that integrals which would only cover the half space $z > 0$ now cover the entire space. We can therefore use (5.235) as it stands in (5.234), with which we can study E_x (5.233) and E_{xc} (5.232).

In order to study E_x we write (5.233) as

$$E_x = -\frac{1}{2\pi}\int_0^\infty d\omega \,\mathrm{Im} \sum_{\kappa} \int dz\, R_0(\boldsymbol{\kappa}; z, z; \omega^+) - \frac{1}{2}\int \frac{e^2}{|\boldsymbol{r}-\boldsymbol{r}'|}\delta(\boldsymbol{r}-\boldsymbol{r}')n_0(\boldsymbol{r}')\,\mathbf{d}\boldsymbol{r}\,\mathbf{d}\boldsymbol{r}'. \tag{5.236}$$

Taking $\boldsymbol{\kappa}$ and ω^+ as implicit we have

$$R_0(z, z') = \int R_0(q, q_1)\,e^{iqz}\,e^{-iq_1z'}\frac{dq}{2\pi}\frac{dq_1}{2\pi}. \tag{5.237}$$

Using (5.237), (5.234) and (5.235) in (5.236), and using again the same device as in § 5.2 to pick out the V or A scaling factors, we find, taking the dependence on ω^+ as understood,

$$\begin{aligned} E_x = &-\frac{V}{2\pi}\int_0^\infty d\omega\,\mathrm{Im}\sum_{\boldsymbol{k}}[1-\varepsilon(\boldsymbol{k})] \\ &-\frac{1}{2}\int\frac{e^2}{|\boldsymbol{r}-\boldsymbol{r}'|}\delta(\boldsymbol{r}-\boldsymbol{r}')n_0(\boldsymbol{r}')\,\mathbf{d}\boldsymbol{r}\,\mathbf{d}\boldsymbol{r}' \\ &+\frac{A}{2\pi}\int_0^\infty d\omega\sum_{\kappa}\mathrm{Im}\left\{\frac{\varepsilon(\boldsymbol{\kappa},0)-1}{4}+\int\frac{4\pi}{k^2}\frac{\kappa}{2\pi k^2}\chi_b(\boldsymbol{k})\,dq\right. \\ &-\int\frac{2\pi}{k^2}[\delta\chi(q,q)+\delta\chi(q,-q)]\,dq \\ &\left.+\int\frac{2\pi}{k^2}dq\int\frac{\kappa}{2\pi k_1^2}[\delta\chi(q_1,q)+\delta\chi(q_1,-q)]\,dq_1\right\}. \end{aligned} \tag{5.238}$$

In order to extract from here the surface exchange energy it is important to remember the analysis of chapter 3, where we saw that charge neutrality requires us to remove the jellium edge a distance $3\pi/8k_F$ away from the infinite barrier. Thus the volume contribution corresponding to the area A multiplied by this length must be separated from the first term on the r.h.s. of (5.238) and counted as part of the surface energy. We then find for the surface exchange energy, expressed as energy per unit area,

$$\begin{aligned} E_x^s = &-\frac{3}{16k_F}\int_0^\infty d\omega\,\mathrm{Im}\sum_{\boldsymbol{k}}[1-\varepsilon(\boldsymbol{k})] \\ &+\frac{1}{2\pi}\int_0^\infty d\omega\sum_{\kappa}\mathrm{Im}\left\{\frac{\varepsilon(\boldsymbol{\kappa},0)-1}{4}+\int\frac{4\pi}{k^2}\frac{\kappa}{2\pi k^2}\chi_b(\boldsymbol{k})\,dq\right. \\ &-\int\frac{2\pi}{k^2}[\delta\chi(q,q)+\delta\chi(q,-q)]\,dq \\ &\left.+\int\frac{2\pi}{k^2}dq\int\frac{\kappa}{2\pi k_1^2}[\delta\chi(q_1,q)+\delta\chi(q_1,-q)]\,dq_1\right\}. \end{aligned} \tag{5.239}$$

The evaluation of E_{xc} proceeds in much the same manner, but starting from (5.232). Before doing this it is convenient to rewrite $1-R_0$ as a product of two factors, one of which is easy to work out. The idea is to return to (5.235) and make there the semiclassical approximation $\delta\chi = 0$. Repeating the above arguments we then find

$$R_0'(\boldsymbol{k}, \boldsymbol{k}_1) = \tfrac{1}{2}(2\pi)^3\Big\{[1-\varepsilon(\boldsymbol{\kappa})][\delta(q-q_1)+\delta(q+q_1)] - \frac{\kappa}{\pi k^2}[1-\varepsilon(\boldsymbol{k}_1)]\Big\}. \tag{5.240}$$

If we used this as the value of R_0 the formulae of the present r.p.a. analysis would reproduce the semiclassical results of § 5.2. Let us define the first term of (5.240) as

$$R_0''(\boldsymbol{k}, \boldsymbol{k}_1) = \tfrac{1}{2}(2\pi)^3[1-\varepsilon(k)][\delta(q-q_1)+\delta(q+q_1)]. \tag{5.241}$$

If this approximation for R_0 were used in (5.232) then E_{xc} would be very easy to evaluate. In fact this would reproduce the band-edge effects of the semiclassical analysis. The idea is then to define the factorisation

$$(2\pi)^3\delta(q-q_1) - R_0(q, q_1) \equiv 1 - R_0(q, q_1) = [1-R_0''(q, q')] \times [1-R_1(q', q_1)]\, \mathrm{d}q'. \tag{5.242}$$

This defines R_1, which is what we now want to find. To this effect we write (5.242) in concise notation as

$$1-R_0 = (1-R_0'')(1-R_1). \tag{5.243}$$

We furthermore define δR_0 through

$$R_0 = R_0'' + \delta R_0. \tag{5.244}$$

We can always find δR_0, since we know R_0 and R_0''. From (5.243) and (5.244) we find

$$R_1 = (1-R_0'')^{-1}\delta R_0, \tag{5.245}$$

i.e. remembering (5.241),

$$R_1(q, q_1) = \delta R_0(q, q_1)/\varepsilon(q). \tag{5.246}$$

We can then write

$$\ln \operatorname{Det}(1-R_0) = \ln \operatorname{Det}(1-R_0'') + \ln \operatorname{Det}(1-R_1),$$

R_1 being given by (5.245) or (5.246). Using all this in (5.232) we find –

dependence on ω^+ against understood –

$$E_{xc} = \frac{V}{2\pi}\int_0^\infty d\omega \sum_{\boldsymbol{k}} \operatorname{Im} \ln \varepsilon(\boldsymbol{k})$$
$$-\frac{1}{2}\int \frac{e^2}{|\boldsymbol{r}-\boldsymbol{r}'|}\delta(\boldsymbol{r}-\boldsymbol{r}')n_0(\boldsymbol{r}')\,\mathbf{d}\boldsymbol{r}\,\mathbf{d}\boldsymbol{r}'$$
$$+\frac{A}{2\pi}\int d\omega \sum_{\boldsymbol{\kappa}} \operatorname{Im} \frac{\ln \varepsilon(\boldsymbol{\kappa}, 0)}{4} + \int_0^\infty \frac{d\omega}{2\pi} \operatorname{Im} \ln \operatorname{Det}(1-R_1). \tag{5.247}$$

Remembering again the distance from the jellium edge to the infinite barrier we have the r.p.a. result, per unit area,

$$E^s_{xc} = \frac{3}{16k_F}\int_0^\infty d\omega \sum_{\boldsymbol{k}} \operatorname{Im} \ln \varepsilon(\boldsymbol{k})$$
$$+\frac{1}{8\pi}\int_0^\infty d\omega \sum_{\boldsymbol{\kappa}} \operatorname{Im} \ln \varepsilon(\boldsymbol{\kappa}, 0)$$
$$+\int_0^\infty d\omega \operatorname{Im} \ln \operatorname{Det}(1-R_1). \tag{5.248}$$

The last term must be evaluated numerically by using the result (5.246) found for R_1.

The interest of formulae (5.239) and (5.248) lies in that one can now study the exchange and correlation contributions separately, and this proves helpful in comparing the results of different approaches to the surface-energy problem. Table 10 shows the results obtained on evaluating E_x and E_{xc} for three different electron densities according to (5.239) and (5.248). The same quantities could be calculated by an entirely different approach, by calculating in the *Lang–Kohn model* (§ 3.3) the

TABLE 10. *Surface energy terms obtained as explained in the text*

r_s	2.07 (Al)	4	6
E^s_x†	700	100	30
E^s_{xc}‡	1388	203	63

† Calculated by J. Harris & R. O. Jones, *J. Phys.* F: *Metal Phys.* **4**, 1170 (1974).
‡ Calculated by D. C. Langreth & J. Perdew, *Phys. Rev.* B **15**, 2884 (1977).

charge density corresponding to the infinite-barrier case and then using this charge density profile to evaluate the contributions to the surfa energy by the methods discussed in § 3.3. The results obtained by carrying out this process for the same values of r_s as those of table 10 are shown in table 11. On comparing these tables it appears that the local density functional approach proves rather poor for the exchange energy E_x^s *separately*, but it works quite well for E_{xc}^s. This point will be more fully discussed in the following section.

TABLE 11. *Surface energy terms obtained in the local density functional approach*

r_s	2.07 (Al)	4	6
E_x^s	1107	153	45
E_{xc}^s	1241	184	58

Calculated by D. C. Langreth & J. Perdew, *Phys. Rev.* B **15**, 2844 (1977).

Finally we stress that the results derived here, including interference effects, can be used to extract from them a normal-mode analysis which contains all the key features stressed in § 5.2, but is a generalisation of the semiclassical results derived there.

5.4. The surface energy of simple metals: local versus non-local approach

It is interesting to delve further into the comparison between the local approach of § 3.3 and the non-local approach to the surface energy problem expounded in this chapter. We have just seen that both agree fairly well for E_{xc}^s, but they disagree quite conspicuously for E_x^s alone. The point is that in calculating E_x^s by the gradient expansion method of § 3.3 we have incorrectly made an uncritical use of this formalism. It turns out that, while such an expansion can be justified for the exchange and correlation energy – of a system in general – this is not the case for the exchange energy *alone*. This is why the local approach works fairly well for E_{xc}^s. An illuminating view on the mutual relationship between the local and non-local approaches is provided by the *Langreth–Perdew wavevector analysis*, which on general lines runs as follows.

Let us go back to the general formula (5.23) which gives E_{xc} for an arbitrary – homogeneous or inhomogeneous – system. Concentrate on the integral

$$\int \mathbf{d}\boldsymbol{r}' \frac{1}{|\boldsymbol{r}-\boldsymbol{r}'|}\left[-\frac{1}{\pi}\int_0^\infty \mathrm{d}\omega \,\mathrm{Im}\,\chi(\boldsymbol{r},\boldsymbol{r}';\omega^+) - e^2\,\delta(\boldsymbol{r}-\boldsymbol{r}')n_0(\boldsymbol{r}')\right]. \tag{5.249}$$

In an homogeneous system this would be independent of $\boldsymbol{r}$ and, on performing the further integration (over $\boldsymbol{r}$) indicated in (5.23), we would find that E_{xc} is proportional to the volume V. It is convenient to introduce the structure factor $S(\boldsymbol{k})$ defined by

$$\begin{aligned} S(\boldsymbol{k}) &= -\frac{V}{Ne^2}\int \mathrm{d}(\boldsymbol{r}-\boldsymbol{r}')\,\mathrm{e}^{\mathrm{i}\boldsymbol{k}\cdot(\boldsymbol{r}-\boldsymbol{r}')}\int_0^\infty \frac{\mathrm{d}\omega}{\pi}\,\mathrm{Im}\;\chi(\boldsymbol{r}-\boldsymbol{r}';\omega) \\ &= -\frac{V}{Ne^2}\int_0^\infty \frac{\mathrm{d}\omega}{\pi}\,\mathrm{Im}\,\chi(\boldsymbol{k};\omega). \end{aligned} \tag{5.250}$$

Then (5.249) can be cast in the form

$$(5.249) = \frac{N}{V}\sum_{\boldsymbol{k}} \frac{4\pi e^2}{k^2}[S(\boldsymbol{k})-1]. \tag{5.251}$$

Thus, for a homogeneous system

$$E_{xc} = \frac{1}{2}\int_0^1 \frac{\mathrm{d}\lambda}{\lambda} N \sum_{\boldsymbol{k}} \frac{4\pi e^2\lambda}{k^2}[S(\boldsymbol{k};\lambda)-1]. \tag{5.252}$$

For an inhomogeneous system we define

$$\begin{aligned} S(\boldsymbol{k},\boldsymbol{k}') &= -\frac{1}{Ne^2}\int \mathbf{d}\boldsymbol{r}\,\mathbf{d}\boldsymbol{r}'\,\mathrm{e}^{\mathrm{i}\boldsymbol{k}\cdot\boldsymbol{r}}\,\mathrm{e}^{-\mathrm{i}\boldsymbol{k}'\cdot\boldsymbol{r}'}\int_0^\infty \frac{\mathrm{d}\omega}{\pi}\,\mathrm{Im}\,\chi(\boldsymbol{r},\boldsymbol{r}';\omega) \\ &= -\frac{1}{Ne^2}\int_0^\infty \frac{\mathrm{d}\omega}{\pi}\,\mathrm{Im}\,\chi(\boldsymbol{k},\boldsymbol{k}';\omega), \end{aligned} \tag{5.253}$$

which is the extension of (5.250). Using this in (5.23) yields, after some algebra,

$$E_{xc} = \frac{1}{2}\int_0^1 \frac{\mathrm{d}\lambda}{\lambda} N \int \frac{4\pi e^2\lambda}{k^2}[S(\boldsymbol{k},\boldsymbol{k};\lambda)-1]\frac{\mathrm{d}^3k}{(2\pi)^3}, \tag{5.254}$$

where we have changed from a summation to an integration over $\boldsymbol{k}$. This formula holds for an inhomogeneous system. The question now is which approach is to be used. Within the framework of the local density functional approach we would expect (5.254) to take the form

$$E_{xc} = \frac{1}{2}\int_0^1 \frac{\mathrm{d}\lambda}{\lambda}\int \mathbf{d}\boldsymbol{r}\, n(\boldsymbol{r})\int \frac{4\pi e^2\lambda}{k^2}\{S_b[\boldsymbol{k};\lambda;n(\boldsymbol{r})]-1\}\frac{\mathrm{d}^3k}{(2\pi)^3}, \tag{5.255}$$

where $S_b[n(\boldsymbol{r})]$ would be the bulk structure factor of a homogeneous system with electron density equal to the local value of $n(\boldsymbol{r})$. If $n(\boldsymbol{r})$ varies sufficiently smoothly we may expect to obtain (5.254) on performing the integration over $\boldsymbol{r}$ in (5.255). Suppose $n(\boldsymbol{r})$ does not vary smoothly. Even then we must consider that there is also an integration over $\boldsymbol{k}$ to be performed. Now, high values of $\boldsymbol{k}$ represent the contributions of high wavevectors and these encounter the system as locally homogeneous. Thus we expect the contributions of both integrands to be very similar in the high $\boldsymbol{k}$ range. If d is a distance which characterises the scale of variation of $n(\boldsymbol{r})$ we have

$$S(\boldsymbol{k}, \boldsymbol{k}) - 1 \approx \int \frac{n(\boldsymbol{r})}{N} \{S_b[\boldsymbol{k}; n(\boldsymbol{r})] - 1\} \, \mathrm{d}\boldsymbol{r}; \; k^{-1} \ll d. \qquad (5.256)$$

The error introduced on using a local approximation like (5.255) then comes from the energy associated with low-wavevector fluctuations. It is now convenient, for the discussion of the surface problem, to introduce the terms $e_{xc}^{V}(\boldsymbol{k})$ and $\delta e_{xc}^{A}(\boldsymbol{k})$ defined by

$$\begin{aligned} E_{xc} &= \int e_{xc}(\boldsymbol{k}) \frac{\mathrm{d}^3 k}{(2\pi)^3} \\ &= V \int e_{xc}^{V}(\boldsymbol{k}) \frac{\mathrm{d}^3 k}{(2\pi)^3} + A \int \delta e_{xc}^{A}(\boldsymbol{k}) \frac{\mathrm{d}^3 k}{(2\pi)^3}. \end{aligned} \qquad (5.257)$$

From (5.254),

$$e_{xc}(\boldsymbol{k}) = \frac{N}{2} \int_0^1 \mathrm{d}\lambda \, \frac{4\pi e^2}{k^2} [S(\boldsymbol{k}, \boldsymbol{k}, \lambda) - 1], \qquad (5.258)$$

and from (5.255),

$$e_{xc}^{V}(\boldsymbol{k}) = \frac{n_b}{2} \int_0^1 \mathrm{d}\lambda \, \frac{4\pi e^2}{k^2} [S_b(\boldsymbol{k}; \lambda; n_b) - 1]. \qquad (5.259)$$

Hence, from (5.257), since $V n_b = N$,

$$A \delta e_{xc}^{A}(\boldsymbol{k}) = \frac{N}{2} \int_0^1 \mathrm{d}\lambda \, \frac{4\pi e^2}{k^2} [S(\boldsymbol{k}, \boldsymbol{k}; \lambda) - S_b(\boldsymbol{k}; \lambda; n_b)]. \qquad (5.260)$$

The surface exchange and correlation energy expressed as energy per unit area is then

$$E_{xc}^{s} = \int \delta e_{xc}^{A}(\boldsymbol{k}) \frac{\mathrm{d}^3 k}{(2\pi)^3}. \qquad (5.261)$$

So far (5.258)–(5.260) are exact. Suppose we now introduce a local approximation l_{xc} and δl_{xc}^{A} to the exact values e_{xc} and δe_{xc}^{A}. Then we have

$$l_{xc}(\mathbf{k}) = \frac{1}{2}\int_0^1 d\lambda \int \mathbf{dr}\, n(\mathbf{r}) \frac{4\pi e^2}{k^2}\{S_b[\mathbf{k}; \lambda; n(\mathbf{r})] - 1\}, \tag{5.262}$$

whence

$$A\delta l_{xc}^{A}(k) = \frac{1}{2}\int_0^1 d\lambda \int \mathbf{dr} \frac{4\pi e^2}{k^2}$$
$$\times \{n(\mathbf{r}) S_b[\mathbf{k}; \lambda; n(\mathbf{r})] - n_b \theta(z) S_b(\mathbf{k}; \lambda; n_b)\}, \tag{5.263}$$

and this is what we use to evaluate E_{xc}^{s} in the local approximation. Notice that $\delta l_{xc}^{A}(\mathbf{k})$ depends only on the modulus of $\mathbf{k}$. Thus we can write

$$E_{xc}^{s} = \int_0^\infty \frac{k_F k^2}{\pi^2} \delta l_{xc}^{A}(k)\, d\left(\frac{k}{2k_F}\right). \tag{5.264}$$

This integrand is plotted in fig. 117 as a function of $k/2k_F$ by using the density profile $n(z)$ corresponding to an infinite-barrier model for the case $r_s = 2.07$. The surface exchange and correlation energy per unit surface area is given, in the local approximation, by the area enclosed by the full line curve.

Now, we know from the above argument that $\delta l_{xc}^{A}(k)$ is a good approximation to $\delta e_{xc}^{A}(k)$ for high values of k, i.e. for $k \gtrsim k_F$. The differences between these two functions come from the low-wavevector

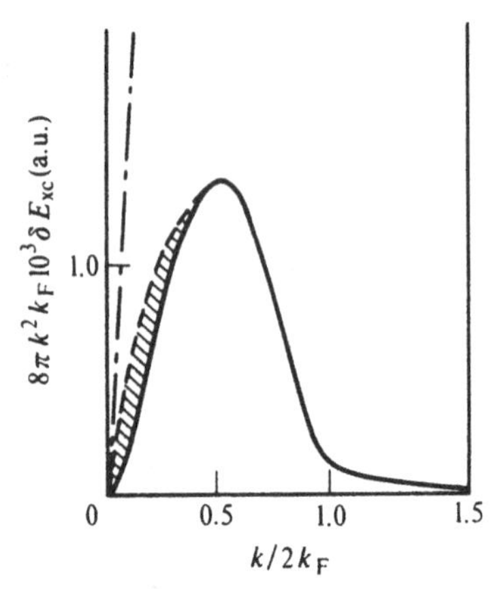

Fig. 117. The integrand of (5.264) (vertical axis) as a function of $(k/2k_F)$ for $r_s = 2$ and infinite barrier. Solid line: the integrand of (5.264) evaluated in the local approximation. Dash-dotted line: limiting behaviour of this function, for $k \to 0$, when evaluated from the exact formula (5.260). Dotted line: interpolation between both. D. C. Langreth & J. Perdew, *Solid St. Commun.* **17**, 1425 (1975).

range. It is therefore interesting to try and see whether one can study the behaviour of δe^{A}_{xc} for small k and this can be done at least to lowest order in the limit $k \ll k_F$. The physical idea behind this is that the rapid changes across the interface do not significantly affect the main contributions to $S(\boldsymbol{k}, \boldsymbol{k})$ for $k \to 0$ because in this limit the lengths of physical significance are much longer than the typical thickness of the interface. It is a well-known fact, for example, that the limiting value $\omega_p/\sqrt{2}$ of the surface plasmon frequency for $\kappa \to 0$ is a common feature of all models, no matter how crude, and is insensitive to the details of the interface. Let us then look at $S(\boldsymbol{k}, \boldsymbol{k}; \lambda)$ for $k \to 0$ in the semiclassical model, but treat it exactly. It follows from (5.13) that

$$\int \frac{\delta\rho^{i}(\boldsymbol{r})}{\delta\rho^{e}(\boldsymbol{r}')} e^{i\boldsymbol{k}\cdot(\boldsymbol{r}-\boldsymbol{r}')} \, \mathbf{d}\boldsymbol{r} \, \mathbf{d}\boldsymbol{r}' = \frac{4\pi}{k^2}\chi(\boldsymbol{k}, \boldsymbol{k}).$$

From this and (5.253) we have

$$S(\boldsymbol{k}, \boldsymbol{k}) = -\frac{k^2}{4\pi N e^2}\int_0^\infty \frac{d\omega}{\pi} \operatorname{Im} \int \frac{\delta\rho^{i}(\boldsymbol{r})}{\delta\rho^{e}(\boldsymbol{r}')} e^{i\boldsymbol{k}\cdot(\boldsymbol{r}-\boldsymbol{r}')} \, \mathbf{d}\boldsymbol{r} \, \mathbf{d}\boldsymbol{r}'. \tag{5.265}$$

A semiclassical evaluation of the functional derivative appearing here was obtained in § 5.2. Thus we can combine (5.265) and § 5.2 to study $S(\boldsymbol{k}, \boldsymbol{k})$, and hence $e_{xc}(\boldsymbol{k})$ in the limit $k \to 0$. We are therefore interested in

$$\lim_{k\to 0} e_{xc}(\boldsymbol{k}) = \lim_{k\to 0} -\frac{1}{2}\left\{\int_0^1 \frac{d\lambda}{\lambda}\int_0^\infty \frac{d\omega}{\pi} \operatorname{Im} \int \frac{\delta\rho^{i}(\mathbf{r})}{\delta\rho^{e}(\boldsymbol{r}')} e^{i\boldsymbol{k}\cdot(\boldsymbol{r}-\boldsymbol{r}')} \mathbf{d}\boldsymbol{r} \, \mathbf{d}\boldsymbol{r}' + \frac{4\pi N e^2}{k^2}\right\}. \tag{5.266}$$

In order to compare this with the result of the local approximation it is convenient to define the angular average – over the solid angle $\mathbf{d}\Omega_{\boldsymbol{k}}$ in $\boldsymbol{k}$ space –

$$\begin{aligned} e_{xc}(k) &= \int e_{xc}(\boldsymbol{k})\frac{d\Omega_{\boldsymbol{k}}}{4\pi} \\ &= -\frac{1}{2}\int_0^1 \frac{d\lambda}{\lambda}\int_0^\infty \frac{d\omega}{\pi}\int \frac{\delta\rho^{i}(\boldsymbol{r})}{\delta\rho^{e}(\boldsymbol{r}')} \\ &\quad \times e^{i\boldsymbol{k}\cdot(\boldsymbol{r}-\boldsymbol{r}')}\frac{d\Omega_{\boldsymbol{k}}}{4\pi} \, \mathbf{d}\boldsymbol{r} \, \mathbf{d}\boldsymbol{r}' - \frac{2\pi N e^2}{k^2}. \end{aligned} \tag{5.267}$$

Using now the results of § 5.2 and the definition (5.257) we find

$$\lim_{k\to 0} \delta e^{A}_{xc}(k) = \frac{\pi}{4k}\left(\frac{\omega_p}{\sqrt{2}} - \frac{\omega_p}{2}\right). \tag{5.268}$$

Notice that in the integrand of (5.261) this is still multiplied by k^2, which yields a straight line in the plot of fig. 117. We stress that this asymptotic result is in this limit exact although it is based on the semiclassical model. Fig. 117 shows the extent to which the local approximation becomes inadequate in this limit. This suggests that a reasonable approximation could be provided by an interpolation as shown (dashed line) in the figure. The dashed area represents an estimate of the error one may expect when using the local approximation to calculate E^s_{xc}.

With this we can understand why the local density functional approach works fairly well for the surface exchange and correlation energy. Formally speaking the gradient expansion is justified – although, we stress, not for the exchange or the correlation energy alone – while on the practical side we see that the expected error is not too serious. Table 12 shows a practical example. Three different evaluations of E^s_{xc} for the

TABLE 12. *The surface exchange and correlation energy in* erg/cm^{-2} (*infinite-barrier model and* r_s = 2.07 (Al)) *from three different calulcations*

	(a)	(b)	(c)
E^s_{xc}	1241	1365	1388

(a) Local density functional approach (see table 11). (b) Interpolation scheme based on the wave-vector analysis calculated by D. C. Langreth & J. Perdew, *Phys. Rev.* B **15**, 2884 (1977). (c) non-local approach of § 5.3.4 (see table 10).

same system – infinite barrier and $r_s = 2.07$ – yield results in quite reasonable agreement. The local approximation is within 10% of the others. The density functional formalism is not as accurate as the non-local approach discussed in this chapter, but we have seen that the errors are small and it is in a form ready for direct application to other physical problems – e.g. those of chapter 3 – and also to more realistic models of non-abrupt barriers by using the interpolation scheme discussed here. It turns out that typical discrepancies are also of order 10% and the general conclusions as regards the comparison with the local approximation are similar to those reached on the basis of an infinite-barrier model.

5.4. *Surface energy*

The latest developments in the application of the scheme discussed in this section indicate (see the last reference for this chapter) that this yields better results for the surface energy than those obtained by using only the first term in the gradient expansion method for exchange and correlation, which seems to overestimate the corrections by a factor of 2. In other words, the values shown in table 4 (chapter 3) would seem to be too large by about a factor of 2. In conclusion, it seems that the scheme discussed in this section provides one of the best methods currently available for the calculation of the surface energy.

References

Chapter 1.
General background:
Jackson, J. D. (1962). *Classical electrodynamics.* New York: Wiley & Sons.
Landau, L. D. & Lifshitz, E. M. (1963). *Electrodynamics of continuous media.* Oxford: Pergamon Press.
Pines, D. & Nozières, P. (1966). *The theory of quantum liquids,* vol. 1. New York: W. A. Benjamin.
Ginzburg, V. L. (1964). *The propagation of electromagnetic waves in plasmas.* Oxford: Pergamon Press.
Agranovich, V. M. & Ginzburg, V. L. (1966). *Spartial dispersion in crystal optics and the theory of excitons.* New York: Interscience.

§§ 1.3–1.8 are based on the following papers:
García-Moliner, F. & Flores, F. (1977). *J. Physique* **38**, 851.
Flores, F. & García-Moliner, F. (1977). *J. Physique* **38**, 863.
Flores, F., García-Moliner, F. & Navascués, G. (1971). *Surface Sci.* **24**, 61.
Flores, F. & Navascués, G. (1973). *Surface Sci.* **34**, 773.
García-Moliner, F. (1969). *Electron transport in solids.* Lecture notes from the 2nd Nordic Summer School on Solid State Physics, Copenhagen.
Flores, F. & García-Moliner, F. (1972). *Solid St. Commun.* **11**, 1295.

Chapter 2.
This chapter is mostly based on the papers:
García-Moliner, F. & Rubio, J. (1971). *Proc. R. Soc. Lond.* A**324**, 257.
Flores, F. (1973). *Nuovo Cim.* **14**B, 1.
García-Moliner, F. (1977). *Ann. Physique* **2**.

Further general background material:
(i) For elastic surface waves in solids
Love, A. E. M. (1944). *A treatise on the mathematical theory of elasticity.* New York: Dover.
Wallis, R. F. (1972). Lecture notes in *Atomic structure and properties of solids.* Proc. LII Int. School 'E. Fermi', Varenna (ed. E. Burstein). New York and London: Academic Press.
(ii) For thermodynamics of solid surfaces
García-Moliner, F. (1975). 'Surface thermodynamics', in *Surface science,* vol. 1. Proc. Int. Winter College on Surface science, I.C.T.P., Trieste, 1974. I.A.E.A.
Linford, R. G. (1973). 'Surface thermodynamics of solids', in *Solid state surface science,* vol. 2 (ed. M. Green). New York: Marcel Dekker.

(iii) For viscous fluids

Landau, L. D. & Lifshitz, E. M. (1959). *Fluid mechanics.* Oxford: Pergamon Press.

Lucassen-Reyndeers, E. H. & Lucassen, J. (1969). *Adv. Colloid Interface Sci.* **2**, 347.

Chapter 3.

General background:

Lang, N. D. (1973). 'The density-functional formalism and the electronic structure of metal surfaces', in *Solid state physics*, vol. 28. New York and London: Academic Press.

Hedin, L. & Lundqvist, S. (1969). 'Effects of electron–electron and electron–phonon interactions on the one-electron states of solids', in *Solid state physics*, vol. 23. New York and London: Academic Press.

§ 3.1 is partly based on:

Paasch, G. & Wonn, H. (1975). *Phys. St. Sol. (b)* **70**, 555.

Flores, F. & Tejedor, C. (1977). *J. Physique* **38**, 949.

§ 3.2 is based on:

Hohenberg, P. & Kohn, W. (1964). *Phys. Rev.* **136**, 864.

Lang, N. D. & Kohn, W. (1971). *Phys. Rev.* B **3**, 1215.

Budd, H. F. & Vannimenus, J. (1973). *Phys. Rev. Lett.* **31**, 1218.

Vannimenus, J. & Budd, H. F. (1974). *Solid St. Commun.* **15**, 1739.

Smith, J. R. (1969). *Phys. Rev.* **181**, 522.

Kohn, W. & Sham, L. J. (1965). *Phys. Rev.* **140**A, 1133.

Lang, N. D. & Kohn, W. (1970). *Phys. Rev.* B **1**, 4555.

Lang, N. D. (1969). *Solid St. Commun.* **7**, 1047.

Singwi, K. S., Sjoländer, A., Tosi, M. P. & Land, R. H. (1970). *Phys. Rev.* B **1**, 1044.

Vashishta, P. & Singwi, K. S. (1972). *Phys. Rev.* B **6**, 875.

Rose, Jr, J. H., Shore, H. B., Geldart, D. J. & Rasolt, M. (1976). *Solid St. Commun.* **19**, 619.

§ 3.3 is based on:

Lang, N. D. & Kohn, W. (1970). *Phys. Rev.* B **1**, 4555.

Lang, N. D. & Kohn, W. (1971). *Phys. Rev.* B **3**, 1215.

Perdew, J. P. & Monnier, R. (1976). *Phys. Rev. Lett.* **37**, 1286; and private communication.

Tejedor, C. & Flores, F. (1976). *J. Phys.* F: *Metal Phys.* **6**, 1647.

Chapter 4.

§ 4.1 is based on:

Flores, F. & Tejedor, C. (1977). *J. Physique* **38**, 949.

Tejedor, C., Flores, F. & Alvarellos, J. E. (1977). *Phys. Lett.* **62**A, 99.

§ 4.2 is based on:

Kohn, W. (1959). *Phys. Rev.* **115**, 809.

Heine, V. (1963). *Proc. Phys. Soc. Lond.* **81**, 300.

Heine, V. & Jones, R. O. (1969). *J. Phys.* C: *Solid St. Phys.* **2**, 719.

Elices, M., Flores, F., Louis, E. & Rubio, J. (1974). *J. Phys.* C: *Solid St. Phys.* **7**, 3020.

Flores, F. & Rubio, J. (1973). *J. Phys.* C: *Solid St. Phys.* **6**, L258.

Louis, E. Yndurain, F. & Flores, F. (1976). *Phys. Rev.* B **13**, 4408.
Flores, F., García-Moliner, F., Louis, E. & Tejedor, C. (1976). *J. Phys.* C: *Solid St. Phys.* **9**, L429.
García-Moliner, F. & Flores, F. (1976). *J. Phys.* C: *Solid St. Phys.* **9**, 1609.
Flores, F., Tejedor, C. & Martín-Rodero, A. (1978). *Phys. St. Sol.* (*b*) **88**, 591.

§§ 4.3 and 4.4 are based on:
Masri, P. & Lannoo, M. (1975). *Surface Sci.* **52**, 377.
Lannoo, M. & Decarpigny, J. N. (1973). *Phys. Rev.* B **8**, 5704.
Djafari-Rouhani, B., Dobrzynski, L., Flores, F., Lannoo, M. & Tejedor, C. (1978).
Rajan, V. T. & Falicov, L. M. (1976). *J. Phys.* C: *Solid St. Phys.* **9**, 2533.
Djafari-Rouhani, B., Dobrzynski, L. & Lannoo, M. (1978).

§ 4.5 is based on:
Louis, E., Yndurain, F. & Flores, F. (1976). *Phys. Rev.* B **13**, 4408.
Tejedor, C., Flores, F. & Louis, E. (1977). *J. Phys.* C: *Solid St. Phys.* **10**, 2163.

Chapter 5.
General background:
Hedin, L. & Lundqvist, S. (1969). 'Effect of electron–electron and electron-phonon interactions on the one-electron states of solids', in *Solid state physics*, vol. 23. New York and London: Academic Press.

§ 5.1 is based on:
Wikborg, E. & Inglesfield, J. E. (1977). *Phys. Scripta* **15**, 37.
Harris, J. & Jones, R. O. (1974). *J. Phys.* F; *Metal Phys.* **4**, 1170.
Langreth, D. C. & Perdew, J. P. (1977). *Phys. Rev.* B **15**, 2884.

§ 5.2 is based on:
Wikborg, E. (1974). *Phys. St. Sol.* (*b*) **63**, K151.
Harris, J. & Griffin, A. (1975). *Phys. Rev.* B **11**, 3669.
Griffin, A., Kranz, H. & Harris, J. (1974). *J. Phys.* F: *Metal Phys.* **4**, 1774.
Zaremba, E. & Kohn, W. (1976). *Phys. Rev.* B **13**, 2270.
Chan, D. & Richmond, P. (1976). *J. Phys.* C: *Solid St. Phys.* **9**, 163.
Flores, F. & García-Moliner, F. (1979). *J. Phys.* C: *Solid St. Phys.* **12**, 1.

§ 5.3 is based on:
Beck, D. E., Celli, V., Lo Vecchio, G. & Magnaterra, A. (1970). *Nuovo Cim.* **68B**, 230.
Wikborg, E. & Inglesfield, J. E. (1975). *Solid St. Commun.* **16**, 335.
Inglesfield, J. E. & Wikborg, E. (1974). *Solid St. Commun.* **14**, 661.
Harris, J. & Jones, R. O. (1974). *J. Phys.* F: *Metal Phys.* **4**, 1170.
Wikborg, E. & Inglesfield, J. E. (1977). *Phys. Scripta* **15**, 37.
Beck, D. E. & Celli, V. (1970). *Phys. Rev.* B **2**, 2955.
Garrido, L., Flores, F. & García-Moliner, F. (1979). *J. Phys.* F: *Metal Phys.* **9**, in press.

§ 5.4 is based on:
Langreth, D. C. & Perdew, J. P. (1975). *Solid St. Commun.* **17**, 1425.
Lang, N. D. & Sham, L. J. (1975). *Solid St. Commun.* **17**, 581.
Perdew, J. P., Langreth, D. C. & Sahni, V. (1977). *Phys. Rev. Lett.* **88**, 1030.

Index

Index

Index